Handbook of
ION
SOURCES

Handbook of
ION
SOURCES

Edited by

Bernhard Wolf, Ph.D.

GSI Center for Heavy Ion Research
Darmstadt, Germany

CRC Press
Taylor & Francis Group
Boca Raton London New York

CRC Press is an imprint of the
Taylor & Francis Group, an **informa** business

CRC Press
Taylor & Francis Group
6000 Broken Sound Parkway NW, Suite 300
Boca Raton, FL 33487-2742

First issued in paperback 2020

© 1995 Taylor & Francis Group, London, UK
CRC Press is an imprint of Taylor & Francis Group, an Informa business

No claim to original U.S. Government works

ISBN 13: 978-0-367-57970-8 (pbk)
ISBN 13: 978-0-8493-2502-1 (hbk)

Visit the Taylor & Francis Web site at
http://www.taylorandfrancis.com

and the CRC Press Web site at
http://www.crcpress.com

Library of Congress Cataloging-in-Publication Data

Handbook of ion sources / edited by Bernhard Wolf.
    p.   cm.
    Includes bibliographical references and index.
    ISBN 0-8493-2502-1
    1. Ion sources--Handbooks, manuals, etc.   I. Wolf, Bernhard,
1939–
QC702.3.H36 1995
539.7'3--dc20

95–3189
CIP

Library of Congress Card Number 95-3189

PREFACE

This Handbook on Ion Sources was designed and written to support developers, users and operators of ion sources by concentrating the data they need in their daily work in one volume and to give beginners in the field of ion sources a comprehensive overview of the variety of ion sources and related issues.

The emphasis is on the technical and practical aspects of ion sources and ion beams and less on the physical background which has already been described in other publications (see introduction to chapter 2).

Each section of the book is written by specialists on the topics described below and gives state of the art information and references to the latest publications and some of historic interest, which will give the reader further information on special issues.

I designed the book to be on hand in the workplace of people involved in ion sources and accelerators; it contains all the information I have collected over the years from a wide variety of books and publications. During the process of composing this book it became obvious, however, that not every detail of interest could be included without extending the size (and price) of the book significantly. So many ingenious ion source designs could not be included despite their success in specialized areas (see proceedings of International Conferences on Ion Sources, Ion Implantation, Mass separators and various Accelerators).

The first chapter, written by Ian Brown, gives a short introduction to the plasma physics related to ion sources and explains many terms and formulas used throughout the book.

The second chapter, written by several authors (G. Alton, R. Becker, J. Ishikawa, T. Jolly, K. Leung, B. Sharkow and B. H. Wolf), presents various ion sources with details of their history, performance, operation data, special techniques used, area of application, and names or places of some users and deliverers and references to the latest and the most important publications for each type of ion source.

The third chapter, written by the editor, discusses various methods of ion production from non-gaseous feeding materials applicable to most of the ion sources. Physical and technical data on those materials as well as on construction materials for ion source parts, especially for high temperatures, are summarized in this chapter in various tables.

Chapter four by Peter Spädtke is on ion beam formation and transport. It discusses design criteria for ion extraction and acceleration systems and the importance of space charge compensation for efficient beam transport. Characteristics for beam quality such as emittance, brilliance energy spread and the conditions for their conservation or growth are given. Transformation matrices for several electrostatic and magnetic ion optical elements are also summarized.

Peter Strehl gives in chapter five a comprehensive review of beam diagnostic methods and presents a variety of practical designs for Faraday cups, beam transformers, emittance measurement and time structure of ion beams. He discusses the advantages and limits of devices and materials and gives formulas to estimate cooling of devices and power limits.

Chapter six offers selected examples of the modern design of electronics for ion sources by Hans Horneff and an introduction to microwave equipment used in connection with ECR and μ-wave ion sources by Francois Bourg.

The final chapter by Peter Spädtke gives a summary on computer codes used for the stimulation of extraction and various elements of ion beam systems.

I would not have been able to prepare this book without the joint effort of all the authors and the overwhelming response of the ion source community to my request for information on their work. I have to thank the staff of my group at GSI for their steady support and constructive criticism during the preparation of the manuscripts. I would also like to express my appreciation to CRC Press for their support and the opportunity to write this book. Last, I would like to thank my wife for her support in proof reading the manuscripts.

Brensbach 3.20.1995

EDITOR'S BIOGRAPHY

Bernhard H. Wolf, Ph.D., is Senior Physicist and Head of the Ion Source and Injector Group at GSI the German heavy ion research center, Darmstadt, Germany.

Born in December, 1939, he studied physics, chemistry, mathematics, astronomy and radiation biology at the University of Heidelberg. His Diploma thesis was on the investigation of laser interferometers and he received his Ph.D. in applied physics with a thesis on cold cathode Penning ion sources.

In 1970 Dr. Wolf was appointed to GSI for ion source development and in 1973 became head of the ion source group at GSI. He has investigated Duoplasmatron-, Duopigatron-, Penning-, CHORDIS, ECR- and Mevva-ion sources and was involved in research projects on surface modification of materials by ion beams.

He has worked in several international cooperations on accelerator projects and was chairman of the 4th International Conference on Ion Sources in Bensheim, Germany (1991).

CONTRIBUTORS

Gerald D. Alton
Oak Ridge National Laboratory
Oak Ridge, Tennessee

Reinhard Becker
Intitute for Applied Physics
University of Frankfurt
Frankfurt, Germany

François Bourg
Centre d'Etudes Nucleaires
  de Grenoble
Grenoble, Cedex, France

Ian G. Brown
Lawrence Berkeley Laboratory
University of California
Berkeley, California

Hans Horneff
GSI Center for Heavy Ion Research
Darmstadt, Germany

Junzo Ishikawa
Department of Electronics
Kyoto University
Kyoto, Japan

Timothy W. Jolly
Oxford Instruments
Bristol, Avon, United Kingdom

Ka-Ngo Leung
Ion Beam Technology Program
Lawrence Berkeley Laboratory
Berkeley, California

B. Sharkov
Institute of Theoretical and
Experimental Physics
Moscow, Russia

Peter Späedtke
GSI Center for Heavy Ion Research
Darmstadt, Germany

Peter Strehl
GSI Center for Heavy Ion Research
Darmstadt, Germany

Bernhard Wolf
GSI Center for Heavy Ion Research
Darmstadt, Germany

TABLE OF CONTENTS

Chapter 1

FUNDAMENTAL PROCESSES

Ian Brown

CONTENTS

1 INTRODUCTION

The plasma state is the fourth state of matter, following the solid, liquid, and gaseous states. This common description of plasma refers to its energy content — as the temperature of a substance is increased, the material changes first from solid to

0-8493-2502-1/95/$0.00+$.50

liquid, then liquid to gas, and finally from gas to plasma. In a plasma, some of the orbital electrons are stripped from their nuclei and are free to participate as individual particles. While in a gas the individual particles are molecules of the gas species (or atoms, for the noble gases), in a plasma the individual particles that make up the plasma are, in general, three different kinds — ions, electrons, and neutrals. Because some of the particles are now charged, as opposed to the neutral particles of an ordinary gas, the kinds of interactions that can take place between the particles, and between the plasma as a whole and external fields, are much more diverse than for the gaseous state.

Plasmas exist in nature in those environments where the temperature is adequately high, such as in the sun, stars and in the ionosphere, and on the earth in transient forms such as lightning. Man-made plasmas have become commonplace and are part of the modern world in forms such as fluorescent lamps, neon signs, and high-voltage sparks. Laboratory plasmas can be created in a wide variety of ways, most commonly as electrical discharges of one kind or another. Industrially, plasmas are used in various forms for semiconductor processing, materials modification and synthesis, and other purposes. In the 1970s and 1980s the controlled-fusion research programs in many countries around the world dominated the plasma physics research scene internationally. In ion sources, plasmas are the medium from which the ions are extracted.

Here we survey and summarize the basic parameters and characteristics of the plasma medium. Particle density and temperature are defined and related to some of the different kinds of plasmas encountered in ion sources. Ionization phenomena — the ways in which the ionized plasma may be formed from the neutral medium — are discussed next. Finally, we summarize some of the characteristics of the plasma boundary and plasma behavior in magnetic fields. The presentation is condensed; for more detail than is presented here the reader might consult any of a number of excellent texts on the subject (e.g., References 1 to 7). The review presented here is limited to those parts of the very broad field of plasma physics that are important for an understanding of ion source fundamentals.

2   BASIC PLASMA PARAMETERS

2.1   Plasma Density, Degree of Ionization, and Temperature

In the simplest case a plasma contains positively charged ions and negatively charged electrons. The ion density, particles per centimeter cubed or per meter cubed, is commonly given the symbol n_i and the electron density n_e. If the ions are all singly ionized, $q_i = +1$, and because the plasma is overall charge-neutral, the ion density equals the electron density, $n_i=n_e$. In the more general case the plasma may contain multiply charged ions, say $q_i = 1+, 2+, ...,$ or even negatively charged ions, $q_i = -1$, and there may also be neutral particles of density n_n; then the ion and electron particle densities need not be the same. Charge neutrality is still preserved, however, as expressed by the general condition

$$\Sigma q_i n_i = n_e \tag{1}$$

The term *plasma density* is often used to mean the ion density or electron density of the plasma, but note that the term is ill-defined except for the case of a plasma of singly charged positive ions. A more precise way of describing the plasma particle density is to specify the electron density and the distribution of ion charge states.

The fractional ionization or percentage ionization of the plasma is defined as the ratio of ion density to total density of ions and neutral particles,

$$\text{fractional ionization} = \frac{n_i}{n_i + n_n} \tag{2}$$

If there are no neutral particles in the plasma then the plasma is said to be fully ionized; the term *highly ionized* is loosely used to describe plasmas with percentage ionization greater than about 10% or so. Confusion often occurs with the use of the term *highly ionized*, since it can be used to refer either to a plasma with a high-percentage ionization or to ions that have several electrons removed. The solution is to be aware of this possibility and to use other terminology. The latter might be referred to, for example, as highly stripped or multiply ionized.

Most plasmas encountered in the laboratory, especially in ion source applications, have densities in the broad range of 10^{10} to 10^{14} cm^{-3}. More commonly, because of the conditions imposed on the ion source plasma for good beam formation (good ion optics, no breakdown in the extractor grids, etc.) and because the physical dimensions of the beam-formation electrodes (grid spacing and beamlet aperture size) are typically in the range of several millimeters to about 1 cm, the density of the plasma at the extractor is generally in the 10^{12} cm^{-3} range. As a reference, recall that a room temperature gas at a pressure of 1×10^{-4} torr (1.3×10^{-2} Pa) has a particle density of 3.3×10^{12} cm^{-3}.

The temperature of the plasma is also a basic and important parameter. It is usual to specify the plasma temperature in units of electron volts (eV), where

$$1\,\text{eV} = 11,600°\text{K} \tag{3}$$

The ion temperature T_i and the electron temperature T_e are not necessarily equal, and if the plasma is in a magnetic field, an anisotropy is introduced and the particle temperatures parallel to and perpendicular to the field may be different; then we can have four particle temperatures: $T_{i\parallel}$, $T_{i\perp}$, $T_{e\parallel}$, and $T_{e\perp}$. The neutral component will likely be at a different temperature too, T_n. The term *plasma temperature* is thus not very meaningful in the most general case. Furthermore, the concept of temperature is valid only for Maxwellian energy distributions, as described below. Although this covers many kinds of plasmas, the concept of temperature is also often loosely extended to describe plasmas that are not in thermal equilibrium. It is important not to confuse kinetic drift energy of a plasma with temperature. For example, a plasma gun might produce a puff of plasma with drift energy hundreds or even thousands of electron volts, but the temperature would likely be relatively cold, perhaps on the order of 1 eV; in this case one speaks of a drifting or shifted Maxwellian. A cold plasma, such as could be produced in a surface ionization plasma source, for example, might have a temperature of, say, 0.2 eV, which is over 2000°K, however. Arc-produced plasmas might have ion temperatures ~1 eV and electron temperatures somewhat higher, perhaps several electron volts, depending on the kind of arc and the discharge parameters. In microwave-produced plasmas the energy is input directly into the electrons at low pressure; for example, for electron cyclotron resonance (ECR) discharges,[7] T_e can be over 1 keV while the ions remain cold, $T_i < 1$ eV.

2.2 Distribution Functions; Means

Here we briefly summarize the various distribution functions that describe plasmas in Maxwellian equilibrium; they lead to some useful and important derived parameters.

The velocity distribution function f(v) describes the number of particles in a given velocity interval, $dn = f(v)dv = f(v_x, v_y, v_z)dv_x dv_y dv_z$. For a plasma in thermal equilibrium the distribution function is Maxwellian, given by

$$f(v_x, v_y, v_z) = n\left(\frac{m}{2\pi kT}\right)^{3/2} e^{-\frac{m(v_x^2+v_y^2+v_z^2)}{2kT}} \tag{4}$$

where m is the particle mass and kT is the temperature in energy units. The three-dimensional Maxwellian velocity distribution function is a product of three independent one-dimensional velocity distribution functions, each of which is of the form

$$f(v_x) = n\left(\frac{m}{2\pi kT}\right)^{\frac{1}{2}} e^{-\frac{mv_x^2}{2kT}} \tag{5}$$

The distribution functions of speed F(v) and of energy f(E) can be obtained from the velocity distribution function,

$$F(v) = 4\pi v^2 f(v)$$

$$= 4\pi v^2 n\left(\frac{m}{2\pi kT}\right)^{\frac{3}{2}} e^{-\frac{mv^2}{2kT}} \tag{6}$$

and

$$f(E) = F(v)\frac{dv}{dE}$$

$$= n\sqrt{\frac{4}{\pi}}(kT)^{-\frac{3}{2}}\sqrt{E}e^{-\frac{E}{kT}} \tag{7}$$

The mean particle speed and mean energy then follow,

$$\bar{v} = \sqrt{\frac{8kT}{\pi m}} \tag{8}$$

and

$$\bar{E} = \frac{3}{2}kT \tag{9}$$

In an isotropic plasma the kinetic energy is divided equally between the three degrees of freedom:

$$\overline{E}_x = \overline{E}_y = \overline{E}_z = \frac{1}{2}kT \tag{10}$$

The mean thermal speeds of ions and electrons can be written as

$$\overline{v}_e = 67\sqrt{T_e} \quad \text{cm } \mu s \tag{11}$$

and

$$\overline{v}_i = 1.57\sqrt{\frac{T_i}{A}} \quad \text{cm } \mu s \tag{12}$$

where T_e and T_i are in electron volts and A is the ion mass in atomic mass units.

2.3 Collisions

Because the plasma particles are charged, their interactions are fundamentally different from the interactions between the neutral particles of an ordinary gas. In a plasma, energy and momentum are exchanged mostly via a large number of distant encounters rather than by close encounters, and the velocity vector of a test particle is changed in direction and magnitude by a random-walk process of many small steps; particle momentum and energy suffer a gradual change rather than a sudden change at each collision. The concepts of discrete billiard-ball-like collisions, and the associated collision times and mean free paths, as we are accustomed to thinking about for molecules in a gas are no longer appropriate, and need now to be replaced by the concepts of random walk and relaxation time. For example, the angular relaxation time τ_θ can be defined as the time for a particle to be deflected through an angle of 90° by the sum effect of many distant encounters; this might also be called the 90° deflection time. Depending on the kind of plasma (constituent particles, anisotropies, different species at different temperatures, etc.), there can be many different kinds of relaxation times — between like and unlike particles (ions, electrons), change in speed or exchange of energy or momentum, perhaps between species with different properties in different directions (parallel or perpendicular to an applied magnetic field) and perhaps between hot and cold components, etc. For simplicity, and as a residue from the more familiar case of neutral particle encounters, it is, nevertheless, common to speak loosely of collision time, with the implicit understanding that what is really meant is relaxation time.

With this terminology simplification we can use many of the relationships that are familiar from the kinetic theory of gases. The mean free path λ for a particular process is related to the cross section σ for that process by

$$\lambda = \frac{1}{n\sigma} \tag{13}$$

where n is the appropriate particle density. The collision time τ and its reciprocal ν, the collision frequency, are related by

$$\tau = 1/\nu = \lambda/v = 1/n\sigma v \tag{14}$$

where v is the particle velocity.

Collision times in plasmas span many orders of magnitude. For the kinds of plasmas encountered in ion sources, they generally vary from nanoseconds to milliseconds. Some handy formulas (taken from Reference 3) follow.

Electron-ion collision frequency:

$$\nu_{ei} = 1.5 \times 10^{-6} \frac{q_i^2 n_i \ln \Lambda}{T_e^{3/2}} \tag{15}$$

Electron-electron collision frequency:

$$\nu_{ee} = 3 \times 10^{-6} \frac{n_e \ln \Lambda}{T_e^{3/2}} \tag{16}$$

Ion-ion collision frequency:

$$\nu_{ii} = q_i^4 \left(\frac{m_e}{m_i} \right)^{1/2} \left(\frac{T_e}{T_i} \right)^{3/2} \nu_{ee} \tag{17}$$

Collision mean free path:

$$\lambda_{ei} = 4.5 \times 10^{13} \frac{T_e^2}{q_i^2 n_i \ln \Lambda} \approx \lambda_{ee} \approx \lambda_{ii} \tag{18}$$

where in these expressions n is the particle density in inverse centimeters cubed, T is the particle temperature in electron volts, $\ln \Lambda$ is a parameter called the Coulomb logarithm[1-5] whose value can, with good approximation, be taken as 10, and the frequencies are in inverse seconds, and the mean free paths in centimeters.

2.4 Plasma Frequency

Plasmas have a number of natural modes of oscillation, perhaps the most fundamental of which are the electron plasma oscillations — the response of the plasma to small departures from complete charge neutrality. These are oscillations of the plasma electron component, and the oscillation frequency is called the electron plasma frequency. The plasma frequency is denoted by the symbol ω_{pe} and is given by

$$\omega_{pe}^2 = \frac{e^2 n_e}{\varepsilon_0 m_e} \tag{19}$$

where ε_0 is the permittivity of free space. The ions can also oscillate at their own natural frequency, as, for example, in standing acoustic waves. This frequency is called the ion plasma frequency, ω_{pi}, and is given by

$$\omega_{pi}^2 = \frac{q^2 e^2 n_i}{\varepsilon_0 m_i} \tag{20}$$

where q is the ion charge state.
These expressions can be written in the convenient forms

$$f_{pe} = 8980 \sqrt{n_e} \quad (Hz) \tag{21}$$

and

$$f_{pi} = 210 q \sqrt{\frac{n_i}{A}} \quad (Hz) \tag{22}$$

where n_e and n_i are the electron and ion densities in inverse centimeters cubed and A is the ion mass in atomic mass units.

3   IONIZATION PHENOMENA

Ions are formed from neutrals by ionization mechanisms of one kind or another. In all plasma devices, including ion sources, the means whereby the plasma is formed is pivotal. The various ways in which the ionization can be done include electron impact ionization, photoionization, field ionization, and surface ionization. Often the names of different kinds of ion sources refer to the specific ways in which the plasmas are formed, as, for example, field emission ion sources, laser ion sources, vacuum arc ion sources, etc. Here we review some of the features of ionization processes that are relevant to ion source behavior.

3.1   Electron Impact Ionization

Ionization of neutrals by collisions with electrons in a gas is the most fundamental kind of ionization mechanism; this is called electron impact ionization.[8] In this case the free electrons in the gas, preestablished by one means or another, are accelerated by an applied electric field to an energy sufficient to cause ionization when they collide with a neutral. In the collisions more free electrons are created, and the discharge grows. A certain minimum energy is needed from the electron-neutral collision for ionization to occur. The electron energy must exceed the energy needed to remove the outermost bound electron from the neutral atom, called the ionization potential, or $E_e > e\phi_i$; more specifically, this is the first ionization potential, referring to removal of the first electron. Actually, the probability of ionization increases with electron energy, rising from zero for energy just below $e\phi_i$ up to a maximum for an electron energy about three to four times $e\phi_i$. In a plasma there is a distribution of energies and the mean electron energy is $3/2 \, kT_e$, where T_e is the plasma electron temperature. Electron impact ionization within a plasma is maximum, therefore, when the electron temperature is several times the ionization potential of the gas being ionized. This condition is rarely met; the ionization potentials for most gases are in the vicinity of 15 eV (helium is the highest at 24.6 eV) while typical gaseous discharge plasma electron temperatures are 1 to 10 eV, and, usually, most ionization events are due to electrons in the tail of the distribution. At the other extreme, note that the electron temperature should not be too great or the ionization efficiency will again decrease. It is usual to refer to the energetic electrons that cause the ionization as primary electrons, and the colder electrons that are produced as part of the ionization of neutrals as secondary electrons. Power must be coupled to the discharge at a rate sufficient to compensate for the energy loss of the primaries due to their ionizing collisions with neutrals.

Gaseous ionization can be enhanced in a number of ways. One way is to use the electrons as efficiently as possible. The electrons can be "reused" by causing them to reflect backward and forward many times in the potential well established between negatively biased electrodes, usually also in the presence of an axial magnetic field. This configuration is called a reflex discharge or a PIG (after the Philips, or Penning, Ionization Gauge), and the electrons are said to be reflexing or pigging. For this to work, clearly the gas pressure, or rather, the electron mean free path for ionizing collisions, must be in the right range with respect to the Paschen curve. (Note also that electron reflexing can occur in high-voltage vacuum devices when it is *not* wanted, leading to electrical breakdown; the solution is to disrupt the reflexing geometry.) Another way of enhancing the ionization process is to provide good confinement, both for the primary electrons doing the ionization as well as for the positive ions that are created. This usually means that the plasma formation process is done within a magnetic field, such as a magnetic mirror or magnetic multipole plasma confinement geometry. The operational parameters and potential applications of an ion source are determined in large part by the source's efficiency as measured in a number of ways — electrical efficiency, ionization efficiency, mass utilization efficiency, etc. Thus, it is that ion sources often embody sophisticated magnetic plasma confinement systems.

3.2 Multiple Electron Removal; Single-Step and Multistep Ionization

When just a single electron is removed from the atom, the ion is said to be singly ionized and the ion charge state is unity, $q_i = 1$. This is the most usual situation encountered in gas discharge phenomena, simply because of the energetics involved. It is also possible for more than one electron to be removed from the atom, and in this case the ion is said to be multiply ionized, multiply stripped, or multiply charged; the ion charge state is then greater than unity, $q_i = 2, 3, ...$ (or 2+, 3+, ...). The ion may also be referred to as being highly ionized, but note that this is the same term used to describe the fractional ionization of the plasma; to avoid confusion it is better not to use the term *highly ionized* to refer to multiply stripped ions. The production of multiply stripped ions has been considered in some detail by a number of authors.[9-11]

Just as a minimum electron energy is needed for removal of the first electron to form the singly charged ion, a similar condition applies for formation of the more highly stripped ionization states. The electron must have an energy at least equal to the n'th ionization potential for formation of ions of charge state q = n+. A chart showing the calculated ionization potentials for all charge states of all the elements is shown in Figure 1.[12] The energy needed to create very highly stripped states is impressive, over 100 keV for fully stripped uranium, for example; a very hot plasma indeed is required to produce such ions.

One can envision two different ways in which multiply stripped ions can be formed: by a single electron-atom encounter in which many electrons are removed in a single event, or by multiple electron-atom/ion encounters in which but a single

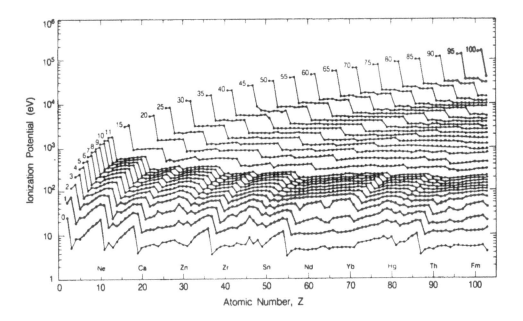

FIGURE 1
Calculated ionization potentials for all charge states of all the elements. (Plotted from data presented in Reference 12.)

electron is removed at a time and the high-charge state is built up by a sequence of many ionizing events. It turns out that both of these processes are possible and do occur, but for most conditions encountered, it is the stepwise ionization process that dominates. This simple picture can be further complicated by the formation of metastable states.[13]

Consider a plasma in which ionization is taking place and high-charge state ions are being created by successive electron impact. The electron temperature T_e is clearly a pivotal parameter in determining the equilibrium ion charge state distribution. Just as for the case of singly charged ions where the ionization cross section maximizes at an electron temperature of about three or four times the first ionization potential, so also for creation of ions of charge state $q = n+$ the optimal T_e is about three or four times the n'th ionization potential. For example, if we were contemplating the possibility of making an ion source for which the production of helium-like Xe, i.e., Xe^{52+}, is to be maximized, then by reference to Figure 1, we see that the 52nd ionization potential is approximately 10 keV, and we can say immediately that a plasma with electron temperature of about 30 to 40 keV is needed (or an electron beam of about this energy). If T_e is much less than this then the ion charge state distribution will maximize at less than $q = 52+$, and if the temperature is much greater then the production rate of Xe^{52+} ions will fall off. On the other hand, and as another example, if we would like just the lowest few charge states of Xe, say with the Xe^{2+} current maximized, then (Figure 1 and a factor of three to four again) an electron temperature of about only 60 to 80 eV is needed. Of course, it may not be easy, or even possible, to establish the optimum plasma conditions; for example, the plasma may oscillate or be extinguished. A major part of the science (art?) of ion source design lies in finding simple ways of using practical plasma techniques to approximate the theoretical requirements.

The electron temperature is only part of the whole story. Consider a plasma with an adequately high, even optimized, electron temperature for the production of some arbitrary multiply charged ion species. If the ion confinement time (or residence time) within the plasma region where the stripping is taking place is too short to allow the step-by-step ionization process to take place, then the targeted charge states will not be produced in spite of the right T_e. It turns out that the important parameter is $n_e\tau_i$ — the product of the plasma electron density n_e and ion confinement time τ_i. Note that a plasma of electron density n_e and electron temperature T_e has an equivalent random electron current j_e given by $j_e = n_e v_e$, where v_e is the mean electron speed as given by Equation 8 or 11, and the parameter $j_e\tau_i$ is often used instead of $n_e\tau_i$. The use of $n_e\tau_i$ seems natural for the case when the stripping medium is a plasma of thermal electrons, and $j_e\tau_i$ seems appropriate for the case when the stripping medium is an electron beam; both situations occur, and, in any case, the two systems are interconvertible. (See Chapter 2.11.)

The time $\tau_i(q)$ needed for stripping to charge state q by successive electron impact and stepwise ionization in a plasma of electron density n_e is given by

$$\tau_i(q) = \sum_{k=0}^{q-1} \frac{1}{n_e\langle\sigma_{k,k+1}v_e\rangle} \tag{23}$$

where $\sigma_{k,k+1}$ is the cross section for ionization from charge state k to charge state k+1, v_e is the electron velocity, and the average $\langle\sigma v\rangle$ is taken over the distribution of electron velocities, usually Maxwellian. A semiempirical calculation of the cross sections for multiple ionization has been presented by Lotz.[14] An example of the kinds of results that are obtained from this model is shown in Figure 2. Here the required $j_e\tau_i$ is plotted as a function of electron energy for the stripping to various charge states of Xe.[15] In an alternative presentation of the results, the evolution of the charge state distributions as a function of the parameter $j_e\tau_i$ is shown in Figure 3. Here the species being stripped to progressively higher charge states is argon, and the (monoenergetic) electron energy is 10 keV.[16]

3.3 Surface Ionization

Atoms can be ionized by contact with a hot metal surface. This is called contact ionization or surface ionization, and was first considered by Langmuir and Kingdon.[17,18] As long as the residence time of particles on the surface is long enough for them to come into thermal equilibrium with the hot surface (typically 10^{-5} to 10^{-3} s), the probability of ionization is given by (a form of) the Langmuir-Saha equation[17]

$$P_i = \frac{n_i}{n_0 + n_i} = \left(1 + \frac{g_0}{g_i}e^{e(\varphi_i - W)/kT}\right)^{-1} \tag{24}$$

where g_i and g_0 are the statistical weights of the ion and atom states (for the alkali metals, $g_i/g_0 = 1/2$), φ_i is the ionization potential of the atom, W is the work function of the metal, and T is the hot-plate temperature. The fractional ionization that can be produced for most φ_i-W combinations is, in general, very low. But for the low-work-function alkali metals and alkali earths (Li, Na, K, Rb, Cs, Ca, Sr, Ba, etc.) on refractory metal hot plates (Ta, W, Re, Ir, Pt, etc.) the phenomenon can provide a useful laboratory tool for ionization. For example, for some typical cases: P_i(K on Pt at 1500 K) = 1.0, P_i(Cs on W at 1500 K) = 0.99, and P_i(Ba on Re at 2200 K) = 0.12. Even for other

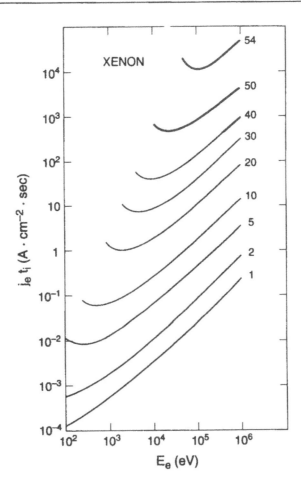

FIGURE 2

$j_e\tau_i$ needed to produce various charge states of Xe as a function of electron energy E_e. (Replotted from data presented in Reference 15.)

elements having higher work functions, surface ionization can sometimes be used to provide a low-current ion beam, though the ionization efficiency is tiny; the method can be useful, for example, in ion beam analysis instrumentation where extremely low-intensity beams are adequate (perhaps involving single particle counting). The ionization can be increased greatly by coating the hot plate with an aluminosilicate of the metal. Zeolite-A is a synthetic alkali aluminosilicate with the composition $6X_2O\cdot6Al_2O_3\cdot12SiO_2$, where X is an alkali metal. Sources of this kind are referred to as aluminosilicate type sources, β-eucryptite sources, spodumene sources, zeolite sources, or Blewett-Jones sources.[19-22]

Electrons are also generated by thermionic emission from the hot plate that supplies the ion flux, and the electron current density j_e is given by the Richardson-Dushman equation,[23]

$$j_e = AT^2e^{-\frac{eW}{kT}}\tag{25}$$

where A is the Richardson constant (120 A cm^{-2} K^{-2}) or an empirical A_{emp} that can vary from 32 to 160 A cm^{-2} K^{-2} for pure (polycrystalline) metals and over a much greater range for oxide and composite surfaces.[24] Surface ionization in combination with an

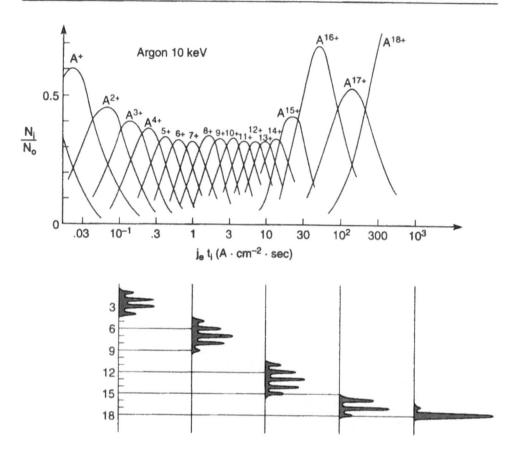

FIGURE 3

Evolution of charge states as a function of $j_e\tau_i$ for electron impact ionization of Ar by 10-keV electrons. (Reproduced with permission from the *Annu. Rev. Nuclear and Particle Sci.*, Vol. 31, ©1981, by Annual Reviews Inc.)

accompanying electron emission can be used as a way of synthesizing a plasma, with more or less separate control over the ion generation and the electron generation. Plasma devices of this kind are called Q-machines (because the plasma can be especially quiescent), and have been well described in the literature.[25,26]

3.4 Field Ionization

Field emissions from sharp points at which very intense electric fields are created can be used to extract either electrons or ions from the solid (or liquid) state. The field ionization phenomenon is used in liquid-metal ion sources to produce beams of species of low-melting-point metals such as Ga, In, Bi, Al, Sn, etc. with extraordinarily high brightness — the ions are formed at a point with effective diameter on the order of a few hundred angstroms, and the beam can be transported and focused to a spot size of this same order. Beam currents are typically a few tens of microamperes and the current density at the focal spot can be over 1 A cm^{-2}.

3.5 Ion Impact Ionization

Ion-atom collisions that involve the transfer of an electron between the interacting particles are called charge-transfer or charge-exchange collisions. In the simplest

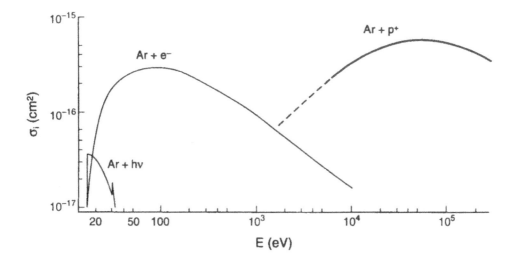

FIGURE 4

Ionization cross sections as functions of energy for ionizing collisions with fast electrons, protons, and photons. (From Winter, H., in *Experimental Methods in Heavy Ion Physics*, Springer-Verlag, Berlin, 1978. With permission.)

case an energetic ion collides with a low-energy neutral atom to produce a cold ion and fast neutral. When the two particles involved are of the same atomic species, the ionization states of the incident fast ion and the resultant slow ion are the same and the process is called resonant charge exchange. Charge exchange can be an important loss mechanism in hot plasmas, for example, in experimental fusion plasmas, but it is rarely a useful ion production mechanism in ion sources.

Ionization can be caused by impact of energetic ions with neutral atoms, but the ion energies required are high compared to electron impact ionization. This is because of the ion-electron mass ratio and the high ion energy required for the ion to have the same speed as a lower-energy electron; the ionization cross section maximizes when the fast particle has a speed equal to that of the orbital electron to be removed. A comparison of the ionization cross sections for Ar with incident electrons, protons, and photons (see Section 3.6) is shown in Figure 4.

3.6 Photoionization

A gas or vapor can be ionized by passage through it of an intense beam of high-energy photons. The absorption of the photon by the atom will result in ejection of an electron, i.e., photoionization, if the photon energy exceeds the ionization energy of the atom,

$$h\nu > e\varphi_i \tag{26}$$

and any excess photon energy is carried off as electron energy. Because ionization potentials are generally in the range 5 to 15 eV and a 1-eV photon has a wavelength of 12,000 Å, the critical wavelength needed for photoionization lies in the broad range of ~800 to 2500 Å, i.e., the vacuum ultraviolet or soft X-ray region. The photoionization cross section maximizes sharply at a photon energy just slightly greater than the minimum required and falls off rather quickly as the photon energy rises.

High-current spark-gap discharges have been used as intense sources of UV radiation for the photoionization of low-pressure (~10^{-3} to 10^{-2} torr) gas[27] and plasmas of density in the 10^{10} cm^{-3} range, and xenon flashtubes have been used to produce fully ionized alkali metal plasmas of density $\sim 10^{11}$ cm^{-3}. Note that laser-produced plasmas involve a completely different phenomenon; in this case, absorption and energy transfer is from the electromagnetic field directly to free electrons.

3.7 Negative Ions

Negative ions can be produced by double-charge exchange of positive ions. The method is often used to form negative ion beams from positive ion beams, for example, for some accelerator injection applications and for the formation of intense beams of energetic neutrals (usually D^0) for fusion plasma heating. The positive ion beam is passed through a vapor cell of low-ionization-potential neutral atoms where negative ions are formed either in a single-step process (for the alkaline earth vapors, e.g., Mg, Ca, Sr, Ba) or a two-step process (for the alkali metal vapors, e.g., Na, K, Rb, Cs). For example, D$^-$ ions can be formed by the processes

$$D^+ + Ba^0 \rightarrow D^- + Ba^{2+}$$

or

$$D^+ + Cs^0 \rightarrow D^0 + Cs^+$$

$$D^0 + Cs^0 \rightarrow D^- + Cs^+$$

These and related ionization phenomena have been investigated and reported by a number of authors.[28-30]

Negative ions are also created within the plasma volume; this is referred to as volume production.[31-33] Of the various atomic processes that can occur in the plasma, it has been found that the dominant process that accounts for the generation of negative ions is the dissociative attachment of thermal electrons to vibrationally or rotationally excited molecules. Quite some skill has been developed in the optimization of the plasma in order to maximize the volume negative-ion production, and as with all negative-ion sources, the extraction of beam is complicated by the need to separate the electron component (using a transverse magnetic field in the extraction region).[34,35]

Another way in which negative ions can be formed is by surface production processes. A beam of positive ions is bombarded against a ceslated metal surface and a fraction of the backscattered flux is returned as negative ions. The positive-ion flux can be provided either by an energetic beam or by appropriate biasing of the target immersed in a plasma. Negative-ion sources based on surface conversion were initially developed by Middleton and Adams.[36]

The production of intense beams of negative hydrogen ions was widely pursued in the late 1970s and 1980s at a number of laboratories as part of the world effort to develop high-energy neutral beams for controlled-fusion application.[37]

4   THE PLASMA BOUNDARY

The electrically charged plasma particles exert an influence on each other at a distance via their electric fields, but the influence of a field particle on a test particle extends only a certain distance through the plasma. This distance is called the shielding distance or screening distance. For distances greater than the shielding length, the field of distant particles is excluded by the (other) plasma particles themselves. This same phenomenon occurs when an electric field is imposed on a plasma from an external source — the field is shielded from the interior region, quite similarly to the way a metal excludes electric fields from its interior. What actually happens is that the plasma particles are redistributed and the field is attenuated exponentially over a distance determined by the plasma electron density and temperature.

A boundary layer is formed at the interface between a plasma and a material wall surrounding the plasma or placed in the plasma. The faster moving (usually) electrons leave the plasma at a greater rate than the more massive ions and the wall acquires a negative charge with respect to the plasma; viewed the other way around, the plasma acquires a positive potential with respect to the wall that confines it. Thus, it is usual (but not universal) that the equilibrium potential of a plasma is positive. The boundary layer is called the plasma sheath. Its physical basis is the same as that which determines the shielding distance described above, and the sheath thickness is the same as the shielding distance. These phenomena were first investigated by Peter Debye some 75 years ago,[38] and the characterizing scale length is commonly called the Debye length, λ_D, given by the expression

$$\lambda_D^2 = \frac{\varepsilon_0 k T_e}{e^2 n_e}$$ (27)

or

$$\lambda_D = 743 \sqrt{\frac{T_e}{n_e}}$$ (28)

where in the second expression T_e is the electron temperature in electron volts, n_e is the electron density in inverse centimeters cubed, and λ_D is the Debye length in centimeters. The magnitude of the sheath drop, i.e., the plasma potential relative to the wall, depends on the precise plasma configuration, but is typically several (three or four) times the electron temperature (sheath drop in volts, electron temperature in electron volts). A floating probe or electrode inserted into a plasma will normally assume a potential, called the floating potential, that is negative with respect to the plasma by approximately 3 to 4 times kT_e; one can look on this as being due to an excess electron flux to the probe, with respect to the ion flux, because of the higher electron velocity. Similarly, floating electrodes within an ion source will assume a potential that depends on the surrounding ion and electron flow.

Within the plasma sheath the plasma is not charge neutral. This is the region where the plasma particles (mostly the electrons, because of their higher mobility) assume such a distribution as to cancel out the external field or to establish the equilibrium transition layer between the plasma and its boundary. The sheath does not have a sharp boundary, but is an exponential fall-off of uncompensated charge density. The $1/e$ width of this transition region is the Debye length λ_D. These concepts of plasma self-shielding, field penetration, and plasma/electrode sheaths are critical to the consideration of ion beam formation at the ion source extractor grids, where one wants a well-defined plasma boundary to form and the electric field applied by the grids to accelerate the ions into a usable beam.

In yet another way of picturing the significance of the Debye length, one can consider this as the scale over which local space-charge fluctuations cancel out. For dimensions within the plasma greater than λ_D, charge neutrality is preserved to a high degree and local fluctuations in potential are small, whereas for dimensions small compared to λ_D, there can be spatial and temporal fluctuations in charge neutrality that are not small.

The Debye length λ_D is the appropriate scale of the boundary or transition region for the case when the wall or electrode assumes its potential only via the plasma, and no additional potential is applied to it. But it is intuitively obvious that if a high voltage is applied to an electrode in a plasma then the sheath must be thicker than for the unbiased case when the plasma-electrode potential (i.e., the sheath drop) is just a few times kT_e. In this case, the high-voltage sheath, the sheath thickness is greater than the Debye length by the root of the ratio of applied voltage to electron temperature:

$$d_{sheath} = \lambda_D \sqrt{\frac{V_{appl}}{kT_e}} \qquad (29)$$

where V_{appl} and kT_e are both expressed in the same units (e.g., volts). This has considerable application to laboratory experimentation through such things as probe sizes, dimensions of wire meshes, extractor hole sizes and separation, etc.

5 MAGNETIC FIELD EFFECTS

The ions and electrons in a plasma are charged particles in motion and experience an interaction with a magnetic field — the plasma can be magnetized. The ions and electrons move in orbits around the magnetic field lines and, apart from collisions with other plasma particles, act as though they are tied to the field lines. The behavior of a plasma in a magnetic field can be profoundly different from a plasma in the absence of a magnetic field.

5.1 Gyromotion

A particle with charge Q and velocity v moving in a magnetic field of flux density B experiences a force F given by

$$F = Qv \times B \qquad (30)$$

which translates as an inward force on the particle as it moves with perpendicular velocity $v_\perp$ in a circular motion about the parallel component of magnetic field $B_\parallel$. The parallel velocity component $v_\parallel$ has no interaction with the magnetic field ($v_\parallel \times B_\parallel = 0$), and the general trajectory of a charged particle in a magnetic field is helical. Equating the Lorentz force of Equation 30 to the centripetal force, an expression for the radius of the circular orbit is obtained:

$$\rho = \frac{mv_\perp}{QB} \qquad (31)$$

called the particle cyclotron radius or gyro-radius. This can be expressed in terms of the particle temperature via the equality $E_\perp = \frac{1}{2} mv_\perp^2 = kT_\perp$, but note that temperature is not a single-particle parameter but describes a Maxwellian distribution of velocities, so the new gyro-radius is an average over the distribution. To reasonable accuracy the ion gyro-radius can be written as

$$\rho_i = 0.0014 \frac{\sqrt{AT_\perp}}{qB} \quad (cm) \tag{32}$$

where A is the ion mass in atomic mass units, $T_\perp$ is the perpendicular ion temperature in electron volts, q is the ion charge state, and B is the magnetic flux density in Teslas. Similarly, the electron gyro-radius can be written as

$$\rho_e = 0.00033 \frac{\sqrt{T_{e\perp}}}{B} \quad (cm) \tag{33}$$

where $T_{e\perp}$ is the perpendicular electron temperature in electron volts and again B is in Teslas. For the parameters generally encountered in ion-source plasmas, ρ_i is typically several millimeters and ρ_e is on the order of a tenth of a millimeter.

The frequencies with which the ions and electrons gyrate around the field lines are important parameters. Equating the centripetal force in the form $m\omega^2 r$ to the Lorentz force of Equation 30 in the form $Q\omega rB$ yields an expression for the cyclotron frequency, or gyro frequency, as

$$\omega_c = \frac{QB}{m} \tag{34}$$

where for ions $Q = q_i e$ and m is the ion mass m_i, and for electrons $Q = e$ and m is the electron mass m_e. One can write the ion cyclotron frequency in simplified form as

$$f_{ci} = 15.2 \, qB/A \quad (MHz) \tag{35}$$

and the electron cyclotron frequency as

$$f_{ce} = 28B \quad (GHz) \tag{36}$$

where again q is the ion charge state and B is in Teslas. For magnetic fields on the order of a kilogauss (100 mT) the ion cyclotron frequency is typically in the low to fractional megahertz range and the electron cyclotron frequency is in the low-microwave band.

The cyclotron frequency as given by Equation 34 carries a sign corresponding to the sign of the charge Q. This signifies the sense of the particle rotation about its magnetic field line. The electron gyromotion is right handed, and the ion gyromotion is left handed. These senses become important for the coupling of some kinds of waves into the plasma — axially propagating waves (of the right frequency) can couple power into the plasma via the electrons if the wave is right-hand polarized or into the ions if the wave is left-hand polarized.

The electron component of a plasma can be heated by coupling into it microwave power at the electron cyclotron frequency — the power transfer is resonant, and the scheme is referred to as electron cyclotron resonance heating (ECRH). Plasmas formed and heated in this way are called ECR plasmas. The electrons can be heated to very high energies, and hot-electron temperatures in ECR plasmas can readily reach the 10 keV range (the term *temperature* here is used loosely, since the hot-electron energy distribution in ECR plasmas is not Maxwellian). For this to be possible it is necessary that the electron collision frequency be small compared to the electron cyclotron frequency, i.e., that the residual gas pressure be not too high; typically, the pressure is in the 10^{-6} to 10^{-5} torr range. Because of the high electron energy, very highly stripped ions are produced in these kinds of plasma. This approach can be embodied in an ion source configuration and the ions extracted as a beam, so forming an ECR ion source.

Microwave power is often used for plasma formation at higher gas pressures also, say in the broad range of 10^{-3} to 10^2 torr. In this case the coupling is not resonant, since collisions effectively prohibit any ordered electron cyclotron motion. Nevertheless, a microwave-produced plasma can be formed, with or without magnetic fields present. The terminology for this kind of plasma is still evolving, but one description that has been used and that fits quite well is *high pressure microwave discharge*. Whereas ECR plasmas are typified by high electron temperature and high ion charge states, high-pressure microwave plasmas are usually of low electron temperature and low ion charge state. ECR plasmas are also in general of lower density than high-pressure microwave plasmas.

In the simplest case it can be shown that the resonant absorption of microwave power at the electron cyclotron frequency can occur only for plasmas with electron density lower than a certain critical density, n_{crit}, also called the cutoff density: $n < n_{crit}$, otherwise the microwave power is not transported into the plasma but is reflected. This condition can also be expressed as a condition on the microwave frequency, which must be greater than the electron plasma frequency of the plasma: $\omega > \omega_{pe}$. The wave frequency that is equal to the plasma frequency for a given density is called the cutoff frequency or critical frequency. For a given microwave frequency the critical density is given by

$$n_{crit} = 1.25 \times 10^{10} f'^2 \quad \left(cm^{-3}\right) \tag{37}$$

where the microwave frequency f' is expressed in gigahertz. In spite of this limit, however, it has been found that under some conditions (having to do with mode conversion, nonlinear effects, and/or finite plasma effects) microwave-produced plasmas can be formed with density greater than the critical density. Such plasmas are called overdense. Since a higher-density plasma in an ion source translates into a higher current of the extracted ion beam, there is clearly some advantage in being able to produce overdense plasmas in ion source configurations. While the production of highly overdense microwave-produced plasmas of the high-pressure type has been demonstrated, attempts to produce overdense ECR plasmas have not met with a great deal of success.

5.2 Magnetic Confinement

That the ions and electrons in a plasma are tied in their orbital motion to the field lines provides a means of confining the plasma, at least in the direction transverse to the field. Plasma loss along the field can be reduced by increasing the field strength at the ends of the confinement region; this configuration is called a magnetic mirror. In another magnetic geometry that has become widespread in recent years, a surface array of individual permanent magnets, usually of the high-field rare earth kind (Sm-Co or Nd-Fe), is used to establish a field that is appreciable (typically approaching 1 kG [100 mT] or thereabouts) only near the wall and is negligible in most of the interior region; this is called a magnetic multipole confinement geometry, or more vividly, a magnetic bucket.

The essence of magnetic mirroring of plasma particles lies in the transfer of their kinetic energy between the perpendicular (to the magnetic field) and parallel directions. In the absence of collisions or electric fields, the particle energy is given by $E = E_\perp + E_\parallel$, where $E_\perp = \frac{1}{2} mv_\perp^2$ and $E_\parallel = \frac{1}{2} mv_\parallel^2$. As a particle, either ion or electron, moves in its helical trajectory upward into a region of higher field strength, energy is transferred from $E_\parallel$ to $E_\perp$; the parallel velocity is decreased until, at the mirror point, the parallel velocity is reduced to zero and then the particle reflects. It is as if the particle experiences a force on it, and indeed the situation can be treated in this way; in a nonuniform magnetic field the particle experiences a force given by $\mu \mathrm{grad}B$, where μ is the magnetic moment of the particle, defined as $\mu = \frac{1}{2} mv_\perp^2/B$.

Note that all these ideas are valid only so long as the particles are free to execute the gyromotion. Often this condition is not met because of particle collisions. If an ion suffers a collision on the average in a time short compared to the ion cyclotron period, then it is clear that no gyromotion can be sustained; ion transport is dominated in this case by collisional processes and the magnetic field will have little effect. The ions can experience the field only if the ion collision frequency is small compared to the ion cyclotron frequency, $v_i \ll \omega_{ci}$. Similarly, for the electrons to be magnetized (experience the effect of the magnetic field and execute gyromotion) their collision frequency must be small compared to the electron cyclotron frequency, $v_e \ll \omega_{ce}$. Nevertheless, it can sometimes occur that the primary magnetic confinement mechanism is through the electrons (electrons are mirrored), with the ions being confined electrostatically (potential well established by the electrons), even for the case of collisional ions.

5.3 Magnetic and Plasma Pressure

The magnetic field pressure in the direction transverse to the field is given by the standard expression

$$P_{mag} = \frac{B^2}{2\mu} \tag{38}$$

while the kinetic pressure of the plasma (transverse to the field) is given by

$$P_{plasma} = n_e k T_{e\perp} + n_i k T_{i\perp} \tag{39}$$

The kinetic pressure is the sum of contributions from the electrons and the ions, and it is the perpendicular components of temperature that are relevant here. In general, there will be some magnetic field in the interior, plasma region, B_{int}, which will be different from the field external to the plasma, B_{ext}. In equilibrium the (magnetic plus kinetic) pressure in the interior region must equal the magnetic pressure in the external region,

$$\frac{B_{int}^2}{2\mu} + \left(n_e kT_{e\perp} + n_i kT_{i\perp}\right) = \frac{B_{ext}^2}{2\mu} \tag{40}$$

The ratio of the plasma pressure to the confining (external) magnetic field pressure is sometimes of interest; this parameter is called the plasma β,

$$\beta = \frac{P_{plasma}}{P_{mag}} = \frac{2\mu\left(n_e kT_{e\perp} + n_i kT_{i\perp}\right)}{B_{ext}^2} \tag{41}$$

For most confinement systems, $\beta \ll 1$. It is often helpful to examine the magnetic and particle pressures, i.e., the β, when considering possible magnetic confinement schemes.

Gas discharge plasmas as used in most ion sources are usually low pressure, with or without confining magnetic field. Note, however, that plasmas can be created more or less directly from the solid state, in which case they are "born" at extremely high pressures, since the density will in the very first place be near to solid density ($\sim 10^{23}$ cm^{-3}) and the temperature at least a few electron volts ($>10^4$ K). Such plasmas include, for example, those produced by the interaction of focused, short-pulse, high-power laser beams with solid surfaces (laser plasmas for short) where the power density can be as high as $\sim 10^{10}$ to 10^{15} W cm^{-2}, and at the cathode spots in vacuum arcs where the arc current constricts down to micrometer-sized regions and the current density can reach $\sim 10^6$ to 10^8 A cm^{-2}. Then, the plasma dynamics are governed by the intense pressure gradients in the initial high-pressure plasma, and the plasma expands rapidly away from the point of creation as a plume, decreasing in density and temperature under the influence of the expansion. These plasmas also can be embodied within ion sources, and the ion source design must accommodate (or utilize!) the plasma expansion. Laser ion sources (Chapter 2.10) and vacuum arc ion sources (Chapter 2.12) are of this type.

REFERENCES

1. D. J. Rose and M. Clark, Jr., *Plasmas and Controlled Fusion*, Wiley, New York, 1961.
2. S. C. Brown, *Introduction to Electrical Discharges in Gases*, Wiley, New York, 1966.
3. F. F. Chen, *Introduction to Plasma Physics*, Plenum, New York, 1974.
4. J. A. Bittencourt, *Fundamentals of Plasma Physics*, Pergammon, New York, 1986.
5. A. Anders, *A Formulary for Plasma Physics*, Akademie-Verlag, Berlin, 1990.
6. G. Fuchs, *IEEE Trans. Nucl. Sci.*, NS-19, 160, 1972.
7. G. Shirkov, *Rev. Sci. Instrum.*, 63, 2896, 1992; *Nucl. Instrum. Meth. Phys. Res.*, A302, 1, 1991; *Proc. 11th Workshop ECR Ion Sources*, C. Lyneis, Ed., Gronigen.
8. T. D. Maerk and G. H. Dunn, Eds., *Electron Impact Ionization*, Springer, New York, 1985. See, in particular, T. D. Maerk, Partial ionization cross sections, chap. 5 therein, and G. H. Dunn, Electron-ion ionization, chap. 8 therein.
9. H. Winter, Production of multiply charged heavy ions, in *Experimental Methods in Heavy Ion Physics*, K. Bethge, Ed., Lecture Notes in Physics series, Springer-Verlag, Berlin, 1978.
10. H. Winter and B. H. Wolf, *Plasma Phys.*, 16, 791, 1974; *Nucl. Instrum. Methods*, 127, 445, 1975.
11. R. K. Janev, L. P. Presnyakov, and V. P. Shevelko, *Physics of Highly Charged Ions*, Springer series in Electrophysics, Vol. 13, G. Ecker, Ed., Springer, Berlin, 1985. See, in particular, Electron-impact ionization of highly charged ions, chap. 5 therein.
12. T. A. Carlson, C. W. Nestor, Jr., N. Wasserman, and J. D. McDowell, *At. Data*, 2, 63, 1970.
13. F. Aumayr and H. Winter, *Phys. Scr.*, T28, 96, 1989.
14. W. Lotz, *Z. Phys.*, 216, 241, 1968.

15. Report SFEC T 10 — Cryebis II, Laboratoire National Saturne, Center for Nuclear Studies, Saclay, France, 1981.
16. J. Arianer and R. Geller, *Annu. Rev. Nucl. Part. Sci.*, 31, 19, 1981.
17. I. Langmuir and K. H. Kingdon, *Proc. R. Soc. London*, A107, 61, 1925.
18. E. Y. Zandberg and N. I. Ionov, *Sov. Phys. Usp.*, 67(2), 255, 1959; *Usp. Fiz. Nauk*, 57, 581, 1959.
19. J. L. Hundley, *Phys. Rev.*, 30, 864, 1927.
20. K. T. Bainbridge, *J. Franklin Inst.*, 212, 317, 1931.
21. J. P. Blewett and E. W. Jones, *Phys. Rev.*, 50, 465, 1936.
22. S. K. Allison and M. Kamegai, *Rev. Sci. Instrum.*, 32, 1090, 1961.
23. S. Kuhn, *Plasma Phys.*, 21, 613, 1979; Plasma Phys., 23, 881, 1981.
24. G. F. Smith, *Phys. Rev.*, 94, 295, 1954.
25. R. Schrittwieser, R. Koslover, R. Karim, and N. Rynn, *J. Appl. Phys.*, 58, 598, 1985.
26. R. W. Motley, *Q Machines*, Academic Press, New York, 1975.
27. J. E. Robin and K. R. MacKenzie, *Phys. Fluids*, 14, 1171, 1971.
28. R. E. Olson, *IEEE Trans. Nucl. Sci.*, NS-23, 971, 1976.
29. A. S. Schlachter, K. R. Stalder, and J. W. Stearns, *Phys. Rev.*, A22, 2494, 1980.
30. R. Geller, B. Jacquot, C. Jacquot, and P. Sermot, *Nucl. Instrum. Methods*, 175, 261, 1980.
31. M. Bacal, A. M. Bruneteau, H. J. Doucet, W. G. Graham, and G. W. Hamilton, *Proc. Int. Symp. Production and Neutralization of Hydrogen Ions and Beams*, Brookhaven National Laboratory, Upton, NY, October 1980.
32. G. W. Hamilton, M. Bacal, A. M. Bruneteau, H. J. Doucet, and M. Nachman, *Proc. Int. Symp. Production and Neutralization of Hydrogen Ions and Beams*, Brookhaven National Laboratory, Upton, NY, October 1980.
33. F. G. Baksht, G. A. Djuzhev, L. I. Elizarov, V. G. Ivanov, A. A. Kostin, and S. M. Shkolnik, *Plasma Sources Sci. Technol.*, 3, 88, 1994.
34. K. N. Leung and K. W. Ehlers, *Proc. Int. Symp. Production and Neutralization of Hydrogen Ions and Beams*, Brookhaven National Laboratory, Upton, NY, November 1983; *Rev. Sci. Instrum.*, 54, 56, 1983.
35. R. McAdams, A. J. T. Holmes, M. P. S. Nightingale, L. M. Lea, M. D. Hinton, A. F. Newman, and T. S. Green, *Proc. Int. Symp. Production and Neutralization of Hydrogen Ions and Beams*, Brookhaven National Laboratory, Upton, NY, October 1986.
36. R. Middleton and C. T. Adams, *Nucl. Instrum. Methods*, 118, 329, 1974.
37. K. N. Leung, Negative ion sources, in *The Physics and Technology of Ion Sources*, I. G. Brown, Ed., Wiley, New York, 1989.
38. P. Debye and W. Hückel, *Phys. Z.*, 24, 185, 1923.

Chapter 2

CHARACTERIZATION
OF ION SOURCES
INTRODUCTION

B. H. Wolf

0-8493-2502-1/95/$0.00+$.50
© 1995 by CRC Press Inc.

The development of ion sources started with the investigation of the canal rays by Goldstein in 1886.[1] Later, electron-atom collisions were used to design low-current ion sources. Higher ion currents became available with the investigation of the arc discharge ion sources in the 1930s. In the 1940s the rf- and microwave-driven discharges were investigated for ion beam production. Historical reviews and bibliographies can be found in the works of von Ardenne,[2] L. Vályi,[3] and R. Wilson and G. Brewer.[4] Various books are available on the physical and technological aspects of ion sources in general or with chapters on specific ion sources.[5-13]

In the following sections, various ion sources are presented in the order of their working principle with a general description at the beginning of each section. Typical designs are shown in a standard form to make information on source data quickly available and comparison of different ion sources easy. The selection of specific sources of one type follows personal criteria of the authors and are not a general quality mark. The distribution of a source, its originality, and its special features are such criteria, and also the information that was delivered to the authors by designers and users. Ion source performance data for one specific ion source may be collected from various publications over some years and may not always show the source potential to date because of the fast development in the field of ion sources.

Each section has its own specific references. This chapter concentrates more on the technical aspects of the ion sources and less on the physics background, which is discussed in detail in the above-mentioned books on ion sources. The first section, however, gives information on cathodes that belongs to all types of ion sources.

REFERENCES

1. E. Goldstein, *Berl. Ber.*, 39, 691, 1896.
2. M. v. Ardenne, *Tabellen zur Angewandten Physik I*, VEB Deutscher Verlag der Wissenschaften, Berlin, 1956 and 1962.
3. L. Vályi, *Atom and Ion Sources*, John Wiley & Sons, London, 1977.
4. R. G. Wilson and G. R. Brewer, *Ion Beams with Applications to Ion Implantation*, R. E. Krieger, Malabar, FL, 1973 and 1979.
5. I. G. Brown, Ed., *The Physics and Technology of Ion Sources*, John Wiley & Sons, New York, 1989.
6. A. T. Forrester, *Large Ion Beams*, John Wiley & Sons, New York, 1987.
7. G. Dearnaly, J. H. Freeman, R. S. Nelson, and J. Stephen, *Ion Implantation*, North Holland, Amsterdam, 1973.
8. M. D. Gabovich, *Physics and Technology of Plasma Ion Sources*, Atomprint, Moscow, 1972; Translated by Foreign Technology Division WO-AFB, Ohio, 1973.
9. A. Septier, *Focusing of Charged Particles*, Academic Press, New York, 1967.
10. A. Guthrie and R. K. Wackerling, *The Characteristics of Electrical Discharges in Magnetic Fields*, McGraw-Hill, New York, 1949.
11. H. S. W. Massey, *Negative Ions*, Cambridge Press, London, 1950.
12. H. S. W. Massey, Ed., *Ionization Phenomena and Ion Sources*, Academic Press, New York, 1983.
13. J. Koch, Ed., *Electromagnetic Isotope Separators and Application of Electromagnetically Enriched Isotopes*, North Holland, Amsterdam, 1958.

CHARACTERIZATION OF ION SOURCES

B. H. Wolf

1 CATHODES

Primary electrons delivered by cathodes of various design play a vital role in many ion sources. Therefore, some data on cathode design and material are summarized.

1.1 Thermionic Cathodes

Electrons are emitted from metal and other surfaces if the temperature is higher than the threshold potential. The maximum current density for a specific temperature is given by the saturation current density j_{eS}, which is given by the Richardson formula

$$j_{eS} = AbT^2 \exp(-e\Phi/kT) \quad A/cm^2 \qquad (1)$$

where ϕ is the work function in volts, T the temperature in kelvin, and A a universal constant, $A = 4\pi mek^2/h^3$ or 120.4 A/cm^2K^2; b is a material-related factor. In many publications A is material dependent, which means it contains b. In Table 1.1 the work function and the factor b for typical cathode materials are summarized and in Figure 1.1 the saturation current densities of these materials are shown.

The increase of the saturation current with temperature is very strong; a 10% change in temperature corresponds to a 10-fold increase of j_{eS}, 20% to a 100-fold increase, and just 1% yields a 26% change in the saturation current.

1.1.1 Metal Cathodes

The simplest and most common cathode is a wire of a high-melting-point material in hairpin, spiral, or spring shapes (see Figure 1.2). These cathodes are usually heated directly by an ac or dc current.

Figure 1.3 shows temperature dependency of the resistivity for various cathode materials. For different wire diameters d and length l the current is a function of $d^{3/2}$ and is independent of l, and the voltage is a function of $\sqrt{d}$ and increases linearly with the length l. Figure 1.4 shows the voltage and the current needed to reach a specific temperature depending on the wire diameter for tungsten and tantalum.

0-8493-2502-1/95/$0.00+$.50

TABLE 1.1

Work Function and Factor A × b for Various Materials [1]

Material	ϕ(V)	A × b (A cm^{-2} K^{-2})
Molybdenum	4.15	55
Nickel	4.61	30
Tantalum	4.12	60
Tungsten	4.54	60
Barium	2.11	60
Cesium	1.81	160
Iridium	5.40	170
Platinum	5.32	32
Rhenium	4.85	100
Thorium	3.38	70
Ba on W	1.56	1.5
Cs on W	1.36	3.2
Th on W	2.63	3.0
Thoria	2.54	3.0
BaO + SrO	0.95	~10^{-2}
Cs-oxide	0.75	~10^{-2}
TaC	3.14	0.3
LaB$_6$	2.70	29

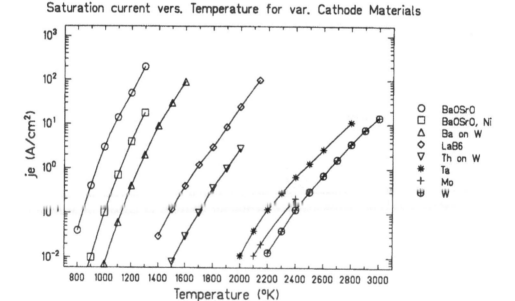

FIGURE 1.1

Saturation current density vs. temperature for various materials.[2,3]

Most of the power put into a filament is radiated and very little is lost through the ends. Most high-temperature metals show similar radiation behavior (~20 W/cm² at 2000°K). The increase of temperature for higher electron output is limited by the increasing evaporation of cathode material, which decreases the cathode lifetime. Tables 1.2 and 1.3 give the respective data for W and Ta.

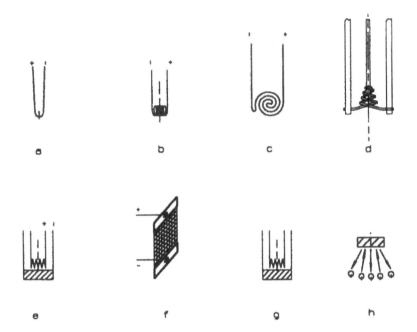

FIGURE 1.2
Common shapes of wire cathodes: (a) hair pin, (b) spiral, (c) helix, (d) helix with middle pin, (e) block cathode, (f) oxide cathode on Ni mesh, (g) indirectly heated oxide cathode, and (h) secondary emission cathode.

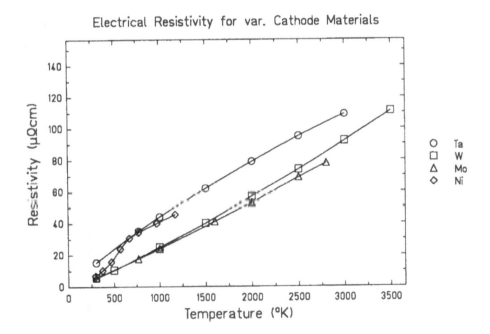

FIGURE 1.3
Resistivity vs. temperature for various cathode materials.[2,3]

In ion sources the cathode lifetime, however, is mostly limited by sputtering, especially in high-current, heavy-ion sources. To extend the lifetime, block cathodes (Figure 1.2e), which are heated from the reverse side by electron bombardment, are used especially in PIG and Calutron type ion sources. Metal cathodes operate at high

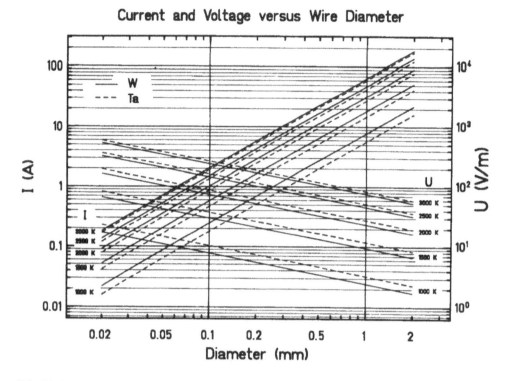

FIGURE 1.4
W and Ta filament voltage and current vs. wire diameter.[4]

temperatures, which implies high radiation heat for the surrounding walls with the need for intensive cooling. They also are easily corroded by oxygen and other reactive gases.

1.1.2 Oxide Cathodes

Alkaline metals or their oxides show high electron emissions at moderate temperatures (~1000°C). Because the metals have a low melting point, and the oxides a low mechanical stability, one has to use a carrier material with a metal surface film or an oxide layer (see Figure 1.1). The carrier material also influences the emissivity of the film and is included in Formula 1 in factor b. The carrier may be heated to the

TABLE 1.2

Electron Emission and Evaporation Data for Tungsten[4]

Temperature (K)	Electron emission (A cm⁻²)	Evaporation rate (g cm⁻² s⁻¹)	Lifetime ⌀ 0.1 mm (h)	Lifetime ⌀ 1 mm (h)	Lifetime ⌀ 1 cm (h)
2100	3.9×10^{-3}	1.6×10^{-12}	1.0×10^{6}	1.0×10^{7}	1.0×10^{8}
2200	1.3×10^{-2}	1.2×10^{-11}	1.4×10^{5}	1.4×10^{6}	1.4×10^{7}
2300	4.1×10^{-2}	7.8×10^{-11}	2.0×10^{4}	2.0×10^{5}	2.0×10^{6}
2400	1.2×10^{-1}	4.4×10^{-10}	3.0×10^{3}	3.0×10^{4}	3.0×10^{5}
2500	3.0×10^{-1}	2.0×10^{-9}	8.3×10^{2}	8.3×10^{3}	8.3×10^{4}
2600	7.0×10^{-1}	8.8×10^{-9}	1.8×10^{2}	1.8×10^{3}	1.8×10^{4}
2700	1.6	3.2×10^{-8}	50	500	5000
2800	3.5	1.1×10^{-7}	14	140	1400
2900	7.3	3.5×10^{-7}	4.6	46	460
3000	14.0	9.7×10^{-7}	1.7	17	170

TABLE 1.3

Electron Emission and Evaporation Data for Tantalum[2,4]

Temperature (K)	Electron emission (A cm^{-2})	Evaporation rate (g cm^{-2} s^{-1})	Lifetime		
			ϕ 0.1 mm (h)	ϕ 1 mm (h)	ϕ 1 cm (h)
2000	1.1×10^{-2}	1.6×10^{-12}	1.0×10^6	1.0×10^7	1.0×10^8
2200	1.2×10^{-1}	9.8×10^{-11}	7.5×10^4	7.5×10^5	7.5×10^6
2400	6.5×10^{-1}	3.0×10^{-9}	7.0×10^2	7.0×10^3	7.0×10^4
2600	2.7	5.5×10^{-8}	35	350	3500
2800	11	6.6×10^{-7}	3	30	300

necessary temperature directly by current flow or indirectly by a separate heater (Figure 1.2f and g). Table 1.4 shows emission and evaporation data for oxide cathode materials.

Some recipes for oxide mixtures and for the preparation of oxide cathodes can be found in Reference 4.

Some oxide cathodes are stable in oxygen discharges. They have to be activated in vacuum in accordance with the instructions given above or by the deliverer of the materials. Some oxide cathodes are destroyed or have to be reactivated after being put back in air. Most oxide cathodes deteriorate under heavy-ion bombardment.

1.1.3 Dispenser Cathodes

Dispenser cathodes maintain a thin film of the emitting material out of a reservoir behind or within a porous carrier, usually tungsten. The thoriated tungsten cathode belongs to this category (tungsten wire doped with $\leq$2% thorium; see Figure 1.1). Figure 1.5 shows typical examples of dispenser cathodes. Most dispenser cathodes have to be activated similarly to oxide cathodes, but they are more stable against ion bombardment. They show high emission at low temperatures, $\leq$1000°C. Table 1.5 shows emission and evaporation data for a BaO-dispenser cathode.

Dispenser cathodes can be purchased from a specialized company.[5]

1.1.4 LaB$_6$ Cathodes

Lanthanum hexaboride is an excellent electron emitter at medium (1600°C) temperatures. It is, in principle, a dispenser type cathode since the lanthanum evaporates slowly out of the boron frame. LaB$_6$ is usually sintered from powder and can be

TABLE 1.4

Electron Emission and Evaporation Data
for BaO-SrO Oxide Cathodes[4]

Temperature (K)	Electron emission (A cm^{-2})	Evaporation rate (g cm^{-2} s^{-1})	Lifetime of 20-μm layer (h)
500	1×10^{-5}		
600	4×10^{-4}		
700	7×10^{-3}		
800	4×10^{-2}		
900	0.4		
1000	3.0	5×10^{-11}	10^4
1100	12	2×10^{-9}	230
1200	50	8×10^{-8}	6
1300	200	1×10^{-6}	0.5

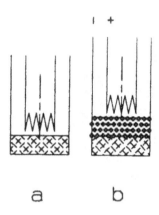

FIGURE 1.5
Dispenser cathodes: (a) porous carrier, (b) separate reservoir.

bought in rods or as pills. It can be shaped by grinding or with a diamond cutter (Figure 1.6).[6] LaB$_6$ has a very low resistivity and can be directly heated or used soldered (special recipe[4]) to a molybdenum foil heater. Data on LaB$_6$ are summarized in Table 1.6.

LaB$_6$ cathodes are stable against ion bombardment and well suited for ion sources. Because of their more difficult preparation techniques they are less used than metal cathodes despite their excellent emission data (see Figure 1.1).

1.2 Liquid-Metal Cathodes

Liquid-metal cathodes (Figure 1.7) are generally known as mercury pool cathodes and are rarely used in ion sources.[7] The metal vapor vacuum arc cathode is, in principle, a liquid-metal cathode. In the cathode spots (see Chapter 2.12) molten material evaporates atoms and emits electrons.[8]

1.3 Hollow Cathodes

Hollow cathodes are usually cold cathodes, which have much higher electron emissions than planar cathodes. Electrons inside a narrow tube (diameter of some millimeters and several times the diameter in length) are trapped and undergo more

TABLE 1.5

Electron Emission and Evaporation Data
for BaO-Dispenser Cathodes[4]

Temperature (K)	Electron emission (A cm^{-2})	Evaporation rate (g cm^{-2} s^{-1})	Lifetime of 30 mg/cm^2 reservoir (h)
800	1×10^{-5}		
900	3×10^{-4}		
1000	5×10^{-3}		
1100	6×10^{-2}		
1200	0.4		
1300	2.0	1.7×10^{-10}	5×10^4
1400	10	2×10^{-9}	4300
1500	30	1.7×10^{-8}	500
1600	90	1×10^{-7}	75

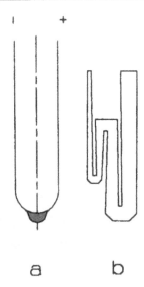

FIGURE 1.6
LaB$_6$ cathodes: (a) point cathode, (b) directly heated cathode cut from solid LaB$_6$.

TABLE 1.6

Electron Emission and Evaporation Data for LaB$_6$[4]

Temperature (K)	Electron emission (A cm^{-2})	Evaporation rate (g cm^{-2} s^{-1})	Lifetime Point cath. (h)	Solid cath. (h)
1600	0.4	1×10^{-11}	2×10^4	7×10^5
1700	1.0	1×10^{-10}	2×10^3	7×10^4
1800	3.0	1×10^{-9}	215	7000
1900	8.5	9×10^{-9}	25	800
2000	25	7×10^{-8}	3	100
2100	100	6×10^{-7}	0.35	12

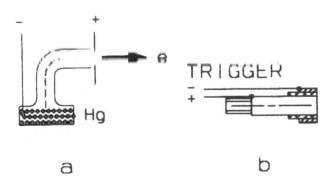

FIGURE 1.7
Liquid-metal cathodes: (a) mercury pool cathode, (b) vacuum arc cathode.

ionizing collisions, creating a denser plasma compared to outside the tube. This dense plasma releases electrons for the main discharge. Hollow cathodes may have coated or heated walls to increase electron emission. Figure 1.8 shows typical arrangements of hollow cathodes.[9]

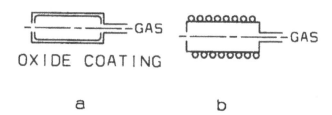

FIGURE 1.8
Hollow-cathode arrangement: (a) with oxide coating, (b) with separate heater.

1.4 Discharge Cathodes

A modification of a hollow cathode is the gas discharge cathode. A separate discharge at higher pressure (usually a Penning discharge in a magnetic field) serves as electron emitter (Figure 1.9).[10] Both hollow cathodes and discharge cathodes can be used in oxygen or corrosive gases and are also relatively stable against ion bombardment.

1.5 Microwave Cathodes

Microwave cathodes are a relatively new type of discharge cathode. Instead of a cold cathode discharge, a microwave discharge is generated and delivers a high electron current for the main discharge. The microwave generators for the usual frequency of 2.4 GHz are easy to build from parts from microwave ovens and low in cost. Because the microwave antennas or windows can be made from a wide variety of materials, corrosion can be kept low and the microwave cathode lifetime is high. Figure 1.10 shows a typical microwave cathode.[11]

1.6 Secondary Emission Cathodes

Secondary emission by ion bombardment can be quite sufficient to maintain a discharge, especially in the presence of a magnetic field, which profoundly increases the ionization efficiency of the electrons in the discharge. Secondary emission or cold cathodes (Figure 1.2h) are mainly used in Penning ion sources of various design (see Chapter 2.5). Table 1.7 gives secondary emission factors of various materials. Surface layers can influence the secondary emission factor quite dramatically in both directions. For example, the emission of pure aluminum or titanium is low, but cathodes

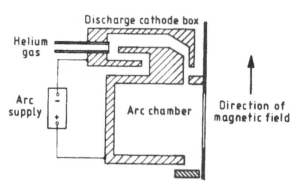

FIGURE 1.9
Discharge cathode arrangement for a calutron ion source.

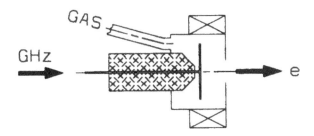

FIGURE 1.10
Microwave cathode.

TABLE 1.7

Secondary Emission Factors for Various Metals and Oxides[1,12]

Material	e⁻	(eV)	Ion (1 keV)				
			He	Ne	A	Kr	Xe
Carbon	1.0	(300)					
Aluminum	0.97	(300)			0.09	0.045	0.015
Silicon	1.1	(600)	0.18	0.17	0.04	0.02	0.002
Copper	1.35	(500)		0.16	0.09	0.07	0.035
Germanium	1.2	(400)	0.19	0.16	0.05	0.02	0.002
Molybdenum	1.23	(400)	0.25	0.23	0.09	0.07	0.02
Tantalum	1.35	(600)	0.28	0.3	0.1		
Titanium	0.9	(300)					
Tungsten	1.43	(700)	0.25	0.25	0.1	0.6	0.016
Al_2O_3	2.6	(250)					
TiO_2							

from these materials give lower discharge voltages or have higher emission than tantalum or tungsten due to oxide layers that are permanently maintained. If one spot of the surface is cleaned by sputtering, the oxygen goes immediately to the next free atom. The secondary emission increases with voltage up to several tens of kiloelectron volts. Multiply charged ions have a higher secondary emission factor than singly charged ions.

REFERENCES

1. C. J. Smithells, *Metals Reference Book*, Vol. III, Butterworths, London, 1967, 737ff.
2. W. Espe, *Werkstoffe der Hochvacuumtechnik*, VEB Deutscher Verlag der Wissenschaften, Berlin, 1959.
3. Y. S. Touloukian, *Thermophysical Properties of High Temperature Solid Materials*, MacMillan, New York-London, 1967.
4. M. v. Ardenne, *Tabellen zur Angewandten Physik I*, VEB Deutscher Verlag der Wissenschaften, Berlin, 1956 and 1962.
5. Spectra-Mat, Inc., Highway 1, Watsonville, CA 95076, U.S.A.
6. K. N. Leung, *Vacuum*, 36, 865, 1986.
7. R. Dawton, The mercury pool cathode applied to ion sources, *Proc. 1st Int. Conf. Ion Sources*, Saclay, 1969, 303.
8. I. G. Brown et al., *Rev. Sci. Instrum.*, 57, 1069, 1986.
9. A. T. Forrester, *Large Ion Beams*, John Wiley & Sons, New York, 1987.
10. A. Guthrie and R. K. Wackerling, *The Characteristics of Electrical Discharges in Magnetic Fields*, McGraw-Hill, New York, 1949.
11. Y. Matsubara et al., *Rev. Sci Instrum.*, 61, 541, 1990.
12. G. Carter and I. S. Colligon, *Ion Bombardment of Solids*, Heinemann, London, 1968, 38.

CHARACTERIZATION
OF ION SOURCES

B. H. Wolf

2 ELECTRON BOMBARDMENT ION SOURCES

Advantages: Low energy spread of the ions; easy and cheap design possible.
Disadvantages: Low ion current; filament problems with oxygen and corrosive elements.

Electron bombardment ion sources are mainly used in areas where a well-defined electron energy is requested and a moderate ion current is sufficient. This is the case in mass spectrometry and in investigations of metastable states in ions and atoms. The design of an electron bombardment ion source can be simple and cheap.

2.1 Ion Source History

Electron bombardment ion sources have been investigated by many people starting with Dempster in 1916.[1] Various designs have been published; the most important for application are the Nier,[2] Bernas,[3] and Nielsen[4] types.

2.2 Working Principle of the Electron Bombardment Ion Source and Description of the Discharge (Figure 2.1)

Electrons are emitted by a cathode, usually by thermionic emission, and accelerated to an anode. Some of these primary electrons have collisions with gas atoms and ionize them. Secondary electrons from these collisions can be accelerated toward the anode to energies depending on the potential distribution and the starting point of the electron. The potential distribution depends on electron current density and gas pressure. At low pressure the potential is linear between cathode and anode without an electron flow and is bent down to negative values and a strong increase at the anode (see Figure 2.2) under the influence of the electron space charge. At higher pressures a discharge occurs, started by the avalanche of ionizing secondary electrons, leading to a plasma on a specific potential with potential drops at both the cathode and the anode. Typical potential distributions are shown in Figure 2.2.[5,6]

If the anode potential is increased to the ionization potential of the gas used, ions are created in an increasing number until the energy reaches the maximum of the ionization cross section (see Figure 2.3).[7] With increasing anode potential, more and more secondary electrons can gain energies above the threshold energy and contribute to the ion production. The ion energy spread can be as large as the potential

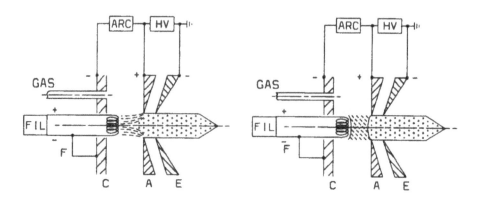

FIGURE 2.1
Electron bombardment ion source scheme and circuitry: (a) low-pressure mode, (b) discharge mode.

difference between the threshold energy and anode potential, depending on the point
of ionization and the position of the extraction aperture. The ion energy spread in the
low-pressure regime can be drastically reduced by the introduction of a grid close to
the cathode, which acts as anode, and a drift space to the extraction aperture (former
anode), which is kept a few volts above or at grid potential. In this way, energy of the
ionizing electrons is uniform and the energy spread of the ions small. This is called
the forced electron beam-induced arc discharge (FEBIAD) mode (Figure 2.4). For the
high-intensity arc mode, this method is not applicable because the grid would be
destroyed by the plasma ions, and the plasma potential is, in any case, uniform.

   In the case of the arc discharge mode, the plasma potential defines the starting
energy of the ions and the ion temperature and the plasma oscillations define their

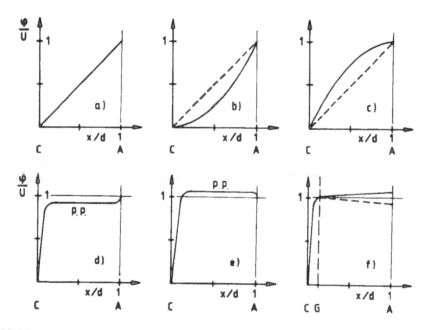

FIGURE 2.2
Potential distribution between anode A and cathode C: (a) virgin potential, (b) with electron flow,
(c) with electrons and ions, (d) with plasma potential below anode, (e) with plasma potential above
anode, (f) FEBIAD with positive or negative potential of A with respect to grid G.

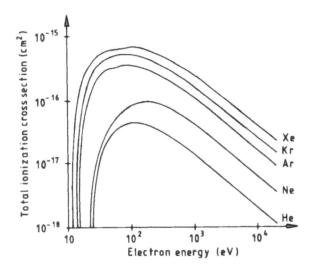

FIGURE 2.3
Ionization cross section vs. electron energy for noble gases.[7]

energy spread. Table 2.1 gives typical energy-spread values for various ion sources. The cathode design also influences the energy spread of the primary electrons through the potential difference along the directly heated cathodes. This leads to low-voltage, high-current cathodes, which, on the other hand, may cause instabilities by the magnetic field induced by the high filament current.

With increasing electron energy, higher charge states may also be created by single or multiple impacts. The latter is more likely at higher ion densities in the discharge mode. Figure 2.5 shows the dependence of the ionization cross section for Ar^+, Ar^{2+}, and Ar^{3+} on the electron energy.[7,10,11]

By careful analysis of the threshold behavior of the ion current, the appearance of metastable states also may be investigated.[12,13]

Ions can be extracted through the anode, perpendicular to it, or through the cathode area. For higher currents cathode lifetime is limited by ion bombardment.

To increase the (poor) ionization efficiency of the electrons in electron bombardment ion sources, several modifications have been introduced. An additional small

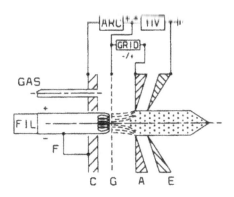

FIGURE 2.4
FEBIAD ion source.[8]

TABLE 2.1

Energy Spread of Different Ion Sources[9]

Ion Source	Energy spread (V)	Typical ion current (mA)
Prim. Electron collision	~1	10^{-6}
Electron bombardment	1–10	10^{-3}–1
Oscill. electron discharge	10–50	1
FEBIAD ion source	1–50	10^{-6}–1
Low-voltage arc	0.2–5	0.1–1
High-voltage arc	10–100	1–10
Duoplasmatron	5–20	1–10^3
PIG ion source	10–50	5–500
Bucket type ion sources	5–50	10–10^5
RF ion sources	10–100	0.1–10^3
Microwave ion sources	5–50	1–100
ECR ion sources	10–100	1–10^3
Surface ionization	0.2–0.5	10^{-2}–100
Liquid-metal ion sources	5–50	10^{-3}–1
Gas-phase field ionization	1–5	10^{-4}–10^{-2}

magnetic field confines the electrons inside the anode and lets them spiral along the magnetic field lines, multiplying on their way to the anode and increasing the ionization efficiency of the ion source. By using a cylindrical anode and a reflector electrode connected to the cathode, the electron path is further enlarged. Many mass separator ion sources are of this type, such as the Nier,[2] Bernas,[3] Nielsen,[4] and other[14,15] sources. These are presented separately in more detail with their specific operating data in Section 2.3.

Increasing the magnetic field increases the plasma density and the current density in the extraction aperture, and also the instabilities and the energy spread of the extracted ions. The calutron ion sources used for isotope production are of this type[16] and are the PIG sources (Penning ion sources); both are described in Chapter 5.

The well-known Freeman ion source with its coaxial cathode corresponds more to the magnetron ion sources[17] and is discussed in Chapter 4.

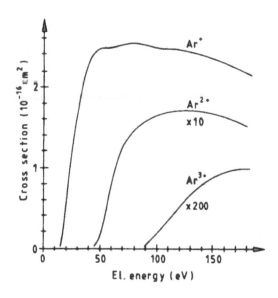

FIGURE 2.5
Cross section for Ar charge states vs. electron energy.[7]

2.2.1 Metal Ions

All methods for metal ion production, described in Chapter 3, can be applied to electron bombardment ion sources. In particular, chemical synthesis have been used extensively.[14,18-21]

2.3 Specific Ion Source Design

2.3.1 Nier and Bernas Types Ion Sources (Figure 2.6)

- Special design and construction details of the source

  The Bernas ion source has rectangular or cylindrical arc chamber positioned in an external magnetic field of some 10 mT. The source contains a single-turn helical filament at one side of the arc chamber (anode) and a reflector at the other side. Electrons from the cathode are confined inside the anode cylinder by the magnetic field and can oscillate between filament and reflector resulting in a high ionization efficiency. Ions are extracted perpendicular to the anode axis through a slit of ~2-mm width and about 40-mm length, depending on the specific design (gap of the analyzing magnet). ≤10% ++ and ≤1% +++ ions achievable.

- Ion source material and vacuum conditions

  The ion source anode is made of Cu, Mo, Ta, steel, or graphite, depending on the ion specimen and the temperature necessary. The filament is made of Ta or W. The gas flow is small so the source will operate in a vacuum environment of ≤10⁻⁴ Pa.

- Application area of the source

  Mass separation, medium current ion implantation.

- Deliverer or user

  Varian Ion Implant Systems, 35 Dory Road, Gloucester, MA 01930–2297, U.S.A.

  IEF Universite Paris Sud, 91405 Orsay, France

  Soreq Nuclear Research Center, Yavne, Israel

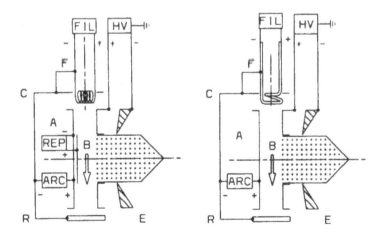

FIGURE 2.6
(a) Nier type ion source with ion repeller, (b) Bernas type ion source.

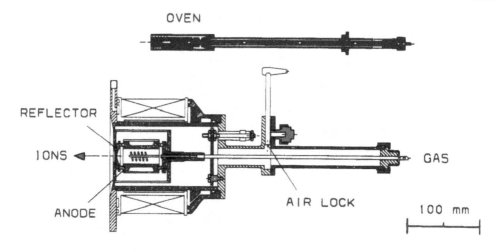

FIGURE 2.7
Nielsen ion source with air lock and oven. (Courtesy of Danlysik, A/S, Jyllinge, Denmark.)

2.3.2 Nielsen Ion Source (Figure 2.7)

- Special design and construction details of the source

  The Nielsen ion source operates on the principle of oscillating electrons emitted from a coaxial W or Ta filament into the cylindrical anode and confined by the external magnetic field. The ion source can operate at temperatures up to 1000°C for a steel chamber and 1200°C for a graphite one. The exchange of the cathode filament or ovens for various materials and methods (CCl_4) is easily managed through air locks.

- Ion source material and vacuum conditions

  The source parts are made of steel or graphite. The ion source works at low pressure and with high efficiency.

- Application area of the source

  Surface bombardment, target preparation, ion implantation, isotope separation.

- Deliverer or user

  Danfysik A/S, DK-4040 Jyllinge, Denmark

  Many research laboratories.

2.3.3 Hollow Cathode Ion Source (Figure 2.8)

- Special design and construction details of the source

  The hollow-cathode ion source can operate at high temperatures of up to 1700°C due to the special design of extraction from the hollow cathode, which is formed by two Ta pieces with a Ta or W filament placed in between. The outlet plate is made of W and so is the anode. Insulators are made of boron nitride. The magnetic coil is connected in such a way that the field of the filament is compensated. The source body is air-cooled. Ovens can be easily exchanged through an air lock. Since the arc voltage and the gas pressure, the two parameters important for the production of multiple charged ions, can be adjusted independently, the percentage of doubly charged ions can be controlled. Doubly charged ions can be generated to up to 10% of the singly charged fraction.

- Ion source material and vacuum conditions

  The vapor pressure inside the source is about 1 Pa, but the gas load is small so the vacuum outside the source can easily be kept below 10^{-4} Pa.

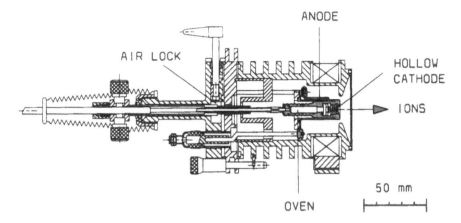

FIGURE 2.8
Hollow-cathode ion source. (Courtesy of Danlysik, A/S, Jyllinge, Denmark.)

- Application area of the source

  Ion implantation, isotope separation, ion beam mixing, RBS, RNRA, PIXE.

- Deliverer or user

  Danfysik A/S, DK-4040 Jyllinge, Denmark
  High Voltage Engineering Europe, 3800 AB Amersford, Netherlands
  Many research laboratories.

2.3.4 Wilson Ion Source (Figure 2.9)

- Special design and construction details of the source

  Electrons emitted by the thermionic filament are accelerated radially toward a cylindrical anode and the axial magnetic field causes them to revolve, thus forming an ionizing electron cloud. The ends of the cylindrical anode are closed by disks on cathode potential. Ions are extracted axially from an area close to the cathode tip. Low-vapor-pressure material is introduced in a container that is heated by the cathode filament. Its temperature is controlled by its position inside the source.

- Ion source material and vacuum conditions

  The ion source parts are made of steel, Mo, Ta, or graphite, depending on the elements used. The pressure inside the source has to be 10 Pa, but the pressure in the vacuum chamber is below 10^{-4} Pa.

- Application area of the source

  Ion implantation, ion beam modification of materials, mass spectroscopy, RBS; can be used for all elements.

- Deliverer or user

  R. G. Wilson, Hughes Research Lab., Malibu, CA, U.S.A.

2.3.5 Metal Ion Source of Wilbur and Wei (Figure 2.10)

- Special design and construction details of the source

  Metal vapor required for operation of the source is supplied by drawing sufficient electron current from the filament to the positively biased crucible. The discharge is initiated between the cathode, the anode, and the crucible within the high-temperature graphite chamber by supplying argon support gas, which can be turned off once a stable metal vapor plasma exists. The entire source is at high potential and

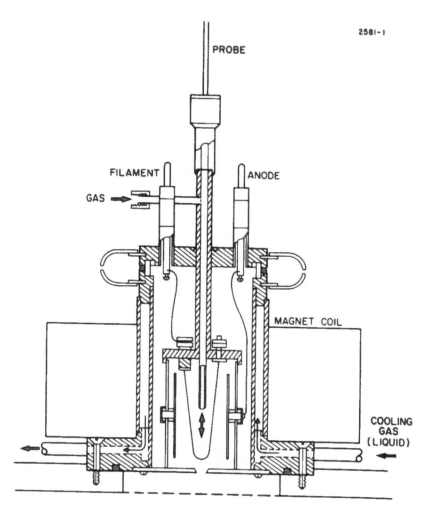

FIGURE 2.9
Wilson ion source. (Courtesy of R. G. Wilson, Hughes Res. Lab.)

a sheath containing mesh facilitates ion beam focusing over a wide range of beam currents and energies. The high-temperature graphite chamber minimizes metal condensation. Vertically upward beam; simple, inexpensive construction with components held together by gravity; easy access for maintenance.

- Ion source material and vacuum conditions

Graphite except for W cathode (dual 0.25-mm Ø wires 5 cm long), alumina insulators, stainless steel screen plate, copper base, and Ta radiation shield. Mesh made of 0.25-mm Ø W wire.

- Application area of the source

High-current-density, broad-beam, metal-ion implantation.

- Deliverer or user

Department of Mechanical Engineering, Colorado State University, Fort Collins, CO 80523, U.S.A.
Colorado Engineering Research Lab., 1500 Teakwood Ct., Fort Collins, CO 80525, U.S.A.

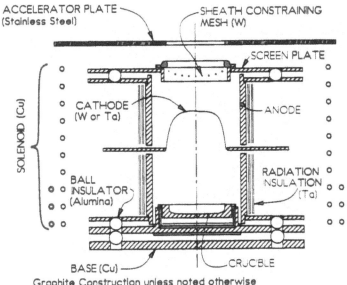

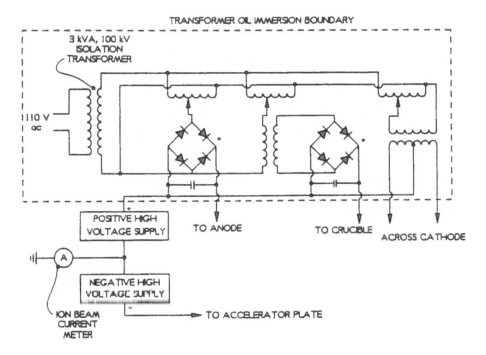

FIGURE 2.10
Metal ion source of Wilbur and Wei. (From Wilber, P. J. and Wei, R., *Rev. Sci. Instr.*, 63, 2491, 1992.
With permission.)

TABLE 2.2

Operating Data of Electron Bombardment Ion Sources

		Bernas	Nielsen	FEBIAD	Hollow cath.	Wilson	Wilbur
Arc voltage	(V)	30–120	20–200	20–200	20–250	50–100	50–250
Arc current	(A)	0–10	0–5	0–5	0–5	0–0.5	1–10
Arc power	(kW)	≤1	≤1	≤1	≤0.5	≤0.03	≤1
Filament							
Current	(A)	200	60	60	5–30	12	15
Voltage	(V)	5 dc	25 dc	20 dc	2–5 dc	10 dc	12 ac
Magnetic field	(mT)	0–40	0–100	0–50	0–40	0–10	0–10
Magnet supply							
Voltage	(V)	0–20	0–120	0–30	0–10	0–5	0–10
Current	(A)	0–50	0–30	0–8	0–30	0–2	0–100
Gas pressure							
From	(Pa)	10^{-3}	10^{-2}	10^{-3}	10^{-4}	1	
To	(Pa)	10^{-1}	1	10^{-1}	10^{2}	10	$\leq 10^{-2}$
Gas consumption	(sccm)	≤1	~0.1	$\leq 0.1^{-4}$		$\leq 10^{-3}$	
Ion current	(mA)	0–15	0–1	0–0.5	0–0.2	0–0.2	0–50
Extraction area	(cm²)	≤5	0.02	0.01	0.01	0.2	7
Extraction voltage	(kV)	2–70	5–50	0–30	0–30	–2	10–50
Noise level	%	≤20					
Gas efficiency	%	≤50	≤50	≤50			
Grid voltage	(V)			0–300			

2.3.6 Operating Data (Table 2.2)

REFERENCES

1. A. J. Dempster, *Philos. Mag.*, 31, 438, 1916; *Phys. Rev.*, 18, 415, 1921.
2. A. Nier, *Rev. Sci. Instrum.*, 18, 398, 1947.
3. R. Bernas and I. Chavet, *Nucl. Instrum. Methods*, 47, 77, 1967; *Nucl. Instrum. Methods*, 51, 77, 1967.
4. K. O. Nielsen, *Nucl. Instrum. Methods*, 1, 289, 1957.
5. Th. Wasserab, *Gaselectronic I*, Bibliographisches Institut, Zurich, 1971, 180 ff.
6. M. v. Ardenne, *Tabellen zur Angewandten Physik I*, VEB Deutscher Verlag der Wissenschaften, Berlin, 1956 and 1962.
7. E. W. McDaniel, *Collision Phenomena in Ionized Gases*, John Wiley & Sons, New York, 1964.
8. R. Kirchner and E. Röckel, *Nucl. Instrum. Methods*, 133, 187, 1976; Nucl. Instrum. Methods, B26, 235, 1987.
9. R. G. Wilson and G. R. Brewer, *Ion Beams with Applications to Ion Implantation*, R. E. Krieger, Malabar, FL, 1973 and 1979.
10. T. D. Maerk and G. H. Dunn, *Electron Impact Ionization*, Springer, New York, 1985, chaps. 5 and 8.
11. R. K. Janev et al., *Physics of Highly Charged Ions*, Springer Series in Electrophysics, Vol. 13, Springer, Berlin, 1985, chap. 5.
12. H. D. Hagstrum, *Phys. Rev.*, 96, 325, 1954; *Phys. Rev.*, 104, 309, 1956; *Phys. Rev.*, 104, 317, 1956.
13. F. Aumayr and H. P. Winter, *Phys. Scr.*, T28, 96, 1989.
14. R. G. Wilson, *Ion Mass Spectra*, John Wiley & Sons, New York, 1974, 1.
15. P. J. Wilbur and R. Wei, *Rev. Sci. Instrum.*, 63, 2491, 1992.
16. A. Guthrie and R. K. Wackerling, *The Characteristics of Electrical Discharges in Magnetic Fields*, McGraw-Hill, New York, 1949.
17. J. H. Freeman, *Nucl. Instrum. Methods*, 22, 306, 1963.
18. J. H. Freeman and G. Sidenius, *Nucl. Instrum. Methods*, 107, 477, 1973.
19. G. Dearnaly, J. H. Freeman, R. S. Nelson, and J. Stephen, *Ion Implantation*, North Holland, Amsterdam, 1973.
20. G. Alton, *Nucl. Instrum. Methods*, 189, 15, 1981.
21. B. H. Wolf, *Rev. Sci. Instrum.*, 65, 1248, 1994.

Chapter 2/Section 3

CHARACTERIZATION OF ION SOURCES

B. H. Wolf

3 PLASMATRON ION SOURCES

Advantages: Compact source for intensive ion beams; good beam quality; moderate costs.

Disadvantages: Not well-suited for nonvolatile materials; filament life limited with heavy elements, oxygen, and corrosive gases.

3.1 Unoplasmatron

3.1.1 Ion Source History

The unoplasmatron ion source was developed by v. Ardenne (1948)[1] to improve the plasma density and the ion current of an electron bombardment ion source.

3.1.2 Working Principle of the Unoplasmatron Ion Source and Description of the Discharge

An intermediate electrode (IE) is positioned between cathode and anode of an electron bombardment ion source running in an arc mode (Figure 3.1). The IE has a conical shape and a center bore of a few millimeters in diameter and length. This electrode is connected through resistors to the cathode and/or to the anode or has a separate power supply. The cathodic plasma is concentrated by a double layer to be able to pass the IE channel and to reach the anode. In this way the ion density in front of the anode is increased and also the extracted ion current.

3.2 Duoplasmatron

3.2.1 Ion Source History

The duoplasmatron ion source was developed by v. Ardenne (1956)[1] as a powerful source for gas ions. Further investigations of the duoplasmatron was carried out by Demirkanow (1962),[2] Fröhlich (1959),[3] and Kistemaker (1965).[4] Many investigations came from the group around Septier[5] and Gautherin.[6]

Negative ions were extracted by Moak (1959)[7] and Collins (1965).[8] Braams (1965)[9] reported on multiply charged ions from a duoplasmatron. Krupp (1968)[10] and Illgen (1968)[11] investigated multiply charged ion production. Winter and Wolf explained some of the processes involved in multiply charged ion production.[12]

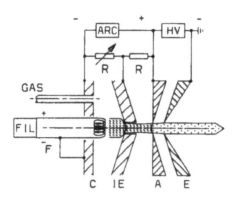

FIGURE 3.1
Unoplasmatron ion source scheme and circuitry.

3.2.2 Working Principle of the Duoplasmatron Ion Source and Description of the Discharge

The duoplasmatron ion source consists of two plasma regions: the lower-density cathode plasma between the cathode and the IE and the high-density plasma between the IE and the anode (Figure 3.2). The cathode plasma is compressed by a double layer into the IE channel and then further compressed by an axial magnetic field. In this way a very high plasma density ($10^{14}/cm^3$) can be reached.

Ions (positive or negative) are extracted on axis through the anode. To reduce the ion beam density for easier beam formation and transport, an expansion cup is commonly used at the extraction side of the anode. In case of negative-ion extraction, electrons can be restricted from extraction by a magnetic filter or by extraction off axis (see section 2.16).

The duoplasmatron discharge has a rather complex structure as shown in Figure 3.3. In addition to the double layer already mentioned, there is one or more layer inside the IE channel to adjust the electron and ion flows to the needs of a stable discharge. The positive potential maximum between IE and anode accelerates ions toward the anode.

The IE is connected to the anode via a resistor of about 1 kΩ to ease the ignition of the discharge. After ignition, the IE potential adjusts close to cathode potential, depending on the electron flow from the cathode. Low electron emission increases the potential drop between the IE and the cathode; high cathode emission reduces the

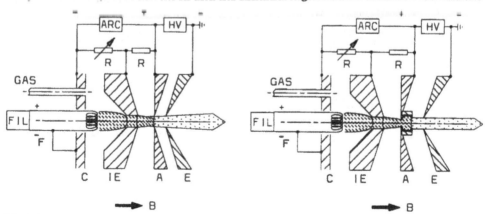

FIGURE 3.2
Duoplasmatron ion source scheme and circuitry: (a) direct extraction, (b) expansion cup extraction.

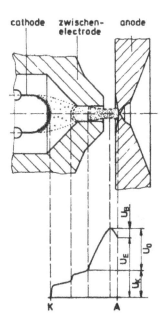

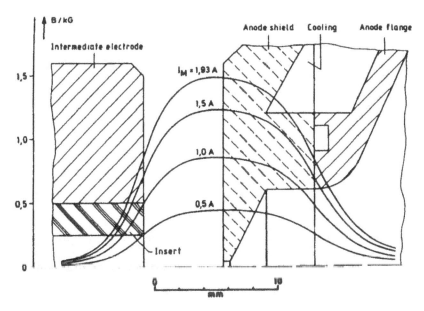

FIGURE 3.3
Duoplasmatron discharge, potential, and magnetic field distribution.

potential and can even change it to negative values. Low electron emission from the cathode increases the discharge voltage and also sputtering of the cathode, especially by heavy ions. Too high an electron emission creates high-frequency instabilities of the discharge (hash).[4] The best operational mode for a duoplasmatron is when the cathode is adjusted to an emission at which the intermediate electrode balances, or is slightly above, the cathode potential.

The cathode is usually a filament made out of a tantalum or tungsten wire of various shapes (see Chapter 2.1, Figure 1.2). For proton and H⁻ sources, oxide cathodes give longer lifetimes and high currents. Because filament lifetime is short, hollow cathodes are commonly used in cases where the ion source runs with oxygen or corrosive gases. The hollow cathode can be made of a tantalum cylinder of 0.1- to 0.5-mm wall thickness or of a cylinder coated on the inside with some alkaline oxide.[13,14,15]

The arc power in a duoplasmatron is limited to 1 to 2 kW by the heat dissipation at the anode. In dc operation, about 30 A is the limit, but in pulsed operation more than 100 A can be applied and higher ion currents can be extracted correspondingly. Increasing the magnetic field increases the ion current and also plasma instabilities.[4] To keep instabilities small the magnetic field should be below 0.1 T.[12]

3.2.3 Charge State Distribution

To reach a high ion current with limited power in the ion source, the arc voltage is kept low in duoplasmatron ion sources. Naturally, the energy of the primary electrons is too low to produce a significant amount of higher-charged ions. In a pulsed mode of operation, however, it is possible to run the ion source with much higher arc voltages by reducing the gas pressure, which further improves the conditions for multiply charged ion production.[11,16] A high magnetic field improves the yield of multiply charged ions and a careful selection of the IE geometry is necessary for optimum conditions.[16] A spectrum for xenon with charge states up to 11^+ is shown in Figure 3.4.

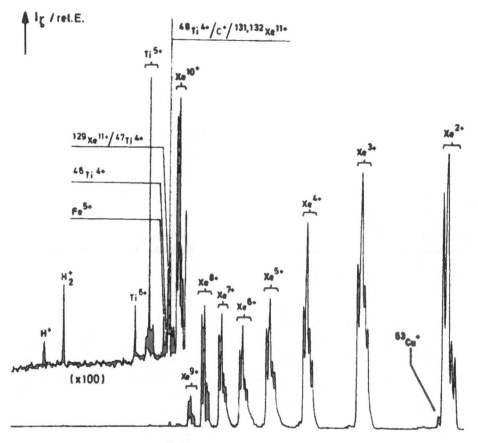

FIGURE 3.4
Xe charge spectrum of a pulsed Duoplasmatron.

Multiply charged ions, however, can also be generated in dc operation of the duoplasmatron[10,19] by proper adjustment of the magnetic field and IE geometry. The best conditions for multiple-ion production are reached if the product of ion lifetime and electron density ($\tau_i \times n_e$) is a maximum. Figure 3.5 shows the charge state distribution of a Duoplasmatron in dc operation. The optimum magnetic field is just below the appearance of instabilities and depends on the gas used (Table 3.1).[12] The admixture of a heavy gas may also improve the amount of multiply charged ions.[12]

Duoplasmatrons can deliver high ion currents of ions from gases, but they are not well suited for ion production from solids. There have been attempts to incorporate an oven between the IE and the anode[17] or in the expansion cup[18] and also to design a sputter version of a duoplasmatron,[20] but none have been used for accelerators or ion implantation.

3.3 Duopigatron

The duopigatron is a modification of the duoplasmatron with an additional reflector electrode following the anode (see Figure 3.6). The reflector electrode is connected to the cathode potential and reflects electrons coming through the anode hole, and in this way the ionization efficiency of the electrons is increased.

3.3.1 Ion Source History

The duopigatron was invented by Demirkanow (1964)[2] and further developed by O. B. Morgan[21] as a high-current proton source. Shubaly developed a high-current source for heavy ions.[22] A duopigatron as a metal ion source for ion implantation was developed by Winter[23] and Wolf.[24]

3.3.2 Working Principle of the Duopigatron Ion Source and Description of the Discharge

By proper design large-area extraction of high ion currents is possible, especially if a permanent magnet multipole is applied in the reflector region (see Chapter 2.6).[21,22,25] Table 3.2 gives ion currents of this ion source.

In the case of this high-current duopigatron ion source, the cathode-IE part acts like a plasma cathode for the multipole source between anode and reflector. For large-area H_2^+ duopigatron ion sources, for fusion devices, more than one cathode-IE part have been connected to the main plasma generator.[27]

Because the reflector electrode is on cathode potential, intensive sputtering occurs in the case of reflector holes of diameters smaller than the main discharge plasma (similar to the cathode erosion in PIG discharges). The effect of reflector electrode erosion can be used to produce metal ions of the reflector electrode material.[23,24,28] To get a high sputter rate of metal atoms the ion source should run on a heavy gas like Ar, Kr, or Xe. The metal ion yield increases with the magnetic field and with the arc voltage adjusted by lowering the gas pressure. A large variety of materials can be ionized in the duopigatron metal ion source by just changing a small insert in the reflector electrode. The potential and magnetic field distribution of the duopigatron metal ion source are shown in Figure 3.7. The magnetic field is symmetric to the anode and has a flat maximum.

Multiply charged ions, especially of heavier elements, can also be found in the extracted ion beam (see Table 3.3).

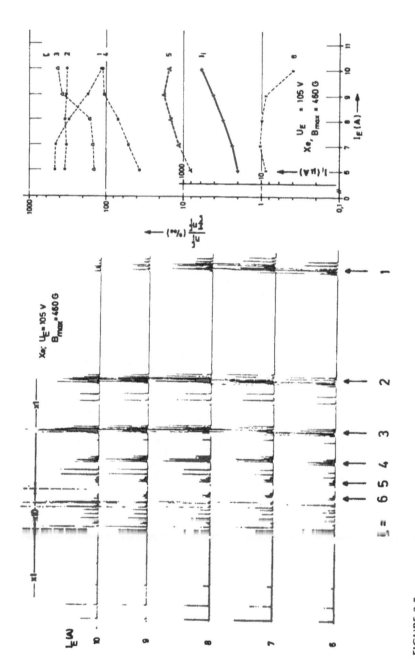

FIGURE 3.5
Xe charge state distribution of a dc Duoplasmatron.

TABLE 3.1

Magnetic Field Optimum for Multiply
Charged Ions in dc Operation

Element	B_{OP} (mT)
Ar	80
Kr	60
Xe	45

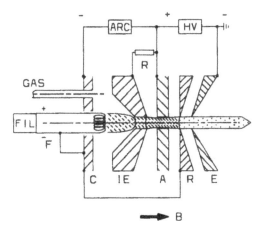

FIGURE 3.6
Duopigatron scheme and circuitry.

TABLE 3.2

Data of High-Current Duopigatron Ion Sources[22,26]

		H^+	H_2^+	H_3^+	He	N	O	Ar	Xe	Li	P	Ca	Bi
Ion current	(mA)	60	85	50	120	190	138	140	90	16	26	27	8.2
Arc voltage	(V)	70	70	70		60–70	100	50–65	25–40		50–100		
Arc current	(A)	12	13	12		12	20	12	12		10–20		
Extr. holes	5-mm ϕ	3	3	3	3	7	7	7	7		3		
Extr. voltage	(kV)	50	50	50	40	35	52	35	35	33	31	38	27
Support gas	—	—	H_2	H_2	H_2	—	—	Ar	—	—	Ne	He	Ne	Ne

3.4 Specific Ion Source Design

3.4.1 The GSI Duoplasmatron for Multiply Charged Heavy Ions (Figure 3.8)

- Special design and construction details of the source

  Anode cooling optimized for high arc power and easy exchange of the anode center
  with the extraction hole. Magnetic field shaped for production of higher charge
  states (maximum at the anode).

- Ion source material and vacuum conditions

  Cobalt steel for intermediate electrode, ferromagnetic steel for anode. Copper-
  tungsten alloy for anode insert, Teflon insulator between anode and intermediate
  electrode. Cathode of Ta or W wire 0.8- to 1.2-mm ϕ about 100 mm long.

- Application area of the source

  Accelerators, tandem accelerators, ion implantation, heavy ions, high charge states
  (in particular, in pulsed operation), sputter deposition, IBAD, ion beam modification of materials, ion beam analysis, SIMS, RBS, NBS, PIXE.

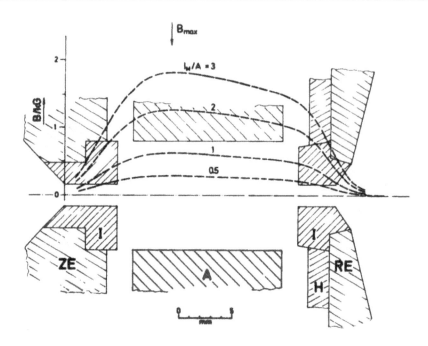

FIGURE 3.7
Magnetic field distribution of the Duopigatron.

- Deliverer or user

  GSI Planckstr.1, 64291 Darmstadt, Germany
  N.T.G. Neue Technologien, Im Steinigen Graben 12, 63571 Gelnhausen-Hailer, Germany
  Other duoplasmatrons are used at many places such as:
  CERN, BNL, Universite Paris Sud, Orsay, France
  High Voltage Engineering Europa, PO Box 99, 3800 AB Amersford, Netherlands
  National Electrostatic Corp. PO Box 310, Middleton, WI 53791–9990, U.S.A.
  ULVAC Ltd., Japan, and others.

3.4.2 Hollow Cathode Duoplasmatron (Figure 3.9)

- Special design and construction details of the source

  The ion source hollow cathode has a length-to-diameter ratio of 5:1 and is surrounded by an extension of the intermediate electrode made of soft iron to shield it from the magnetic field. The magnetic field is maintained by insulating permanent magnet rings. The extractor electrode is also made of mild steel to extend the field lines to the extraction region.

  The dimensions with respect to the diameter (d) of the IE canal are cathode diameter, 11 d; distance between anode and IE, 0.5 d; IE-canal, 2.5 d; extraction aperture, 0.1 d.

  The ion source uses just one power supply for the hollow-cathode discharge and the anode arc. The IE resistor helps to start the discharge and let the IE float close to cathode during source operation.[14]

- Ion source material and vacuum conditions

  Mild steel for IE body and extractor; nonmagnetic anode.

- Application area of the source

  Mass spectroscopy for all elements, isotope separation, ion beam analysis SIMS.

TABLE 3.3

Metal Ion Currents from a Duopigatron Ion Source[24]

Element	Electrical current (μA)				Element	Electrical current (μA)			
	1+	2+	3+	4+		1+	2+	3+	4+
H	250				Mo	10	10	1	
D	150				Ru	—			
He	1000				Rh	—			
Li	10				Pd	21	7.5	0.5	
Be	—				Ag	36	19	0.5	
B	30				In	65	13	—	0.1
C	50				Sn	40	40	5	
N	500	10			Sb	15	5		
O	50				Te	—			
F	50				J	—			
Ne	1000	100	1		Xe	300	70	5	
Na	30				Cs	—			
Mg	25	1.6			Ba	20	15	5	
Al	60	20	1		La	28	18	6	0.1
Si	10	2			Ce	15	10	3	
P	40				Pr	10	5	1	
S	50	5			Nd	20	10	2	
Cl	15				Sm	10	5	1	
K	—				Eu	5	5	1	
Ar	1000	350	50		Gd	5	5	1.5	
Ca	10	3			Tb	27	20	2.5	
Sc	5	2			Dy	20	10	1	
Ti	10	5			Ho	10	5	1	
V	5	1.7			Er	120	120	22	0.5
Cr	50	15	0.5		Tm	20	10	2	
Mn	39	3.5			Yb	20	10	2	
Fe	65	14	0.3		Lu	20	10	1	
Ni	54	11	0.1		Hf	20	10	1	
Co	65	13			Ta	5	5	0.5	
Cu	75	25	0.5		W	22	10	1.5	0.1
Zn	104	40	0.2		Re	—			
Ga	5	2			Os	—			
Ge	10	5	2		Ir	10	15	5	0.5
As	—				Pt	10	15	5	0.5
Se	30	10			Au	38	43	9	0.5
Br	—				Hg	15	20	5	0.5
Kr	400	60	3	0.1	Tl	—			
Rb	—				Pb	30	70	10	0.6
Sr	30	10			Bi	14	22	14	0.3
Y	10	5			Th	—			
Zr	4	2.2	0.1		U	30	40	30	1.5
Nb	32	25	1.1						

- Deliverer or user

  Hasler Research Center, Applied Research Lab., Goleta, CA, U.S.A.

3.4.3 UHV Duoplasmatron (Figure 3.10)

- Special design and construction details of the source

  All-metal vacuum seals, source body bakeable to 300°C, magnet coil removable, hollow-cathode operation possible.[29]

- Ion source material and vacuum conditions

  Stainless steel, Cu, or ceramic insulators electron-beam welded to copper or steel, bakeable to 300°C; can be used with UHV apparatus.

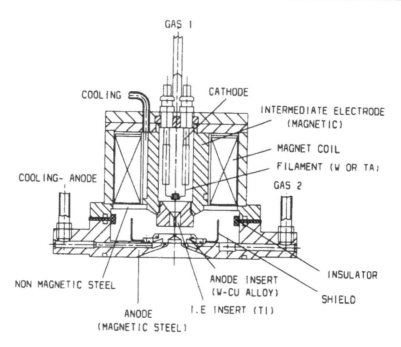

FIGURE 3.8
GSI Duoplasmatron for heavy ions.

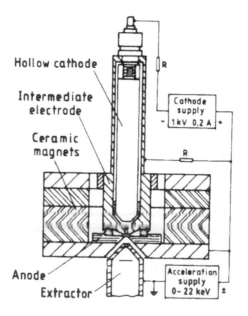

FIGURE 3.9
Hollow-cathode Duoplasmatron.[14]

- Application area of the source

    On-line ion beam analysis and SIMS in UHV arrangements.

- Deliverer or user

    Research and Development Center of Vacuum Electronics (OBREP) ul. Dulga 44/50, PL-00–241 Warszawa, Poland

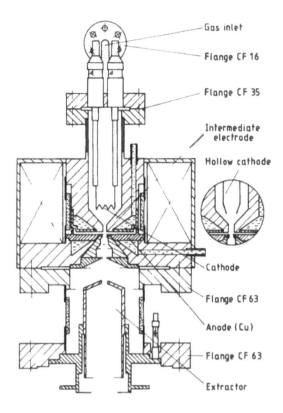

FIGURE 3.10
UHV Duoplasmatron. (From Konanki, P., *Rev. Sci. Instr.*, 63, 2397, 1992. With permission.)

3.4.4 Operating Data (Table 3.4)

3.4.5 The GSI Duopigatron for Metal Ions (Figure 3.11)

- Special design and construction details of the source

  In general, the same design as the GSI duoplasmatron. The anode is at a distance of just 2 mm from the IE and the reflector electrode to avoid parasitic discharges. The Cu anode contains a set of iron bars at the outer diameter to guide the magnetic field to the reflector electrode. The reflector electrode is water-cooled and has inserts made of the material to be ionized. Power supplies and discharge data correspond to those of the duoplasmatron.

- Ion source material and vacuum conditions

  Cobalt steel for intermediate electrode, ferromagnetic steel for reflector electrode, sputter material for reflector electrode insert, Teflon insulator between anode and intermediate electrode. Cu anode, cathode of Ta or W wire 0.8- to 1.2-mm ϕ about 100 mm long. The pressure outside the source is about 10^{-3} Pa.

- Application area of the source

  Accelerators, metal ion implantation, ion beam modification.

- Deliverer or user

  GSI Planckstr.1, 64291 Darmstadt, Germany

  N.T.G. Neue Technologien, Im Steinigen Graben 12, 63571 Gelnhausen-Hailer, Germany

  Most duoplasmatrons can be changed to a duopigatron for metal ions.

TABLE 3.4

Operating Data of Plasmatron Ion Sources

		Unopl.	GSI-dc	GSI-pulse	Hollow cath.	UHV
Arc voltage	(V)	30–80	20–150	20–300	300–400	50–100
Ignition	(V)		150–250	150–400	1000	150–250
Arc current	(A)	0.3–2	0–20	0–200	0–0.2	1–2
Arc power	(kW)	160	0–2	0–2	0–0.1	0–0.3
Filament						
Current	(A)	35	35–60	35–60	—	5–20
Voltage	(V)	10	7.5–12	7.5–12	—	5–10
Magnetic field	(mT)	—	0–300	0–300	400	
Magnet supply						
Voltage	(V)	—	0–30	0–30	Perm.	
Current	(A)	—	0–10	0–10	magnet	2–7
Gas pressure	(PA)	2–5	1–100	1–100	1–30	2–5
Gas consumption	(sccm)		10^{-2}–10	10^{-2}–10		
Ion current	(mA)	0–10	0–100	0–500	0–0.2	0–1
Extraction hole	(mm)ϕ	0.1–2	0.3–1.5	0.3–1.5	0.4	1.5
Extraction voltage	(kV)	10–50	5–50	5–50	0–22	2–10
Noise level	%		1–20	1–20	Low	
Gas efficiency	%	10–20	$\leq$50	—		
Charge states	Xe		$\leq7^+$	$\leq11^+$	—	

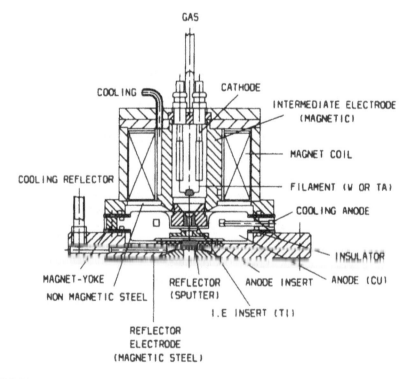

FIGURE 3.11
GSI Duopigatron for metal ions.

3.4.6 *High-Current Duopigatron (Figure 3.12)*

• Special design and construction details of the source

The source uses an oxide-coated tantalum cathode. The IE is made of mild steel, the anode of copper, and the reflector of mild steel again. The insulators are alumina

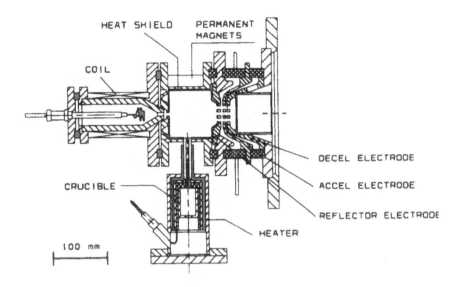

FIGURE 3.12
High-current Duopigatron.[26]

rings and the extraction is a multiaperture accel-decel system. The magnetic field is maintained by 3000 turns operated at about 1 A. Twelve NeFeB magnets provide a multipole field in the main discharge region, improving ion beam uniformity. The discharge is connected to a 250-V, 20-A power supply. The extraction can have up to 13 holes, 5-mm ϕ each; the largest area for extraction is of 50-mm ϕ.

- Ion source material and vacuum conditions

The gas flow for argon is typically 2.5 sccm and the pressure in the vacuum chamber is about 5×10^{-3} Pa.

- Application area of the source

Proton accelerators, fusion devices, high-current ion implantation.

- Deliverer or user

Accelerator Physics Branch, Chalk River Nuclear Lab., Chalk River, Ontario, K0J 1J0, Canada

Similar duopigatron: Oak Ridge National Lab., Oak Ridge, TN 37831–6368, U.S.A.

REFERENCES

1. M. V. Ardenne, *Technik*, 11, 65, 1956; *Tabellen der Elektronenphysik, Ionenphysik, und Übermikroskopie*, VEB Deutscher Verlag Wissenschaften, Berlin, 1956.
2. R. A. Demirkanow et al., *Proc. 1st Int. Conf. High Energy Acc.*, BNL Rep. 767, Brookhaven, 1962, 218.
3. H. Fröhlich, *Nukleonik*, 1, 183, 1959.
4. J. Kistemaker, *Nucl. Instrum. Methods*, 38, 1, 1965.
5. A. Septier, *Focusing of Charged Particles*, Academic Press, New York, 1967.
6. G. Gautherin, *Nucl. Instrum. Methods*, 59, 261, 1968.
7. C. D. Moak et al., *Rev. Sci. Instrum.*, 30, 694, 1959.
8. L. E. Collins and R. H. Gobbet, *Nucl. Instrum. Methods*, 35, 277, 1965.
9. C. M. Braams et al., *Rev. Sci. Instrum.*, 36, 1411, 1965.
10. H. Krupp, Diploma thesis, Univ. Heidelberg Unilac-report 1 - 1968, GSI Darmstadt, 1968.
11. J. Illgen, Ph.D. thesis, Univ. Heidelberg Unilac-report 4 - 1968, GSI Darmstadt, 1968.

12. H. Winter and B. H. Wolf, *Nucl. Instrum. Methods*, 127, 445, 1975; *Plasma Phys.*, 16, 791, 1974.
13. V. A. Batalin et al., *IEEE Trans. Nucl. Sci.*, 23, 1097, 1976.
14. H. J. Roden, A hollow cathode duoplasmatron for reactive gas ions, *Int. Symp. Ion Sources and Formation of Ion Beams*, Rep. LBL 3399, Berkeley, 1974, VII-7-1.
15. W. Aberth, Design of a general purpose duoplasmatron, *Int. Symp. Ion Sources and Formation of Ion Beams*, Rep. LBL 3399, Berkeley, 1974, I-7-1.
16. R. Keller and M. Müller, GSI-PB-2-75, GSI Darmstadt, 1975; Proc. Int. Conf. Heavy Ion Sources, *IEEE Trans. Nucl. Sci.*, NS 23, 1049, 1976.
17. J. Schulte in den Bäumen et al., *Proc. 2nd Int. Conf. Ion Sources*, Vienna, 1972.
18. R. Masic et al., *Nucl. Instrum. Methods*, 71, 339, 1969.
19. H. Krupp and H. Winter, *Nucl. Instrum. Methods*, 127, 459, 1975.
20. R. Keller and M. Müller, *Inst. Phys. Conf. Ser.*, Vol. 38, Bristol, 1978, 40.
21. O. B. Morgan et al., *Rev. Sci. Instrum.*, 38, 467, 1967.
22. R. M. Shubaly, *Inst. Phys. Conf. Ser.*, Vol. 54, Bristol, 1980, 333.
23. H. Winter, *Proc. 7th Yugoslav. Symp. Physics of Ionized Gases*, Rovinji, 1974, 79; Report GSI-PB-1, GSI Darmstadt, 1974, 15.
24. B. H. Wolf, Nucl. Instrum. Methods, 139, 13, 1976.
25. J. E. Osher, *Proc. Symp. Ion Sources and Formation of Ion Beams*, Brookhaven Nat. Lab., 1971; *Proc. 2nd Symp. Ion Sources and Formation of Ion Beams*, Lawrence Berkeley Lab., LBL-3399 1974, IV-7-1.
26. T. Taylor et al., *Rev. Sci. Instrum.*, 61, 454, 1990.
27. M. M. Menon et al., *Rev. Sci. Instrum.*, 56, 242, 1985.
28. R. Keller and H. Winter, *Part. Accel.*, 7, 77, 1976.
29. P. Konarski et al., *Rev. Sci. Instrum.*, 63, 2397, 1992.

Chapter 2/Section 4

CHARACTERIZATION OF ION SOURCES

B. H. Wolf

4 MAGNETRON AND FREEMAN TYPES ION SOURCES

Magnetron
Advantages: Compact ion source for light ions, sputter source for coating processes.
Disadvantages: Short cathode life for heavier elements, noisy beam.

Freeman
Advantages: Most commonly used in commercial ion implanters, high beam quality,
   stable operation, for all elements.
Disadvantages: Cathode lifetime limited for oxygen and corrosive gases.

4.1 Magnetron

4.1.1 Ion Source History

The magnetron ion source got its name from the similarity to the magnetron microwave tube design. It was first presented by Van Voorhis (1934).[1] A magnetron with radial extraction through a slit was built by Cobić (1963).[2] Other versions had radial extraction (Piotrowsky, 1973).[3] A very compact hollow-cathode magnetron was developed by Miljević (1982).[4]

4.1.2 Working Principle of the Magnetron Ion Source and Description of the Discharge

Similarly to the calutron or the Nier and Bernas types ion sources, the magnetron (Figure 4.1) has a hollow anode, but the cathode is a straight wire parallel to the anode axis and the magnetic field, which is 0.1 T or more. The primary electrons have to spiral along the magnetic field lines, which gives them a good ionization efficiency. The performance of a magnetron ion source is similar to a calutron or a hot cathode Penning source. Similarly to other discharges in strong magnetic fields, magnetron ion sources tend to develop oscillations. Cathode lifetime is limited by sputtering and very short when the source is used with heavier gases. This, quite likely, is the reason for the disappearance of this type of ion sources despite their simple design. For coating processes, however, the cathode erosion is used in a planar magnetron sputter source of different geometry, which is extensively used to generate low-energy particles for uniform film production, but not for ion beam formation.[5]

0-8493-2502-1/95/$0.00+$.50
© 1995 by CRC Press Inc.

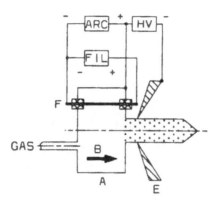

FIGURE 4.1
Magnetron ion source scheme and circuitry.

4.2   Freeman Ion Source

4.2.1   Ion Source History

A similar ion source design to that of the magnetron was developed at Harwell by Freeman in 1963.[6] Freeman put the cathode of a Nier type mass separator ion source inside the hollow anode, as in a magnetron, but left the magnetic field low. Several review articles have been published on the Freeman source.[7-9] The excellent performance of the Freeman ion source has made it the most successful source for ion implantation and industrial application, especially for semiconductor purposes. Many companies delivering ion implantation machines supply them with Freeman ion sources (see Section 4.3.2). Optimization of the Freeman ion source for high currents was done by Aitken in the 1980s.[14]

4.2.2   Working Principle of the Freeman Ion Source and Description of the Discharge

The Freeman ion source (Figure 4.2) has a similar design to a magnetron, but it uses just a low external magnetic field of about 100 G. The arc current is 1 to 3 A and the arc voltage just 40 to 70 V. A quite massive cathode rod, usually 2 mm in diameter and made of tantalum or tungsten, is heated with about 130 A and a few volts to the right temperature. The axial position of the cathode in the Freeman ion source shows several advantages.

1. The inherent magnetic field of the cathode forces the primary electrons to move around the cathode, concentrating the electron density in this area.
2. The high electron density next to the extraction slit produces a high ion density in this area, and, thus, a high ion beam current.
3. The straight-filament rod fixes the plasma parallel to the magnetic field lines and the extraction aperture, which is the reason for the excellent beam quality of Freeman ion sources.
4. The low magnetic field does not force instabilities like the high field in a magnetron.

The lifetime of the source is given by the lifetime of the filament, which is between 10 and 100 h, depending on the elements used and the arc current. The erosion of the filament is not uniform, but stronger at the positive end due to electron movement and higher plasma density. Changing the polarity of the filament after some time of operation improves cathode lifetime. Heating by ac has the same effect but increases plasma instabilities and the energy spread of the extracted ions.

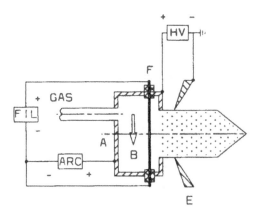

FIGURE 4.2
Freeman ion source scheme and circuitry.

The Freeman ion source is especially designed to deliver ion beams from nongaseous materials. There are many versions with ovens for various temperatures and for the application of chemical compounds and *in situ* chemical synthesis of the required material.[8,9,15] With chemical compounds, corrosion problems occur not only at the cathode, but also at the anode and oven. Carbon instead of tantalum is often used as anode material. There are very few current values of metal ions published; Table 4.1 gives currents for some elements that are of interest for ion beam modification of materials.[10] A list of elements, methods, and temperatures used can be found in Reference 9. Methods for ion source operation with nonvolatile materials are described in detail in Chapter 3.

Ion currents of several milliamps can be produced for most elements and more than 20 mA for a few elements like arsenic and phosphorus under favorable circumstances and using large extraction areas. The extraction slit is usually about 2 mm wide and about 40 mm long. Larger slits are possible, such as 90 mm,[10] but there are some disadvantages because the current density is not uniform along the long slit due to the bigger voltage drop along the cathode. By careful design of the anode and the magnetic field, however, it was possible to overcome this problem.[7]

The ion current density is controlled by the arc current, which is controlled by the filament, the gas pressure, and the magnetic field. Figure 4.3 shows the dependencies of the ion beam current on arc voltage, filament current, and magnetic field, and also arc current on arc voltage for three different gas flows.[6] The correspondence of arc current and ion beam current is obvious. For higher ion currents or arc currents above about 2 A, however, a fine control of the arc current with a current-controlled power supply is necessary to keep the discharge stable. The discharge voltage is then slowly readjusted to its optimum value by readjusting the filament emission.

4.3 Specific Ion Source Design

4.3.1 *Magnetron Ion Source with Axial Extraction (Figure 4.4)*

- Special design and construction details of the source

  The cathode is positioned off the axis inside the hollow anode. The magnetic field of 10 to 20 mT is maintained by a solenoid. Ions are extracted through one end flange on cathode potential and the gas or vapor from an oven enter the source through the other end flange. The discharge chamber is surrounded by a heat shield and can reach 1000°C. Along the ceramic rod behind the discharge chamber there is a

TABLE 4.1

Typical Metal Ion Beams with a Standard Freeman Source

Element	Method	Reactive gas	Ion current (mA)
Beryllium	a	$BF_3 + CCL_4$	0.08
Aluminium	b	Cl_2	0.54
Titanium	a	$CCl_4 + N_2$	0.1
	b	Cl_2	0.5–1
Chromium	a	$BF_3 + CCl_4$	0.27
	b	Cl_2	0.7–1.2
	c	$CrCl_3$	1.4
Nickel	a	$BF_3 + CCl_4$	0.07
Copper	b	Cl_2	0.085
Strontium	c	Sr	3.5
Yttrium	b	Cl_2	0.02
Molybdenum	a	CCl_4	0.1
	b	Cl_2	0.05
Palladium	a	$BF_3 + CCl_4$	0.005
Samarium	c	$SmCl_3$	1.8
Europium	c	$EuCl_2$	1.5
Tantalum	a	$BF_3 + CCl_4$	0.1
Tungsten	Filament	BF_3	0.03
Platinum	a	arsine + BF_3	0.003–0.03

a With foil liner in anode.
b With graphite furnace[16]
c Similar to b but[6]

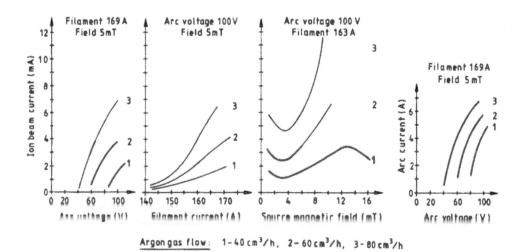

Argon gas flow: 1–40 cm³/h,   2–60 cm³/h,   3–80 cm³/h

FIGURE 4.3
Dependence of ion current on various source parameters for a Freeman ion source. (From Freeman, J. H., *Nucl. Instr. Meth.*, 22, 306, 1963. With permission.)

temperature gradient from 100 to 1000°C. The sample to be evaporated is moved to a position with the right temperature.

- Ion source material and vacuum conditions

The anode is made of graphite or molybdenum and the other parts of the discharge chamber of molybdenum, the insulators of alumina, and the cathode of 1-mm ∅ tungsten wire.

- Application area of the source

Mass separators, ion implantation.

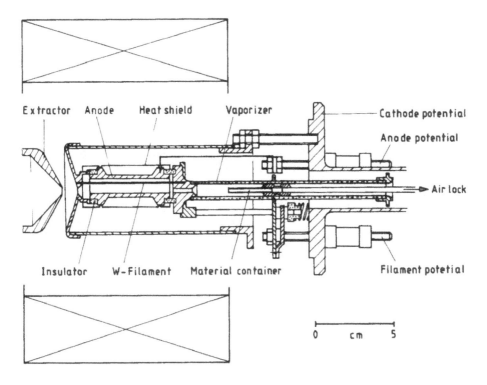

FIGURE 4.4
Magnetron ion source with axial extraction.[3]

4.3.2 Hollow-Cathode Magnetron (Figure 4.5)

- Special design and construction details of the source

  The hollow-cathode magnetron consists of a diode with two coaxial cylinders placed in an axial magnetic field. A cylindrical anode is around the cylindrical mesh cathode and leaves a free optical axis through the ion source. Anode (18-mm Ø × 60 mm) and cathode (5.5-mm Ø × 60 mm) are insulated to the base flange with the extraction aperture in its center. The discharge plasma is established inside the hollow cathode. When the discharge is established and the base flange connected to the anode, an ion current is obtained even at low accelerating voltages.

- Ion source material and vacuum conditions

  The discharge chamber is a glass tube (30-mm Ø), the anode cylinder Al or stainless steel, the cathode mesh stainless steel wire (0.4-mm Ø), eight lines per centimeter, and the insulators are made of lava. The base flange is nonmagnetic.

- Application area of the source

  Accelerators, ion implantation, SIMS, ion beam analysis, optical spectroscopy.

- Deliverer or user

  V. Miljević, Institute of Nuclear Science VINČA Atomic Physics Lab., PO Box 522, 11001 Belgrade, Yugoslavia

4.3.3 Freeman Ion Source (Figure 4.6)

- Special design and construction details of the source

  The ion source is mounted inside the vacuum chamber. All feed-throughs are mounted on a base flange perpendicular to the source axis. All source parts are selected to withstand aggressive elements. The extraction slit is usually 40 × 2 mm but designs up to 100 × 5 mm have been realized.

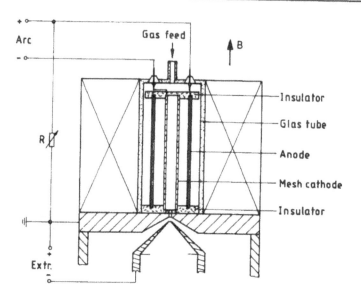

FIGURE 4.5
Hollow-cathode magnetron ion source. (Courtesy of V. Miljević, VIVČA, Belgrade.)

- Ion source material and vacuum conditions

  Arc chamber Mo, Ta, or graphite; W or Ta filament, 1.5- to 2.5-mm $\varnothing$; alumina or BN insulators; vacuum in chamber $\leq 10^{-3}$ Pa.

- Application area of the source

  Medium current ion implantation, ion beam modification of materials, isotope separation.

- Deliverer or user

  AEA Industrial Technology, Harwell Lab., Ditcot OX11 0RA, U.K.

  Applied Materials, Foundry Lane, Horsham, West Sussex RH13 5PY, U.K.

  EATON Corp. 108 Cherry Hill Drive, Beverly, MA 01915, U.S.A.

  VARIAN Ion Implant Systems, 35 Dory Road, Gloucester, MA 01930–2297, U.S.A.

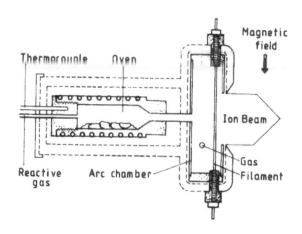

FIGURE 4.6
Freeman ion source.

NISSIN Electric, Applied and Thin Film Dept., 47, Umezu-takase-Cho, Ukyo-Ku, Kyoto 615, Japan

Naval Research Lab. Washington, D.C. 20375, U.S.A.

Used at many industrial ion implanters.

4.3.4 Operating Data

TABLE 4.2

Operating Data of Magnetron Ion Sources

		Axial extr. mag.	Hollow cath. mag.	Freeman source
Arc voltage	(V)	50–200	300–400	30–120
Arc current	(A)	0.5–3	0–0.08	0–8
Arc power	(kW)	0.1–0.5	0–0.03	≤ 1
Filament				
Current	(A)	45–60	Cold	100–200
Voltage	(V)	5–10	cath.	2–5 dc
Magnetic field	(mT)	10–20	0–50	0–20
Magnet supply				
Voltage	(V)		0–24	0–20
Current	(A)		0–5	0–50
Gas pressure				
From	(Pa)	0.1	1	10^{-2}
To	(Pa)	1	100	1
Gas consumption	(sccm)		—	0.2–5
Ion current	(mA)	10^{-3}–1	0–1	0–40
Extraction area	(mm)	0.5–3 $\varnothing$	1 $\varnothing$	2×45
Extraction area	(cm²)	0.002–0.07	8×10^{-3}	0.2–4
Extraction voltage	(kV)	30	0–3	2–70

REFERENCES

1. S. N. Van Voorhis et al., *Phys. Rev.*, 45, 492, 1934.
2. B. Cobić et al., *Nucl. Instrum. Methods*, 24, 358, 1963.
3. A. Piotrowsky et al., *Instrum. Exp. Tech.*, 15, 323, 1972.
4. V. Miljević, *Rev. Sci. Instrum.*, 55, 121, 1984.
5. R. K. Waits, *J. Vac. Sci. Technol.*, 15, 179, 1978.
6. J. H. Freeman, *Nucl. Instrum. Methods*, 22, 306, 1963.
7. D. Aitken, In *The Physics and Technology of Ion Sources*, I. G. Brown, Ed., John Wiley & Sons, New York, 1989, 187.
8. G. Dearnaly, J. H. Freeman, R. S. Nelson, and J. Stephen, *Ion Implantation*, North Holland, Amsterdam, 1973.
9. D. J. Chivers, *Rev. Sci. Instrum.*, 63, 2501, 1992.
10. Lintott Engineering/Applied Materials, U.K.
11. Extrion/Varian, U.S.A.
12. Eaton, U.S.A.
13. Nissin Electric, Kyoto, Japan.
14. D. Aitken, *Vacuum*, 36, 953, 1986; *Nucl. Instrum. Methods*, B21, 274, 1987.
15. J. H. Freeman and G. Sidenius, *Proc. 2nd Int. Conf. Ion Sources*, Vienna, 1972; *Nucl. Instrum. Methods*, 107, 477, 1973.
16. J. K. Hirvonen et al., *Nucl. Instrum. Methods*, 189, 103, 1981.

CHARACTERIZATION OF ION SOURCES

B. H. Wolf

5 PENNING ION SOURCES

Advantages: High current of multiple charged ions (Xe^{16+}), easy metal ion production by sputtering, internal source for cyclotrons.

Disadvantages: Noisy beam, medium beam quality, beam quality changes during source operation time, short lifetime for highly charged heavy ions, expensive if separate ion source magnet is needed.

5.1 Ion Source History

The Penning discharge was first investigated by L. R. Maxwell in 1931.[1] The Penning ion source got its name, PIG, from F. M. Penning who invented the Penning or Philips Ionization vacuum Gauge in 1937.[2] Penning ion sources were first used as internal sources in cyclotrons in the 1940s.[3-7] Later they were adjusted to linear accelerators as ion sources for multiply charged ions (1956).[8-10] A very powerful Penning ion source with heated cathodes was developed in Russia by Morozov and Makov (1957)[11] for use in cyclotrons. A modified version of this source was developed for the Unilac linear accelerator by Schulte and Wolf.[12] Several investigations of the PIG discharge have been published,[13-15] but still not all aspects of the PIG discharge are well understood. Extended review articles on Penning ion sources have been published.[10-20]

The calutrons used as internal ion sources in 180° mass separators are, in principle, Penning ion sources, but their discharge is running in a high-pressure mode where no electron oscillation occurs, indicated by the fact that the polarization of the reflector electrode to cathode or anode potential has no influence on the ion output of this source.[13,21]

5.2 Working Principle of the Penning Ion Source and Description of the Discharge

5.2.1 General Description

The Penning ion source (Figure 5.1) consists of a hollow anode cylinder with one cathode on each end. A strong axial magnetic field confines the electrons inside the anode and keeps them oscillating between the cathodes. This gives a high ionization

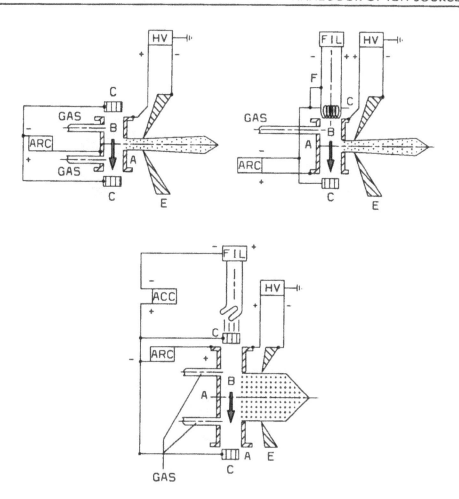

FIGURE 5.1
Penning ion source scheme and circuitry: (a) cold cathodes, (b) filament cathode, (c) heated cathode.

efficiency. The cathodes can be cold or hot (Figure 5.1a) or one filament and a cold anticathode (Figure 5.1b) or an indirectly heated block cathode and a cold anticathode (Figure 5.1c). The anticathode is usually connected to the cathode, but sometimes just floating because of space problems in some cyclotrons.

Ions can be extracted from PIG ion sources axially (Figure 5.2a) through one cathode or, more commonly, radially through a slit in the anode (Figure 5.2b). Gas is fed to the discharge usually through the anode close to the cathode(s) to ease ignition of the arc and to keep the neutral gas flow through the extraction slit in the anode low.

PIG sources can be operated in dc or pulsed mode, depending on the operational mode of the accelerator and on the power level needed to generate a specific ion beam. Depending on the cathode (hot or cold) and on the gas pressure, the arc voltage of a PIG ion source can vary from a few hundred volts to several kilovolts and the arc current from some milliamperes to tens of amperes. The magnetic field is between 0.1 and 1 T and is usually homogeneous, but also a mirror shaped field[22] and opposite a weaker field in the cathode area[23] have been tried for specific advantages.[14]

5.2.2 Characterization of the PIG Discharge

The Penning discharge shows two main regimes: a low-pressure regime between 10^{-6} and 1 Pa and a high-pressure regime between 10^{-1} and 10^2 Pa.[13,24] In the low-

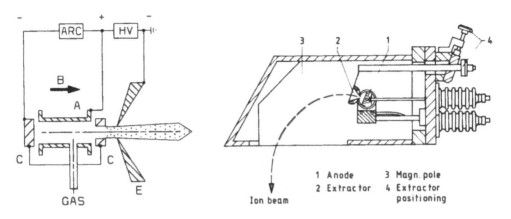

1 Anode   3 Magn. pole
2 Extractor 4 Extractor
             positioning

FIGURE 5.2
PIG ion source with axial (a) and radial (b) extraction.

pressure regime a high-voltage, low-current discharge occurs with various modes and potential distributions in the ion source.[13,24] More references can be found in Reference 10.

More important for ion sources is the high-pressure regime, which is divided into two main types: the cold cathode PIG source with arc voltages above 1 kV and currents between 0.5 and 5 A, and the hot cathode PIG source with arc voltages below 1 kV and currents between 1 and 50 A (Figure 5.3).

The arc plasma is a few volts negative in respect to the anode potential and nearly the full arc voltage drops along the narrow cathode sheath. Electrons emitted by the cathode(s) have an energy distribution dependent on the cathode temperature or the secondary emission process and are accelerated in the cathode sheath (Figure 5.4). More than half of the electrons can reach the opposite cathode and are lost for the ionization process. The others may oscillate many times between the cathodes or become thermalized in the dense plasma within a few oscillations.[25,15,14] Secondary electrons make a background of low-energy electrons, which may gain enough energy by local oscillating fields to take part in the ionization process (Figure 5.4). One electron can produce about eight ions or charges on average.[25] Electrons can reach the anode across the magnetic field lines, accelerated by local fields of plasma oscillations (Bohm diffusion).[26,16]

The discharge is controlled by the arc current and the arc voltage increases with decreasing gas flow or particle density in the discharge chamber until the discharge becomes unstable (Figure 5.5). There is little influence of the magnetic field on the discharge parameters as long as it reaches a certain minimum of roughly 0.1 T. Ignition of the arc in pulsed operation, however, is easier at higher magnetic field values, especially if minimum gas flow conditions have to be adjusted for multiply charged ion production.

In cold cathode PIG ion sources the ion current is the same to each cathode,[25] and in heated cathode PIG sources, a higher percentage of the ion current goes to the cathode with increasing cathode heating or emission compared to the cold anticathode.[15] The ion current to the anode has about the same value as to the cathodes,[25] which means that the ion current density at the cathodes is five to ten times the density at the anode surface, and, consequently, the extracted current densities show the same relation. The total extracted current of a PIG ion source is proportional to the arc current (Figure 5.6), and for extraction through the anode, about 10 to 100 (mA/cm²)/A, depending on the mass of the element used and the total anode surface.

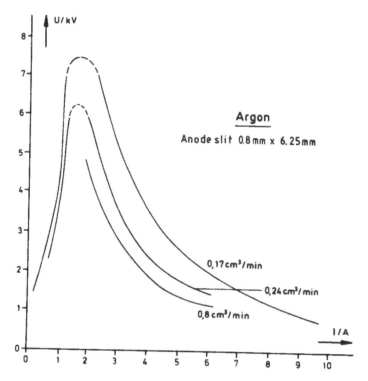

FIGURE 5.3
Discharge characteristic of a PIG source in a cold cathode and a hot cathode mode.[23]

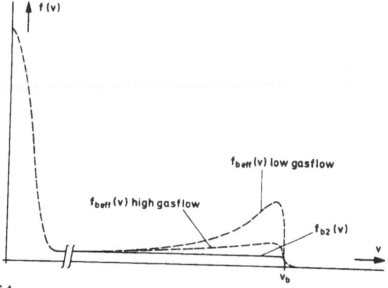

FIGURE 5.4
Electron energy distribution in a PIG ion source.[15]

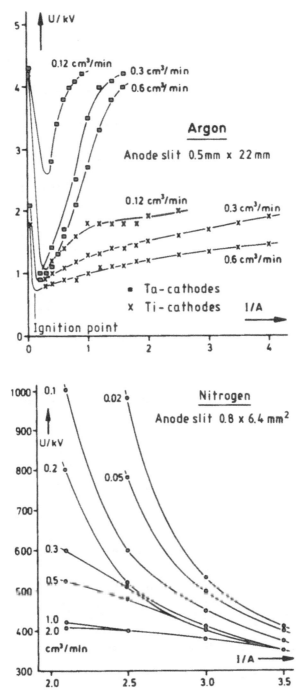

FIGURE 5.5
Variation of arc characteristic for different gas flows; (a) cold cathodes,[10] (b) hot cathodes.[17]

Cold Cathode PIG

The cold cathode discharge depends on the secondary electron emission from the cathodes and, hence, from the cathode material as well as from the gas used. Figure 5.7 shows arc characteristics for various cathode materials and gases. The arc power for cold cathode operation is limited to about 1 kW per cathode, because of the start

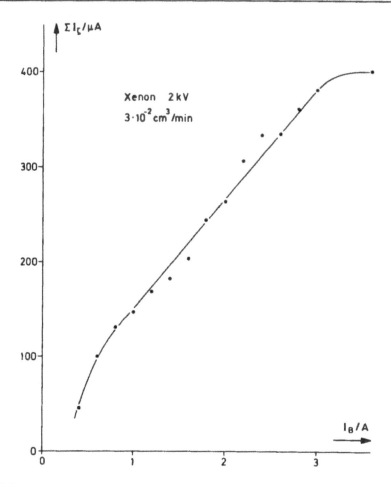

FIGURE 5.6
Total extracted ion current depending on arc current.[10]

of thermal electron emission, and depends on the cathode material and the ion source design.

For the high-current operation needed for the production of highly charged ions, the cold cathode source has to be pulsed to limit the mean power of the arc.

The lifetime of the source is limited by cathode erosion by sputtering, which depends on the cathode material and the gas used. Titanium has been selected as the best compromise.[8–10] If the cathodes are allowed to run hot, tantalum has been shown to be a good choice.[23,27] The cold and hot cathodes are worn out when the erosion crater's depth reaches around the anode bore radius. The discharge becomes unstable under these conditions. Rotating disks or rods have been tried to increase cathode lifetime.[28,29] Cathode lifetime is shorter in the cold cathode mode due to the higher arc voltage. The main advantage of cold cathode PIG sources is their simplicity, and for low-duty cycle operation, as for synchrotrons, they are still a good choice.

Hot Cathode PIG

As mentioned previously, cathodes run hot at higher arc power levels due to ion bombardment, especially in dc operation. The arc characteristic shows a strong negative tendency (see Figure 5.5b and Reference 17), which allows high-current operation with relatively low arc voltage possible. High arc current means high ion current can be extracted from the source. Most hot cathode PIG ion sources are used in cyclotrons with their limited space in the center region.[17,30,31] Tantalum and tungsten are mostly used as cathode materials, but tungsten carbide and hafnium carbide are also used in some cases.[17,32]

Heated Cathode PIG

The first PIG ion source had a filament cathode to feed electrons to the arc plasma and an anticathode just on floating potential. These ion sources worked reasonably well in cyclotrons for protons or helium, but for heavier ions the lifetime of the filament was too short because of sputtering. During the development of the calutron ion source, indirectly heated block cathodes were developed.[13]

Electrons emitted from a filament and accelerated to about 1 kV heat the cathode from the rear side. In that way the cathode temperature can be controlled independently of the ion flux from the plasma to the cathode, and optimum arc conditions can be adjusted for the ion species needed (Figure 5.8). Extensive research on heated cathode PIG ion sources has been done by Makow and the Dubna group.[14,33,16] Their research shows that the electrons are thermalized after a short way through the dense anode plasma and also shows that the performance of the ion source can be explained by an electron temperature of some 10 eV.

The anticathode is kept cold in a heated cathode PIG ion source and the electron flow to the anticathode exceeds the secondary electron emission from the anticathode plus the ion flow to it (Figure 5.9). More than –20 V is necessary to suppress this electron current. The electron flow to the anticathode is also influenced by the cathode heating power (Figure 5.9c).

The lifetime of the heated cathode exceeds that of cold or hot cathodes because it is possible to burn them totally down without plasma instabilities occurring. The cathode lifetime depends less on the element used because the cathode erosion happens mainly by self-sputtering with the tungsten or tantalum cathode materials.[33,34] Hot cathode PIG ion sources deliver the highest currents of highly charged ions of all conventional ion sources (see Table 5.1).

5.2.3 Multiply Charged Ions

Multiply charged ions can be efficiently produced with PIG sources. This is due to the high plasma density ($n_e \leq 10^{13}/cm^3$), the high primary electron energy, low gas pressure, or high degree of ionization and a reasonable ion confinement time. The ion confinement time may be estimated as $\tau_i = r^2 B / T_e$, with r being the plasma radius in centimeters, B the magnetic field in teslas, and T_e the electron temperature in V electron volts, and is about 10 to 100 μs for the usual PIG source parameters for multiply charged ion production. Figure 5.10 shows the time development of xenon charge states 1 to 8 together with the arc voltage U_B and the arc current I_B for a pulsed cold cathode discharge. After ignition it needs more than 100 μs to reach the equilibrium ion current for the higher charges. A pumping effect of the discharge may also have an effect in this case. The afterglow shows a immediate decrease of the higher charges, but an increase of charge 1 and 2, which show a maximum after 40 μs. These are ions that have caught electrons by charge-exchange reactions and must live longer than 40 μs.

The yield of multiply charged ions increases with ion current or plasma density (Figure 5.11) and with decreasing gas pressure or neutral particle density (Figure 5.12). The arc adjusts to the missing particles by increasing the arc voltage and producing multiply charged ions to be able to carry the current requested by the external circuit.

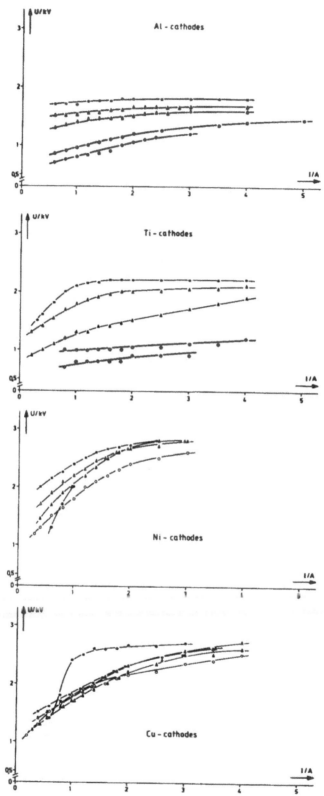

FIGURE 5.7
Cold cathode arc characteristics for various cathode materials.[10]

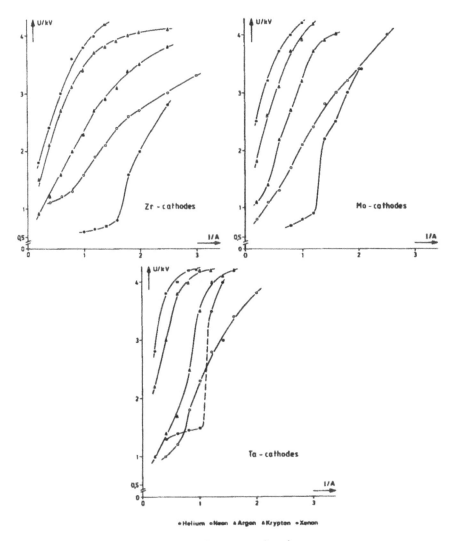

FIGURE 5.7 (continued)

When this compensation is no longer possible the arc becomes unstable and is extinguished.

The highest charge state achieved for xenon has been 13+ for cold and 16+ for hot cathode PIG ion sources (Figure 5.13).[36]

There is no significant influence of the arc voltage and the magnetic field (despite the dependence of τ_i on B) on the charge state distribution if they are above the minimum of around 1 kV or 0.1 T, respectively. In Table 5.1 ion currents of different charge states and gases are summarized for a PIG ion source with heated tantalum or tungsten cathodes.

5.2.4 Metal Ion Production

Metallic ions or ions from nongaseous materials can be generated in PIG sources using the techniques described in Chapter 3 (gaseous compounds, chemical synthesis, evaporation, and sputtering). Sputtering is the best choice in most cases as long as the melting point of the substance used is not too low. Alloys can be used to increase the melting point of an electrode or to get a nonferromagnetic one.[38] The

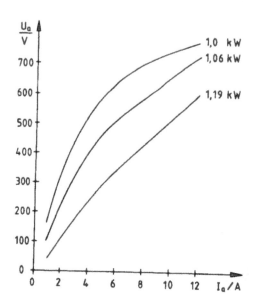

FIGURE 5.8
Effect of cathode heating power on the arc characteristic.[14]

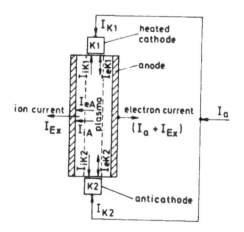

FIGURE 5.9
Electron and ion currents of cathode and anticathode: (a) principal current flow between plasma and electrodes, (b) vs. anticathode potential, (c) vs. cathode heating power.[15]

sputter electrode is usually positioned in a slit in the anode wall opposite or beside the extraction slit (Figure 5.14a)[12,39,40] and connected to a potential of several hundred volts, negative with respect to the anode. Half- or full-cylindrical sputter electrode shapes have also been successfully used (Figure 5.14b).[12,41]

In Table 5.2 metal ion currents for a heated cathode PIG ion source are summarized for charge states up to 12+.[37]

PIG Sources with Axial Extraction

As mentioned previously, the ion current density at the cathodes of a PIG source is five to ten times the density at the anode, which means that one can extract the same

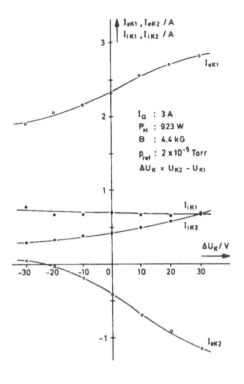

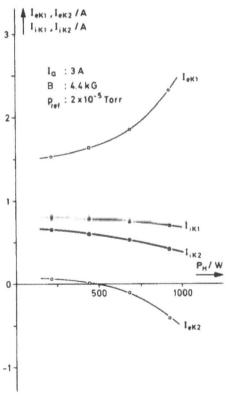

FIGURE 5.9 (continued)

TABLE 5.1

Ion Yields of Multiply Charge Ions from PIG Ion Source Ta or W Heated Cathodes[35]

Ion	Cath. material	Arc kV	Arc A	Charge state (mA-peak) 1	2	3	4	5	6	7	8	9	10	11	12	13	14	15
Ar	Ta	1100	13	25	28	35	12	10	2.8	1.2	0.2							
	W	900	6	20	37	42	25	7	1.6	0.4	0.1							
Kr	Ta	1000	10	—	—	25	14	10	—	6	3.7	1.1	0.12	0.015				
	W	450	13	11	17	40	47	32	20	7	2.5	0.2	0.02					
Xe	Ta	1100	13	—	—	27	18	16	14	16	—	9.1	4.9	2.5	0.8	0.14	0.06	0.01
	W	950	13	—	—	12	13	11	10	11	11	5	2.8	0.9	0.12			

Note: Extraction voltage $\leq$ 25 kV.

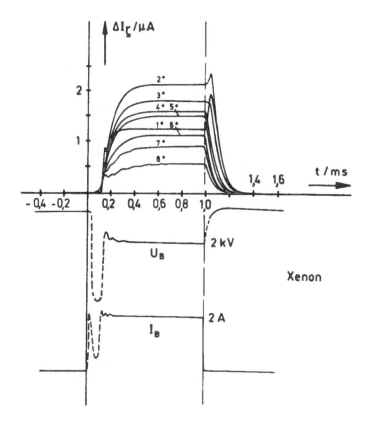

FIGURE 5.10
Time development of various xenon charge states.[10]

ion current from a fivefold smaller area. On the other hand, the area available for extraction is limited to less than 0.1 cm², which is not the case for radial extraction where several centimeters squared can be used for extraction, depending on the ion source size and design. PIG sources with axial extraction can be constructed very compactly and can deliver milliamp beams of singly charged ions (Figures 5.20 and 5.21).[42,43]

PIG sources with axial ion extraction give lower currents of multiply charged ions for several reasons. The multiply charged ion production is larger in the center of the anode[14,33] and the charge-exchange losses are high in the long extraction channel. A high yield of multiply charged ions needs a high arc current, which causes

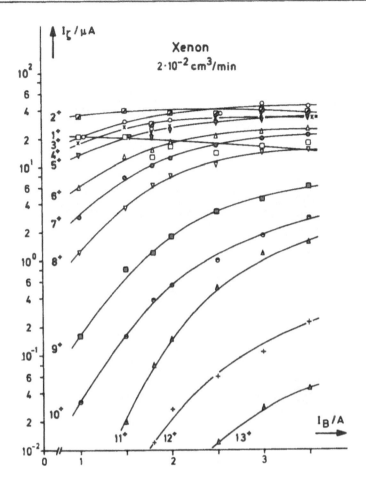

Arc voltage ≈ 2 kV, magnetic field 5 kGauss, anode
diaphragm 10^{-2} cm^2, Titanium cathodes, duty factor 5%

FIGURE 5.11
Variation of xenon charge states with arc current.[10]

high cathode erosion and in this way changes the extraction conditions during the
source lifetime. Therefore, PIG sources with axial extraction are used for low charge
states only. The current efficiency of axial extraction is higher than for radial extraction, especially in low-current operation.[42] If a high-current mode of operation is used
in a PIG with axial extraction, an expansion cup improves the beam formation and
quality (Figure 5.22).[44]

5.3 Specific Ion Source Design

5.3.1 Cyclotron Source with Discharge-Heated Cathode (Figure 5.15)

- Special design and construction details of the source

  The source is designed to be introduced through the center bore of a cyclotron. All
  electrical and cooling connections have to be inside the support cylinder. Here the
  anticathode connection is done through a contact inside the cyclotron. Similar
  designs have an internal connection between the two cathodes. Ion extraction is

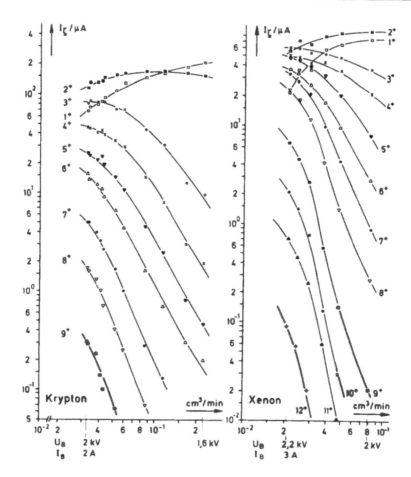

FIGURE 5.12
Variation of xenon charge states with gas flow.[10]

done by the rf field of the cyclotron or by a separate extraction electrode (puller). The cathodes are thermally insulated to enable them to run hot. Like most cyclotron sources, this PIG source is designed to run in dc operation.

• Ion source material and vacuum conditions

The anode is water-cooled and made of Cu; the cathode support is a water-cooled Cu rod that holds the BN cathode insulator; the anticathode has a Mo support and also a BN insulator and is not cooled except for the thermal contact to the anode. The extraction slit (1 × 25 mm) is made of W.

• Application area of the source

Internal ion source for cyclotrons; separate arrangements can be used for high-energy ion implantation using the high charge states.

• Deliverer or user

AEA Harwell Laboratory, Didcot OX11 0RA, U.K.

Oak Ridge Nat. Lab., P.O. Box 2008, Oak Ridge, TN 37831–6368, U.S.A.

Texas A&M University, College Station, Texas 77843–3366, U.S.A.

Lawrence Berkeley Lab., Berkeley, CA 94720, U.S.A.

Many other cyclotron laboratories.

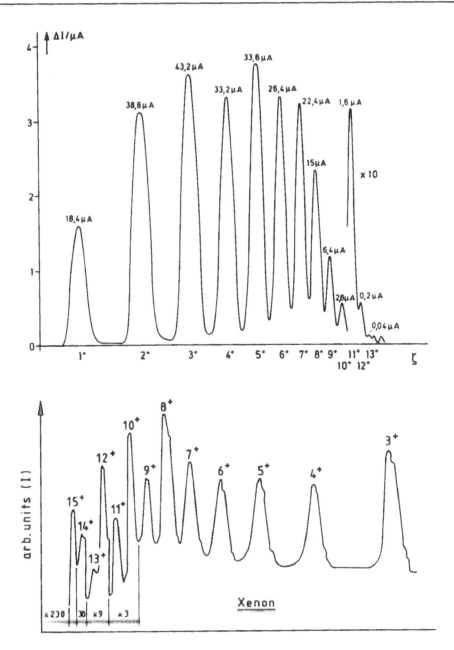

FIGURE 5.13
Xenon charge spectrum of: (a) cold cathode PIG,[10] (b) hot cathode PIG.[36]

5.3.2 Penning Source with Cold Cathode (Figure 5.16)

- Special design and construction details of the source

  Simple and compact design; can be used as internal cyclotron ion source or with a compact separate magnet for linear accelerators. In low-duty-cycle operation, high ion current is achievable. Intensive cooling of the cathodes only.

- Ion source material and vacuum conditions

  Cathodes, Ti; cathode supports, Cu; anode, stainless steel; extraction slit, Ta; extractor slits, W; cathode insulators, alumina or BN.

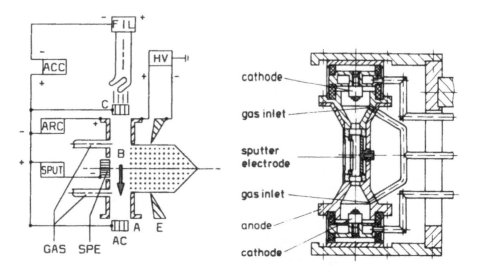

FIGURE 5.14

PIG sputter electrode arrangements: (a) block shape, (b) cylindrical shape.

TABLE 5.2

Ion Yield of Metal Ions from PIG Sputter Ion Sources[37]

		Arc		Sputter		Charge state (mA-peak)											
Ion	Gas	kV	A	kV	A	1	2	3	4	5	6	7	8	9	10	11	12
Mg	Ar	660	7.5	520	2.9	18	85	32	3	0.4	0.06						
Al	Xe	300	8.2	980	1.9	22	59	20	3.4	0.35	0.04						
Si	Xe	620	11	1900	2.5	12	46	35	13	1.6	0.18	0.02					
P	Ar	1000	5.3	1000	1.0	3	6.3	5.5	1.2	0.26	0.03						
Ca	Xe	600	9.5	540	1.8	3	23	22	14	4.5	1	0.18	0.04				
Sc	Xe	500	8	1700	0.64	1.3	18	19	4.9	0.8	0.05	0.01					
Ti	Xe	750	16	1250	2.4	2.1	8.5	14	14	8.8	1.1	0.33	0.04				
V	Xe	900	15	1400	1.6	6.3	19	12	14	6.4	1.1	0.15	0.03				
Cr	Xe	1200	15	850	1.0	3.7	11	5.2	10	8.8	3.3	0.55	0.06				
Mn	Xe	650	8.5	650	0.7	5	16	18	11	6.1	1.6	0.39	0.09	0.03			
Fe	Xe	480	13	600	1.6	3	13	14	6.4	1.7	0.33	0.12	0.04				
Co	Xe	500	15	790	1.3	13	19	15	10	2.5	0.53	0.06	0.01				
Ni	Xe	800	6	1000	1.0	—	23	24	16	6.7	1.5	0.44	0.08	0.03			
Cu	Ar	540	10	400	2.5	—	30	32	35	26	6.6	1.9					
Zn	Xe	400	7.5	560	1.6	—	70	55	29	9.5	3.95	0.76	0.2	0.02			
Ge	Ar	900	5.5	2000	0.5	8.4	17	9.5	11	6.4	1.35	0.5	0.09				
Se	Xe	950	10	500	0.65	3.9	9.5	7.6	5.8	1.35	0.56	0.18	0.1				
Zr	Ar	1100	5	1100	0.8	—	6.9	4.1	5.9	3.7	1.4	1	0.47	0.15	0.04		
Nb	Xe	700	20	880	1.6	0.7	4.8	6.8	11	17	3.8	1.1	0.33	0.04			
Mo	Xe	380	9.5	940	1.8	—	24	25	23	17	9.4	2	0.4				
Cd	Ar	900	10	200	1	—	14	18	28	37	22	11	3	0.74	0.11	0.025	
In	Ar	540	11	1250	0.65	—	1.7	3.3	6.4	7.4	3.2	1.2	0.33	0.02			
Sn	Xe	840	11	550	0.75	—	1.6	2.6	3.3	3	3.7	1.3	0.13				
La	Xe	760	11	850	1.1	—	—	8.5	7.7	7.9	5.5	3.3	2	0.9	0.04		
Hf	Ar	650	7	750	1	3.9	4.1	3.6	4.5	4.7	2.3	0.6	0.21	0.05	0.01		
Ta	Ar	470	9.8	970	1.5	—	—	—	11	19	13	8.4	3				
W	Xe	360	9	980	1.4	—	—	20	17	13	6.8	3.3	0.7	0.12			
Re	Ar	580	20	540	2.8	—	15	14	12	8.6	6	0.9	0.6				
Pb	Xe	950	10	500	0.65	—	—	1.3	4.9	5.5	9	8.8	7	3.9	2.1		
Bi	Ar	860	8.5	480	0.5	—	—	2.4	2.8	5.6	6.3	4.2	2	1.2	0.9	0.22	0.03
Th	Ar	1200	10	750	0.9	—	—	1.6	4.3	8.6	6.8	15	18	18	12	4.8	

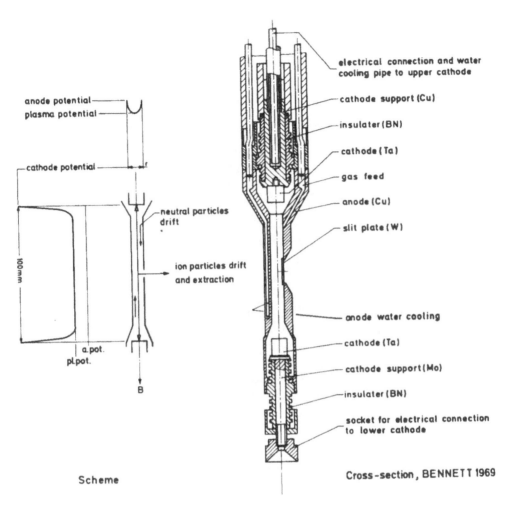

FIGURE 5.15
Cyclotron source with discharge-heated cathode.

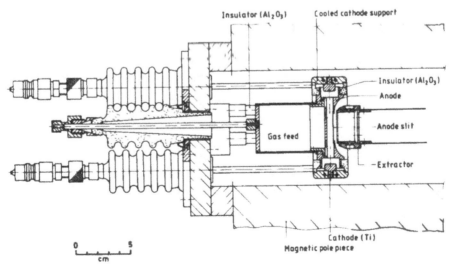

FIGURE 5.16
Cold cathode PIG with radial extraction.

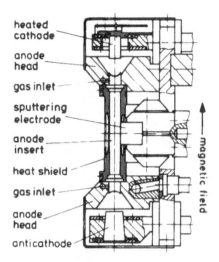

FIGURE 5.17
Penning source with heated block cathode.

- Application area of the source

  Internal ion source for cyclotrons, linear accelerators; separate arrangements can be used for high-energy ion implantation using the high charge states.

- Deliverer or user

  Lawrence Berkeley Lab., Berkeley, CA 94720, U.S.A.

  GSI Planckstr.1, 64291 Darmstadt, Germany

5.3.3   Penning Source with Heated Block Cathode (Figure 5.17)

- Special design and construction details of the source

  The source exists in different designs for use as a cyclotron ion source or in a separate magnet. Cyclotron versions usually have longer anodes because of the larger gap of cyclotron magnets. The source in Figure 5.17 is designed for a gap of 14 cm in a separate magnet for injection into a linear accelerator. To save space the filament had to be flat and the anode short (8 cm). The extraction aperture is 45 mm long and 0.5 to 2 mm wide, depending on the need for isotope separation. The cathode (W or Ta) is mounted in a chimney that serves as a heat shield. Cathode holder and anode are water cooled and so is the sputter electrode, which can be positioned opposite the anode aperture. The source is at positive potential with respect to the extractor and the vacuum chamber.

- Ion source material and vacuum conditions

  Anode, Cu or stainless steel; cathode and anticathode holder, Cu; cathode shield and anticathode, Mo; filament, W; extraction slit, Mo; and extractor aperture, W. The base flange of the source is mounted with an epoxide insulator to the vacuum chamber. The pressure in the chamber during source operation is 2 to 4×10^{-3} Pa.

- Application area of the source

  Internal ion source for cyclotrons, linear accelerators; separate arrangements can be used for high-energy ion implantation using the high charge states.

- Deliverer or user

  GSI Planckstr.1, 64291 Darmstadt, Germany

  FLNR, JINR, 141980 Dubna, Moscow Region, Russia

  Several cyclotron laboratories in Eastern Europe and in Japan.

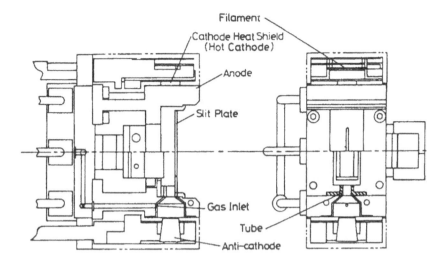

FIGURE 5.18
Short-pulse Penning source with heated block cathode. (Courtesy of Y. Sato, Ciba, Japan.)

5.3.4 Short-Pulse Penning Source with Heated Block Cathode (Figure 5.18)

- Special design and construction details of the source

  The design is mainly the same as the source described previously with a few changes to optimize the high current of short ion pulses (160 μs) with low repetition rate (1 Hz) for synchrotron injection. The source uses cylindrical cathodes and a reduced diameter at the ends of the anode to reduce the gas flow to the anode center. The source is operated in a starvation mode, which gives a maximum of the high-charge state current about 200 μs after ignition. The lifetime of the source under these operating conditions is more than a week.

TABLE 5.3

Ion Currents from the Low-Duty-Cycle PIG Source[45]

Element	Charge state (mA)							
	1	2	3	4	5	6	7	8
He	3.5	3.0						
C	1.0	3.5	—	0.6	0.02			
N	—	2.0	2.5	1.2	0.2			
O	—	2.0	2.3	—	0.3	0.03		
Ne	—	2.0	2.0	0.8	0.4	0.02		
Si[a]	—	—	0.4	0.6	0.3	0.05	0.01	
Ar	—	—	1.5	1.9	1.8	0.8	0.4	0.2

[a] Silicon was sputtered with argon from a single crystal.

- Ion source material and vacuum conditions

  Same as for Penning source with heated block cathode (Section 5.3.3).

- Application area of the source

  Synchrotron accelerator for medical application.

- Deliverer or user

  Quantum Equipment Division, Sumitomo Heavy Industry, 2–1-1 Yato, Tanashi 188, Japan

  National Institute of Radiological Science, 4–9-1 Anagawa, Chiba 260, Japan

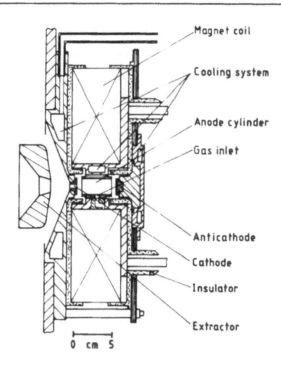

Magnet coil

Cooling system

Anode cylinder

Gas inlet

Anticathode

Cathode

Insulator

Extractor

0   cm   5

FIGURE 5.19
Baumann axial extraction PIG ion source. (Courtesy of H. Baumann, Univ. of Frankfort, Germany.)

5.3.5 Penning Source with Cold Cathode, Axial Extraction (Figure 5.19)

- Special design and construction details of the source

  The source is a compact design that delivers low-charged ions. The total extracted current is several milliamps even when the arc current is just 50 mA (Table 5.4). The source is operated in a special low-pressure (0.1 Pa) dc mode that focuses the plasma to the extraction hole in one of the cathodes. This enables cathode lifetimes of more than 60 h because of the low arc power of about 200 W.

- Ion source material and vacuum conditions

  Anode, stainless steel; cathode, Ta; anticathode, Ta for operation with gases, or the specific metal if metal ions are required. The base flanges are water-cooled iron and the solenoid Cu wire. The extractor is made of Ni.

- Application area of the source

  Tandem Graaf accelerators, ion implantation, ion beam analysis.

- Deliverer or user

  Institut für Kernphysik, Universität, Frankfurt, Germany

  Several research laboratories.

5.3.6 Penning Source with LaB$_6$ Cathode, Axial Extraction (Figure 5.20)

- Special design and construction details of the source

  This source is just 8 cm in diameter and 8 cm high. The magnetic field of about 0.1 T is delivered by a SmCo permanent magnet. The LaB$_6$ cathodes give low discharge voltages ($\leq$500 V) and a correspondingly low discharge power of just 30 W, ten times lower than the previous source (Section 5.3.5). Cathodes made from single crystals give a more stable arc and lower arc voltage.

TABLE 5.4

Ion Currents from the PIG Source with Axial Extraction[42]

Element	U_{arc} (kV)	I_{arc} (mA)	Charge state (mA)							
			1	2	3	4	5	6	7	8
He	3.2	15	3.1	0.017						
N	4	11.5	0.5	0.016						
O	4	19	0.7	0.023						
Ne	4.5	8.4	2.4	0.17	0.01					
Cl	4.2	4.8	1.5	0.36	0.004					
Ar	4.5	7	1.7	0.17	0.02	0.004				
Kr	4.2	4.4	1	0.19	0.06	0.015	0.005	0.002		
Xe	5.4	3.7	0.84	0.23	0.12	0.04	0.012	0.004	0.002	0.001

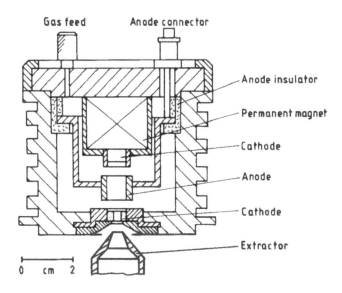

FIGURE 5.20
Penning source with LaB_6 cathode, axial extraction. (From Song, Z., *Rev. Sci. Instrum.*, 61, 463, 1990. With permission.)

- Ion source material and vacuum conditions

  The anode is made of graphite and so are the cathode sockets to prevent chemical reactions with the hot LaB_6. The source body is made of steel and just air-cooled; the magnet is SmCo.

- Application area of the source

  Small accelerators, ion implantation.

- Deliverer or user

  Dept. of Technical Physics, Peking University, Beijing 100871, China

5.3.7 Operating Data

TABLE 5.5

Operating Data of Penning Ion Sources

		Hot cath.	Cold cath.	Heated cath.	Low-duty cycle	Axial extract.	LaB₆ cath.
Arc voltage	(kV)	0.3–1.3	1–5	0.3–1.3	1–1.5	1–5	0.3–0.5
Arc current	(A)	5–10	1–5	5–20	3.5	0.02–0.1	0.02–0.1
Arc power	(kW)	1–10	0–1	1–25	10^{-3}	0–0.3	0–0.03
Ignition volt.	(kV)	3	5	3	3	6	1
Duty cycle	(%)	dc	≤25	≤100	≤1	dc	dc
Filament							
Current	(A)	—	—	50–150	55	—	—
Voltage	(V)	—	—	2–8	—	—	—
Electron acceleration							
Voltage	(kV)	—	—	0–2	500–700	—	—
Current	(A)	—	—	0–2.5	0.8–1.5	—	—
Magnetic field	(T)	≤1.2	≥0.4	0.2–2	~0.4	~0.1	~0.1
Magnet supply	(kW)	Cycl.	5	25	20	≤1	—
Gas pressure	(Pa)	1–10	1–10	1–10	1–2	0.1–1	0.1–1
Gas consumption	(sccm)	0.2	0.2	0.2–0.6	0.2–1.1	—	—
Ion current	(mA)	≤25	≤5	≤100	≤10	≤3	≤2
Charge states	max.	Ar^{9+}	Xe^{13+}	Xe^{16+}	Ar^{8+}	Xe^{9+}	N^{2+}
Extraction area	(cm²)	0.25	0.4	0.7	0.4	0.07	0.07
Extraction voltage	(kV)	5–25	5–35	5–25	20–30	10–50	5–30
Anode aperture	(mm)	1 × 25	1.5 × 25	1.5 × 45	2 × 20	3 Ø	3 Ø
Anode canal	(mm) Ø	8	6	10 × 10	10	20	6.5
Cathode	(mm) Ø	12	9	7 × 7	7	20	6
Cathode distance	(cm)	10	6.5	8–25	8	5	1.8

REFERENCES

1. L. R. Maxwell, *Rev. Sci. Instrum.*, 2, 129, 1931.
2. F. M. Penning, *Physica*, 4, 71, 1937.
3. A. T. Finkelstein, *Rev. Sci. Instrum.*, 11, 94, 1940.
4. R. S. Livingston, *J. Appl. Phys.*, 15, 2, 1944.
5. H. Heil, *Z. Phys.*, 120, 212, 1943.
6. R. S. Livingston and R. J. Jones, *Rev. Sci. Instrum.*, 25, 552, 1954.
7. R. J. Jones and A. Zucker, *Rev. Sci. Instrum.*, 25, 562, 1954.
8. C. E. Anderson and K. W. Ehlers, *Rev. Sci. Instrum.*, 27, 809, 1956.
9. B. Gavin, *Nucl. Instrum. Methods*, 64, 73, 1968.
10. B. H. Wolf, *IEEE Trans. Nucl. Sci.*, 19, 74, 1972; Ph.D. thesis, Univ. Heidelberg Rep. GSI 73–13, GSI, Darmstadt, 1993.
11. P. M. Morozov et al., *Atom. Energ.*, 3, 272, 1957.
12. H. Schulte et al., *IEEE Trans. Nucl. Sci.*, 23, 1042, 1976.
13. A. Guthrie and R. K. Wackerling, *The Characteristics of Electrical Discharges in Magnetic Fields*, McGraw-Hill, New York, 1949.
14. B. N. Makov, *IEEE Trans. Nucl. Sci.*, 23, 1035, 1976.
15. H. Schulte et al., *IEEE Trans. Nucl. Sci.*, 23, 1053, 1976.
16. M. D. Gabovich, *Physics and Technology of Plasma Ion Sources*, Atomprint, Moscow, 1972; Translated by Foreign Technology Division WO-AFB, Ohio, 1973.
17. J. R. J. Bennett, *IEEE Trans. Nucl. Sci.*, 19, 48, 1972.
18. T. S. Green, *Rep. Prog. Phys.*, 37, 1257, 1974.
19. L. Vályi, *Atom and Ion Sources*, John Wiley & Sons, London, 1977.
20. B. F. Gavin, in *The Physics and Technology of Ion Sources*, I. G. Brown, Ed., John Wiley & Sons, New York, 1989.

21. J. Koch, Ed., *Electromagnetic Isotope Separators and Applications of Electromagnetically Enriched Isotopes*, North Holland, Amsterdam, 1958.
22. M. Müller, Annual Rep. 1982 GSI-83–1, GSI, Darmstadt, 1983.
23. J. R. J. Bennett and B. F. Gavin, *Part. Accel.*, 3, 85, 1972.
24. W. Schuurman, *Physica*, 36, 136, 1967.
25. J. Backus, *J. Appl. Phys.*, 30, 1866, 1959.
26. D. Bohm in Reference 13.
27. J. R. J. Bennett, *Proc. 5th Int. Cyclotron Conf.*, Oxford, 1969, 499; *Proc. 1st Int. Conf. Ion Sources*, Saclay, 1969, 571.
28. M. L. Mallory and D. H. Crandell, *IEEE Trans. Nucl. Sci.*, 23, 1069, 1976.
29. B. F. Gavin, private communication.
30. M. L. Mallory et al., *IEEE Trans. Nucl. Sci.*, 19, 118, 1972.
31. D. J. Clark et al., *IEEE Trans. Nucl. Sci.*, 19, 114, 1972.
32. H. Kuhn and F. Schulz, Rep. KFK-1783, KFK, Karlsruhe, 1972.
33. B. N. Makov, *Proc. 2nd Int. Conf. Ion Sources*, Vienna, 1972, 527.
34. A. S. Pasjuk et al., *Sov. Phys. Tech. Phys.*, 3, 42, 1965.
35. A. S. Pasjuk et al., Rep. JINR P9–11914, Dubna, 1978; German translation, Rep. GSI-tr-79/4, GSI, Darmstadt, 1979.
36. V. Kutner, *Rev. Sci. Instrum.*, 65, 1039, 1994.
37. V. Kutner et al., *Rev. Sci. Instrum.*, 61, 487, 1990.
38. M. Müller et al., Rep. GSI-82–11, GSI, Darmstadt, 1982; K. Leible and B. H. Wolf, Conference series No. 38, Inst. of Physics, London, 1978, 96.
39. A. S. Pasjuk and Yu. P. Tretjacov, *Proc. 2nd Int. Conf. Ion Sources*, Vienna, 1972, 512.
40. A. S. Pasjuk et al., JINR Rep. P7–4488, Dubna, 1969.
41. B. Gavin, *IEEE Trans. Nucl. Sci.*, 23, 1008, 1976.
42. H. Baumann et al., *IEEE Trans. Nucl. Sci.*, 19, 88, 1972; *Nucl. Instrum. Methods*, 122, 517, 1974.
43. Z. Song et al., *Rev. Sci. Instrum.*, 61, 463, 1990.
44. F. I. Mineev and O. Kovpik, *Prib. Tekh. Eksp.*, 4, 33, 1963; *Instr. Exp. Technol.*, 8, 1072, 1964.
45. Y. Sato et al., *Rev. Sci. Instrum.*, 63, 2904, 1992.

Chapter 2/Section 6

CHARACTERIZATION
OF ION SOURCES

K. N. Leung

6   MULTICUSP ("BUCKET") TYPE) ION SOURCE

6.1   History of Multicusp Ion Sources

During the early 1980s, ion sources capable of generating high current density and high-power beams of energetic particles over large areas were required for fusion research. The permanent magnet-generated line-cusp type plasma source developed at UCLA is able to fulfill most of these requirements.[1] Subsequently, this type of ion source has been developed to produce various positive and negative ion beams for accelerator and industrial applications.[2-9]

6.2   Working Principles of the Multicusp Ion Source

In the multicusp ion source, the primary ionizing electrons are normally emitted from tungsten-filament cathodes. The source chamber walls form the anode for the discharge. The surface magnetic field generated by rows of permanent magnets, typically of samarium-cobalt or neodymium-iron, can confine the primary ionizing electrons very efficiently. As a result, the arc and gas efficiencies of these sources are high.

The multicusp ion source is relatively simple to operate. There are three main components in the source, the filaments (cathode), the chamber (anode), and the first, or plasma, electrode. A schematic diagram of the multicusp source is shown in Figure 6.1. The source chamber can be rectangular, square or circular in cross section. The permanent magnets can be arranged in rows parallel to the beam axis (Figure 6.1). Alternatively, they can be arranged in the form of rings perpendicular to the beam axis[7,10] as illustrated in Figure 6.2. The back plate also contains rows of the same permanent magnets. Grooves are generally milled on the external wall so that the magnets are mounted within approximately 3 mm from the vacuum. These magnets generate multicusp magnetic fields for primary electron as well as for plasma confinement.

The open end of the source chamber is closed by a set of extraction electrodes. The source can be operated with the first (plasma) electrode electrically floating or connected to the negative terminal of the cathode.[11] The background gas is introduced

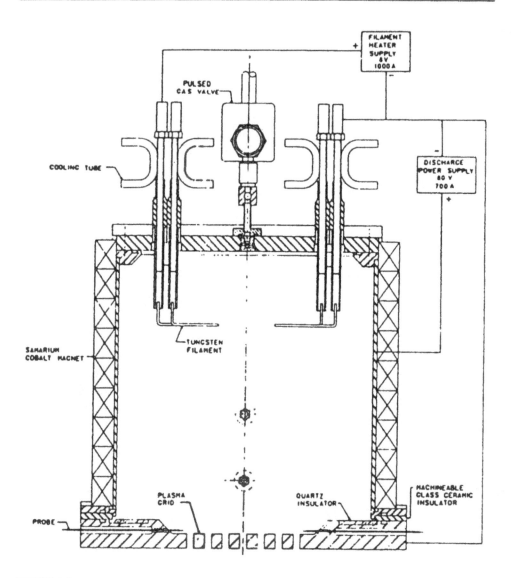

FIGURE 6.1
Schematic diagram of a multicusp ion source.

into the source chamber through a needle or a pulsed valve. The plasma is produced by primary ionizing electrons emitted from one or more tungsten filaments, which are normally biased negatively with respect to the chamber wall (anode). These filaments are located in the field-free region of the ion source chamber and they are mounted on molybdenum holders. The plasma density in the source, and, therefore, the extracted beam current, depends on the magnet geometries, the discharge voltage and current, the biasing voltage on the first extraction electrode, and the length of the source chamber.

For low discharge power, the ion source chamber can be made of thin stainless steel.[3] If the ion source is designed for high-power discharge operation, then the chamber walls are made of copper for fast heat conduction.[11] The source chamber is normally pumped down to about 10^{-7} torr before gases are introduced for the discharge.

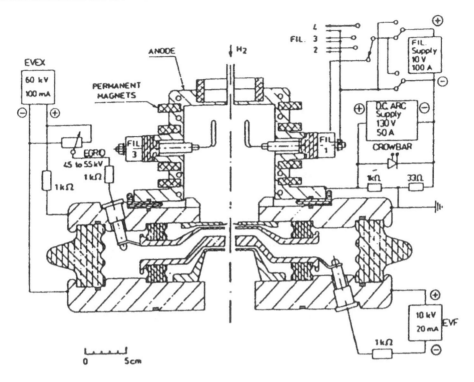

FIGURE 6.2
A multicusp source equipped with rings of magnets perpendicular to the beam axis.

6.2.1 Positive and Negative Ion Beam Production

The multicusp source is generally used to produce positive or negative hydrogen/deuterium ion beams for particle accelerators and for neutral beam heating of tokamak plasmas. It can also be used to generate positive ion beams of the inert gases (He, Ne, Ar, Kr, and Xe). For ion implantation and for surface modification purposes, beams of B^+, P^+, As^+, and N^+ ions have been extracted from some multicusp ion sources.[12,13] By operating the source at higher discharge voltages, ions with multiply charged states have been generated.[14]

It has been found that atomic species >90% can be obtained for the diatomic gases such as hydrogen and nitrogen if a magnetic filter is incorporated into the multicusp ion source.[15] This filter, generated either by inserting small magnets into the source chamber or by installing a pair of dipole magnets on the external surface of the source chamber, provides a narrow region of transverse B-field that is strong enough to prevent the primary electrons from reaching the extraction region, but is weak enough to allow the plasma to leak through. The absence of energetic electrons will prevent the formation of H_2^+ or N_2^+ in the extraction region, but dissociation of the molecular ions can still occur. As a result, the atomic ion species (H^+ or N^+) percentage in the extracted beam is enhanced. Figure 6.3 shows the dependence of the deuterium ion species with and without the presence of a magnetic filter.[16]

H^- ions are used extensively in particle accelerators such as cyclotrons, tandem accelerators, and proton storage rings. In order to heat plasmas and to drive currents in future fusion reactors, multiamperes of very high-energy neutral beams will be required. The high neutralization efficiency of H^-/D^- ions enables them to form atomic beams with energies in excess of 200 keV. Two different types of multicusp

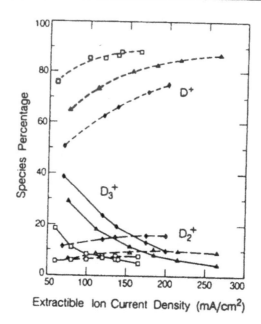

FIGURE 6.3
Species variation with extractible ion current density; □, strong filter; ▲, weak filter; ◆, filterless.

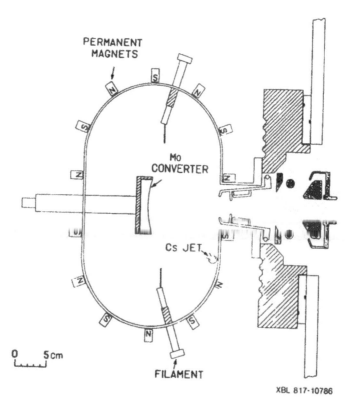

XBL 817-10786

FIGURE 6.4
A schematic diagram of the multicusp surface conversion H⁻/D⁻ ion source.

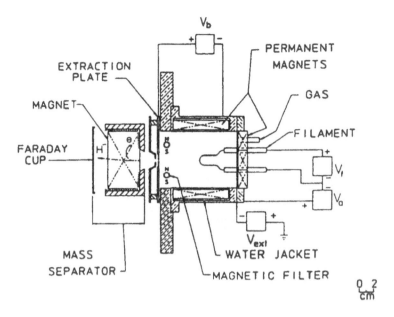

FIGURE 6.5
Schematic diagram of the small multicusp ion source equipped with a magnetic filter for negative ion extraction.

H⁻ ion sources have been developed at Lawrence Berkeley Laboratory. Figure 6.4 shows a LBL type surface conversion source. In this H⁻ source, the negative ions are formed on a negatively biased converter surface. They are then accelerated back through the sheath by the same potential. The bias voltage on the converter thus determines the energy of the H⁻ ion leaving the source (self-extraction). The converter surface is normally curved to geometrically focus the H⁻ ions through the plasma to the exit aperture.

The advantages of the surface conversion H⁻ source are low source operating pressure (~1 mtorr) and very small electron content in the H⁻ beam. The source can be operated either in the steady-state or pulsed mode. However, source operation always requires the presence of cesium, which can cause voltage breakdown in the accelerator column. Nevertheless, steady-state H⁻ beams with current greater than 1 A have been generated by this type of surface conversion sources.[17,18]

A hydrogen plasma contains not only positive ions and electrons, but also H⁻ ions. In 1983, a novel method of extracting volume-produced H⁻ directly from a multicusp source was reported by Leung et al.[19] In this H⁻ source, a permanent magnet filter divides the source chamber into discharge and extraction regions (Figure 6.5). Excitation and ionization of the gas molecules are performed by primaries in the discharge region. In the extraction region, the low electron temperature makes it favorable for the production and survival of H⁻ ions. It has been demonstrated that H⁻ current density greater than 250 mA/cm² can be extracted from a compact multicusp volume source.[20] Surprisingly, an experimental study at LBL on a small-volume H⁻ source indicates a large increase (more than a factor of five) in H⁻ output current when cesium vapor is added to a hydrogen discharge, resulting in an extractable current density exceeding 1 A/cm².[21,22] It is also found that the improvement in H⁻ yield is accompanied by a large reduction in the extracted electron current as well as the optimum source operating pressure.

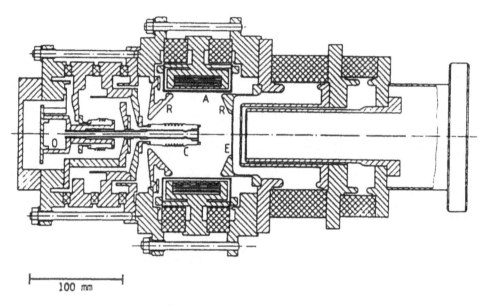

|———————————|
   100 mm

Oven/cathode chamber   Discharge chamber   Extraction system

FIGURE 6.6
Cold or hot reflex discharge ion source (CHORDIS) scheme: A, anode; C, cathode; E, first extraction
electrode; R, reflector electrode; O, oven. Cross-hatched, insulators; densely horizontally hatched,
permanent magnets. This figure shows the most elaborate, hot-running source version.

6.2.2 Metallic Ion Beam Production

The multicusp source has been modified at GSI (Germany) for metallic ion beam
production. The so called CHORDIS source was designed as a modular system of
high-current ion sources, separating the functions of particle feed, plasma generation,
and ion beam extraction.[23] The source uses a thermionic cathode to sustain a cw or
pulsed discharge. The plasma is confined radially by an array of permanent magnets
that form a multicusp field and axially by two biased reflector electrodes. Particles are
fed through the rear reflector in the form of gases or vapors. The front reflector forms
the first electrode of the extraction system. This outlet electrode can be equipped with
a sputtering target surrounding the extraction aperture.[24,25]

Figure 6.6 shows a hot-running version of the CHORDIS with an internal oven
chamber. The discharge chamber contains 18 directly cooled permanent magnets
enclosed in a vacuumtight shell. The inner wall of the chamber is fitted with a hot
liner whose surface temperature typically reaches 1500°C. The internal oven can be
operated at up to 1000°C. The vapor generated is conducted through the cathode
holder into the discharge or ionization chamber. Single- or multiple-aperture extraction systems with total emitting area up to 11 cm² have been used with the CHORDIS.
Beam energy ranges from 5 to 100 keV.

The source modules are normally fabricated with nonmagnetic stainless steel
with incorporated cooling channels. Hot-running parts are typically made from
molybdenum or tungsten/copper sinter metal. All flanges are sealed by viton gasket;
the flange at the end of the extraction system is UHV-compatible. Normally, six
tantalum filaments form the cathode, and eight tungsten filaments heat the internal
oven.

The CHORDIS source has been operated successfully to produce high currents of
bismuth (>30 mA).[26] By using a sputtering target installed on the back flange of the

source chamber, it has also been demonstrated by the Ishikawajima-Harima Heavy Industries Co., Japan that a 20-mA tungsten or tantalum and a 55-mA copper ion beam can be obtained from a multicusp ion source.

6.3 Ion Source Design

6.3.1 Operating Data

- Arc voltage: 15 to 350 V
- Arc current: several milliamps to tens of amperes for cw operation and hundreds of amperes for pulsed operation
- Arc power: as high as tens of kilowatts
- Cathode data: ~10 V/100 A (for a 1.5-mm-diam tungsten filament)
- Duty cycle: any, up to cw
- Time structure of the discharge: cw or square pulses, noise level <1%
- Magnetic field data: typically 4 kG for SmCo magnets
- Gas pressure: 10^{-5} torr to tens of 10^{-3} torr

6.3.2 Ion Source Performance

- Ion current: from milliamps to tens of amperes
- Charge state distribution: as high as 7+ for Ar or Xe for 250-V arc
- Extraction area: from millimeters to centimeters in diameter or multiple aperture
- Extraction voltage: from hundreds of volts to hundreds of kilovolts
- Emittance: depends on aperture size (~0.1 $\pi \cdot$mm$\cdot$mrad for 3-mm-diam aperture)
- Noise level: <1%
- Gas efficiency: >50% for hydrogen and higher for other gases

6.3.3 Ion Source Application

- Research: for cyclotrons or synchrotrons and in neutral beam systems for fusion research
- Industry: for ion implantation, material surface modification, ion beam lithography, neutron tubes for medical and oil-well logging
- Analysis, etc.: for fusion plasma diagnostics, for analyzing energetic electron beams

6.3.4 Deliverer or User

Lawrence Berkeley Laboratory (U.S.A.), Culham Laboratory (U.K.), JAERI (Japan), NIFS (Japan), GSI (Germany), DANFYSIK (Denmark), ROKION (Germany), Nissin Co. (Japan), Hitachi Co. (Japan), Ishikawajima-Harima Heavy Industries Co. (Japan), and IBA (Belgium).

Users include TFTR, JET, JT-60, Doublet-III tokamaks, NIFS (Japan), TRIUMF (Canada), PSI (Switzerland), DESY (Germany), GSI (Germany), CERN (Switzerland), Los Alamos Natl. Lab. (U.S.A.), Sandia Lab. (U.S.A.), Grumman Corp. (U.S.A.), IMS GmbH (Austria), Varian Corp. (U.S.A.), Mission Research Corp. (U.S.A.), IBA (Belgium), University of Frankfurt (Germany), KFA (Germany), the Ecole Polytechnique (Paliseau, France), Tohoku University (Japan), Exxon Corp. (U.S.A.), City College of New York (U.S.A.), and FOM Institute (Netherlands).

REFERENCES

1. R. Limpaecher and K. R. MacKenzie, *Rev. Sci. Instrum.*, 44, 726, 1973.
2. R. L. Stenzel and B. H. Ripin, *Rev. Sci. Instrum.*, 44, 617, 1973.
3. K. N. Leung, R. D. Collier, L. B. Marshall, T. N. Gallaher, W. H. Ingham, R. E. Kribel, and G. R. Taylor, *Rev. Sci. Instrum.*, 49, 321, 1978.
4. W. L. Stirling, P. M. Ryan, C. C. Tsai, and K. N. Leung, *Rev. Sci. Instrum.*, 50, 102, 1979.
5. A. Goede, T. S. Green, and B. Singh, *Proc. 8th European Conf. Controlled Fusion and Plasma Physics*, Prague, 1977.
7. M. Olivo, *Proc. 11th Int. Conf. Cyclotrons and Their Applications*, Ionics, Tokyo, 1986, 519.
8. For negative ion applications, see for example: *Proc. Int. Symp. Production and Neutralization of Negative Ions and Beams*, Brookhaven, NY.
9. For positive ion applications, see for example: *Proc. Int. Ion Source Conference*, at Berkeley, CA, 1989; Bensheim, Germany, 1991; and Beijing, China, 1993.
10. K. N. Leung and K. W. Ehlers, *Rev. Sci. Instrum.*, 52, 1452, 1981.
11. K. W. Ehlers and K. N. Leung, *Rev. Sci. Instrum.*, 50, 1353, 1979.
12. S. R. Walther, K. N. Leung, and W. B. Kunkel, *Rev. Sci. Instrum.*, 61, 315, 1990.
13. S. R. Walther, *Ion Implantation Technology - 92*, D. F. Downey, M. Farley, K. S. Jones, and G. Ryding, Eds., Elsevier, New York, 1993, 503.
14. K. N. Leung and R. Keller, *Rev. Sci. Instrum.*, 61, 333, 1990.
15. K. W. Ehlers and K. N. Leung, *Rev. Sci. Instrum.*, 52, 1452, 1981.
16. P. A. Pincosy, K. W. Ehlers, and A. F. Lietzke, *Rev. Sci. Instrum.*, 57, 2387, 1986.
17. K. N. Leung and K. W. Ehlers, *Rev. Sci. Instrum.*, 53, 803, 1982.
18. J. W. Kwan et al., *Rev. Sci. Instrum.*, 57, 831, 1986.
19. K. N. Leung, K. W. Ehlers, and M. Bacal, *Rev. Sci. Instrum.*, 54, 56, 1983.
20. K. N. Leung et al., *Rev. Sci. Instrum.*, 59, 453, 1988.
21. S. R. Walther, K. N. Leung, and W. B. Kunkel, *J. Appl. Phys.*, 64, 3424, 1988.
22. K. N. Leung, C. A. Hauck, W. B. Kunkel, and S. R. Walther, *Rev. Sci. Instrum.*, 60, 531, 1989.
23. R. Keller, P. Spadtke, and F. Nohmayer, *Proc. Int. Ion Engineering Congr.*, Kyoto, Japan, 1983, 25.
24. R. Keller, B. R. Nielsen, and B. Torp, *Nucl. Instrum. Methods*, B37/38, 74, 1989.
25. B. Torp et al., *Rev. Sci. Instrum.*, 61, 595, 1990.
26. R. Keller, IEEE No. 87CH2387-0, 382, 1987.

CHARACTERIZATION OF ION SOURCES

K. N. Leung

7 RF ION SOURCES

7.1 Ion Source History

The use of rf voltage to create a plasma dates back to the late 1940s.[1,2] An extensive review of rf ion sources and their operating characteristics was reported by Blanc and Degeilh in 1961.[3] Rf ion sources offer the advantages that they can operate with any type of background gases, in particular, gases such as oxygen that can easily poison tungsten filament cathodes. For this reason, rf sources have found important applications in plasma and reactive ion beam etching and ion beam doping. Rf ion sources are also useful when long-life operation or clean plasma production is required. Today they are widely employed by the semiconductor industry.

7.1.1 Working Principle of the RF Ion Source and Description of the Discharge

In practice, an rf discharge is formed in a vacuum vessel filled with a gas at a pressure of about 10^{-3} to 10^{-2} torr. A few hundred watts of rf power is typically required to establish a suitable discharge. The rf frequency can vary from a megahertz to tens of megahertz. There are two ways in which a low-pressure gas can be excited by rf voltages: (1) a discharge between two parallel plates across which is applied an alternating potential (capacitively coupled discharge) and (2) a discharge generated by an induction coil (inductively coupled discharge). Most rf ion sources are operated with the second type of discharge. In this case, an azimuthal rf electric field is generated by the alternating magnetic field in the discharge region. Electrons present in the gas volume are excited into oscillation by the rf electric field. They quickly acquire enough kinetic energy to form a plasma by ionizing the background gas particles. The ions are then extracted from the source chamber in a manner similar to a dc discharge source.

0-8493-2502-1/95/$0.00+$.50
© 1995 by CRC Press Inc.

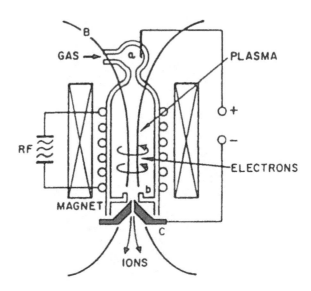

FIGURE 7.1
A Thonemann type rf ion source.

7.2 RF-Driven Ion Source Design and Operation

7.2.1 *Inductively Coupled Discharges with RF Coil Outside the Plasma Chamber*

Figure 7.1 shows a schematic diagram of a Thonemann type inductively coupled rf ion source.[4] It consists of a quartz discharge chamber surrounded by the rf-induction coil from the outside. There are four external variables that affect the character of the discharge and the resulting ion beam: the gas pressure in the chamber, the rf field (its magnitude and coupling to the plasma), the external magnetic field, and the extraction voltage. These parameters are not all independent. For instance, both the pressure and the magnetic field will affect the rf fields by influencing the electrical properties of the plasma, which in turn affect the rf coupling. As a consequence of these complex relationships, it is not possible to examine the source characteristics as a function of only one of these variables.

Typical operating parameters for this source are gas pressure, 10^{-3} torr; rf power, 350 W; oscillator frequency, 13 MHz; extraction voltage, 3 kV; magnetic field, 4 m T; and estimated ion density, 10^{11} ions/cm^3. The energy spread of the ion beams was much larger (by a factor of 10 to 100) than that expected from the thermal distribution of ion velocities in the plasma. Ero[5] and Levitskii[6] independently argued that the observed large energy anomalies are due to modification and variation of the plasma potential by energetic electrons that have come under the influence of strong rf electric fields existing in the region between the plasma and the electrode, or between the plasma and the walls.

Researchers in the University of Giessen, Germany later developed larger cross-sectional area (up to 10 cm in diameter) rf-driven ion sources for ion thrusters and neutral beam injector applications.[7-9] The schematic diagram of their RIM 10 model rf source is illustrated in Figure 7.2. The large quartz chamber is surrounded externally by the induction coil of the rf generator (1 to 30 MHz, <500 W). Beam extraction and formation is accomplished by a three-grid accel-decel extraction system. The rf power transfer to the discharge plasma is very efficient (98%). The RIM 10 source had

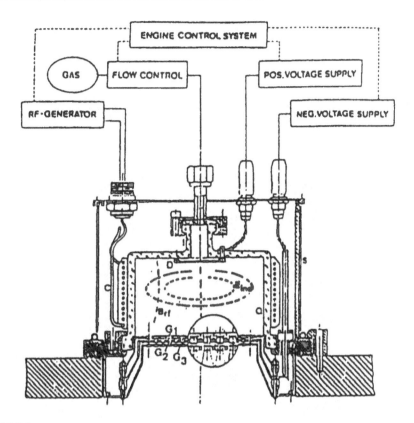

FIGURE 7.2
Cross section of RIM 10 and block diagram of the supply units (C: rf-induction coil; D: gas distributor; Q: quartz discharge chamber; S: rf screen; G_1, G_2, G_3: 1st, 2nd, and 3rd grid).

been operated with many different gases including oxygen, nitrogen, bromotrifluoromethane, and freon-12.[7,8] For cw operation without cooling the extraction grids, ion beam currents up to 300 mA has been achieved.

7.2.2 Inductively Coupled Discharges with RF Coil Inside Plasma Chamber

Generation of an rf discharge by placing the induction coil (or antenna) inside a multicusp source chamber was tested at Berkeley and Garching for neutral beam applications.[10] In 1991, a new rf driven H⁻ source (Figure 7.3) has been developed at Lawrence Berkeley Laboratory for use in the injector unit of the Superconducting Super Collider (SSC).[11] The source chamber is a copper cylinder (10-cm diam and 10-cm length) surrounded by 20 columns of samarium-cobalt magnets that form a longitudinal linecusp configuration for plasma confinement. The discharge is generated through inductive coupling of the plasma via a two-turn antenna coil. The antenna is fabricated from 4.7-mm-diam copper tubing and is coated with a thin layer of hard porcelain material. The thin coating is slightly flexible and resistant to cracking. It has maintained a clean plasma in cw operation for periods of a week or more. The antenna coil is connected to a matching network and isolation transformer (Figure 7.4), matching the 50 Ω impedance of the amplifier with the impedance of the plasma. The rf signal is generated by a digital synthesizer. The signal (~2 MHz) is sent to a preamplifier, and then to the rf amplifier. Peak performance allows a maximum pulsed rf input power of 50 kW. The rf power can be controlled by changing the

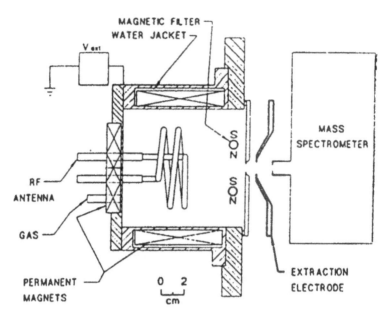

FIGURE 7.3
Schematic diagram of the rf multicusp ion source.

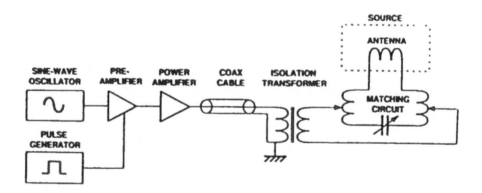

FIGURE 7.4
Schematic diagram of the complete rf power system.

amplitude and frequency of the synthesizer signal. Maximum efficiency is achieved when the output voltage and current from the rf amplifier are in phase and operating at a 50-Ω impedance. For pulse operation, a small hairpin tungsten filament is normally used as a starter for the rf induction discharge.

Operation of the rf-driven multicusp H⁻ ion source has been reported.[11,12] The source is able to generate volume-produced H⁻ beams with extracted current densities higher than 200 mA/cm². A total current of ~40 mA can be obtained from a 5.6-mm-diam aperture with the source operating at a pressure of about 12 mtorr and a rf input power of 50 kW. The same rf multicusp source has also been tested with inert gas plasmas such as He, Ne, Ar, Kr, and Xe.[13] Figure 7.5 shows the extractable positive ion current (and current density) as a function of rf power. The optimum source pressure is typically below 1 mtorr. In most cases, the extractable ion current density can be as high as 1 A/cm² at approximately 50 kW of rf input power. Source operation with gases such as N_2, O_2, and BF_3 have proved to be equally successful.

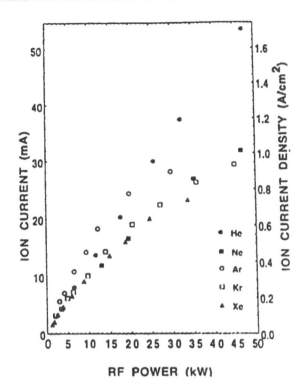

FIGURE 7.5
Extracted beam current and density as functions of rf power for various inert gas plasmas.

The advantages of the rf-driven multicusp source are that there is no cathode lifetime limitation and that the discharge is cleaner than most filament source discharges. However, there is always some concern that the plasma parameters (e.g., plasma potential and electron density and temperature) of this type of source may oscillate significantly, which is undesirable for most positive and negative ion source applications. Langmuir probe characteristics were sampled at different phases of the rf cycle.[14] Results indicate that the plasma potential, electron density, and temperature do not vary significantly during the rf cycle.

The multicusp rf source has also been tested for metallic ion beams such as copper by the sputtering technique.[13] In this case, neutral copper atoms are first sputtered from an electrode by the background Ar⁺ ions. These atoms are subsequently ionized by the plasma electrons. Mass analysis shows that the extracted beam contains about 20% of Cu⁺ ions in cw operation. Other metallic ions can also be generated in a similar manner or by using other techniques such as electron beam evaporation.[15]

7.2.3 Capacitively Coupled Discharge

Figure 7.6 shows a capacitively coupled rf ion source developed by Plasma Consult GmbH, Germany.[16] The rf discharge is created between the centrally placed rf cathode and the cylindrical anode. This radial excitation is especially efficient in the presence of an axial magnetic field generated by the solenoid coils. The high-energy electrons move in the cathode plasma sheath on cycloidal trajectories around the cathode (magnetron effect) and can ionize neutrals or impinge on the cathode surface, causing secondary electron emission. The axial magnetic field suppresses the

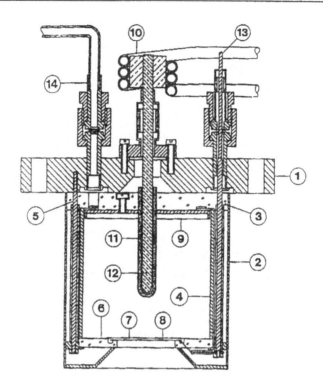

FIGURE 7.6

Capacitively coupled rf ion source. Ion source cross section: (1) mounting flange; (2) ion source case; (3) source base of Al_2O_3; (4) plasma chamber walls of stainless steel — rf anode; (5) mounting slabs; (6) ceramic grid holder; (7) screen grid; (8) acceleration grid; (9) electrostatic end confinement made of Al_2O_3; (10) water cooling for the rf cathode; (11) cathode cover made of Al_2O_3; (12) rf cathode; (13) grid polarization; (14) gas inlet.

loss of electrons on the anode. To avoid electron loss in the direction parallel to the magnetic field, electrostatic end confinement is used. The floating screen grid and the floating ceramic plate provide the essential confinement geometry. This ion source operates at an excitation frequency of 13.5 MHz and rf power less than 250 W. An integrated extraction system allows one to extract ion beams with current density higher than 0.8 mA/cm² and beam energy between 100 eV and 2 keV.

7.3   Ion Source Application

- Research: for cyclotrons or synchrotrons and in neutral beam injectors for fusion research
- Industry: for ion implantation, ion beam etching, ion beam lithography, material surface modification

7.3.1   Deliverer or User

Lawrence Berkeley Laboratory (U.S.A.), Grumman Corporation (U.S.A.), Hauzer Techno Coating (Netherlands), University of Giessen (Germany), Nordiko Limited (U.K.), Veeco Instrument, Inc. (U.S.A.), Plasma Consult GmbH (Germany), High Voltage Engineering (Europe), and University of Kaiserslautern (Germany), Max-Planck Institute for Plasma Physics (Germany), and Lam Research (U.S.A.).

Users include Superconducting Super Collider Laboratory (U.S.A.), DESY (Germany), CERN (Switzerland), Sandia Laboratory (U.S.A.), IMS (Austria), Grumman Corporation (U.S.A.), Max-Planck Institute for Plasma Physics (Germany), Institute of Nuclear Research (Hungary), and Princeton Plasma Physics Laboratory (U.S.A.).

REFERENCES

1. P. C. Thonemann, J. Moffatt, D. Roaf, and J. H. Saunders, *Proc. Phys. Soc. London*, 61, 483, 1948.
2. R. N. Hall, *Rev. Sci. Instrum.*, 19, 905, 1948.
3. D. Blanc and A. Degeilh, *J. Phys. Radium*, 22, 230, 1961.
4. P. C. Thonemann, *Progress in Nuclear Physics*, Pergamon Press, London, 1953, 219; P. C. Thonemann and E. R. Harrison, AERE Report Gp. R1190, 1955.
5. J. Ero, *Acta Phys. Hung. Acad. Sci.*, 5, 391, 1956; *Nucl. Instrum.*, 3, 303, 1958.
6. S. M. Levitskii, *Sov. Phys. Tech. Phys.*, 2, 5, 913, 1958.
7. J. Freisinger et al., *Kerntechnik*, 51, 125, 1987.
8. H. W. Loeb et al., Preprint IAF-88–258, Bangalore, India, 1988.
9. J. Freisinger et al., IEPC-Paper 88–117, Garmisch-Partenkirchen, FRG, 1988.
10. M. C. Vella, K. W. Ehlers, D. Kipperhan, P. A. Pincosy, R. V. Pyle, W. F. DiVergilio, and V. V. Fosnight, *J. Vac. Sci. Technol.*, A 3, 1218, 1985; J. H. Feist et al., *14th Symp. Fusion Technol.*, Avignon, 1986, 1127.
11. K. N. Leung, G. J. DeVries, W. F. DiVergilio, R. W. Hamm, C. A. Hauck, W. B. Kunkel, D. S. McDonald, and M. D. Williams, *Rev. Sci. Instrum.*, 62, 100, 1991.
12. K. N. Leung, D. A. Bachman, C. F. Chan, and D. S. McDonald, *Proc. 15th Int. Conf. High Energy Accelerators*, Hamburg, Germany, July, 1992, 200.
13. K. N. Leung, D. A. Bachman, P. R. Herz, and D. S. McDonald, *Nucl. Instrum. Methods Phys. Res.*, B74, 291, 1993.
14. D. S. McDonald, D. A. Bachman, W. F. DiVergilio, W. B. Kunkel, K. N. Leung, and M. D. Williams, *Rev. Sci. Instrum.*, 63, 2741, 1992.
15. J. Waldorf and H. Oechsner, *Rev. Sci. Instrum.*, 63, 2570, 1992.
16. D. Korzec, *Fortschrittberichte VDI, Reihe 9: Elektronik*, Nor. 160, IBSN 3–18–146009–5, 1993.

Characterization of Ion Sources

B. H. Wolf

8 MICROWAVE AND ECR ION SOURCES ≤2.4 GHZ

Advantages: No hot filament, can run with aggressive gases, long lifetime, easy to operate, low emittance.

Disadvantages: Lifetime of microwave window limited by electron bombardment and metal vapor deposition, runs just with gases or vaporized material.

8.1 Ion Source History

Microwave (f > 1 GHz) ion sources were first developed for multiply charged ion production at the end of the 1960s by R. Geller and his group[1] using electron cyclotron resonance heating of the electrons and magnetic confinement for the ions (see Chapter 2.9). Microwave ion sources discussed in this section are designed to produce high currents of singly charged ions. The wavelength of microwaves is, in contrast to radio frequency (rf) (f ≤ 500 MHz) ion sources (Chapter 2.7), of similar dimension or smaller than the ion source. The first microwave ion source was built by Sakudo (1977)[2] using an antenna to couple the microwaves into the ion source chamber. In 1978[3] he developed a microwave ion source with slit extraction for mass separation using a microwave window to couple the microwaves into the ion source. A compact microwave ion source using permanent magnets was developed by Ishikawa (1984).[4] Leung changed a microwave-driven spectroscopy lamp into an ion source (1984)[5] and Asmussen used hybride modes to heat his microwave ion source plasma (1981).[6] Torii (1987) developed a powerful broad-beam microwave ion source and significantly improved the window lifetime.[7,8] Review articles on microwave ion sources for high-current ion beams are presented in References 9 and 10.

8.2 Working Principle of the Microwave Ion Source and Description of the Discharge

A microwave ion source consists of a single or multimode cavity, which is also the discharge chamber (Figure 8.1a), or contains a separate vessel made of insulating material (Figure 8.1b). In most microwave ion sources, a magnetic field is superimposed to improve the microwave heating of the plasma electrons. Microwaves (typical between 0.1 and 1 kW) are coupled into the discharge chamber with an antenna

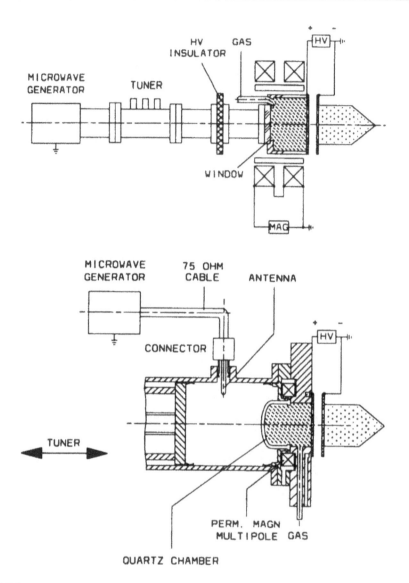

FIGURE 8.1
Microwave ion source scheme and circuitry: (a) metal discharge chamber, (b) quartz discharge chamber.

or through a microwave window (see Chapter 6). Ions are extracted from the microwave-generated plasma in the usual way through single- or multiple-aperture or slit extraction systems. A high-voltage insulator is mounted between two waveguide flanges and the magnet coils are insulated from the discharge chamber. In this way the microwave generator and the magnet with its power supply can remain at ground potential and only the discharge chamber is at high voltage. For voltages higher than 50 kV the magnet coil has to be at high potential and the magnetic field should be shielded from the extraction and acceleration area.[11]

Microwave and high-current ECR ion sources commonly use the frequency of 2.45 GHz by several reasons:

- cheap and reliable magnetron tubes and generators are available, especially since the microwave oven was introduced

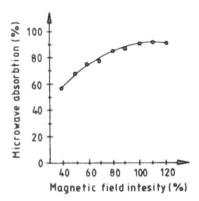

FIGURE 8.2
Microwave absorption vs. magnetic field.[2]

- the size of the waveguides is not too large (about 5×10 cm) and for a microwave power up to about 300 W coaxial cables can be used (see Chapter 6)
- the low frequency needs a relatively low magnetic field of 0.0875 T to reach the ECR condition (see Chapter 1)
- the low frequency creates more low-energy electrons and, thus, less multiply charged ions

Because the impedance of a high-density plasma changes with magnetic field, gas pressure, and the degree of ionization, a tuner has to be placed between the microwave generator and ion source to prevent high reflection of microwave power due to mismatch of the plasma impedance.

Microwaves entering the discharge chamber can propagate through the plasma and transfer their energy to free electrons, which further ionize the gas atoms. In the absence of a magnetic field there is an upper limit for the electron density in the plasma

$$n_{ep} \leq 1.11 \times 10^{10} \ f^2 \ cm^{-3}$$

where f is in GHz, which for 2.45 GHz gives $n_{ep} = 6.66 \times 10^{10}$ cm^{-3}. This is about two orders of magnitude lower than needed for high-current ion sources. If n_{ep} reaches its maximum value, microwaves cannot enter the plasma anymore and can only be weakly absorbed at the plasma surface, and most of the microwave power will be reflected under those circumstances. In the presence of a sufficiently high magnetic field ($B_c = 0.0357 \times f$ or 0.0875 T for 2.45 GHz), however, electron cyclotron heating (see Chapter 1) occurs and the right-hand circular polarized fraction of the microwave can enter the plasma, again heating up electrons.[12] If the magnetic field exceeds the electron cyclotron resonance value and the neutral particle density is sufficiently high (gas pressure >1 Pa), the energy of the microwave is effectively absorbed by off-resonance heating processes[13] (see Figure 8.2). In this way electron densities of 10^{13}/cm^3 can be generated and high ion currents can be extracted from the plasma (Figure 8.3).

The position and design of the microwave window is very important for the functioning of the ion source and for the lifetime of the window. The highest plasma densities are generated when the magnetic field at the microwave window is slightly above the ECR resonance field.[14] Usually, the plasma can reach the window in high-current microwave ion sources. The main danger for the window are metal deposits

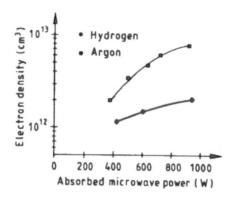

FIGURE 8.3
Electron density vs. microwave power.[2]

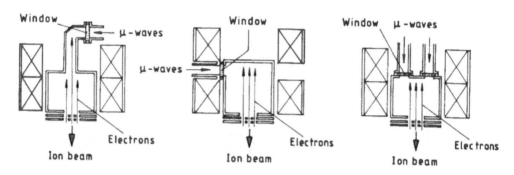

FIGURE 8.4
Possible window positions out of the range of backstreaming electrons.[8]

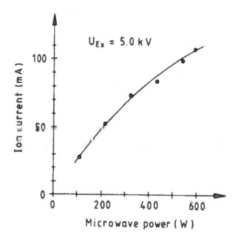

FIGURE 8.5
Ion beam current vs. microwave power.[15]

on the window and high-energy electrons accelerated by the extraction potential into the ion source. Torii[14] developed a three-layer window that absorbs the electrons without destruction of the third layer, which serves as a vacuum seal. A further improvement was made by splitting the waveguide and coupling the microwaves parallel through two windows with a solid middle part into the source (see Figure 8.4).[8]

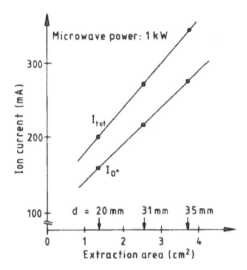

FIGURE 8.6
Ion beam current vs. extraction area.[14]

The ion temperature in microwave heated plasmas, with densities used in ion sources, is low (less than 1 eV) because only the electrons are heated by the microwaves at the magnetic fields used. The ion beam current density increases almost linearly with microwave power (Figure 8.5) and ions can be extracted over a large area of constant plasma density (Figure 8.6).

By coating the discharge chamber with an insulating material that discourages recombination of hydrogen molecules, up to 90% of H^+ ions can be generated in a microwave ion source (Figure 8.7),[10]

Dielectric material inside the ion source and/or in the waveguide can help to reduce the size of the ion source[8] or to reduce the plasma volume and increase the power efficiency of the ion source.[3] Rare earth permanent magnets are used to design

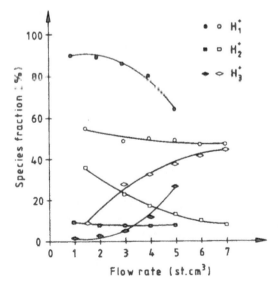

FIGURE 8.7
Proton fraction vs. gas flow with and without BN liner, Microwave power, 1 kW. (From Taylor, T., Rev. Sci. Instrum., 63, 2507, 1992. With permission.)

various magnetic field geometries, such as ring magnets, providing nearly parallel field lines,[4] multipole fields,[16] or cusp fields (see below).[17] Broad-beam microwave ion sources are well suited for low-energy ion extraction used for ion etching and deposition techniques.

8.2.1 Metal Ions

Microwave ion sources run well with oxygen and corrosive gases. Metal ions can be produced using volatile gaseous compounds. Evaporation is less suited because the deposition of metallic films can destroy the microwave window or reflect the incoming microwaves. Heating of the window above the evaporation point of the material used may help for those materials with low evaporation point, and shielded positions of the microwave window behind knees can help to reduce the metal vapor deposits. Sputtering of antenna material, as in rf-ion sources (see Chapter 2.7), is not sufficient for high-current ion beam generation. In Table 8.1 ion currents for various elements are summarized for the Hitachi microwave source[11,19] and compared with values from a Freeman ion source.[20]

8.3 Specific Ion Source Design

8.3.1 Microwave Ion Source with Slit Extraction (Figure 8.8)

- Special design and construction details of the source

  The ion source uses a rectangular waveguide as a discharge chamber. To reduce the discharge volume, a ridged waveguide is used, parts of which are filled with boron nitride as dielectric so that the microwave power is concentrated in the gas-filled part close to the extraction slit. The 2.45-GHz microwave power is coupled from a regular waveguide through a vacuum window into the discharge chamber. A solenoid, electrically insulated from the high potential of the source, provides a magnetic field of up to 0.15 T. Two iron pieces increase the magnetic field in front of the window, protecting it from the plasma. The entire wall of the discharge is covered with BN to reach a high wall temperature and to reduce condensation of metals introduced by evaporation or by chemical compounds. An improved version of the source with a two-stage extraction and acceleration can deliver 100-keV ion beams. A similar microwave ion source was built by Walter (Varian) using permanent magnets instead of a solenoid.

TABLE 8.1

Metal Ion Currents of a Microwave Source Compared
to Values from a Freeman Source in mA

Element	Microwave	Compound	Freeman	Compound
B	6	BF_3	4	BF_3
O	10	O_2	5	O_2
Al	4	$AlCl_3$	0.5	$Al + Cl_2$
P	15	PH_3	7	P
Sc	0.5	$Sc_2O_3 + CCl_4$		
Ti	3.5	$TiCl_4$	0.5	$Ti + Cl_2$
Ga	1.8	$GaCl_3$		
Ge	2	GeS		
As	10	AsH_3	15	As
Sb	3.1	Sb		
Hf	2	$HfCl_4$		

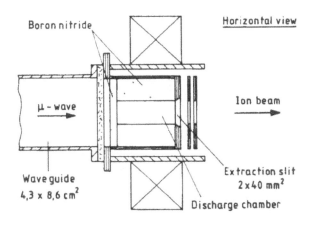

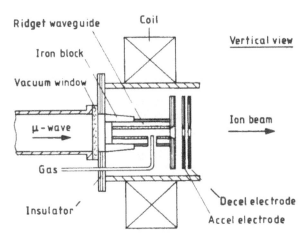

FIGURE 8.8
Hitachi ion source.

- Ion source material and vacuum conditions

  Discharge chamber, Cu covered with BN; extraction aperture, graphite; accel-decel electrodes, stainless steel.

- Application area of the source

  Ion implantation into semiconductors, surface modification of materials.

- Deliverer or user

  Hitachi Research Lab. 1–1–1 Kokubu-cho, Hitachi-shi, Ibaraki-ken 316, Japan

  Varian Ion Implant Systems, 35 Dory Rd., Glouchester, MA 01930, U.S.A.

8.3.2 Microwave Ion Source with Metal Discharge Chamber (Figure 8.9)

- Special design and construction details of the source

  Microwaves of 2.45 GHz are coupled from a rectangular waveguide through a triple-layer window to a multimode cylindrical metal discharge chamber. The magnetic field is slightly above the ECR condition of 0.0875 and, thus, produces a dense plasma in the center of the discharge chamber. The window consists of a quartz vacuum window, an alumina microwave window, and a boron nitride protector against damage to the window by backstreaming fast electrons. Ions are extracted by a multiaperture three-electrode system. A modified version of this source improved the lifetime by using a split microwave path into the discharge chamber so

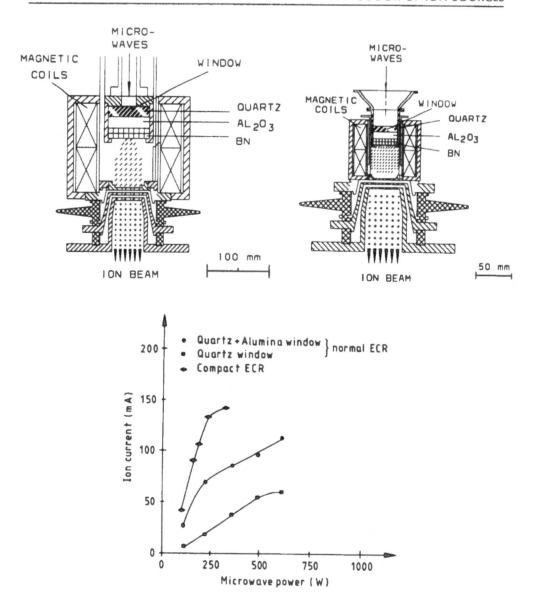

FIGURE 8.9

ECR ion source of Torii. (a) normal size, (b) compact size, (c) ion current vs. microwave power for sources (a) and (b).[8,14]

that backstreaming electrons cannot reach the windows. A third version has a reduced volume, increasing the microwave efficiency. To get a proper coupling of the microwaves to the small chamber, the end of the waveguide is filled with alumina in this case.

- Ion source material and vacuum conditions

  Plasma chamber, stainless steel; extraction grid, mild steel; window, Al_2O_3, BN, and quartz.

- Application area of the source

  High-current oxygen (and other gases) implantation.

- Deliverer or user

  NTT LSI Laboratories, 3–1 Morinisato Wakamiya, Atsugi-shi 243–01, Japan

  Eaton Semiconductor Equipment Division, Beverly, MA 01915, U.S.A.

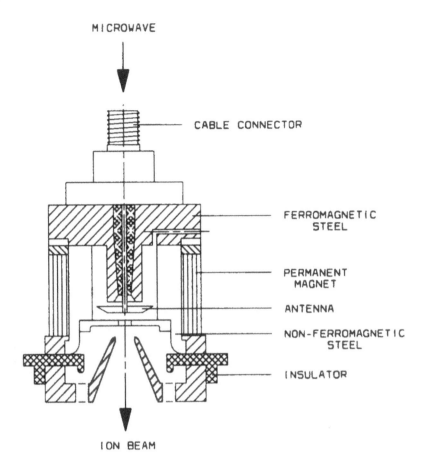

FIGURE 8.10
Compact microwave ion source.[4]

8.3.3 Compact Microwave Ion Source (Figure 8.10)

- Special design and construction details of the source

  This ion source uses a permanent ring magnet to produce a strong magnetic field for an ECR plasma. The magnetic field is optimized by the shape of the pole pieces. A helical antenna couples the microwaves into the plasma. A cable and a sealed N-connector transport the 2.45-GHz microwave power to the antenna. Ions are extracted through a hole in the base pole plate.

- Ion source material and vacuum conditions

  Permanent magnet, CoSm or NdFe; discharge chamber, Al, Cu, or stainless steel; antenna, Mo or stainless steel; pole pieces, mild steel.

- Application area of the source

  Ion implantation and plasma cathode for high-current ion sources.

- Deliverer or user

  Department of Electronics, Kyoto University, Yoshida-honmachi. Sakyo-ku, Kyoto 615, Japan

  Institut Electronique Fundamentale, CNRS UAR 022, Universite Paris Sud, 91405, Orsay, France

  Institute of Nuclear Research, Lanzhou University, Lanzhou, China

  Department of Technical Physics, Peking University, Beijing 100871, China

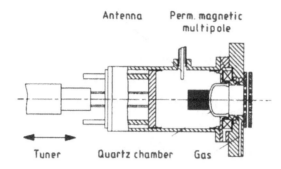

FIGURE 8.11
Microwave ion source with quartz discharge chamber.

8.3.4 Microwave Ion Source with Quartz Discharge Chamber (Figure 8.11)

- Special design and construction details of the source

  The quartz discharge chamber is part of one end plate of an adjustable microwave resonator outside the vacuum system. A set of permanent magnets generate a multipolar field inside the discharge chamber, which provides ECR resonant zones for heating a plasma disk. The plasma has a uniform density along the multiaperture extraction system allowing broad ion beam formation. The source can run at 2.45 GHz or at 915 MHz.

- Ion source material and vacuum conditions

  Resonator, brass; discharge chamber, quartz; base flange and extractor grids, stainless steel. The permanent magnets are rare earth magnets.

- Application area of the source

  Ion beam and plasma processing.

- Deliverer or user

  Dept. of Electrical Engineering, Michigan State University, East Lansing, MI 48824–1226, U.S.A.

  Similar sources FMT, Universität Wuppertal, Postfach 100127, 42001 Wuppertal, Germany

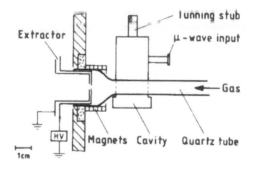

FIGURE 8.12
Small microwave ion source with expansion cup. (From Walther, S. R., Leung, K. N., *Rev. Sci. Instrum.*, 57, 1531, 1986. With permission.)

8.3.5 Small Microwave Ion Source of Leung, LBL (Figure 8.12)

- Special design and construction details of the source

  The source consists of a 10-mm-diam quartz tube, which is extended to 27 mm with the extraction grid (25 holes, 0.8-mm φ) at one side and the gas inlet at the other. The tube goes through an Evanson microwave cavity[21], which is better described as a tunable quarter-wave termination for the coaxial feed line. A set of permanent magnets is positioned around the plasma expansion area where the extraction takes place. The plasma has to be started by rf sparks from a Tesla transformer.

- Ion source material and vacuum conditions

  Resonator, brass; discharge chamber, quartz; base flange and the extractor grids, stainless steel. The permanent magnets are rare earth magnets.

- Application area of the source

  H- and positive ions from reactive gases and metal compounds, small accelerators, ion implantation.

- Deliverer or user

  Lawrence Berkeley Laboratory, Berkeley, CA 91720, U.S.A.

8.3.6 Two-Stage Microwave Ion Source (MIXT; Figure 8.13)

- Special design and construction details of the source

  A high-density plasma is generated in the first chamber (10-cm φ, 30-cm length) using a magnetic field at or above the ECR resonance 0.0875 T generated by a set of solenoid coils. The plasma expands into a second stage, larger in diameter and surrounded by a multipole arrangement of permanent magnets, to be extracted from an area of 27-cm diam by a multiaperture extraction system. Microwaves are coupled from a rectangular, tapered waveguide through a quartz window and a short 1 × 8 cm waveguide into the first chamber.

- Ion source material and vacuum conditions

  Plasma chamber, stainless steel; microwave window, quartz; permanent magnets, SmCo.

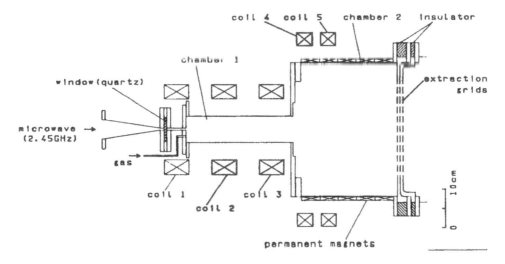

FIGURE 8.13
Two-stage microwave ion source with large-area extraction. (Courtesy of H. Nihei, Univ. of Tokyo.)

- Application area of the source

  Ion implantation, ion beam modification of materials, plasma processing.

- Deliverer or user

  Dept. of Nuclear Engineering, The University of Tokyo, 7-3-1 Hongo, Bunkyo-ku, Tokyo 113, Japan

8.3.7 Operating Data (Table 8.2)

TABLE 8.2

Operating Data of Microwave Ion Sources

		Hitachi	Torii	Compact	Quartz ch.	Leung	MIXT
Rf frequency	(GHz)			2.45			
Microwave power	(W)	0–1000	0–600	0–100	0–6000[a]	400	0–5000
Magnetic field	(mT)	≥87.5	95	87.5	≤300	≤50	≥87.5
Magnet supply	(kW)	1	2–5	—	—	—	0–30A
Gas pressure	(Pa)	0.1–1		0.1–10	0.04–0.4	1–10	0.1–1
Gas flow	(sccm)		2–5	0.1–5		2–10	
Ion current	(mA)	≤40	300	1	≤1000[a]	4.5	1500
Ion beam density	(mA/cm^2)	≤50	100	30	≤10	35	15
Solid material	[b]	Oven	Gas	450°C	Gas	Gas	—
Charge state	max	P^{2+}	1	1	1	Ne^{3+}	1
Extraction area	(cm^2)	0.2 × 4	4	0.07	960	0.12	200
Extraction volt.	(kV)	≤70	45	30	2	3	30

[a] Depending on the diameter of the source.
[b] Gas means chemical compounds.

REFERENCES

1. R. Geller et al., *Proc. 1st Int. Conf. Ion Sources*, Saclay, 1969, 537; *Proc. 2nd Int. Conf. Ion Sources*, Vienna, 1972, 632.
2. N. Sakudo et al., *Rev. Sci. Instrum.*, 48, 762, 1977.
3. N. Sakudo et al., *Rev. Sci. Instrum.*, 49, 940, 1978.
4. J. Ishikawa et al., *Rev. Sci. Instrum.*, 55, 449, 1984.
5. K. N. Leung et al., *IEEE Trans. Nucl. Sci.*, 32, 1803, 1984.
6. J. Asmussen and J. Root, *J. Vac. Sci. Technol.*, B4, 126, 1984.
7. Y. Torii et al., *Nucl. Instrum. Methods*, B 21, 178, 1987.
8. Y. Torii et al., *Rev. Sci. Instrum.*, 63, 2359, 1992.
9. N. Sakudo, in *The Physics and Technology of Ion Sources*, I. G. Brown, Ed., John Wiley & Sons, New York, 1989, 229.
10. T. Taylor, *Rev. Sci. Instrum.*, 63, 2507, 1992.
11. N. Sakudo et al., *Nucl. Instrum. Methods*, B 21, 168, 1987.
12. T. H. Stix, *The Theory of Plasma Waves*, McGraw-Hill, New York, 1962, 32.
13. J. Musil and F. Zacek, *Czech. J. Phys.*, B22, 133, 1972; *Czech. J. Phys.*, B23, 736, 1973; *Plasma Phys.*, 16, 971, 1974; *Plasma Phys.*, 17, 1147, 1975.
14. Y. Torii et al., *Rev. Sci. Instrum.*, 61, 253, 1990.
15. T. Tokiguchi et al., *Vacuum*, 36, 11, 1986.
16. J. Asmussen et al., *Rev. Sci. Instrum.*, 61, 250, 1990.
17. M. Delaunay, *Rev. Sci. Instrum.*, 61, 267, 1990.
18. E. Tojyo et al., *Proc. 11th Workshop Electron Cyclotron Resonance Ion Sources*, KVI-Rep.996, KVI, Groningen, 1993, 234.
19. N. Sakudo et al., *Vacuum*, 34, 245, 1984.
20. D. J. Chivers, *Rev. Sci. Instrum.*, 63, 2501, 1992.
21. F. C. Fehsenfeld et al., *Rev. Sci. Instrum.*, 54, 56, 1983.
22. H. Nihei et al., *Rev. Sci. Instrum.*, 63, 1932, 1992.

CHARACTERIZATION OF ION SOURCES

B. H. Wolf

9 ECR ION SOURCES >2.4 GHZ

Advantages: High currents of high charge states, no filament, long lifetime of the source, long stable operation time, dc or pulse mode possible.

Disadvantages: High power consumption if mirror field is not made with permanent magnets, expensive microwave technology and design, noisy beam, long conditioning time, metal ion production subtle.

9.1 Ion Source History

Electron cyclotron resonance (ECR) ion sources were first proposed by R. Geller in 1965[1] and H. Postma (1970)[2] using their experience with mirror machines for fusion plasma studies. In 1971 Geller and co-workers reported on the first operational ECR ion source.[3] This ECR ion source, called MAFIOS, delivered just moderate charge states as from a PIG ion source because ion lifetime in the plasma was short (~10⁻⁴ s). Other groups studied similar devices.[4,5] In 1975 Geller reported on a new ECR ion source design using a two-stage arrangement called SuperMAFIOS.[6] The first stage was a MAFIOS that created a plasma that was guided into the second stage with a hexapole field for additional radial confinement apart from the axial mirror field. In this way the ion lifetime was improved (~10⁻² s) and very high charge states could be detected. SuperMAFIOS is the basis of all ECR ion sources for multiply charged ion production, but its power consumption of 3 MW was unacceptable for accelerator applications. This changed with the introduction of strong rare earth permanent magnets for the construction of the hexapole. Again, Geller, in cooperation with the cyclotron laboratories in Louvain la Neuve and Karlsruhe, built the MiniMAFIOS source, which needed just 100 kW for the mirror field coils.[7] The first ECR source linked to an accelerator was PikoHISCA of Bechthold (1981)[8] at the Karlsruhe cyclotron using an open hexapole structure that allowed radial access to the second-stage plasma. A larger ECR source was constructed for the LBL cyclotron by Lyneis[9] and a very small one, using for the first time a hexapole that used the whole volume to increase the field strength on the poles, and which also introduced a magnetic yoke to reduce the power consumption of the mirror field, was built by Schulte.[10] Another line of development was the design of superconducting devices by Jongen,[11] Bechtold,[12] and Beucher,[13] who all had problems with the superconducting magnet arrangement, and only Beucher's operated for a long period of time at the Jülich cyclotron. The only

0-8493-2502-1/95/$0.00+$.50
© 1995 by CRC Press Inc.

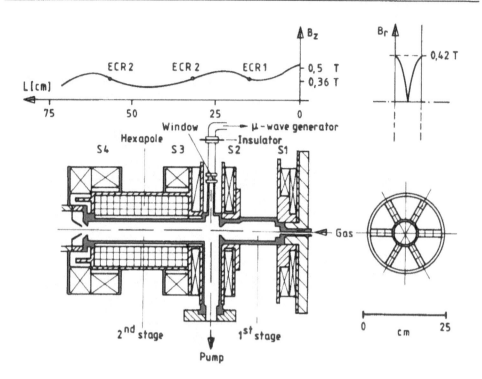

FIGURE 9.1
ECR ion source scheme and axial and radial magnetic field distribution.[7]

remaining operational superconducting ECR ion source was built by Antaya MSU[14] and is able to provide magnetic fields for operation with frequencies up to 35 GHz. The first ECR on a 350-kV platform was built at Argonne in 1987 for injection into a superconducting linac.[15,16] In 1987 Sortais and Geller built the first ECR ion source made entirely of permanent magnets requiring just 1 kW of power for the microwave generator,[17] which made the use of ECR ion sources on high-voltage platforms easy. Nowadays ECR ion sources for the production of multiply charged ions run at many accelerators and in many laboratories for atomic physics experiments. Also, several low-budget designs have been realized.[18–20] There is a series of workshops, started in 1978 in Karlsruhe, reporting on the ongoing development of ECR ion sources; the 11th workshop was held in 1993 in Groningen.[21] Reviews have been presented at various conferences[22–26] and in works of I. Brown[27] and R. Geller.[28]

9.2 Working Principle of the Ion Source and Description of the Discharge

ECR ion sources (Figure 9.1) for the generation of multiply charged ions need the highest possible plasma confinement to reach a high value of the product of electron density and confinement time, $n_e\tau$, necessary for step-by-step ionization of the ions. Since n_e in usual ECR plasmas is $\leq 10^{12}$ cm^{-3}, the confinement time τ should be $\geq 10^{-2}$ s. It is essential to reach this high value of τ; axial and radial confinement of the plasma is essential for a sufficient charge state distribution. Typical field distributions in axial and radial directions are also shown in Figure 9.1. The necessary magnetic field for ECR resonance is given by

$$B_c = 0.0357 \times f \text{ T}$$

where f is the frequency in gigahertz (see Chapter 1). For a 10-GHz ECR ion source

$$B_c = 0.357 \text{ T}$$

9.2.1 General Description

An ECR ion source (see Figure 9.1) consists of a vacuum chamber, which serves also as a microwave cavity, a set of solenoid coils, which provides a magnetic mirror field for the axial confinement, a hexapole (or higher multipole) for the radial confinement, a first or injector stage, which provides a cold plasma, or electrons for the main stage and an axial extractor system to form the ion beam. Microwaves can be coupled into the ion source axially off axis or on axis through a circular waveguide or radially between first and second stage with the microwave window close to the maximum of the magnetic field away from a ECR resonance zone. The ECR zone has to be a closed volume that does not intercept the cavity wall at any point (Figure 9.1).

The vacuum chamber is insulated from the solenoid coils and put on a negative high-voltage potential for ion beam extraction. A high-voltage insulator is also positioned between two flanges of the waveguide system to leave the microwave generator and the magnet power supplies at ground (or platform) potential (Figure 9.1).

Early ECR ion sources had intensive pumping at the first and second stages to provide the pressure gradient between the first and second stages and to keep the pressure in the second stage low to avoid charge-exchange reactions of highly charged ions. In the meantime, pumping is mainly done through the extraction system with sometimes an additional pump at the waveguide entrance to the ion source. Gas is fed to the first stage of the ion source to be able to create a higher pressure at a first resonance zone.

9.2.2 Magnetic Structure

The axial magnetic field is usually generated by at least two sets of solenoid coils for the mirror field of the main stage and additional coils for the first stage, depending on the ion source design. Modern designs use iron yokes to reduce power consumption of the coils and the stray fields around the ion source (Figure 9.2). Without an iron yoke, ferromagnetic materials in the vicinity of the ion source will influence the magnetic field structure and may also influence the extracted ion beam by steering effects. The power consumption of the coils for a usual field of $B_{max} \approx 1$ T is between 20 and 100 kW.

The mirror ratio B_{max}/B_{min} is usually about 2 and B_{max} at least 50% above the ECR resonance field. Sources with iron yokes tend to have higher mirror ratios due to the sharp maxima generated by the ferromagnetic material (Figure 9.3).

For frequencies above 20 GHz, magnetic fields above 1.5 T are needed, which can be provided by superconducting coils. With modern NeFeB permanent magnets it is possible to generate the necessary mirror fields for 10 GHz or even more (Figure 9.4), which reduces the power consumption of the entire ion source to a few kilowatts. The disadvantage of permanent magnets is the frozen field structure, which does not allow further fine tuning of the plasma by changes of magnetic field strength or mirror ratio.

For radial confinement of the plasma, magnetic multipoles are used. The higher the order of the multipole, the larger the loss area at both ends, and, thus, the area usable for ion extraction. A quadrupole has just a loss line at the ends, the orientation

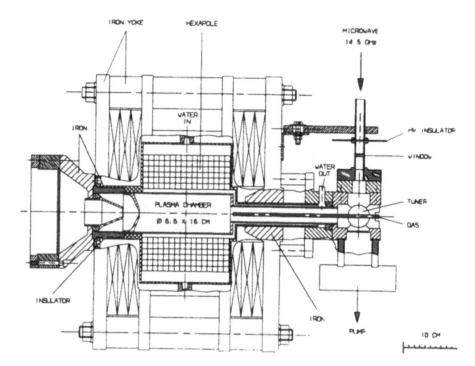

FIGURE 9.2
Caprice ECR with iron yoke.[37,56]

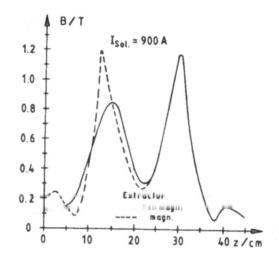

FIGURE 9.3
Axial magnetic field distribution of a Caprice type ECR with and without ferromagnetic extractor. (From Efremov, A., Kutner, V., and Zhao, H., Report JINR, E9-93-441, Dubna, Russia, 1993. With permission.)

of which is determined by the polarity of the superimposed mirror field; a hexapole leaves a triangle and an octupole a square as shown in Figure 9.5. The order of the multipole, however, is limited by the space available and by the field strength necessary for a closed ECR resonance zone, which made the hexapole configuration preferred in most designs.

The multipoles are made of rare earth permanent magnets with a surface field strength of more than 0.8 T, except for the superconducting sources where the

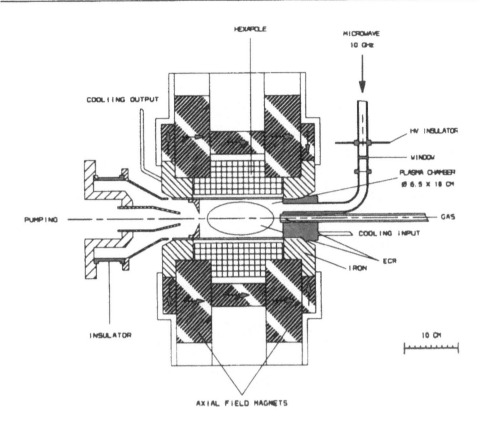

FIGURE 9.4
NeoMAFIOS ECR with permanent magnets.[17]

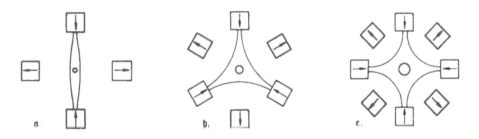

FIGURE 9.5
Axial loss lines of different multipoles.

multipole usually is made with superconducting coils also. There are two designs for the multipole: a closed structure shown in Figure 9.6, which allows the full volume to be filled with magnet material resulting in higher fields, and an open structure shown in Figure 9.7, which allows radial access for pumping, metal particle injection, or plasma diagnostics. Figure 9.8 shows a cross-sectional view on the magnet orientation for a hexapole using the full volume.

9.2.3 ECR Plasma and Scaling Laws

If microwaves are coupled into a multimode cavity (i.e., a metallic vessel with dimensions larger than the wavelength — for 10 GHz, $\lambda \approx 3$ cm) they can generate a

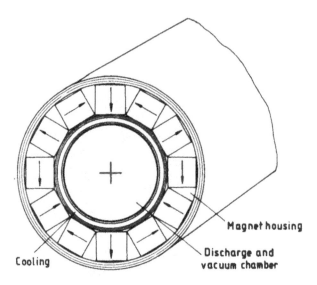

FIGURE 9.6
Closed hexapole structure.

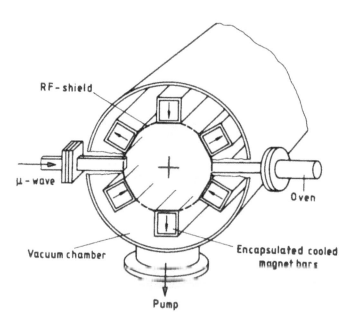

FIGURE 9.7
Open hexapole structure.

plasma. In the presence of a sufficiently high magnetic field, electron cyclotron resonance heating of the electrons will occur. One electron undergoes many passages through the resonance and will gain more and more energy with the help of stochastic or other nonlinear randomizing effects. For a more profound analysis of the heating processes see References 28 and 30 to 34. The electrons reach energies of up to 10 keV in usual modes of operation, but at lower gas pressure, more than 100 keV has been measured.[35,36] The ions, however, are not heated at normal operational conditions and have energies around only 1 eV, which results in a good emittance of the ECR ion source.

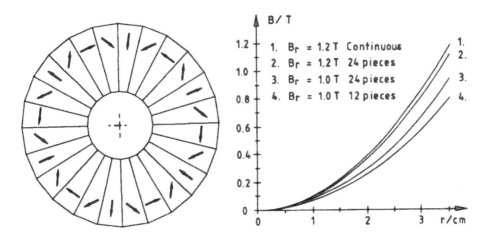

FIGURE 9.8
Magnet orientation for a hexapole with optimized field strength and radial field distribution. (From Efremov, A., Kutner, V., and Zhao, H., Report E9-93-441, Dubna, Russia. With permission.)

There is an upper limit for the plasma density given by

$$n_e \leq 1.2 \times 10^{10} \, f^2 \, cm^{-3}$$

with n_e the electron density and the frequency f in gigahertz, which leads to

$$n_{emax} \approx 10^{12} \, cm^{-3}$$

for 10 GHz.

Although the maximum of the electron density is a function of the square of the frequency, one needs more power to reach this density and the higher power level can cause a severe reduction in lifetime of the ions. It seems to be difficult to avoid instabilities for power levels above 1 W cm^{-3} in present ECR ion source arrangements. This was demonstrated by using the high magnetic field necessary for a 14-GHz source for 10-GHz[37] or 6.45-GHz[38] operation reaching nearly similar results as with the higher frequency. The higher magnetic field improves the confinement of the ions and in this way improves the charge state distribution.

Geller tried to give some scaling laws for the dependency of the optimum charge state on B and f,[25,26] but has been proven to be not very accurate, showing once more the complexity of the ECR ion source plasma. A precise description of the ECR ion source plasma has not been established to date, but great improvements in the understanding have recently have been made.[21,30,31]

9.2.4 First or Injector Stage

As mentioned above, the first stage of an ECR ion source for high charge states acts as an injector of a cold plasma or of electrons for the main stage, which must be operated at the lowest possible gas pressure or lowest possible neutral particle density. The picture R. Geller used defined the first stage as the ion generator and the second stage as a stripper to reach the high charge states. Consequently, the first ECR sources used a higher microwave frequency and a higher magnetic field for the first stage to generate a higher plasma density, which was then diffusing to the lower magnetic field of the second stage to be further ionized.[6,7] Later it was realized that one frequency was sufficient to operate the ion source.

After the operation of the ECR ion source at LBL with silicon, it was discovered that the performance of the source for high charge states had significantly improved.

This effect was due to a wall coating with SiO_2 delivering secondary electrons to the second-stage plasma.[39] If the second stage works better when it receives additional electrons, it should also improve if electrons are directly injected from a filament. This was shown at LBL.[40] If the first stage is just an electron source for the second-stage operation, an electron source should give the same results as a conventional first stage. This also was shown in the experiments at LBL.[39,40]

A similar effect was found using a negatively (~−300 V) biased disk positioned between the first and second stages, reflecting electrons back to the second stage and attracting ions that produce secondary electrons, which are accelerated into the second-stage plasma.[41,42] At RIKEN a biased probe was positioned inside the first stage to push electrons to the second stage,[43] which allowed reduction of the pressure in the second stage by a factor of two. In another experiment, a microwave-driven plasma cathode was used instead of a filament to extract electrons and inject them into the second stage with energies of ~300 eV,[44] which allowed a further reduction of the gas pressure in the second stage (Figure 9.9), improving the yield of high charge states.

9.2.5 Gas Mixing Effects

The production of higher-charged ions can be improved significantly by adding a lighter gas to the plasma (Figure 9.10).[45] For elements up to neon, helium is the best mixing gas and for elements heavier than argon, oxygen usually shows the best results; however, argon can also be used for some heavy elements. Under optimum conditions more than 50% of the particles in the ion source plasma will be of the mixing gas.

If the ion source is running on a getter material such as titanium, oxygen will disappear, so helium has to be used in those cases.[46] There is a strong pumping effect inside the ion source because just 5 to 10% of the ions reach the extraction and the rest hits the wall and most of this is recycled to the plasma as neutral particles. The extracted particles and those absorbed at the wall or pumped away have to be replaced by new particles injected through the first stage.

The most widely accepted model for the gas mixing effect is a cooling of the higher-charged ions by the low-charged mixing gas ions, which increases the confinement time of the high-charged ions, giving them a chance for additional ionizing collisions with energetic electrons.[47] This picture was confirmed by an isotope effect in oxygen where the ratio of ^{18}O to ^{16}O changed in favor of ^{18}O for the high charge states in pure oxygen gas but stayed constant when helium was added as a better mixing gas.[48]

9.2.6 Metal Ions

For the production of ions from nonvolatile materials the methods described in Chapter 3 can be applied in ECR ion sources for high charge states.

Gaseous Compounds

Gaseous compounds can be used in many cases especially if the compound parts can serve as a good mixing gas. One has to look for compounds with light elements and with at least 10% of the needed element. H_2S and SO_2, for example, have been used for sulfur,[49] SiH_4 for silicon,[50] and UF_6 for uranium.[50,51] A long list of metallo-organic volatile compounds also exists for many metals such as ferrocene, $C_{10}H_{10}Fe$ (most of them toxic or inflammable);[52] some are listed in Table 9.1.

Special care has to be taken when using aggressive compounds to prevent damage of the microwave window or other parts of the source. Sometimes an

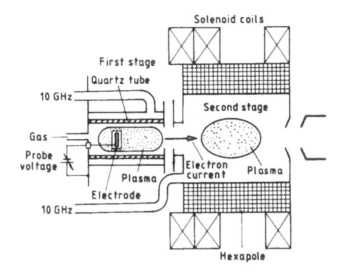

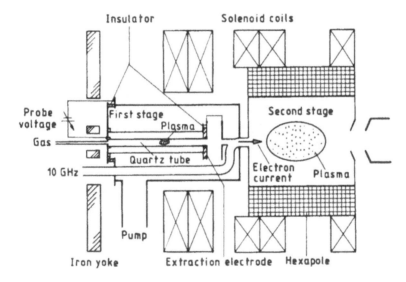

FIGURE 9.9
(a) Biased probe inside the first stage,[43] (b) plasma cathode arrangement.[44]

admixture of an additional element that binds the corrosive element in the source will overcome these problems. This was demonstrated in case of a potassium beam out of KCl with the addition of Ca where the calcium binds the chlorine and just a small amount of free chlorine was found in the spectra.[53]

Low- and High-temperature Oven

Depending on the design of an ECR ion source, an oven can be mounted to feed the second stage directly radially or has to be positioned axially in front of the first stage if there is one. In the first case the oven design does not cause any problems because there is enough space available with open hexapole structures and the oven can be positioned in a way that does not need special care to keep the boiling liquid in it (Figure 9.11).[27,53,54]

Axial ovens have to be of compact design due to the limited space available on the injector side of the ion source. For ECR4 at GANIL a microoven just 6 mm in

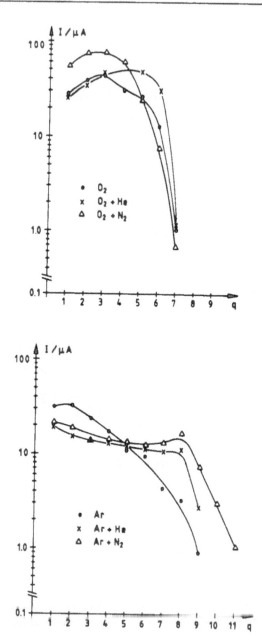

FIGURE 9.10
Yield of multiply charged ions with and without mixing gas: (a) for oxygen, (b) for argon.[45]

diameter has been developed for lead and other materials (Figure 9.12a) that can reach temperatures up to 1500°C.[55] The center pipe has a bore of 1.5 mm and holds the liquid by capilliary forces if the material combination is selected properly. A larger oven has been developed in Grenoble with an outer diameter of 15 mm and an inner bore of 6 mm for a crucible or boat with 3.40×20 mm³ usable space for the material to be evaporated.[37] If the ovens are designed in a way that they can be moved in and out through a vacuum lock, the vacuum in the second stage stays undisturbed by necessary refilling of the oven and long-term operation over weeks is possible. An oven similar to the Grenoble design that can be introduced to the ion source through an air lock is shown in Figure 9.12b.

TABLE 9.1

High-Vapor-Pressure Metalloorganic Compounds

Element	Compound	State	Boiling point (°C)	Comments
Cr	$Cr(CO)_5$		420	Decomposes at 250°C
Fe	$Fe(CO)_5$	liq.	105	Inflammable on air
Ni	$Ni(CO)_4$	liq.	43	Very toxic!!!
Ga	$Ga(CH_3)_3$	liq.	56	Stable
Ge	$Ge(CH_3)_3$	liq.	43	Stable
Sn	$Sn(CH_3)_4$	liq.	78	
Pb	$Pb(CH_3)_4$	liq.		Inflammable
Bi	$Bi(CH_3)_3$	liq.	110	Inflammable

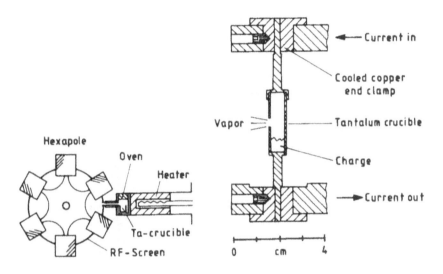

FIGURE 9.11
Oven for radial evaporation into the second stage: (a) low-temperature version (≤1200°C), (b) high-temperature version (≤2000°C).[53,54]

The temperature is controlled with a thermocouple and a feedback circuit that has a good response time because the mass of the oven can be small due to the low material consumption of ~ 1 mg/h. For high temperatures (≥1500°C) the control has to be done optically on axis through the extraction hole or in radial designs through a window opposite the high-temperature oven. Table 9.2 summarizes the ion yield of different charge states for various elements using oven techniques.

The usual oven designs give temperatures up to 1000 or 1500°C. Higher temperatures need well-heat-shielded arrangements of ovens heated by direct current flow or by electron bombardment of high-temperature metal or graphite crucibles. Power dissipation from the oven may also cause severe problems. For the production of ions from refractory materials one of the following methods is commonly used.

Chemical Synthesis

Chemical synthesis should be possible under the same conditions as for chemical compounds, but has not yet been tried.

Internal Evaporation from the Solid Material by Plasma Heating (Insertion Technique)

Samples of refractory material are mounted to a support that can be moved axially into the ion source plasma to a position where it can collect enough energetic

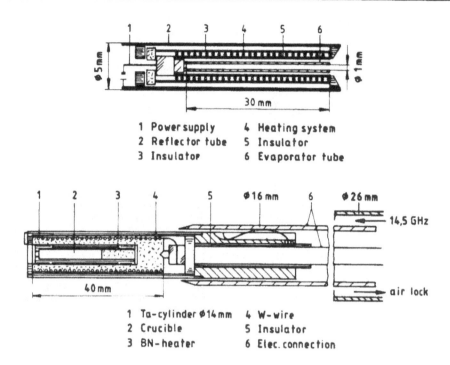

1 Power supply   4 Heating system
2 Reflector tube   5 Insulator
3 Insulator   6 Evaporator tube

1 Ta-cylinder ⌀14mm   4 W-wire
2 Crucible   5 Insulator
3 BN-heater   6 Elec. connection

FIGURE 9.12
Oven for axial evaporation: (a) GANIL microoven, [55](b) GSI oven.

electrons to be heated to a temperature sufficient to evaporate the material from the sample (Figure 9.13).[56,57] The temperature can be controlled optically on axis through the extraction hole. The signal may be used for remote control of the positioning mechanism. High-vapor-pressure materials gain sufficient energy in the area of the first stage; others have to be brought close to the ECR zone to reach the right temperature. In the second case, the charge state distribution may be less favorable for the highest charge states due to the interference of the sample with the plasma reducing the electron density and/or the confinement time of the ions. The size of the samples is 1 to 4 mm in diameter and a length sufficient for the necessary operation time. Not all materials can be purchased in the right dimensions but have to be especially fabricated by powder technology. Table 9.3 shows metal ion yields using insertion techniques.[58] Instead of an axial probe one can also position the sample or a boat with the material inside the plasma chamber so that the electrons can heat it sufficiently. (Figure 9.13). The temperature has to be controlled by the magnetic field, the microwave power, or the auxiliary gas pressure in this case, which means a compromise between optimum evaporation and best charge state distribution.

At MSU a similar technique is used. An alumina rod, covered with powder of the element wanted, is inserted radially into the second stage, producing a typical low material consumption charge state distribution.[38] The material consumption is just 0.1 to 0.5 mg/h.

The insertion technique is applicable to all materials that sublimate at temperatures where their vapor pressure is ~1 Pa. This is valid for many metallic elements and also for many oxides and carbonates. Table 9.4 gives vapor pressure and temperature values for various suitable elements and compounds.

The power to heat the plasma does not need to be increased to heat the probe because the electrons used are otherwise lost to the wall of the plasma chamber. If just 10% of the 300 W typically coupled into the source plasma is used to heat the probe

TABLE 9.2

Metal Ion Yields Using Oven Techniques[37,40,51,53,54]

Charge	Li	Mg	P	K	Ca	Sc	Ti	Fe	Ni	Cu	Se	Ag	Cs	La	Tb	Au	Pb	Bi
											Ion current (el. µA)							
1	5	—	—	—	—	—	—	—	—	—	—	—	—	—	—	—	—	—
2	7	—	—	—	—	—	—	—	—	—	—	—	—	—	—	—	—	—
3	0.5	—	—	4	—	—	—	—	—	—	—	—	—	—	—	—	—	—
4	—	—	—	5	—	—	—	—	—	—	—	—	—	—	—	—	—	—
5	—	5	—	—	—	—	—	—	—	—	2	—	—	—	—	—	—	—
6	—	8	6	8	—	—	—	—	—	—	2.5	—	—	—	—	—	—	—
7	—	—	8	11	—	6	—	—	—	—	2.6	—	12	—	—	—	—	—
8	—	8	12	18	14	7	—	—	—	—	6	—	14	—	—	—	—	—
9	—	6	—	37	22	13	12	—	—	—	7	—	15	—	—	—	—	—
10	—	2	4	—	44	—	10	8	3	5	—	—	10	—	—	—	—	—
11	—	—	—	12	—	—	8	9	—	8	8	—	10	—	—	—	—	—
12	—	—	—	—	14	27	—	—	5	—	7	—	10	—	—	—	—	—
13	—	—	—	—	5	14	1	8	6	10	3	—	10	—	—	—	—	—
14	—	—	—	—	—	—	—	—	5	9	—	—	10	—	—	—	—	—
15	—	—	—	—	—	—	5	3	7	—	2	10	1	—	—	—	—	
16	—	—	—	—	—	—	—	3	—	—	3	6	1.4	—	—	—	—	
17	—	—	—	—	—	—	—	2	3	—	4	6	—	—	—	—	—	
18								—	—	0.5	—	5	2	—	—	25	—	
19								—	—		6	—	3	3	5	24	—	
20											—	—	3	—	—	7	22	—
21											—	3	2	5	5	8	20	—
22											—	1.4	1.3	5	5	10	20	—
23												—	0.7	—	—	12	19	—
24												—	—	4	5	12	19	3
25												—	—	—	4	—	18	4
26												—	—	—	13	—	—	
27												—	2	—	10	17	5.5	
28												—	—	1.2	—	17	6	
29												—	0.4	0.6	4	—	5.7	
30												—	—	2.5	16	—		
31												—	1.5	9	4.5			
32												—	—	6	3.5			
33												—	—	3	2.6			
34												—	—	—	1.5			

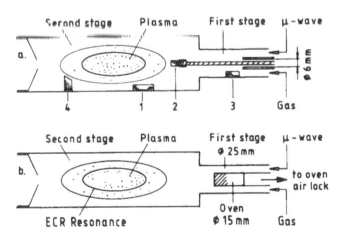

FIGURE 9.13
Possible sample position for the insertion technique (a) and oven (b).[57]

TABLE 9.3

Ion Yield for Various Elements by the Insertion Technique

Ion	Gas	Charge state																						Method
		1	2	3	4	5	6	7	8	9	10	11	12	13	14	15	16	17	18	19	20	21	22	
He	He	500	210																					
C	CO₂,He	75	57	—	18	2.3																		
N	N₂	90	95	63	50	30	2.8																	
O	O₂		85	65	—	15	6	0.25																
Ne	Ne	215	96	82	65	35	6.5																	
Ar	Ar,O₂						30	24	26	10	—													
Kr	Kr,O₂		75		17.5	16	30	17	11.5	9	6.8	0.7	3.7	—	1.2		1		0.5					
Xe	Xe,O₂				10	11	11	13.5	14	10	6.6		3.8	3.1	2.5	2.2		1.3	1.1	—	0.4			
Mg	O₂	30				35	—	30																MgO rod 4 Ø
Al	O₂		22	34	33	—	15	6	2	0.1														Al₂O₃ rod 4 Ø
Si	He			15	—	33				1														SiO₂ rod 3 Ø
Ca	O₂		72	70	55	15	36	27	26	25	—	5.5	1	0.3										CaO rod 5.7 Ø
Ti	O₂		5.2		7.5	15	14	14	10	7.5	6.5	4.4		0.3										Ti rod 2 Ø
V	O₂		2.5		7.5	12	12	12	10	7.5	4.8	2.4	1											V 2 wires 1 Ø
Cr	O₂		9		27	30	33	25	15	8.3	5.8	3.6	2		0.3									Cr rod 2 Ø
Mn	O₂				51	68	50	42	19	9	4	2.6	0.8											Ta crucible
Fe	O₂,He		4	4.8	5.2	15	15	—	14	11	8.2	4.6	3	1										Fe wire 1 Ø
Ni	O₂,He		3.6	7	16	8	12	13	14	11	8	—	1.7	0.75										Ni 2 wires 1 Ø
Co	He			9	16	17	26	19	15	6	2.2	4												Co 2 wires 1 Ø
Cu	O₂						19	17	15	6	6.5	4												Ta crucible
Zn	He		27	32		4.3	34	26	35	8.3	0.4													Ta crucible
Ge	O₂	1.3	1.8	3	2.6	4.3	6.5	6.3	6	4.6	2.6	1.3	0.5	0.27										Ta crucible
Zr	O₂				1	—	17	1.9	2.6	4.2	4.5	5	6.5	—	2.6	—	0.8							Zr rod 2 Ø
Nb	O₂			11	19	23	13	9.5	7	7	7	5	3	1.7	0.6									Nb rod 3 Ø
Mo	O₂		2		3.8	5	—	7.5	7	3.5	3.5	3	3	0.65	0.6									Mo rod 2 Ø
Ag	He		1.2		2.6	5.2	6.5	8.4	8.7	9	7.3	6.5	4.9	2.5	1.6	1.2	0.9	0.4						Ag rod 3 Ø
In	O₂,He			10	21	31	24	22	16	13	10	7.5	5.3	4.5	3	2	1.2	0.6	0.2					Ta crucible
Sn	O₂,He				3	5.2	5.7	6.6	10	7	—	4.2	2.7	1.8	1	—	0.2							Ta crucible
Ta	O₂					8.5	11	18	19	16	14	14	12	—	8.2	—	5.5	5	4.3	3	2	1.1		Ta wire 1 Ø
W	O₂				—	0.9	1	1.4	1.7	2.4	—	—	4	6	3	4.3	1.5	0.9	0.5	0.2	0.4	0.3	0.6	W wire 1 Ø
Au	O₂						11		19	16	14	—	10	6	10		4	2.6	1.5	0.8	0.4	0.3		Ta crucible
Pb	O₂						7.7	10	14	13	11	8.3	—	8.3	5.2	—	3.5	—	1.7	1	0.5	0.3	0.1	Ta crucible
Bi	He					—	3.3	—	8.3	10	8.7	7.6	5.6	7.6	5.6	—	3.6	2.3	1.4	0.9	0.5	0.1	0.1	Ta crucible

Note: Current in electron microamps, diameter in millimeters; 8-GHz NeoMAFIOS source.[56]

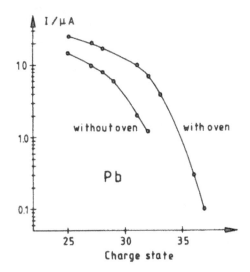

FIGURE 9.14
Yield of Pb charge states using the oven or insertion technique.[37]

(a tungsten wire of 1-mm diameter), it would reach a temperature of ≈3500°C (over a length of 1 cm or a surface of 0.1 cm² if heat conductance is neglected).

Internal Evaporation by Plasma Heating of a Boat

This method is similar to the oven method, but instead by using an independent heater, it is heated by the plasma in the same way as the evaporation from a solid described above. This method is simple, but, in general, an independently heated oven is the better choice because it leaves the plasma of the second stage undisturbed and usually gives a higher yield of high charge states.[37]

TABLE 9.4

Elements and Compounds Suitable for Evaporation from the Solid[59–61]

Element	Compound	1 Pa at °C	Melting point (°C)	Element	Compound	1 Pa at °C	Melting point (°C)
Li	LiF	650	850	Sr	SrO	1260	2430
Be	BeO	2300	2550	Zr	ZrO₂	2500	2715
B	—	1400	2100	Mn	—	2500	2670
Na	NaCl	530	613	Cd	—	260	321
Mg	MgO	2000	2700	Te	—	380	450
Al	Al₂O₃	2000	2050	Ba	BaO	1260	1923
Si	SiO₂	1450	1710	Ce	CeO₂	1900	2600
P	—	5	44	Sm	—	660	1075
K	KCl	510	776	Eu	—	610	822
Ca	CaO	1720	2572	Dy	—	1180	1410
Sc	—	1400	1540	Ho	—	1120	1470
Ti	—	1600	1690	Er	—	1250	1530
V	—	1800	1890	Tm	—	850	1545
Cr	—	1300	1860	Yb	—	460	820
Mn	—	1000	1245	Ta	—	3050	3010
Fe	—	1500	1535	W	—	3200	3410
Ni	NiO	1500	1960	Re	—	3000	3180
Co	CoO	1600	1810	Th	ThO₂	2200	3000
Zn	—	350	420	U	UO₂	1750	2176
As	—	310	817				

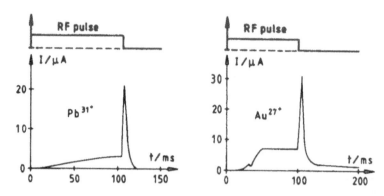

FIGURE 9.15

Ion current of Au^{27+} and of Pb^{31+} with afterglow from 14.5-GHz Caprice source.[37]

9.2.7 Pulse Mode and Afterglow

Operation of the ECR ion source in a pulsed mode with pulse length ≤30 ms and duty cycles ≤10% can increase the yield in high charge states if the ion source is tuned for this mode of operation.[62] The improvement is mainly due to the reduction of average power and, thus, less outgassing from the walls. The ion source discharge is pulsed by pulsing the microwave generator (see Chapter 6).

A different phenomenon was discovered by the Grenoble group when pulsing the ECR ion source at higher microwave power (≥1 kW). An increase of high charge states occurred for a short time of ~1 ms from the afterglow of the plasma and is therefore called the afterglow mode.[63,64] The shape of this afterglow ion current is like a spike with a sharp increase to a maximum and then a slower decay (Figure 9.15). The shape and the pulse length can be influenced by the ion source parameters (microwave power, gas pressure, and magnetic field) in a small range from 0.5 to 2 ms. The fine tuning can be best done by varying the microwave pulse length. This afterglow effect is much more pronounced for heavy ions; for Xe^{17+} the gain compared to dc operation is about a factor two and for Pb^{31+}, about ten. The ion current during the microwave pulse for the highest charge states can be nearly zero and ions are just extracted during the afterglow. Figure 9.16 shows the tuning of an afterglow mode starting from dc operation. The explanation for the afterglow effect is a negative potential well that contains the high-charged ions and disappears when the microwave heating of the plasma is switched off. This effect is also called ECR mirror plugging. More details on the explanation of the afterglow effect are given in References 25, 28, 30, 65, and 66.

Pulse lengths shorter than 0.5 ms can be achieved by pulsing an additional low-inductance coil inside a vacuum chamber that also contains the multipole inside the coil. A test source called PuMa ECR (pulsed magnetic field) was designed at GSI using a 10-GHz MiniMAFIOS source (Figure 9.17) and first experiments gave pulse lengths of 0.15 ms (Figure 9.18).[67-69] To avoid induced surface currents, the multipole has to be constructed in two half-shells.

9.3 Specific Ion Source Design

9.3.1 Minimafios Ion Sources (Figure 9.19)

- Special design and construction details of the source

  The source is characterized by a plasma chamber length (36 cm) close to the axial magnetic bottle length (38 cm). The mirror field is created by a set of coils consuming

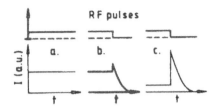

FIGURE 9.16
Tuning of the afterglow: (a) dc, (b) pulsed, (c) afterglow optimized.[65]

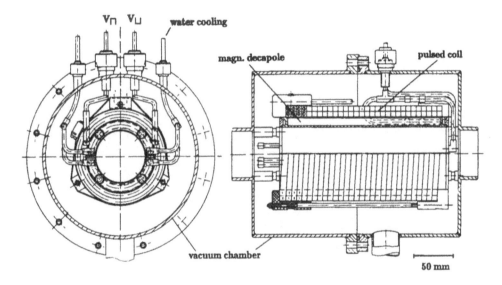

FIGURE 9.17
Pulsed coil and split decapole of PuMa ECR ion source.

about 150 kW and mirror plates but no iron yoke, giving a mirror ratio of ≤ 2.67, adjustable over a wide range. The hexapole gives a radial field of 0.65 T on the plasma chamber wall. A biased probe replaced the separate first stage of earlier versions of the MiniMAFIOS family.

- Ion source material and vacuum conditions

    The materials used are nonmagnetic stainless steel, aluminium, alumina, and SmCo hexapole magnets. The base pressure is 10^{-5} Pa.

- Application area of the source

    Low- and high-energy accelerators, high intensities of high charge states in dc or pulsed operation, atomic physics.

- Deliverer or user

    Centre d'Etudes Nuclear de Grenoble, DRFMC/PSI, 17, rue des Martyrs, 38054 Grenoble Cedex 9, France

    ISN, Grenoble, France

    KVI, Groningen, France

    GSI, Darmstadt, Germany

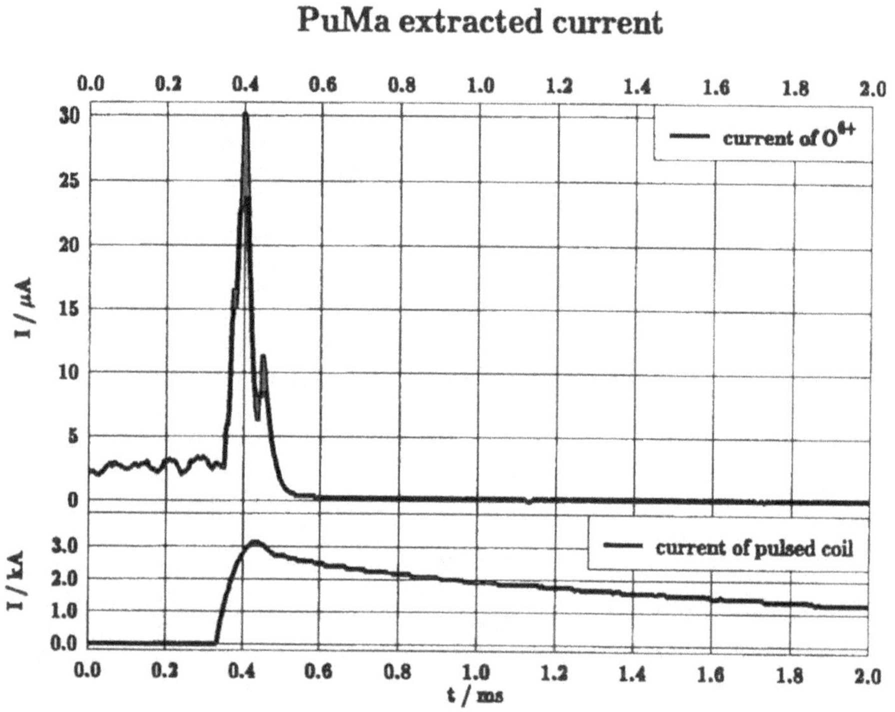

PuMa extracted current

FIGURE 9.18
Pulse shape for total current and O^{6+} current from a PuMa source.

9.3.2 Caprice Ion Source (Figure 9.20)

- Special design and construction details of the source

  The source is compact because of an iron yoke and a small plasma chamber (16-cm length, 6.6-cm Ø). The mirror field is 1.2 T at the extraction because of the extensive use of iron at the extractor side, including the extractor itself. The hexapole field is up to 1.2 T at the plasma chamber wall, depending on the magnets used. The microwave power is launched by a coaxial waveguide, which allows a very efficient first stage providing electrons for the main stage.

- Ion source material and vacuum conditions

  First-stage chamber, Cu; second-stage chamber, stainless steel; high-voltage insulation, polyethylene; extractor insulator, alumina. The hexapole is made with NdFeB permanent magnets.

- Application area of the source

  Accelerators, atomic physics with highly charged ions.

- Deliverer or user

  Centre d'Etudes Nuclear de Grenoble, DRFMC/PSI, 17, rue des Martyrs, 38054 Grenoble Cedex 9, France

  GANIL, Caen, France

  IMP, Lanzhou, China

  ORNL, Oak Ridge, TN, U.S.A.

  PSI, Zürich, Switzerland

  GSI, Darmstadt, Germany

  Univ. of Nevada, Reno, NV, U.S.A.

  Jet Propulsion Lab., Pasadena, CA, U.S.A.

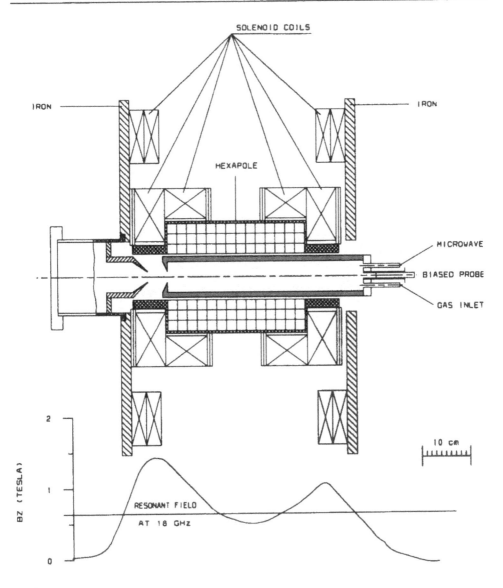

FIGURE 9.19
MINIMAFIOS 10- to 18-GHz ECR ion source.

9.3.3 Neomafios Ion Source (Figure 9.21)

- Special design and construction details of the source in diameter.

  The discharge chamber is just 18 cm long and 6.5 cm in diameter. The axial magnetic field, designed with permanent magnets only, is 0.56 T at the maximum and the mirror ratio is 2.15. The first resonance zone is in a quartz tube supplying the gas and thus acts as the first stage. The microwaves are fed through a rectangular waveguide off axis to allow the introduction of metallic samples on axis.

- Ion source material and vacuum conditions

  Plasma chamber, stainless steel; insulators, polyethylene; magnetic structure, NdFeB permanent magnets. The total power consumption is ≤10 kW.

- Application area of the source

  Low- and high-energy accelerators, atomic physics, ion implantation.

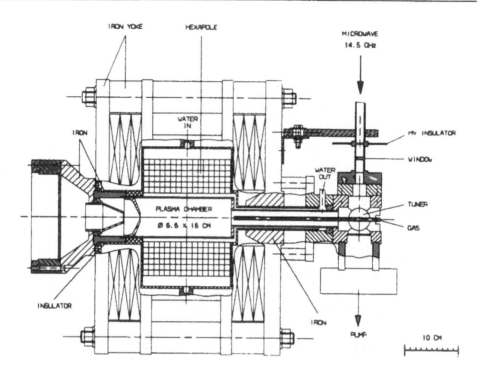

FIGURE 9.20
Caprice ECR ion source.[37]

- Deliverer or user

  Centre d'Etudes Nuclear de Grenoble, DRFMC/PSI, 17, rue des Martyrs, 38054 Grenoble Cedex 9, France

  RILAC, Riken, Japan RCNP, Osaka, Japan

9.3.4 MSU Room-Temperature Ion Source (Figure 9.22)

- Special design and construction details of the source

  The vertical ion source has a large plasma chamber 82 cm long and 14 cm in diameter surrounded by a permanent magnet hexapole and solenoid coils maintaining a 2-B field (twice the ECR resonant field) inside the plasma chamber. The ion source runs with 6.4-GHz microwaves only, giving extremely high charge states.

- Ion source material and vacuum conditions

  Plasma chamber, Cu; hexapole, NdFeB; cylindrical yoke, iron.

- Application area of the source

  Cyclotron injection, heavy-ion accelerators, atomic physics with multiply charged ions.

- Deliverer or user

  Michigan State University, East Lansing, Michigan 48824, U.S.A.

  TSL, Uppsala, Sweden

  University of Jyväskylä, Finland

9.3.5 LBL AECR Ion Source (Figure 9.23)

- Special design and construction details of the source

  The ion source has an open hexapole structure that allows radial access to the plasma chamber. The axial magnetic field reaches 1 T on the injector side and has

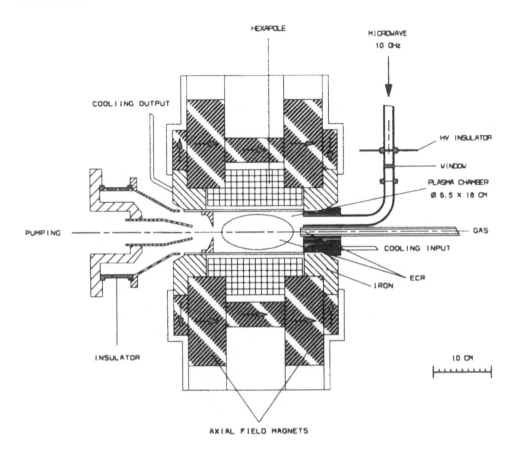

FIGURE 9.21
NeoMAFIOS all-permanent-magnet ECR ion source.[17]

    a high mirror ratio of 2.5 because of the iron plates in the discharge area. Instead of a first stage, an electron gun delivers enough electrons for the main-stage plasma. The extraction gap is adjustable.

- Ion source material and vacuum conditions

  Discharge chamber, stainless steel; hexapole, NdFeB; insulator, alumina; cathode, LaB$_6$.

- Application area of the source

  Cyclotron injection, heavy-ion accelerators, atomic physics with multiply charged ions.

- Deliverer or user

  88″ Cyclotron, Lawrence Berkeley Lab., Berkeley, CA 94720, U.S.A.

  Institut für Kernphysik, Univ. Frankfurt, 60486 Frankfurt, Germany

  Technische Universität Dresden, 01796 Pirna, Germany

9.3.6   GANIL ECR-4 Ion Source (Figure 9.24)

- Special design and construction details of the source

  The axial magnetic field reaches 1 T as maximum, has a mirror ratio of 2, and needs 45 kW. The hexapole reaches its maximum in the middle plane and is weaker towards the ends of the plasma chamber where the axial field is strongest. In this way the solenoid coils can have smaller dimensions compared to the similarly

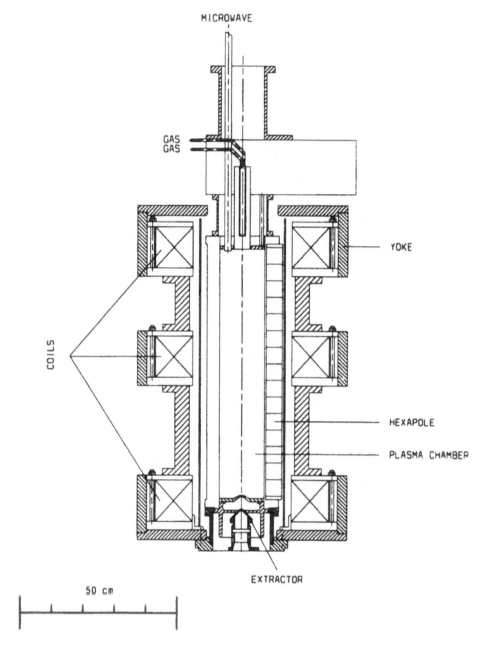

FIGURE 9.22
MSU room-temperature ECR ion source.[14]

designed Caprice source. The entire source is encapsuled in a iron yoke with large pieces of iron at both ends. The vacuum chamber is water-cooled through a double wall and the high-voltage insulation keeps the coils and the outer iron yoke at ground potential. The ECR-4 is easy to operate in the afterglow mode.

- Ion source material and vacuum conditions

Plasma chamber, stainless steel; hexapole, NdFeB; insulation, polypropylene; the vacuum is 10^{-5} Pa.

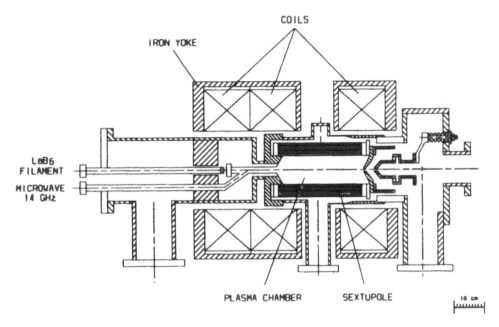

FIGURE 9.23
LBL AECR ion source.[40]

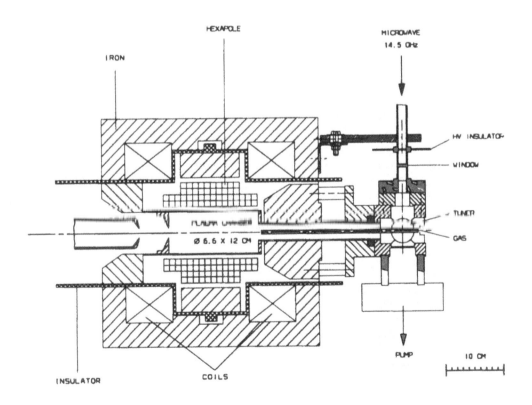

FIGURE 9.24
ECR-4 ion source.[55]

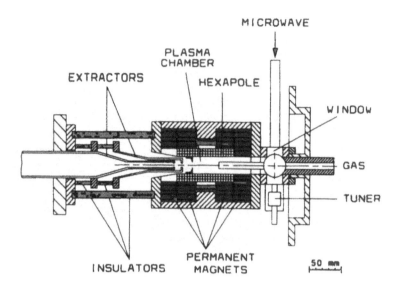

FIGURE 9.25
Nanogan ECR ion source.[70]

- Application area of the source
  Cyclotron injection, linear accelerators.
- Deliverer or user
  GANIL, B.P. 5027, 14021 Caen Cedex, France
  Hahn Meitner Institut, Glienicker str. 100, 14109 Berlin, Germany
  CERN, PS Division, 42 Geneva, Switzerland

9.3.7 Nanogan Ion Sources (Figure 9.25)

- Special design and construction details of the source
  The ion source is just 13 cm in diameter and 22 cm long. The magnetic field is made entirely with permanent magnets. The plasma chamber is only 2.6 cm in diameter but the magnetic field is sufficient for 10-GHz operation. The mirror field has a maximum of 0.74 T. Because of the small plasma volume the source needs less than 100 W of microwave power. The ion source delivers moderate charge states and has a high gas efficiency needed for radioactive beam production.
- Ion source material and vacuum conditions
  Plasma chamber, stainless steel; magnets, all NdFeB.
- Application area of the source
  On-line isotope separation, radioactive beams, accelerators, ion implantation.
- Deliverer or user
  GANIL, B.P. 5027, 14021 Caen Cedex, France
  Pantechnik S.A. Siege Social, 14200 Herouville-Saint-Clair, France

9.3.8 Low-Cost Giessen Type ECR Ion Sources (Figure 9.26)

- Special design and construction details of the source
  The 5-GHz ECR has a strong magnetic field of 0.5 T for the mirror field, a mirror ratio of up to 5, and a hexapole field of 0.5. The source uses a biased disk to enhance multiply charged ion production. All source parts are easy to produce.

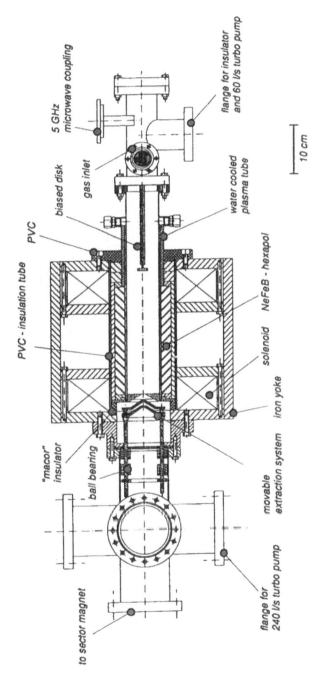

FIGURE 9.26
ECR ion source BERTA. (From Winter, H. P. et al., *Rev. Sci. Instrum.*, 65, 1091, 1994. With permission.)

- Ion source material and vacuum conditions

  Plasma chamber, stainless steel; magnets, NdFeB; magnetic yoke, iron; insulator, machinable ceramics; insulator of source body, PVC.

- Application area of the source

  Atomic physics, low-energy accelerators, ion implantation.

- Deliverer or user

  Institut für Allgemeine Physik, Wiedner Hauptstr.8–10/134, 1040 Wien, Austria

  Institut für Kernphysik, Universität Giessen, 35392 Giessen, Germany

9.3.9 Operating Data

TABLE 9.5

Operating Data of ECR Ion Sources

		MiniMAF.	Caprice	MSU-RTECR	LBL-AECR	ECR-4
Rf frequency	(GHz)	10–18	10–14.5	6.4	14.5	14.5
Microwave power	(kW)	1–5	1	0.5	1.5	≤2
Axial field	(T)	1.5	1.2	0.38	1	1.1
Hexapole field	(T)	0.7	0.8–1.2	0.34	0.8	1
Magnet supply	(kW)	150	1.2	80	90	45
Gas flow	(sccm)	0.1	0.1	0.1	0.1	0.1
Extraction	(mm) Ø	8–12	6–10	8	8–10	6–10
Extr. volt.	(kV)	0–25	0–25	5–20	10	30
Plasma chamber	(cm) Ø	6.6	6.6	14	7.5	6.6
Chamber length	(cm)	36	16	82	30	16
N^{6+}	(μA)	110	150	13	41	—
O^{7+}	(μA)	180	50	5	131	—
Ne^{9+}	(μA)	—	7	1	7.5	—
Ar^{8+}	(μA)	280	550	40	210	500
Kr^{17+}	(μA)	—	42	5	48	50
Xe^{20+}	(μA)	50	42	6	27	41
Ca^{11+}	(μA)	—	18	—	47	—
Au^{26+}	(μA)	—	13	—	—	—
Pb^{25+}	(μA)	—	25	—	—	30
Bi^{25+}	(μA)	50	20	—	10	—
U^{28+}	(μA)	6	6	2.8	—	5
Max.charge state	N	7	7	7	7	—
Max.charge state	Ne	10	9	10	9	—
Max.charge state	Ar	18	18	17	16	16
Max.charge state	Kr	26	23	24	28	20
Max.charge state	Xe	32	29	31	31	29
Max.charge state	Ta	30	29	—	—	31
Max.charge state	U	35	31	36	—	28

		NeoMAFIOS	Nanogan	Low-cost
Rf frequency	(GHz)	8–10	10–14.5	5
Microwave power	(kW)	≤0.6	≤0.1	≤1
Axial field	(T)	0.56	0.75	0.5
Hexapole field	(T)	0.42	0.7	0.5
Magnet supply	(kW)	—	—	30
Gas flow	(sccm)	0.1	0.1	0.1
Extraction	(mm) Ø	6–8	5	6
Extr. volt.	(kV)	0–15	0–15	0–15
Plasma chamber	(cm) Ø	6.5	2.6	7
Chamber length	(cm)	18	16	25
N^{6+}	(μA)	2.8	—	—
O^{7+}	(μA)	0.3	0.3	2
Ne^{9+}	(μA)	—	—	—

TABLE 9.5 (continued)

Operating Data of ECR Ion Sources

		NeoMAFIOS	Nanogan	Low-cost
Ar^{8+}	(μA)	90	65	40
Kr^{17+}	(μA)	1	1	—
Xe^{20+}	(μA)	5	3	—
Al^{5+}	(μA)	28	—	—
Ta^{16+}	(μA)	15	—	—
Max.charge state	N	6	—	6
Max.charge state	Ne	6	—	—
Max.charge state	Ar	12	12	12
Max.charge state	Kr	20	18	—
Max.charge state	Xe	23	22	—
Max.charge state	Ta	22	—	—

REFERENCES

1. R. Geller et al., *Proc. 1st Int. Conf. Ion Sources*, Saclay, 1969, 537.
2. H. Postma, *Phys. Lett.*, 13A, 196, 1970.
3. R. Geller et al., *Proc. 2nd Int. Conf. Ion Sources*, Vienna, 1971, 632.
4. K. Bernhardi and K. Wiesemann, *Plasma Phys.*, 14, 1073, 1972.
5. H. Tamagawa et al., *IEEE Trans. Nucl. Sci.*, 23, 994, 1976.
6. P. Briand et al., *Nucl. Instrum. Methods*, 131, 407, 1975.
7. R. Geller, *Proc. 8th Int. Conf. Cyclotrons and Their Applications*, Bloomington, IN, 1978; *IEEE Trans. Nucl. Sci.*, 26, 2120, 1978; *Proc. 4th Int. Workshop ECR Ion Sources*, CENG Press, Grenoble, 1982, 5.1.
8. V. Bechtold et al., *Proc. 9th Int. Conf. Cyclotrons and Their Applications*, Caen, 1981, 249.
9. D. J. Clark et al., *IEEE Trans. Nucl. Sci.*, 30, 2719, 1983.
10. H. Schulte, Rep. GSI-81-2 p 247, GSI, Darmstadt, 1981; *Proc. Int. Workshop ECR Ion Sources*, 1981, Rep. GSI-81-1, GSI, Darmstadt, 1981, 50.
11. Y. Jongen et al., *IEEE Trans. Nucl. Sci.*, 26, 3677, 1979.
12. V. Bechtold et al., *IEEE Trans. Nucl. Sci.*, 26, 3680, 1979.
13. H. Beuscher et al., *Proc. 9th Int. Conf. Cyclotrons and Their Applications*, Caen, 1981, 285; les Editions de Physique, Les Ulis, France, 1982.
14. A. Antaya et al., *Proc. 6th Int. Workshop ECR Ion Sources*, Berkeley, 1985; Rep. LBL-PUB-5143, p. 126; *Proc. 11th Int. Workshop ECR Ion Sources*, Groningen, 1993, Rep. KVI-996, p. 42.
15. R. C. Pardo and P. J. Billquist, *Proc. Int. Conf. ECR Ion Sources*, East Lansing, 1987, Rep. NSCL-MSUCP-47, p. 279.
16. R. C. Pardo, *Nucl. Instrum. Methods*, B 40/41, 1014, 1989.
17. P. Sortais et al., *Proc. Int. Conf. ECR Ion Sources*, East Lansing, 1987; Rep. NSCL-MSUCP-47, p. 334.
18. H. Beuscher et al., *Proc. 6th Int. Workshop ECR Ion Sources*, Berkeley, 1985, Rep. LBL-PUB-5143, p. 107.
19. M. Liehr et al., *Rev. Sci. Instrum.*, 63, 2541, 1990; *Proc. 10th Int. Workshop ECR Ion Sources*, Knoxville, 1990, 363; ORNL Conf-9011136, 1991.
20. M. Leitner et al., *Proc. 11th Int. Workshop ECR Ion Sources*, Groningen, 1993, Rep. KVI-996, p. 119; *Proc. 5th Int. Conf. Ion Sources*, Beijing, 1993; *Rev. Sci. Instrum.*, 65, 1091, 1994.
21. A. G. Drentje, Ed., *Proc. 11th Int. Workshop ECR Ion Sources*, Groningen, 1993, Rep. KVI-996.
22. C. M. Lyneis and T. A. Antaya, *Rev. Sci. Instrum.*, 61, 221, 1990.
23. A. G. Drentje, *Rev. Sci. Instrum.*, 65, 1045, 1994.
24. R. Geller, *Annu. Rev. Nucl. Part. Sci.*, 40, 15, 1990.
25. R. Geller, *Proc. 5th Int. Conf. Phys. Highly Charged Ions*, Giessen, 1990; *Z. Phys.*, D 21, 117, 1990.
26. R. Geller, *Proc. 1989 Particle Accelerator Conf.*, Chicago, IEEE Conf. Service 89CH2669-0, 1989, 1088.
27. Y. Jongen and C. M. Lyneis, in *The Physics and Technology of Ion Sources*, I. G. Brown Ed., John Wiley & Sons, New York, 1989.
28. R. Geller, *Electron Cyclotron Resonance Ion Sources*, IOP Publishing, Bristol, in preparation (1995/96).
29. V. B. Kutner, private communication.

30. R. Geller, *Proc. 11th Int. Workshop ECR Ion Sources*, Groningen, 1993, Rep. KVI-996, p. 1.
31. K. S. Golovanivsky, *Proc. 11th Int. Workshop ECR Ion Sources*, Groningen, 1993, Rep. KVI-996, p. 78.
32. O. Eldridge, *Phys. Fluids*, 15, 676, 1972.
33. M. A. Liebermann and A. J. Lichtenberg, *Plasma Phys.*, 14, 1073, 1972; *Plasma Phys.*, 15, 125, 1973.
34. M. Bornatici et al., *Nucl. Fusion*, 23, 1153, 1983.
35. K. Bernardi and K. Wiesemann, *Plasma Phys.*, 14, 867, 1981.
36. D. Meyer et al., *Proc. 11th Int. Workshop ECR Ion Sources*, Groningen, 1993, Rep. KVI-996, p. 198.
37. D. Hitz et al., *Proc. 11th Int. Workshop ECR Ion Sources*, Groningen, 1993, Rep. KVI-996, p. 91.
38. T. A. Antaya et al., *Proc. 11th Int. Workshop ECR Ion Sources*, Groningen, 1993, Rep. KVI-996, p. 42.
39. C. M. Lyneis, *Proc. Int. Conf. ECR Ion Sources*, East Lansing, 1987, Rep. NSCL-MSUCP-47, p. 42.
40. C. M. Lyneis et al., *Proc. 10th Int. Workshop ECR Ion Sources*, Knoxville, 1990, ORNL Conf-9011136, 1991, 47; *Rev. Sci. Instrum.*, 62, 775, 1991.
41. G. Melin et al., *Proc. 10th Int. Workshop ECR Ion Sources*, Knoxville, 1990, ORNL Conf-9011136, 1991, 1.
42. S. Gammino et al., *Rev. Sci. Instrum.*, 63, 2872, 1992.
43. T. Nakagawa et al., *Jpn. J. Appl. Phys.*, 31, 1129, 1992.
44. T. Nakagawa et al., *Jpn. J. Appl. Phys.*, 32, 1335, 1993; *Proc. 11th Int. Workshop ECR Ion Sources*, Groningen, 1993, Rep. KVI-996, p. 208.
45. A. G. Drentje and J. Sijbring, *Nucl. Instrum. Methods*, B9, 526, 1985; *Proc. 6th Int. Workshop ECR Ion Sources*, Berkeley, 1985, Rep. LBL-PUB-5143, p. 73.
46. H. Schulte, private communication.
47. T. A. Antaya, *Proc. Int. Conf. ECR Ion Sources*, Grenoble, 1988; *J. Phys. Colloq.*, Cl 50, Cl-707, 1989.
48. A. G. Drentje, *Rev. Sci. Instrum.*, 63, 2875, 1992.
49. R. Geller et al., *Proc. Int. Conf. ECR Ion Sources*, East Lansing, 1987, Rep. NSCL-MSUCP-47, p. 1.
50. F. W. Meyer et al., *Proc. 10th Int. Workshop ECR Ion Sources*, Knoxville, 1990, ORNL Conf-9011136, 1991, 367.
51. R. C. Pardo and P. J. Billquist, *Rev. Sci. Instrum.*, 61, 239, 1990.
52. J. Ärje et al., *Proc. 11th Int. Workshop ECR Ion Sources*, Groningen, 1993, Rep. KVI-996, p. 27.
53. C. M. Lyneis, *Proc. 11th Int. Conf. Cyclotrons and Their Applications*, Tokyo, 1986, Ionics, Tokyo, 1986, 707.
54. D. Clark and C. M. Lyneis, *Proc. Int. Conf. ECR Ion Sources*, Grenoble, 1988; *J. Phys. Colloq.*, Cl 50, Cl-707, 1989.
55. M. P. Bourgarel et al., *Rev. Sci. Instrum.*, 63, 2854, 1992.
56. B. Jacquot et al., *Nucl. Instrum. Methods*, A269, 4, 1988.
57. R. Geller et al., *Rev. Sci. Instrum.*, 63, 2795, 1992.
58. T. Nakagawa et al., *Proc. 10th Int. Workshop ECR Ion Sources*, Knoxville, 1990, ORNL Conf-9011136, 1991 167; *Proc. 11th Int. Workshop ECR Ion Sources*, Groningen, 1993, Rep. KVI-996, p. 208.
59. G. Ranc, *La Technique des Couches Minces et son Application a la Microscopie Electronique*, CEA Rep. 1117, CEA, Saclay, France, 1959.
60. C. J. Smithells, *Metals Reference Book*, Vol. I, Butterworths, London, 1967, 262.
61. S. Dushman, *Scientific Foundations of Vacuum Technique*, John Wiley & Sons, New York, 1962, 691 ff.
62. F. Bourg et al., *Proc. 4th Int. Workshop ECR Ion Sources*, Grenoble, 1982, CENG Press, Grenoble, 1982, 5.1; *Proc. 6th Int. Workshop ECR Ion Sources*, Berkeley, 1985, Rep. LBL-PUB-5143, p. 1.
63. G. Melin et al., *Proc. Int. Conf. ECR Ion Sources*, Grenoble, 1988; *J. Phys. Colloq.*, Cl 50, Cl-673, 1989.
64. P. Briand et al., *Nucl. Instrum. Methods*, A294, 673??, 1990.
65. P. Sortais, *Rev. Sci. Instrum.*, 63, 2801, 1992.
66. G. Shirkov, *Proc. 11th Int. Workshop ECR Ion Sources*, Groningen, 1993, Rep. KVI-996, p. 67.
67. U. Ratzinger et al., *Proc. 1990 Linear Accel. Conf.*, Albuquerque, p. 713.
68. U. Ratzinger et al., *Proc. 11th Int. Workshop ECR Ion Sources*, Groningen, 1993, Rep. KVI-996, p. 204.
69. U. Ratzinger et al., *Rev. Sci. Instrum.*, 65, 1078, 1994.
70. P. Sortais et al. Proc. 11th Int. Workshop ECR Ion Sources, Groninjen, 1993, Rep. kVI-996, p. 97.

CHARACTERIZATION OF ION SOURCES

B. Sharkov

10   LASER ION SOURCE

+ Compact and simple source for intensive pulsed ion beams, ions of any solid material, high, medium and low charge states, moderate costs, no high voltage power supplies necessary.

– short pulse length, low repetition rate.

10.1   Ion Source History

The first two detailed proposals to use a laser produced plasma as a source of ions for a particle accelerator were made independently by Peacock and Pease[1] and Byckovsky et al.[2] both in 1969. Highly charged ions of different materials having high expansion energy (several keV per ion) have been produced by neodymium glass lasers.[3,4] Furthermore, the ions with the higher charges were found to expand into cones with quite narrow angles 20 to 30°.[5] The first operation of a laser ion source based on Nd-glass laser on a cyclotron machine was reported by Anan'in.[6] The application of a laser produced plasma as a source of ions for injection into a high-energy accelerator took place at the 10-GeV synchrotron at HINR (Dubna)[7] 1984. Most of this development used light or medium-mass ions up to chromium (Cr^{13+}).[8]

Laser ion sources for Van-de-Graaf accelerators have been employed at the Technical University of Munich[9] and at ITEP-Moscow[10] 1988, and also at the University of Arkansas a laser ion source has been constructed and intensively investigated.[11]

10.2   Working Principle of the Laser Ion Source and Description of the Operation Principle

The principle of operation of the laser ion source is based on plasma generation by a laser beam focused by a mirror system (or lens) on a solid movable target. The focused laser light is used to evaporate particles from a target which is made out of the material to be ionized. The electrons of the plasma, which is generated during the evaporation process, are heated by the laser radiation to temperatures up to several hundreds of eV. The ions are ionized due to electron-ion collisions. The temperature of the plasma T_e and the final ion charge state distribution strongly depend on the laser power density $P[W/cm^2]$ on the target:[12]

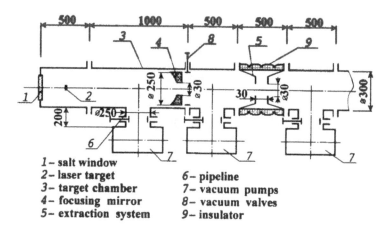

1 – salt window
2 – laser target
3 – target chamber
4 – focusing mirror
5 – extraction system

6 – pipeline
7 – vacuum pumps
8 – vacuum valves
9 – insulator

FIGURE 10.1
Ion source scheme and circuitry.

$$T_e \sim \left(P\bar{Z}\right)^{2/7}$$

The ion pulse duration is determined by the drift space between the target and the extraction plane, where the plasma expands. The ion pulse also expands longitudinally resulting in a longer pulse duration due to the initial energy spread of the ions in the plasma.

10.2.1 General Description

The whole source system consists of three main parts: laser generator, optical focusing system and source vacuum chamber (see Figure 10.1). The vacuum chamber is insulated from the ground; it contains the target mounted on a target positioning mechanism.

Apart from the stepping motor associated with the vacuum chamber to expose a fresh part of the target surface, no power is required at the high voltage platform. The pulsed carbon dioxide laser consisting almost entirely of a pulsed power supply is positioned at ground potential.

10.2.2 Laser Characteristics

A CO_2 laser is found to be the best to deliver the energy to evaporate particles from a target, which is made out of the material to be ionized. The pulse energy of the laser can vary from 1 J to 50 J in a pulse length of ~1 µs and a repetition rate of up to 1 Hz. The divergence of the laser light is about 2 mrad. The laser gas mixture consists of CO_2 to produce the radiation with a wavelength of 10.6 µm, N_2 to keep the CO_2 molecules in an excited state for a relatively long time and He to create a homogeneous discharge in the resonator of the laser. The laser pulse shape, which influences the charge state distribution of the generated plasma, depends critically on the composition ratio of these three gases.[13]

10.2.3 High Current Low Charge State Mode

This mode of source operation can be achieved by using a nitrogen-rich laser gas mixture (for instance CO_2:N_2:He=1:1:3) and (or) by defocusing of the laser beam on

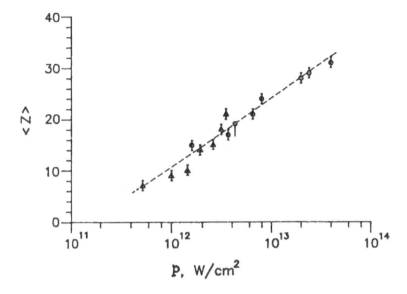

FIGURE 10.2
Dependence of average charge state on laser power density.

the target surface, i.e., by decreasing the laser power density P. The power density level required for efficient production of singly and doubly ionized ions is typically $P \simeq 10^9$ W/cm². The extracted ion current available under these conditions reaches 10 to 100 mA/20 µs pulse.[13]

10.2.4 High Charge State Mode

The higher laser power density $P \geq 10^{12}$ W/cm² required for the production of highly charged ions can be obtained either by increasing the laser output energy and shortening the laser pulse or by decreasing the focal spot diameter. In Figure 10.2 the dependence of the average charge state $\bar{Z}$ in the expanding plasma on the power density P is shown for lead ions. The focal spot diameter is usually limited by the laser beam divergence (and wave length) and is typically about 100 µm.

The laser pulse of a specific shape (Figure 10.3) is provided by using a nitrogen pour mixture (CO_2:N_2:He=4:1:7). The first peak of the distribution in Figure 10.3 is responsible for effective plasma heating and production of highly ionized ions.

The extracted current level for heavy ions with $Z \geq 10+$ ranges typically from 0.5 to 10 mA per 5 + 10 µs pulse.

10.2.5 Time structure of the Ion Beam and Pulse Extension Methods

The pulse duration is determined by the drift space between the target and the extraction plane (typically 70 to 100 cm), and by the initial energy spread generated during the expansion of the order of one keV per charge). The relative momentum spread in the extracted beam can be reduced by increasing the extraction voltage while the absolute momentum spread can be compressed by modulating the extraction voltage during the pulse formation. The ion pulse duration can be adjusted by varying the distance between the target and the extraction system, although limitations on this distance exist: The minimum distance is limited by the maximum plasma density that one can tolerate in the extraction gap without arcing. The maximum

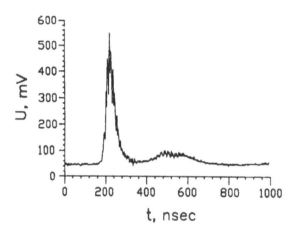

FIGURE 10.3
Laser pulse shape optimized for high charge state production.

distance is limited by the minimum current level required, since the extracted current density decreases with the square of distance (following the solid angle from the extraction hole).

A typical signal of the extracted ion beam is shown in Figure 10.4 for Ta ions. According to the time of flight measurements the first peak contains the mixture of Ta^{11+}, Ta^{10+}, Ta^{9+}, and Ta^{8+} ions. The low charged ions like Ta^{1+}, Ta^{2+}, and Ta^{3+} are mainly represented in the tail of the current signal. A typical current signal for a Ta^{10+} ion beam after charge separation and acceleration in a RFQ accelerator is shown in Figure 10.5. The pulse length generally decreases by increasing of the charge state from about 20 μs for Ta^{1+} to 5 μs for Ta^{10+}.

The pulse duration can be increased when the plasma expands to the extraction plane along the axis of a solenoidal magnetic field. A magnetic field of 0.75 T on the axis of solenoid is reported[14] to increase the total pulse length of a carbon ion beam up to 10 times.

10.2.6 Influence of Magnetic Fields on the Laser Ion Source Plasma

Investigations at the U-200 cyclotron at HINR Dubna and at Moscow Engenering Physics Institut have shown some interesting aspects of the influence of longitudinal and transversal magnetic fields on the laser ion source plasma.[8,15]

A longitudinal magnetic field mainly confines the plasma first narrowing the expansion angle of the plasma plum and finally with increasing of the magnetic field forming a cylindrical plasma along the field axis. The above mentioned increase of pulse length is a result of particle rotations and collective effects in this cylinder. There was no decrease of charge states found at power densities $P \leq 10^{10}$ W/cm^2 and field length of about 10 cm.

A transversal magnetic field has less influence on the plasma expansion angle and can cause instabilities of the expanding plasma for higher power densities and longer drift space of the plasma plum. The plasma expansion is slowed down by a transversal magnetic field and in that way changes the energy spectra of the ions. The transversal magnetic field does not lead to a decrease of the maximum ion charge state of i.e. Si^{8+}, V^{11+}, Zr^{13+}, In^{11+}, Ta^{9+}, and Bi^{8+}.[8] A more detailed description of the influence of transversal and longitudinal magnetic fields on the laser ion source plasma can be found in.[16,17,18]

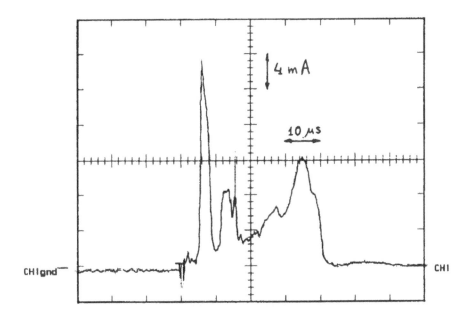

FIGURE 10.4
Typical signal of an extracted ion pulse for Ta.

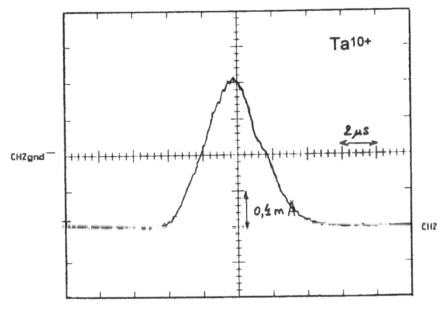

FIGURE 10.5
Current signal of a Ta^{10+} ion beam.

10.2.7 Laser Resonance Ionization for Radioactive Beams

Laser resonance ionization has been applied to ionize radioactive atoms from thick targets at ISOLDE/CERN. Laser light from pulsed dye lasers with high repetition rate (10 kHz) was used for stepwise excitation and ionization.[19] By this method the resonance ionization of elements with different ionization potentials, such as

Yb (E_i = 6.2 eV), Tm (E_i = 5.8 eV) and Sn (E_i = 7.3 eV), was investigated. For Yb, photoionization efficiencies of up to 15% have been measured. An over all efficiency of 0.2% was obtained for Sn at the on-separator ISOLDE-3. A similar laser ion source is under development at GSI(Darmstadt) for γ-spectroscopy of the [102]Sn isotope.

10.2.8 Laser Pumped Polarized Ion Source

In a direct current polarized H[-] ion source[20] a hydrogen plasma is created within an ECR cavity, protons are extracted and pass in an axial magnetic field through a polarized vapor where a fraction of the protons (up to 75%) are neutralized by picking up polarized electrons. As optically pumped medium the vapor of sodium (590 nm), potassium (770 nm), rubidium (795 nm) or cesium (894 nm) are used. The vapor is polarized by laser optical pumping of spectroscopic lines with several watt of laser light directed along the source axis.[21]

10.3 Specific Ion Source Design

10.3.1 The ITEP Laser Ion Source for Multiple Charged Ions

- Special design and construction details of the source

  The cylindrical stainless steel source chamber has a diameter of 350 mm and is 300 mm high. Beside the optical system it contains the target mounted on a target position changing mechanism.

  The beam of a CO_2-laser enters the target chamber through a NaCl (or KCl) window, transparent for λ = 10.6 μm laser radiation. The whole chamber is at a potential of up to 50 kV.

  The optical system consists of a number of gold plated copper mirrors directing the laser light on to a spherical reflector with a focal length F = 170 mm, which focuses the laser light to the surface of the target. The spot size at the focus is about 70 μm.

  The resulting plasma expands hydrodynamically into the vacuum and a part of it is extracted by the extraction system. The distance between the target and the extraction system (700 mm in the case described) has to be chosen to get the necessary particle density in front of the extraction system as well as to provide the required ion pulse duration.

  The diameter of the extraction hole, the shape of the electrodes and the configuration of the matching system are depending on the injection parameters of the following accelerator.

- Ion Source Materials and Vacuum Conditions

  Any solid material can be used as a target in the source. But some of them, having a low temperature melting point, produce a coating on the surface of the focusing optical elements. Therefore the surface of the focusing mirror has to be protected by a thin movable plastic film.

  The target position changing mechanism has to provide 10^6 shots without replacement of the target.

  The source chamber is pumped down to ~10^{-6} Torr after every laser shot with repetition rate 1 Hz.

- Application of the source

  Accelerators, ion implantation, high energy density in matter, heavy ions, high charge states in particular in pulsed operation, SIMS.

- Deliverer ITEP Moscow

- Users ITEP-Moscow, PS division CERN, GSI-Darmstadt, JINR-Dubna

10.3.2 Operating Data

Duty cycle	10^{-3}%
Lifetime	10^6 shots
Laser power density	$10^9 + 10^{13}$ W/cm^2
Wave length of laser radiation:	
usual	10.6 μm
possible	1.06 μm
Laser beam diameter	≥50 mm
Time structure	
total pulse length	≥1 μs
first peak 30% energy	≥100 ns
Power consumption	500 W
Gas consumption	
total	10 st.1/min
Ion Current	0.1 + 20 mA
Ions from solid material	any solid material possible
Charge state distribution	from 1+ to 28+ detected[12]
Extraction area	5 + 30 mm diameter*
Extraction voltage	10 + 50 kV
Emittance	50 + 100 mm·mrad
Pulse shape	gaussian
Material efficiency	up to 50%

* More possible for large ion beams

REFERENCES

1. N. J. Paecock and R. S. Pease. *Br. J. Appl. Phys.* (J. Phys. D), Ser. 2, V. 2, p. 1705, (1969).
2. Yu. A. Byckovsky et. al. Sov. Patent No. 324938, 1969.
3. Yu. A. Byckovsky et. al. *Sov. Phys. JETP*, 33, p. 706, (1971).
4. G. F. Tonon, *IEEE Trans. Nucl. Sci.* 19, 172, (1972).
5. O. B. Anan'in et. al. *Sov. Phys. JETP-Letters*, 16, No. 10, p. 543, (1972).
6. O. B. Anan'in et. al. *Sov. Phys. JETP-Letters*, 17, No. 69, 460, (1973).
7. O. B. Anan'in et. al. *Sov. J. Quantum Electronic*, 7, 873, (1977).
8. V. B. Kutner et. al. *Rev. Sci. Instrum.* 63, 2835, (1992).
9. G. Korshinek and J. Sellmair, *Nucl. Instrum. Meth.* A 268, 473, (1988).
10. L. Z. Barabash et. al. *Sov. J. Atomic Energy*, 64, No. 4, 395, (1988).
11. R. H. Hughes in I. G. Brown Ed. The Physics and Technology of Ion Sources, John Wiley & Sons, New York, 1989, 299.
12. S. V. Laryshev, I. V. Rudskoy. Preprint ITEP, Moscow, No. 2, 1986.
13. B. Yu. Sharkov et. al. *Rev. Sci. Instrum.* 63, 2841, (1992).
14. Yu. B. Byckovsky et. al. Preprint MEPI, Moscow, No. 012-91, (1991).
15. V. B. Kutner et. al. Supplement to Z. *Phys.* D 21, S 165, (1991).
16. O. B. Ananyin et. al. *JETP Lett.* 9, 261, (1983).
17. Yu. A. Byckovsky et. al. *JTPhys.* 58, 1291, (1988).
18. Yu. A. Byckovsky et. al. *Plasma Physics* 13, 1240, (1987).
19. F. Schreer et. al. *Rev. Sci. Instrum.* 63, 283, (1992).
20. C. D. P. Levy et. al. *Rev. Sci. Instrum.* 63, 2625, (1992).
21. L. Buchmann et. al. *Nucl. Instrum. Meth.* A 306, 413, (1991).

CHARACTERIZATION OF ION SOURCES

R. Becker

11 EBIS/EBIT: ELECTRON BEAM ION SOURCE/TRAP[1-10,16,42,43]

Advantages: Highest charge states, excellent beam quality, variable pulse length from µs to dc, no lifetime limitations, UHV-compatible.
Disadvantages: Limited amount of charges per pulse, complex and expensive device.

11.1 Ion Source Principle

11.1.1 Short Description of the Working Principle of EBIS and EBIT (Figure 11.1)

A dense electron beam (length typically in meters in EBIS and centimeters in EBIT), focused by a strong magnetic field of a solenoid, forms by its space charge a radial well for ions, which is closed axially by potentials applied to cylindrical electrodes surrounding the beam. Atoms, once ionized by electron impact inside the electron beam, are trapped and undergo stepwise ionization until the stripping limit of the electron energy is reached or the ions are extracted axially by a change of the axial potential distribution. Typical ionization times in an EBIS are from several milliseconds to several hundred seconds, depending on current density and wanted ion, while trapping in EBIT can last for hours. Extraction times usually vary from 20 to 50 µs, but may be as short as 4 µs,[11] as long as the ionization time (50% duty cycle)[12] or even dc.[13] The time for compensation of the electron beam by trapping ions from the residual gas should be comparable to the ionization time. This requires UHV (10^{-10} mbar for 1 s) for the interior of the trap, which usually is made cryogenic along with the magnet. For high-charge states, ion-by-ion cooling[14] is important to remove the heat from small-angle elastic Coulomb scattering by the ionizing electrons.[15] This is called evaporative cooling in EBIT[16] and known as the gas mixing effect in ECR sources.[17]

11.1.2 Ion Source History

The EBIS was invented by Donets in 1965.[18] In 1969 he reported the production of Au[19+],[19] using a baked-out vacuum tube inside of a normal conducting solenoid. With his co-workers he then developed cryogenic versions KRYON-I to -III, where the electrodes inside a superconducting solenoid were kept at 4.2 K. KRYON-I injected with great success bare nuclei of C, N, and O into the Dubna Synchrophasotron

0-8493-2502-1/95/$0.00+$.50
© 1995 by CRC Press Inc.

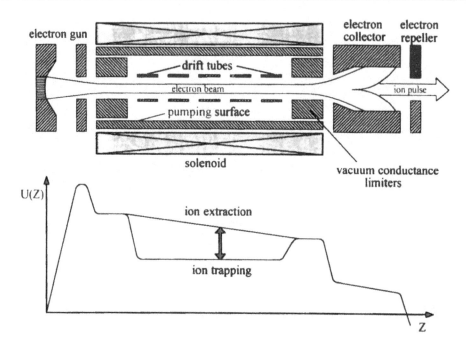

FIGURE 11.1
Scheme of EBIS/EBIT and corresponding axial potential functions.

for 2 years,[20] while KRYON-II was the first EBIS to produce bare ions of Ne, Ar, Kr, and even Xe.[21] Arianer started an EBIS development in 1969 in Orsay.[22] Becker in 1970 in Frankfurt (leading in 1975 to TOFEBIS with dc extraction by Kleinod[23] and to a cryogenic version in 1980[24]), and Hamm in 1973 at Texas A&M.[25] Faure in Saclay obtained the cryogenic CRYEBIS-I from Arianer[26] in 1981 as an injector for the SATURNE synchrotron; later he developed his own EBIS called DIONE,[27] replacing CRYEBIS-I. Arianer built CRYEBIS-II, which became CRYSIS,[28] when moved to Stockholm as a successful injector of CRYRING. Brown started an EBIS development in Berkeley,[29] which was stopped before maturing. Later Schmieder continued the work at SNLL, Livermore and made this EBIS work reasonably well,[30] but his cryogenic SUPEREBIS[31] was not finished and was recently taken over by Hershcovitch at BNL. In the meantime, people became interested in the EBIS as a tool for atomic physics. A Japanese group around Kaneko created the naked ion collision experiment (NICE) in 1901,[32] Kostroun[33] started warm and cold EBIS development at Cornell University, Abdulmanov[34] designed warm devices IMI-I and IMI-II, Stöckli created a cryogenic EBIS[35] dedicated to atomic physics at Manhattan, Kansas, and Okuno from the NICE group completed MINI-EBIS[36] in Tokyo and provided six copies of it to various atomic physics groups. Also in Japan a "small" SMEBIS with permanent magnets was born.[37] The latest push to EBIS development came from accelerators: Faure designed RHEA[38] for a 5-A, 10-kV beam with a warm solenoid and for the future heavy ion colliders RHIC at BNL and LHC at CERN; high-current, medium current density EBIS devices[39] have been proposed for typical charge-to-mass ratio of 0.25 only, but with 10 to 100 times higher ion yields as obtained so far.

The story of EBIT starts with experience from the Berkeley EBIS. Litvin, Vella, and Sessler[40] had concluded that a long electron beam tends to instabilities, and as a way out, a short electron beam ion trap was proposed. One having a Helmholtz coil to allow for radial access for spectroscopy was built and operated very successfully.[41] In the opinion of the author, the reasoning for EBIT was wrong, as demonstrated later by Schmieder,[30] who made the Berkeley EBIS work after some reconstruction. Nevertheless, the EBIT is an interesting variant of EBIS and especially useful for spectroscopic research on highly charged heavy ions. Recently, fully stripped uranium was obtained from SUPEREBIT[42] at LLNL with an electron energy of 198 keV.[43] New EBIT projects have been started in Oxford by Silver,[44] at NBS by Gillaspy,[45] in Troitzk by Antsiferov and Movshev,[46] and in Dubna by Ovsyannikov.[47] It has been shown that the transition between EBIS and EBIT modes is continuous and only governed by the relation of ionization time to compensation time.[48] The distinction between EBIS and EBIT as different devices is therefore arbitrary; to conclude, S stands for source, T stands for trap; a more precise name for a device that may serve both applications could be EBIST.

11.2 Specific Ion Source Design

11.2.1 Special Design and Construction Details of the Source

The basic layout of an EBIS/T essentially looks like a traveling-wave tube amplifier (TWT), waiting for modulation (see Figure 11.2). Plane drift tube ends are excellent coupling gaps to either modulate the electron beam, or decouple amplified rf from the beam; the transition of the connecting wires to the support structure of all feed wires will show an impedance step, where rf may be partially reflected, and the parallel duct of the wires along the EBIS structure provides numerous crosstalk possibilities. Either generation of rf will be observed, if a specific — construction and mode dependent — current is exceeded, or special means have to be provided so that rf feedback will be supressed. Figure 11.2b shows the most common solution by capacitive shunts, which still preserves vacuum conductance. In Figure 11.2c a solution is shown that avoids coherent coupling for wavelengths shorter than the transition length and which has excellent vacuum conductance. Most important, however, is a careful alignment, making sure that the beam is well centered. This also gives minimum electron loss to the electrodes inside the trap and hence minimizes desorption and spoiling of the base vacuum. In order to provide a good centering of a 0.1-mm-diameter electron beam within a 3- to 10-mm drift tube diameter over a length of 1 m (typical for EBIS, not EBIT), straight flux lines of the focusing solenoid are mandatory. The required flux straightness of less than 10^{-4} transverse to longitudinal field imposes serious considerations on the winding technique of solenoids. Normal conducting water-cooled solenoids have too much pitch (or will consume too much power if the inner radius is increased) and superconducting solenoids must be wound with great care to limit the transverse components of magnetic field.

The generation of the electron beam has to follow the design and engineering rules of TWTs, e.g., gun design,[49] magnetic compression,[50] and immersed[51] or Brillouin[52] flow focusing and collector design with emphasis on secondary electron suppression, power distribution on the dissipating surface, and proper ion extraction.[53] Flat cathodes may be located in the uniform or slightly increasing magnetic field to use immersed flow focusing, while the highest current densities are obtained with completely shielded guns with spherical cathodes and magnetic compression into Brillouin flow.[50,54] The disadvantage of highly compressed Brillouin flow is the Gaussian shape of the density profiles, which comes from thermal effects,[55-58] nonuniform electron emission, and aberrations in the gun. While the radial variation of the electron current density is less important for ion production, it should be avoided or carefully considered for atomic physics experiments, where the electron beam is used as a target of known density.

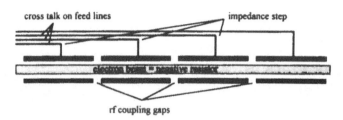

a) simple drift tubes with feed wires seen as rf-circuit

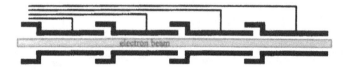

b) capacitive shunts on drift tube ends to short out rf (Donets and Arianer)

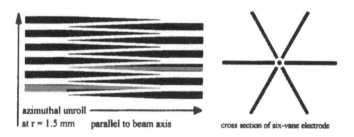

c) gap of tapered fingers to avoid coherent rf coupling [13]

FIGURE 11.2
Electrode structures used in EBIS devices.

11.2.2 Gas Feed and Use of Nonvolatile Materials

The working material may be introduced into the electron beam in several ways: Donets used evaporation from a heated spiral to make ions of gold;[19] then he developed the scheme of a seed trap for gaseous material,[20] the Orsay group evaporated metals nearby to the electron beam by a focused CO_2-laser;[59] Faure[60] injected ions from an external ion source, now also used on EBIT;[42] and trapping from a low-pressure gas mixture has been successful in EBIT[16] as well as in EBIS.[48,61] Since direct feed of nonvolatile working materials is not feasible for an EBIS, the most convenient way is the use of an external ion source. All available techniques can be applied to produce singly charged ions from elements throughout the periodic table. At the same time, the EBIS can be built and operated in a simpler manner without a seed trap and the ion gymnastics required for it. This also removes the need for differential pumping over the short distance between the seed trap and the main ionization trap.

11.2.3 Ion Source Material and Vacuum Conditions

Since UHV is required, the choice of materials is restricted and must follow specific requirements: the gun electrodes should not poison the cathode and their outgassing rate must be low, so only Mo or OFHC-copper should be used. Along with impregnated cathodes, LaB_6 has been found useful[21,22,48] because of its high emission current density. The structure elements of the trap should match the cryogenic considerations, e.g., differential contraction, outgassing at cryogenic temperatures, and electric properties. At 4.2 K, PTFE or saphire can be used as insulating materials with some heat-conduction capability, while nonmagnetic stainless steel or copper can be used for electrodes. The collector needs brazing for cooling channels; therefore, OFHC-copper is best if vacuum brazing is applied; otherwise, degassing will last for a long time. The vacuum conductance limiters on each end of the ionization region are much more effective when kept at 4.2 K. In cryogenic sources a separation of vacua between the cryostat and the trap by means of axial bellows has found wide application in order to prevent He from leaks of the cryostat to enter the trap.

11.2.4 Fields of Application for EBIS and EBIT

While EBIT (without ion extraction) finds its application in atomic physics studies, EBIS is a successful injector of fully stripped light ions into synchrotrons (Dubna, Saclay, Stockholm) but also may be useful for high-charge states of heavy ions (CERN,[62] BNL,[63]). This is due to the fact that the extracted ion pulse can be reasonably short, giving high intensity at low-duty cycle even for single-turn injection. EBIS-like setups also find interest as ion traps[64] or merged beams devices[65,66] for atomic physics studies.

11.3 Theoretical Modeling of EBIS/EBIT — Performance

From the very beginning, EBIS has seduced scientists by its appeal of being theoretically understandable. The models at first were naive — as we know now — but there has always been the attempt to explain all experimental results, including those that were unexpected. Important for the understanding of EBIS/EBIT physics is the space charge compensation by thermal ions, the heating of trapped ions by electrons, the cooling by ion-ion collisions, and the combined influence on the charge evolution under stepwise ionization.

11.3.1 Focusing of Electron Beams

The most efficient magnetic focusing (in terms of focused current density at a given magnetic field) is Brillouin flow,[52] where a uniform beam is rotating as a whole with half of the cyclotron frequency in order to balance the space charge and the centrifugal force by the Lorentz force. The current density of Brillouin flow (with zero field at the cathode) is given by

$$j_{Br} = \varepsilon_0 \frac{e}{2m} B^2 \sqrt{\frac{2e\,U_0}{m}} \tag{1}$$

in MKS units, where B is the magnetic field, U_0 the potential difference on axis between cathode and focused beam, and ε_0 is the dielectric constant, while e and m denote the charge and mass of an electron. Figure 11.3 shows the dependencies of Equation 1 on B and U_0. At 6 T and 10^5 V a current density of 5.2×10^5 A/cm^2 is calculated from Equation 1. This high current density, however, is not accessible because cathodes can deliver only 10 to 100 A/cm^2, needing an area compression in the rising magnetic field of more than a factor of 1000. By this, the transverse temperature of the beam will increase accordingly, resulting in a beam being completely

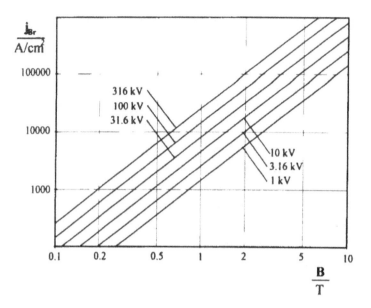

FIGURE 11.3
Dependence of the current density in Brillouin flow without cathode flux and thermal degradation on the magnetic field B and for interesting values of U_0.

dominated by transverse thermal motion. Herrmann[55] has given a formula to calculate the required increase of the magnetic field to overcome the tranverse thermal behavior. Fögen[67] has shown that the optical thermal treatment of Herrmann predicts the same increase of magnetic field as the statistical thermal theory of Pierce and Walker[57] in order to obtain a certain current density despite thermal effects. Amboss[58] used the Herrmann theory to calculate the radius of a thermal beam that contains 80% of the current. His result can be put in a simple form to show the degradation of current density by cathode flux and transverse temperature.

Assuming that the following parameters are known:

I   electron current (A)
U_0 axial potential difference between cathode and focused beam (V)
B_c magnetic field at the cathode (edge) (T)
B   magnetic field in the focused beam (T)
T_r temperature of the cathode (K)
r_c radius of the cathode (m)

then the quantities given below can be calculated.
Potential depression

$$\Delta U = \frac{I}{4\pi\varepsilon_0 \sqrt{\frac{2e}{m} U_0}} \ (V) \tag{2}$$

Brillouin radius

$$r_0 = \frac{8\Delta U}{\sqrt{\frac{e}{m} B^2}} \ (m) \tag{3}$$

Nonthermal flux compression

$$K_0 = \frac{B_c r_c^2}{B r_0^2} \qquad (4)$$

normalized beam temperature

$$\tau = \frac{r_c^2}{r_0^2} \frac{kT_c}{e\Delta U} \qquad (5)$$

The reduction of current density by cathode flux and thermal influence is obtained from the following equation:

$$j_{eff} = j_{Br} \frac{2(1-K_0^2)}{1 + \sqrt{1 + (1-K_0^2)^2 \tau}} \qquad (6)$$

The dependencies on the parameters K_0 and τ are displayed in Figure 11.4. It becomes obvious that due to the limited cathode emission current densities, a high compression will result in thermal degradation of the beam, and the reduction of current density as compared to Equation 1 is considerable. Even further degradation by aberrations in the gun and emittance growth must be considered, however. Fortunately, the influence of cathode flux becomes the less important, the higher the thermal influence.

To continue the example of numbers given before, an impregnated cathode with 5-A/cm^2 emission capability will need a compression of 10^5, hence, at $\Delta U = 10V$ the normalized beam temperature τ will reach 10^3 and the focused current density will be 6% of the Brillouin density or about 3×10^4 A/cm^2, which is still a remarkably large value. A higher perveance will increase ΔU, reduce τ, and increase the current density, as well as increase the emission current density from the cathode.

The advantage of immersed flow is a more uniform current density distribution, while thermally degraded Brillouin flow has Gaussian beam profiles, which may also cause radial loss of electrons. Brillouin flow is supposed to react on space charge compensation by additional beam compression, while immersed flow will not. This additional compression of Brillouin-focused beams by space charge compensation was demonstrated only in the source CRYEBIS-I by Arianer[68] and later by Faure,[69] who deduced current densities in the 10^4 to 10^5 A/cm^2 range from observed ionization times — and even more surprising — these were increasing with the atomic number of the ions.

Besides focusing with an uniform magnetic field, periodic permanent magnet focusing (PPM) has also been proposed[70-72] and realized.[37] The use of SmCo and NdFe magnets gives effective focusing fields in the range of 0.5 to 1 T on axis, leading to interesting current densities (see Figure 11.3). The magnetic mirrors created by the PPM structure are expected to form semipermeable axial mirrors for the ions, being useful for dc extraction of high charge states,[73] which has not yet been proved, however.

A new focusing scheme for the EBIS/T has no magnetic field at all. A short (few millimeters to centimeters) region of high current density is obtained by inertial focusing,[74] e.g., the electrons from an electron gun — carefully designed and built —

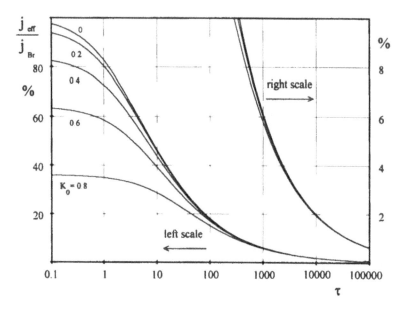

FIGURE 11.4
Reduction of Brillouin current density by flux compression K_0 and thermal influence τ.

are projected to a dense crossover, forming the ion trap. This scheme is interesting for low-perveance, high-energy beams, hence EBIT applications, where the absent magnetic field also can be an advantage for spectroscopic investigations of the trapped ions.

11.3.2 Matching of the Electron Beam to the Rising Magnetic Field

The electron gun may have three distinguished differing positions with respect to the rising magnetic field as shown in Figure 11.5. If the magnetic flux line through the cathode edge remains identical to the beam boundary, the focusing scheme is called immersed flow, with two extremes: fully immersed flow if the cathode is at the highest field, and partially immersed flow if the cathode is sitting in the field buildup and the beam compression develops in proportion to the flux compression. In the third clearly different situation, flux is penetrating during beam compression from outside into the beam. This results in Brillouin flow,[52] which gives the highest beam compression (see Section 11.3.1). In general, the cathode curvature increases from fully immersed flow to Brillouin flow. A fully immersed gun is designed according to the textbook example of Pierce[75] for a planar cathode; a partially immersed gun should have such a cathode curvature that the fluxlines become perpendicular on the cathode surface. Close to the axis and far away from iron, the magnetic field may be well described by series of axial derivatives

$$B_z(z,r) = B_z(z,0) - \frac{r^2}{4}\frac{\partial^2 B_z(z,0)}{\partial z^2} + \dots$$

$$B_r(z,r) = -\frac{r}{2}\frac{\partial B_z(z,0)}{\partial z} + \frac{r^3}{16}\frac{\partial B_z(z,0)}{\partial z} - \dots \tag{7}$$

which allow one to calculate the ratio of B_r to the total field at the cathode edge (z_c, r_c), also given by the ratio of r_c to the spherical cathode radius r_s. From this, r_s is obtained:

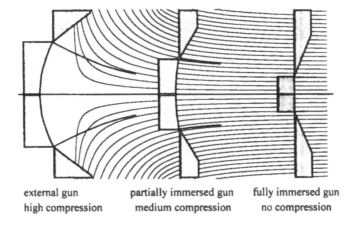

external gun   partially immersed gun   fully immersed gun
high compression   medium compression   no compression

FIGURE 11.5
Locations of cathode and Wehnelt electrode with respect to an increasing magnetic field, represented by converging flux lines. The boundaries of the electron beam are indicated by thick lines.

$$r_s = \sqrt{r_c^2 + \left(\frac{2 \times B_z(z_c, 0)}{\frac{\partial B_z(z_c, 0)}{\partial z}} \right)^2} \qquad (8)$$

Equation 8 shows that independent of the actual magnetic field buildup, an unique spherical radius (independent of r_c) can be found only if the magnitude of the radial field is negligible compared to the axial field. In this case, the spherical cathode radius may become sufficiently large that a flat cathode will be reasonable as well. Otherwise, the spherical radius according to Equation 8 should be adjusted at the cathode edge, where most of the beam current is emitted.

In the simplest magnetic compression system for immersed flow, the flux compression between the location of the gun and the center of a solenoid without iron shielding is used to compress the electron beam. This is shown in Figure 11.6 by the inverse of the axial field decay of solenoids with ratios r_w/b (radius to half-length) from 0.01 to 0.1 The indicated flux compression C is obtained if the cathode is positioned at z_c measured in units of the solenoid's half-length b.

The design procedure for a partially immersed gun then will be to first define the cylindrical cathode radius r_c with respect to the emission capability of the cathode material and the aimed electron current. The axial position of the cathode will define the magnetic compression, ensuring that the Brillouin limit according to Equation 1 will exceed the local current density by at least a factor of five. With the axial position being found in this way, the magnetic field and its gradient will be known and the spherical cathode radius r_s can be calculated from Equation 8. Beyond the anode bore, where the electrostatic focusing is lost, the magnetic field must be strong enough for immersed flow (more than twice the field required for Brillouin flow[51]). Therefore, the calculated current density $j_{Br,a}$ at the anode should be at least fivefold higher than the actual current density $j_{Im,a}$ to be focused in immersed flow:

$$j_{Im,a} \leq \frac{j_{Br,a}}{5} \qquad (9)$$

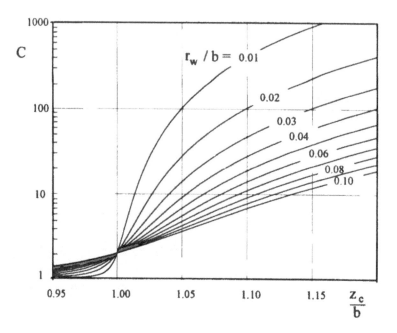

FIGURE 11.6
Flux compression C in thin solenoids with radius r_w over half-length b between 0.01 and 0.1 from
(normalized) axial position z_c/b to the solenoid center.

If the potential difference between cathode and anode is U_a, while the focused beam
in the center of the solenoid has a potential difference of U_b to the cathode, the
Brillouin current density at the anode relates to the Brillouin current density in the
center of the solenoid according to Equation 1 by

$$\frac{j_{Br,a}}{j_{Br,max}} = \frac{B_a^2}{B_{max}^2}\sqrt{\frac{U_a}{U_b}} \tag{10}$$

$$\frac{B_a^2}{B_{max}^2}\sqrt{\frac{U_a}{U_b}} = \frac{j_{Im,a}^2}{j_{Im,max}^2}\sqrt{\frac{U_a}{U_b}} \tag{11}$$

The ratio of flux densities, however, may be expressed by the ratio of immersed flow
current densities, and combining Equations 9, 10, and 11 by eliminating $j_{Br,a}$ finally
gives a simple relation for estimating the maximum obtainable current density in
immersed flow:

$$j_{Im,max} \leq \sqrt{\frac{j_{Br,max} \times j_{Im,a}}{5}}\sqrt{\frac{U_a}{U_b}} \tag{12}$$

In order to obtain a high current density in immersed flow, it is therefore advisable
to use a much higher voltage in the gun than for the finally used beam; however, it
needs $U_a = 16U_b$ in order to win a factor of two — so this seems practical only for the
low electron energies needed for the production of light, fully stripped ions. For fully
immersed flow, $j_{Im,a}$ is identical to the cathode current density, while for partially
immersed flow, all the possible electrostatic beam compression can be used to enhance the cathode current density. Fully immersed flow, using LaB$_6$ cathodes at a

demanding value of 100 A/cm², will result for a beam of 10^5 eV within a 6-T solenoid in a maximum current density of about 3000 A/cm², while partially immersed flow, using an osmium-coated, impregnated cathode of 10-A/cm² emission current density together with a demanding electrostatic compression of 10^3, will be limited at a focused current density of about 30,000 A/cm². This is almost identical to the value possible in Brillouin flow, if thermal degradation is taken into account. In experiments, this limit has been reached in the case of partially immersed flow by Donets, who obtained 650 A/cm² at 18 keV in a 3-T solenoid.[21]

The general Brillouin flow design needs numerical simulations and no recepies are established; only the special case of almost vanishing cathode flux has been described by simple engineering rules.[54] The classical injection into Brillouin flow requires a sudden rise of the magnetic field to maximum value in the position of the electrostatic crossover. In this case, all beam compression has occurred by the electric fields in the gun. This scheme breaks down for dense beams, where the magnetic field buildup cannot be made short enough. Instead, the rising magnetic field gradually starts to focus the beam, overcoming the increasing repulsion of a compressing beam by an appropriate increase of the focusing force. The converging beam therefore has to be matched to the axial gradient of the magnetic field. By numerical analysis it was found that the required gradient of the magnetic field becomes related to gun data such as perveance P ($A/V^{3/2}$), crossover radius r_x (m), and beam voltage U if the additional magnetic beam compression (as compared to the electrostatic beam crossover without magnetic field) exceeds a factor of about ten; namely,

$$\frac{dB_z}{dz} \cong 0.079 \frac{P\sqrt{U}}{r_x^2(1-K_0^2)} (T\ m) \tag{13}$$

where K_0 is defined by Equation 4 for the finally focused beam. The proper magnetic gradient close to the gun and far away from the solenoid can be obtained conveniently by an iron disk of high magnetic flux with a bore radius r of[76]

$$r = \frac{2\ B_{surf}}{\pi \dfrac{dB_z}{dz}} \tag{14}$$

where B_{surf} is the (almost constant) magnetic field on the disk surface facing the solenoid, which depends on the size and strength of the solenoid and the distance between the disk and the solenoid. In addition to the correct matching of the gradient of B_z to gun data, the position of the gun in relation to the rising magnetic field should be such that at the axial position, where the extrapolated highest magnetic gradient hits the axis and the magnetic field buildup seems to start, the beam radius r_{start} (without magnetic field) amounts to

$$r_{start} = 1.795\left(1 - \frac{K_0^2}{2}\right)r_x \tag{15}$$

Extremely important is the local shielding of the spherical cathode to ensure flux lines that are perpendicular on the cathode surface. This may be achieved, in general, by surrounding the cathode of spherical radius r_s with an iron cylinder of appropriate radius.[77] The magnetic field decays exponentially inside the iron cylinder in the axial

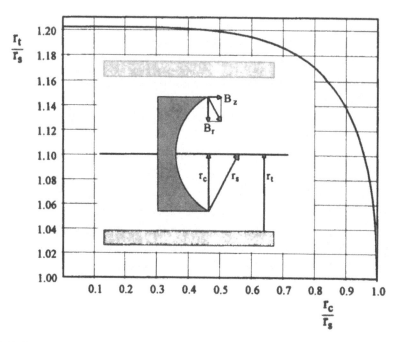

FIGURE 11.7

Use of an iron cylinder with radius r_t to force fluxlines to be perpendicular on the edge of a cathode with sperical radius r_s and cylindrical radius r_c.

direction, giving Bessel functions for the radial dependencies of the magnetic field components. Requiring that the flux line through the cathode edge be perpendicular on the spherical emitting surface then relates the inner radius r_t of the iron tube to the cylindrical radius r_c of the cathode as shown in Figure 11.7. For cathodes with r_c/r_s < 0.5 the radius of the iron tube is about 20% larger than the spherical radius of the cathode. Although this design only applies to the cathode edge exactly, the differences at smaller radius are tolerable.

The outlined design rules have been applied to a magnetic beam compression system, using the Frost gun,[78] but with some changes to improve its laminarity. The electrostatic behavior of the gun gave a (calculated) crossover radius of 0.5 mm at a perveance of 2.2×10^{-6} A/V$^{3/2}$. Because the axis potential was depressed by space charge to about 8600 V, a magnetic gradient of 64.5 T/m was calculated from Equation 13. To realize this, a hole with radius of 5 mm is appropriate with the Frankfurt EBIS solenoid at 3-T central field, giving a surface field of about 0.6 T on the iron flange. The resulting design for the coils and iron parts is shown in Figure 11.8, reproducing a calculation with INTMAG(C).[79] The results of INTMAG(C) can then be used to calculate electron trajectories, including the simulation of the space charge in the beam by the program EGN2e(C), which is the PC version of the well-known electron trajectory optics program EGUN(C) by Herrmannsfeldt.[80] EGN2e(C) can use the readily formatted output of INTMAG(C) in two ways: either the axial field can be used and expanded in the radial direction by Equations 7 up to 6th order, or a map of B_r and B_z can be read in on all mesh points. For the use of the radial expansion it is important that INTMAG(C) integrates magnetic fields and it is therefore free of residual errors in contrast to programs that use the methods of finite differences (FDM) or finite elements (FEM). The results for both methods of using INTMAG(C) output in EGN2e(C) are almost identical, as shown in Figure 11.9, using a small part of the axial magnetic fields of Figure 11.8 ($-80 < Z < -50/0 < R < 10$).

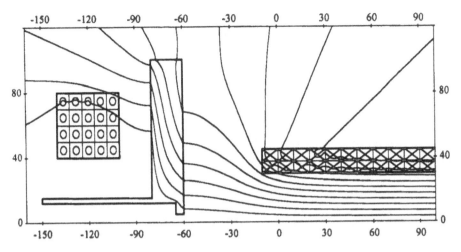

FIGURE 11.8
Numerical simulation of a magnetic beam compression system with INTMAG(C), with a properly shielded cathode to launch a Brillouin focused electron beam. At the right is the end of the superconducting solenoid; at the left is the opposing normal conducting bucking coil. In between is the iron disk with a hole of appropriate radius to form the axial gradient and at the lower left is the iron cylinder that assures perpendicular fluxlines on the spherical cathode surface. The dashed box shows the frame for the electrostatic simulation in Figure 11.9.

The calculated perveance does not change by the magnetic field, which is direct proof of the correct shaping of flux lines in the cathode region by means of the iron tube. At the right side in Figure 11.9 the magnetic field varies only slowly. The formerly matched electron beam will then adjust its compression adiabatically and maintain (local) Brillouin flow conditions. This is true as long as the thermal behavior of the beam by high compression does not use up part of the additional magnetic field, as discussed in Section 11.3.1.

11.3.3 Space Charge Compensation of the Electron Beam

The radial Poisson equation

$$\Delta U(r) = \frac{\partial^2 U}{\partial r^2} + \frac{1}{r}\frac{\partial U}{\partial r} = -\frac{\rho_b + \rho_c}{\varepsilon_0} \tag{16}$$

where U stands for the potential, ε_0 denotes the dielectric constant, and ρ_b and ρ_c give the space charge of beam and compensating particles and are defined by:

$$\rho_b = \frac{I_b}{\pi r_b^2 \sqrt{\frac{2e}{m}U_0}} \tag{17}$$

and

$$\rho_c = -f_c \rho_b \exp\left\{-\frac{e(U - U_0)}{kT_c}\right\} \tag{18}$$

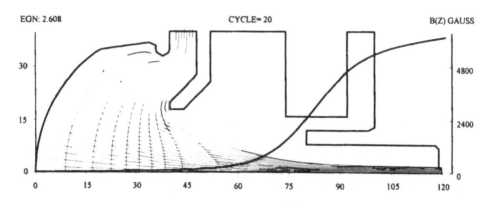

FIGURE 11.9

Numerical simulation of the electron beam compression in the properly shaped magnetic field with EGN2e(C). Shown are electrodes, equipotential lines, trajectories, and the axial function of B_z with a scale at the right side.

where I_b is the beam current, r_b is the beam radius, and U_0 is the potential difference on the axis to the cathode and where f_c is the degree of central compensation, has been solved[81] for various ratios of

$$\mu_b = \frac{e\{U(r_b) = U(r_0)\}}{kT_c} \tag{19}$$

which measures the potential difference in the uniform (uncompensated) beam in units of the thermal temperature of the compensating particles. For any given temperature T_c, there is one distinguished solution where the tube surrounding the beam contains (per unit length) the same amount of positive and negative charges. This is considered as full compensation, although the degree of central compensation f_c has a characteristic value smaller than unity. Another important quantity for the understanding of space charge compensation is the potential difference between the axis and the surrounding tube, also related to the thermal energy of the ions:

$$\mu_t = \frac{e\{U(r_t) - U(r_0)\}}{kT_c} \tag{20}$$

The dependence of f_c and μ_t on μ_b is shown in Figure 11.10 for a ratio of beam to tube radius of 0.1. It may be concluded that the lower the temperature of the compensating particles, the higher the central compensation. A complete compensation is not possible at finite temperature as long as the ion production rate is low enough. The dynamics of compensation provides cooling of the compensating ions, because newly created ions are always colder than existing ones. The remaining potential difference between the axis and the tube, measured in units of kT_c, shows an important result: for colder particles, the repelling potential wall becomes higher and steeper (cf. Figure 11.11). This confines the compensating particles more and more to the beam, the higher the degree of compensation becomes. This is a surprising result at first glance, but very useful for EBIS operation, because compensation does not lead to a loss of containment as could be expected from a more naive point of view. Instead the hotter ions will be lost at the expense of the colder ones, which has been called evaporative cooling.[16]

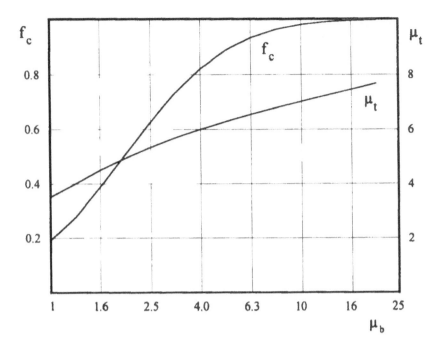

FIGURE 11.10
Dependence of central compensation f_c and normalized well depth μ_t on ion temperature of compensating ions for full compensation.

11.3.4 Stepwise Ionization

By collisions from energetic electrons, the outermost electrons of an atom or ion are ionized with highest probability, although inner-shell ionization may contribute significantly in obtaining high-charge states. Data for cross sections are found in Reference 82. The general dependence on electron and binding energies E_e and E_i is given by a Bethe[83] type cross section formula, reflecting Coulomb ionization in Born approximation. On this basis, Lotz[84] has proposed a semiempirical formula for low-charged ions

$$\sigma_{i \to i+1} = 4.5 \times 10^{-14} \sum_{nl} \frac{\ln\{E_e/E_{i,nl}\}}{E_e \times E_{i,nl}} \; [\text{cm}^2] \tag{21}$$

showing the sudden rise of the cross section for $E_i < E_e$, the $1/E_e$ dependence at high energies, and the total decrease proportional to $1/E_i^2$ for increasing ionization energies. The summation extends over all removable electrons in all orbitals nl, which gives a considerable contribution in the outer shell of heavy ions, where many electrons have almost the same binding energy E_i.

In a comprehensive study, Itikawa and Kato[85] compared various cross section formulae with experimental data for ions with $Z < 20$ and found best overall agreement for the Lotz formula.

The ionization energies to be used with Formula 21 can be taken with sufficient accuracy from HFS calculations[86-88] of the neutral atom. The increase of the binding energies by ionization is then calculated by a simple (spherical) electrostatic shell model[86] of the ion:

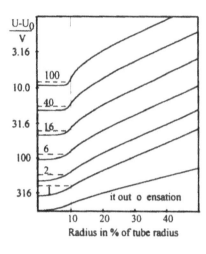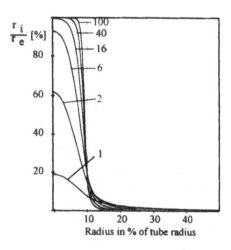

FIGURE 11.11

Radial potential (left) and ion charge (right) distribution in an electron beam of 1 A/10 kV, which is fully compensated by ions of different temperature (μ_b = 1, 2.5, 6, 16, 40, 100 — from bottom to top). The beam-to-tube ratio is 1:10. The dashed lines show potentials, which are kT_i/e above the axis potentials.

$$E_{i,nl}(j) = E_{i,nl}(0) + \left\{ E_{i,(nl)^*}(j) - E_{i,(nl)^*}(0) \right\} \tag{22}$$

where nl stands for the orbital under consideration and (nl)* for the least-bound orbital in the ion of charge state j. $E_{i,(nl)^*}$ (j) then is the ionization energy of this ion. Nonrelativistic scaling of the hydrogen atom gives for the ionization energy of the last electron in an element with Z protons (similar to Mosley's law):

$$E_{i,k}(Z) = 13.6 \times Z^2 \ [eV] \tag{23}$$

Bare argon ions and medium charges of heavy ions can be ionized with 10-keV electrons, helium-like ions of the heaviest elements with 100 keV, and for the production of U^{92+} an electron energy of 150 keV or more is needed. For hydrogen-like ions, Equation 23 can be used with the Lotz formula to estimate the ionization cross section at optimum electron energy

$$\sigma_{z-1 \to z} \approx \frac{10^{-16}}{Z^4} \ \left[cm^2 \right] \tag{24}$$

Ionization times are related to cross sections by

$$\tau = \frac{e}{\sigma j} \tag{25}$$

which gives a similar jτ criterion for the production of highly charged ions as Lawson's criterion for fusion, even of same magnitude for U^{92+}. In "cooking" bare nuclei, most of the time is spent for the last charge state. Therefore, a rough estimate of the required jτ value necessary is obtained by combining Equations 24 and 25 to

TABLE 11.1

Approximate Ionization Energies, Ionization
Cross Sections, and Required $j\tau$ Values
for Bare Ions

Ion	E_i(eV)	σ(cm^2)	$j\tau$(Cb/cm^2)
C^{6+}	490	7.7×10^{-20}	2.1
N^{7+}	666	4.2×10^{-20}	3.8
O^{8+}	870	2.4×10^{-20}	6.5
Ne^{10+}	1360	1×10^{-20}	16
Ar^{18+}	4,400	9.5×10^{-22}	170
Kr^{36+}	17,600	6×10^{-23}	2,700
Xe^{54+}	39,700	1.2×10^{-23}	13,600
Pb^{82+}	91,400	2.2×10^{-24}	72,300
U^{92+}	115,000	1.4×10^{-24}	115,000

$$j\tau \approx \left(\frac{Z}{5}\right)^4 \tag{26}$$

Numerical results from Equations 23 through 26 are given in Table 11.1 for some bare ions throughout the periodic table. Up to Ne^{10+} an electron beam of only a few kiloelectron volts is needed, and with a current density of 100 A/cm^2, these bare ions are obtained in less than 200 ms. To make U^{92+} at least 150 keV is needed and even at 1000 A/cm^2 a containment time of 100 s must be provided.

Besides by ionization, the charge state of an ion can also be changed by charge exchange either with an atom or — less probable — with an ion, by dielectronic and by radiative recombination. While these loss processes certainly will influence the equilibrium distribution in EBIT[89] for long trapping times, much influence has not been seen in EBIS sources, where the trapping time usually is kept smaller than the compensation time under excellent vacuum conditions. Therefore, a simple mathematical model of stepwise ionization must include gain and loss due to ionization to neighboring charge states as well as a thermal loss term, which becomes important when space charge neutralization is reached. Space charge neutralization can then be treated by reducing the residual potential depression until the amount of positive charges equals the negative charge of the beam electrons. While most EBIS sources inject a certain amount of neutrals or singly charged ions, which then become depopulated by ionization to higher charge states, ionization from the working gas may happen all time in EBIT devices. This can be taken into account by a special constant v_n,[90] which describes the neutral influx, but may be set to zero for standard EBIS application. The ionization frequencies $v_{i-1 \to i}$ are the inverse of the ionization times, as defined by Equation 25. This gives for the set of coupled differential equations of charge state abundances with time:

$$\frac{dn_0}{dt} = v_n n_{00} - v_{0 \to 1} n_0$$

$$\frac{dn_i}{dt} = v_{i-1 \to i} n_{i-1} - n_i \left(v_{i \to i+1} + v_{loss} \exp\left\{ -\frac{ieU_{res}}{kT_{ion}} \right\} \right)$$

$$\frac{dn_m}{dt} = v_{m-1 \to m} n_{m-1} - v_{loss} n_m \exp\left\{ -\frac{meU_{res}}{kT_{ion}} \right\}, \tag{27}$$

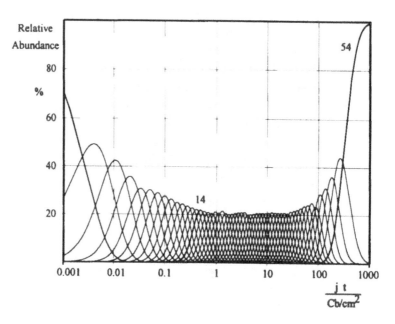

FIGURE 11.12
Relative abundances of charge states by stepwise ionization of Pb at 5-keV electron energy (pulsed Pb injection, no heating, no compensation).

where n_{00} stands for a background gas, while n_0 describes injected neutrals around the electron beam. Without neutral influx ($v_n = 0$) and without thermal loss ($v_{loss} = 0$) the well-known charge state evolution in EBIS, shown in Figure 11.12 for Pb at an energy of 5 keV, is obtained. The time scale applies to a current density of 1 A/cm²; it will be shorter by the actual current density used. Due to the shell structure of atomic binding energies, there is always a considerable increase in binding energy for the last electron in a shell (or subshell).

By using an electron energy just below the ionization energy for the next electron, full abundance may be obtained for this last electron of a shell. Here we chose 5 keV, making use of the relative large change in ionization energy from charge 54 to 55 of 3678 to 5008 eV.[86] Pb^{54+} can therefore be extracted at exclusive abundance in the charge spectrum, because the electron energy is too low to ionize Pb^{54+}. If multistep ionization were taken into account, the charge distribution should broaden and the maximum abundances should decrease, as observed in experiments.[91]

If Pb is fed continuously to the trap, a superposition is seen of stepwise ionization as in Figure 11.12 and of the increase of all charges with time, reflecting the trapping action of the electron beam. The final charge state distribution will depend on the compensation time as shown in Figure 11.13 for $jt = 10$ Cb/cm². By this the residual radial holding voltage U_{res} of the electron beam will be reduced, evaporating off the hot ions and keeping cold ions of such temperature and amount that all ion charges just balance the charge of the electron beam. The charge distribution then becomes time independent, which is the typical EBIT mode of operation. It may also be seen in Figure 11.13 that stepwise ionization is not stopped by the onset of space charge neutralization with thermal ions.

The outlined model of stepwise ionization generally gives satisfactory agreement with measurements; however, there have been found two situations of more complicated processes: Donets[91] has shown that double ionization is important in the outer shell of Ar, and the cross section from Ar^{6+} to Ar^{8+} is even higher than from Ar^{6+} to

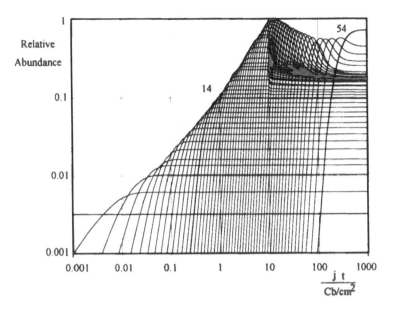

FIGURE 11.13
Stepwise ionization of Pb at 5-keV electron energy (continuous Pb injection, heating, and compensation at $j\tau = 10$ Cb/cm²).

Ar^{7+}. In the measured evolution of charge states, Ar^{7+} therefore never reaches a higher abundance than Ar^{6+}. Dielectronic recombination, which is a resonance process, can depopulate a certain charge state for a specific electron energy, as shown by Stöckli[3] for Ar^{16+} at electron energies between 2 and 4 keV.

11.3.5 Heating of Trapped Ions by Beam Electrons

Electrons performing inelastic ionizing collisions with ions also cause elastic small-angle Coulomb scattering, by which heat is transferred to the ion population in the trap. It has been shown that this heating can be formulated independently of electron current or current density and be mainly attributed to the charge state of a specific ion.[15] This makes the results quite general and also applicable to other ion sources, such as the ECR. The ion temperatures in charge state q by electron heating has been related to a required residual holding voltage to prevent radial escape of the heated ion:

$$U_{res} = \frac{10^{-18}}{qAE_e} \sum_i^{\zeta} \frac{i^2}{\sigma_{i-1 \to i}} \text{ (V)} \tag{28}$$

where A is the atomic mass of the ion, σ the ionization cross section, and E_e the electron energy. U_{res} is shown in Figure 11.14 for representative ion charge states at an electron energy of 6 keV. The ionization of Kr and Xe is stopped at charge states 33 and 44, respectively, due to the choice of the electron energy. However, heating will continue if the highest ion charge states are not extracted. As can be seen from Equation 28, heating becomes less serious for higher electron energies and heavier ions of the same charge state. Other sources of ion heating may be due to time-dependent potentials caused by power supply ripples or charge fluctuations in the electron beam or compensating ion cloud.

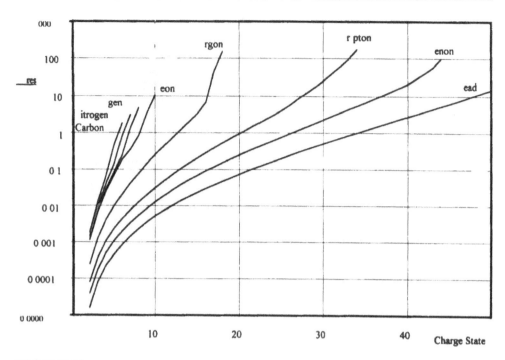

FIGURE 11.14

Radial holding voltages $eqU_{res} = kT_i$ for multiply charged ions heated by electrons of energy 6 keV.

11.3.6 Cooling of Ions by Ion Collisions

In the operation of EBIS sources, the best vacuum conditions have always been the aim. However, the results of EBIT clearly demonstrated that a mixture of different ion species is important to provide evaporative cooling. While the ejection of hot ions occurs in EBIT in the axial direction, in EBIS sources the radial loss after reaching compensation may prevail. Evaporative cooling can start at a low degree of compensation for predominant axial losses, while the radial loss mechanism requires compensation to occur before cooling can start. Let us assume that all ions — different species and different charge states — in the space charge trap of the electron beam interact sufficiently long with each other to provide equipartitioning (there are arguments that strong magnetic fields may break up the collisional coupling).[16] Then all ions will have the same temperature T_{ion}. The loss rate R_{loss} for radial escape of an ion in charge state q will be, according to Boltzmann's law,

$$R_{loss,q} = v_{loss} \exp\{-eqU_{res}/kT_{ion}\} \tag{29}$$

where U_{res} is the residual potential depression in the compensated beam and the loss frequency v_{loss} is the inverse of the compensation time. If we compare the loss rate for ions of different charge state q but same temperature, we see that low-charged ions are lost preferably. This favors light ions as coolant where the maximum charge state is limited and obtained fast enough that these ions are still cold (see Figure 11.14). Residual gas ions — available for free, but difficult to control — have been a reasonable choice, as demonstrated.[48,61] A typical set of charge spectra obtained in this way at different containment times is shown in Figure 11.15. Note that below 320 ms, the EBIS seems to collect only residual gas ions, which suddenly disappear for longer containment times when the neutralization limit is reached and the number of all

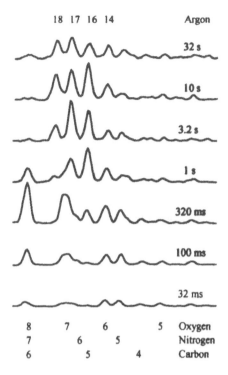

FIGURE 11.15

Charge spectra for a mixture of Ar and residual gas ions at different containment times, showing the transition from EBIS to EBIT mode of operation[48] between 320 ms and 1 s, when space charge compensation occurs.

positive charges does not increase any longer. The stepwise ionization of argon continues nevertheless! This clearly shows the cooling action of residual gas ions and the benefit of this to argon ions. For the classical EBIS operation, where ions are extracted before reaching compensation, this scheme of cooling cannot be applied; instead, a continuous flow of light ions may be useful,[92] but this proposal is missing experimental proof so far.

Another clear evidence for the ion-by-ion cooling process taking place is the isotope enrichment observed for $^{22}Ne^{8+}$ vs. $^{20}Ne^{8+}$ under dc extraction from an EBIS.[13] The heavier isotope is heated less according to Equation 28 and, therefore, survives the radial filtering of the compensated beam better than the lighter one. A similiar effect has been observed in ECR sources,[93] using a gas mixture of different isotopes. This strongly suggests that evaporative cooling in EBIS/EBIT has the same physical origin as the gas-mixing effect observed in ECR sources.

11.3.7 Ion Yield of an EBIS

In EBIS sources, ions are trapped in the potential depression of the electron beam. Therefore, the number of positive charges cannot exceed the number of negative charges within the trap. Although the use of secondary electrons has been discussed for increasing the number of electrons, experience so far has shown that the limit is set by the electrons of the beam. In the nonrelativistic limit this gives

$$N^+ \leq \frac{1}{e}\sqrt{\frac{m}{2e}}\, PU\ell \qquad (30)$$

where e and m are the charge and mass of an electron, P is the perveance of the beam, eU the beam energy, and ℓ the trap length. Taking "typical" numbers

$$P = 10^{-6} \, A/V^{3/2}, \quad U - 10^4 \, V, \quad \ell = 1m \tag{31}$$

a maximum yield of $N^+ \leq 1.05 \times 10^{11}$ positive charges can be trapped within the electron beam, until space charge compensation sets the limit. While this number of trapped charges is independent of the electron beam current density, the number of ions N_q^+ that can be produced per second increases with it, because the ionization time τ for a required charge state q decreases inversely to it, and more cycles of ionization become possible per second:

$$N_q^+ \leq \frac{N^+}{q} \frac{j}{e} \sigma \, \left[s^{-1} \right] \tag{32}$$

Donets[94] extracted 2×10^8 C^{6+} ions per second from a 95-mA, 5-keV beam, corresponding to 18% of the theoretical yield. Arianer[95] achieved about 70% and Faure[96] extracted 2×10^{10} charges from a 550-mA, 9.5-keV beam, which is about 55% of the theoretical yield. So this limit seems to be real if some reduction is taken into account for not matching the acceptance of the detection system. For highly charged ions, where ion-by-ion cooling is needed, an additional reduction will occur, either due to the heating of ions or due to the cooling scheme applied. Despite the fact that the EBIS is a weak ion source with respect to the ions produced per second, it is also an intense one if the extraction is performed in short time. This makes the EBIS less interesting for LINACS and cyclotrons, but attractive for synchrotrons — especially at single-turn injection.[97]

Recently, proposals were made[38,39] to increase the yield of EBIS sources by increasing perveance, electron energy, and length over the classical values in Equation 31 to 5 µA/V³/², 30 keV, and 2 m[98] for the hadron collider facilities to be built at BNL or under consideration at CERN. This could give about 5×10^9 Pb^{50+} ions per extraction pulse. At 200 A/cm² this number of ions is available in 1 s of containment time, and because of the elevated energy, ion heating by Coulomb collisions is not important (cf. Figure 11.13 and Equation 28).

11.3.8 Ion Beam Quality of an EBIS

In general, the ion beam quality of EBIS sources has been adequate for accelerator application as well as for beam-target experiments. It may be widely influenced by the extraction procedure: a short (some 10 µs) extraction pulse will have axial and radial energy spread according to the potential well depth ΔU (Equation 2), while a slow (more than 100 µs) extraction will provide ion beams with improved quality.

Independent of these radial and axial energy spreads, an EBIS will have a significant contribution to the emittance from the magnetic field in which the ions are born. For axial extraction from a magnetic field the conservation of canonical angular momentum predicts skew ion trajectories outside of the magnetic field, which establishes an emittance of[99]

$$\varepsilon = \frac{\pi}{4} \sqrt{\frac{2eq}{M}} \frac{Br^2}{\sqrt{U_0}} \times 10^6 \quad (\pi \cdot mm \cdot mrad) \tag{33}$$

where q is the charge state, e is the elementary charge in coulombs, M is the mass of the ion in kilograms, B is the magnetic field in teslas, r is the beam radius in meters, and U_0 is the ion acceleration voltage. This gives for a "typical" EBIS beam of 0.5 A, 10 keV, 3 T, and 500 A/cm^2 a magnetic contribution to the emittance of $\varepsilon = 7.38$ $\pi \cdot$mm$\cdot$mrad at 10-kV extraction energy for bare nuclei, which is acceptable for most applications. Although the magnetic field used in EBIS/EBIT generally is higher than in ECR sources, the electron beam radius is much smaller than the radius of the extraction electrode in typical ECR sources. Therefore, the emittance of EBIS/EBIT is caused by the extraction procedure in conjunction with the potential depression, while ECR sources seem to be dominated by the magnetic contribution.

11.4 Operating Data, EBIS/EBIT Performances, and Applications

From all known installations of EBIS and EBIT devices, data about the operation and performance as well as the application are listed in Table 2, which updates similar ones in References 1 to 4 and only shows those numbers that were reported in publications as experimental results; however, the numbers do not necessarily belong to the same experiment, which is especially true for the highest-charged ion detected and the number of positive charges extracted per pulse. The trap temperature shown in all cases also gives the operating temperature of the solenoid, e.g., water-cooled (300 K), cooled with liquid nitrogen (77 K) or liquid helium (4.2 K). In the latter case, the solenoids are superconducting.

ACKNOWLEDGMENTS

M. Kleinod, R. W. Hamm, E. D. Donets, and B. Wolf deserve thanks for carefully reading the manuscript and for their important suggestions to improve it. The help of J. Faure, G. Visentin, M. Stöckli, J. Silver, E. D. Donets, A. Hershkovitch, S. Kravis, E. Beebe, and Leif Liljeby has been essential in updating Table 1. M. Kleinod compiled most of these data for the comparative list of different installations and provided the list of references.

TABLE 11.2

Characteristic Data of Operating and Upcoming EBIS/EBIT Devices (Mid 1994)

Location/Lab	Device	Trap T(K)	Trap L(m)	Electron beam I(mA)	Electron beam U(kV)	Focusing B(T)	Focusing j(A/cm²)	τ(s)	Ion	N⁺/pulse	Application, remarks, [reference]
Dubna/JINR	Krion-II-M	4.2	1.2	100	6–50	2.25	400	40	Xe^{53+}	10^3–10^9	Atomic physics, 150 kV beam, H_2 problem, [4]
Dubna/JINR	Krion-III	4.2	0.9	60	8–28	5	1000	0.2	Ar^{18+}	2×10^7	Future injector, [4]
Dubna/JINR	Krion-S	4.2	1.2	130	5–80	1.2	200	2	S^{14+}	10^8	Nuclotron inj., [4]
Frankfurt/IAP	Kryo-EBIS	4.2	0.8	10–200	2–6	1–5	50–500	<100	Ar^{18+}	10^7	Atomic physics, dc extraction, EBIT mode, [48]
Frankfurt/IAP	X-EBIS	300	0.01	20	3	—	2	1	Ar^{13+}	10^4	EBIS physics, inertial focusing, B=0 !, [74]
Frankfurt/IAP	MEDEBIS	300	0.2	169	2	0.8	200				Prototype of Medical Synchrotron injector, [97]
Saclay/LNS	DIONE	4.2		470	9.5	4.5	500	0.16	U^{55+}	2×10^{10}	Saturne injector, [27]
St. Petersburg/IR	IMI-1	300	0.5	5000	6.3	0.9	1000	?	Ti^{12+}	?	B5 Synchrotron, [34]
Novosibirsk/INP	IMI-II	4.2?	0.5	1500	6.3	1	1000	?	Ti^{12+}	?	Cryopump planned, [34]
Nagoya/IPP	NICE-I	4.2		15	2	1.2	?	dc	I^{42+}	?	Atomic physics, [32]
Ithaca/Cornell	CEBIS-I	300	0.5	20	3.6	0.3	150	<1.2	Ar^{16+}	10^5	Atomic physics, [61]
Ithaca/Cornell	CEBIS-II	4.2	1	200	5	1.5	?	<1.2	Xe^{28+}	?	Atomic physics, [33]
Stockholm/MSI	CRYSIS	4.2	1.5	300	12–19	2	300	<22	Xe^{52+}	$<10^{10}$	Cryring injector, closed He system, [28]
Livermore/LLNL	EBIT-II	4.2	0.02	130	<30	3	4000	Hours	Th^{80+}	10^3–10^6	Atomic physics, steady-state charge distrib., [16]
Livermore/LLNL	SuperEBIT	4.2	0.02	200	198	3	5000	Hours	U^{92+}!	4×10^4	Atomic physics, steady-state charge distrib., [42]
Manhattan/KSU	CRYEBIS	4.2	1	150	9.5	5	800	1	Xe^{46+}	$<10^8$	Atomic physics, operated by users, [35]
Tokyo/Metrop. U.	Mini-EBIS	77	0.25	15	2	0.1	10	dc	I^{39+}	10^3–10^4	Atomic physics, LN_2-cooled solenoid, [36]
Oxford/Clarendon	EBIT	4.2	0.02	100	7	2.8		dc	Ba^{46+}		Atomic physics, 50-kV beam, [44]
Gaithersburg/NIST	EBIT	4.2	0.02	150	20	3					Atomic physics, [45]
Wako/RIKEN	REBIS	77	0.3	10	2	0.3	?	2	Ar^{16+}	?	Atomic physics, ibid. [44], p. 1066
Troitzk/ISAN	EBIT	300	0.02	100	10	1					Atomic physics, [46]
Osaka/Kinki U.	SMEBIS	77	0.25	1	2	1					Atomic physics, alternating perm. magnets, [37]
Dubna/JINR	EBIT	4.2	0.02								Atomic physics, [47]
Brookhaven/BNL	SuperEBIS	4.2	1.2	120	20	3					RHIC injector, [63], former SNLL-EBIS, [31]

REFERENCES

The latest review papers on EBIS are listed as References 1 to 4, and the proceedings of a series of six international workshops on EBIS as References 5 to 10.

1. E. D. Donets, in *The Physics and Technology of Ion Sources*, I. G. Brown, Ed., John Wiley & Sons, New York, 1989, 245.
2. R. W. Schmieder, Physics of the EBIS and first ions, in *Proc. NATO Workshop Physics of Highly Ionized Atoms*, Cargese, Corsica 1988, R. Marrus, Ed., Plenum Press New York, 1989.
3. M. Stöckli, The status of the electron beam ion sources, Proc. HCI-90, Giessen, E. Salzborn, P. H. Mokler, and A. Müller, Eds., *Suppl. Z. Phys. D*, 21, 111, 1991.
4. E. D. Donets, The status of electron beam ion sources, Proc. HCI-92, Manhattan, P. Richard, M. Stöckli, C. L. Cocke, and C. D. Lin, Eds., *AIP Conf. Proc.*, 274, 663, 1993.
5. *Proc. EBIS Workshop 1977*, GSI, Darmstadt, B. Wolf and H. Klein, Eds., GSI-P-3-77.
6. *Proc. 2nd EBIS Workshop 1981*, IPN2 Orsay and LNS Saclay, J. Arianer and M. Olivier, Eds.
7. *Proc. 3rd EBIS Workshop 1985*, Ithaca, V. Kostroun and B. W. Schmieder, Eds.
8. *Proc. 4th EBIS Workshop 1988*, BNL, A. Hershcovitch, Ed., *AIP Conf. Proc.*, No. 188, 1989.
9. *Proc. 5th EBIS Workshop 1991*, JINR, Dubna, E. D. Donets and V. Judin, Eds., EBIS-V publishing group, Dubna, 1992.
10. Proc. 6th EBIS Workshop 1994, MSI, Stockholm, C. Herrlander and L. Lilijeby, Eds., *Phys. Scr.*, in press.
11. R. Becker et al., Proc. 4th ICIS, Bensheim, 1991, B. Wolf, Ed., *Rev. Sci. Instrum.*, 63, 2812, 1992.
12. J. Arianer et al., *Proc. Int. Ion Eng. Congr. ISIAT'83 & IPAT'83*, Kyoto, 1983, 176.
13. R. Becker, M. Kleinod, and H. Klein, *Nucl. Instrum. Methods*, B24/25, 838, 1987.
14. E. D. Donets and G. D. Shirkov, Avtorskoe Svidetelstvo U.S.S.R. N1225420 (1984), *Bul. OI*, N44, 1989.
15. R. Becker, ibid. [6], p . 185.
16. M. A. Levine et al., *Phys. Scr.*, T22, 157, 1988.
17. A. G. Drentje and J. Sijbring, KVI Groningen (NL), Annual Report 1983, p. 79.
18. E. D. Donets, U.S.S.R. inventor's certificate No. 248860, 16 March 1967, *Bul. OI*, N24, 65, 1967.
19. E. D. Donets, V. I. Ilushchenko, and V. A. Alpert, Proc. ICIS-1, Saclay 1969, p. 635.
20. E. D. Donets, *IEEE Trans. Nucl. Sci.*, NS-23, 897, 1976.
21. E. D. Donets, *Nucl. Instrum. Methods*, B9, 522, 1985.
22. J. Arianer and C. Goldstein, ibid. [20], p. 976.
23. R. Becker et al., ibid. [20], p. 1017 and M. Kleinod et al., ibid. [20], p. 1023.
24. R. Becker, H. Klein, and M. Kleinod, ibid [5], p. 35.
25. R. W. Hamm, L. M. Choate, and R. A. Kennefick, ibid. [20], p. 1013.
26. J. Arianer, A. Cabrespine, and C. Goldstein, *Nucl. Instrum. Methods*, 193, 401, 1982.
27. B. Gastineau, J. Faure, and A. Courtois, ibid. [21], p. 538.
28. J. Arianer et al., *Phys. Scr.*, T3, 36, 1983.
29. I. G. Brown and B. Feinberg, *Nucl. Instrum. Methods*, 220, 251, 1984.
30. R. W. Schmieder and C. L. Bisson, *Rev. Sci. Instrum.*, 61, 256, 1990.
31. R. W. Schmieder et al., ibid. [11], p. 259.
32. S. Ohtani et al ibid. [28], p. 110.
33. V. O. Kostroun et al., ibid. [28], p. 47.
34. V. G. Abdulmanov et al., Proc. X All Union Conf. Accel. (1987) and ibid. [9], p. 71.
35. M. P. Stöckli et al., *Nucl. Instrum. Methods*, B10/11, 763, 1985.
36. K. Okuno, *Jpn. J. Appl. Phys.*, 28, 1124, 1989.
37. T. Kusakabe, T. Sano, and Y. Nakajima, *At. Coll. Res. Jpn.*, No. 8, 105, 1992.
38. J. Faure et al., ibid. [9], p. 76.
39. H. Haseroth and K. Prelec, ibid. [10].
40. M. Litvin, M. Vella, and A. Sessler, *Nucl. Instrum. Methods*, 198, 189, 1982.
41. M. A. Levine et al., *Nucl. Instrum. Methods*, B43, 431, 1989.
42. D. A. Knapp et al., *Nucl. Instrum. Methods*, A334, 305, 1993.
43. R. E. Marrs, S. R. Elliot, and D. A. Knapp, *Phys. Rev. Lett.*, 75, 26, 4082, 1994.
44. J. D. Silver et al., Proc. ICIS-V (1993) Beijing, *Rev. Sci. Instrum.*, 65, 1072, 1994.
45. J. D. Gillaspy, J. R. Roberts, C. M. Brown, and U. Feldman, Proc. HCI-92, ibid. [4].
46. P. S. Antsiferov and V. G. Movshev, ibid. [9], p. 98.
47. V. P. Ovsyannikov, private communication, 1993.
48. R. Becker, M. Kleinod, H. Thomae, and E. D. Donets, ibid. [4], p. 686.

49. G. R. Brewer, High intensity electron guns, in *Focusing of Charged Particles*, A. Septier, Ed., Vol. 2, Academic Press, New York, 1967, chap. 3.2.
50. K. Amboss, *IEEE Trans. Electron Devices*, ED-16, 897, 1969.
51. G. R. Brewer, Focusing of high density electron beams, ibid. [49], chap. 3.3.
52. L. Brillouin, *Phys. Rev.*, 67, 260, 1945.
53. J. Dietrich and R. Becker, ibid. [9], p. 147.
54. R. Becker, ibid. [5], p. 59.
55. G. Herrmann, *J. Appl. Phys.*, 29, 127, 1958.
56. C. C. Cutler and M. E. Hines, *Proc. I.R.E.*, 43, 307, 1955.
57. J. R. Pierce and L. R. Walker, *J. Appl. Phys.*, 24, 1328, 1953.
58. K. Amboss, *IEEE Trans. Electron Devices*, ED-11, 479, 1964.
59. C. Collart, A. Liebe, P. Nicol, and A. Steinigger, ibid. [6], p. 170.
60. J. Faure, B. Feinberg, A. Courtois, and R. Gobin, *Nucl. Instrum. Methods*, 219, 449, 1984.
61. E. N. Beebe, ibid. [8], p. 166.
62. R. Becker, ibid. [8], p. 359.
63. K. Prelec, ibid. [8], p. 341.
64. S. Kravis et al., ibid. [4] p. 671.
65. P. F. Dittner et al., *Phys. Rev. Lett.*, 51, 31, 1983.
66. R. Becker et al., in *Proc. EPAC, Rome 1988*, S. Tazzari, Ed., World Scientific Publishers, Singapore, 1989, 607.
67. B. Fögen, ibid. [6], p. 119.
68. J. Arianer et al., *IEEE Trans. Nucl. Sci.*, NS-26, 3713, 1979.
69. M. Olivier et al., *IEEE Trans. Nucl. Sci.*, NS-30, 1463, 1982.
70. R. Bikowski, Diploma thesis, Inst. Angew. Physik, University of Frankfurt, 1985.
71. B. Pfisterer, Diploma thesis, Inst. Angew. Physik, University of Frankfurt, 1992.
72. V. P. Ovsiannikov, ibid. [44].
73. R. Becker, H. Klein, and W. Schmidt, *IEEE Trans. Nucl. Sci.*, NS-19, 125, 1972.
74. M. Kleinod, R. Becker, O. Kester, A. Lakatos, H. Thomae, B. Zipfel, and H. Klein, ibid. [44].
75. J. R. Pierce, *Theory and Design of Electron Beams*, D Van Nostrand, New York, 1954, 181.
76. C. R. Moster and J. P. Molnar, unpublished, Bell Labs Memorandum MM-51–290–1 (1951); results reported in V. Bevc et al., *J. Br. I.R.E.*, 18, 696, 1958.
77. R. Becker, ibid. [7], p. 177.
78. R. D. Frost, O. T. Purl, and H. R. Johnson, *Proc. I.R.E.*, 50, 1800, 1962.
79. R. Becker, *Nucl. Instrum. Methods*, A298, 13, 1990.
80. W. B. Herrmannsfeldt, SLAC-331, 1988.
81. R. Becker, Proc. Int. Symp. HIF, Frascati 1993, *Nuovo Cimento*, 106A, 1613, 1993.
82. H. Tawara and T. Kato, Inst. Plasma Physics, Nagoya, IPPJ-AM-37, 1985.
83. H. Bethe, *Z. Phys.*, 76, 293, 1932.
84. W. Lotz, *Z. Phys.*, 216, 241, 1968.
85. Y. Itikawa and T. Kato, Inst. Plasma Phys., Nagoya, IPPJ-AM-17, 1981.
86. T. A. Carlson, C. W. Nestor, Jr., N. Wasserman, and J. D. McDowell, ORNL-4562, 1970.
87. K. D. Sevier, *At. Data Nucl. Data Tables*, 24, 323, 1979.
88. B. Fricke, J. H. Blanke, and D. Heinemann, *Plasmarelevante atomare Daten*, Universität GHS Kassel, FB Physik, 1991.
89. D. M. Penetrante et al., *Phys. Rev.*, A43, 4861, 1991.
90. E. Baron, IPN, Orsay, Rep. IPN-73–05, 1973.
91. E. D. Donets, ibid. [28], p. 11.
92. G. Shirkov, E. Donets, R. Becker, and M. Kleinod, ibid. [9], p. 171.
93. A. G. Drentje, ibid. [11], p. 2875.
94. E. D. Donets, private communication, 1993.
95. J. Arianer and C. Goldstein, IPN, Orsay, IPNO-79–02, 1979.
96. A. Courtois et al., ibid. [11], p. 2815; J. Faure, private communication, 1993.
97. O. Kester, R. Becker, and M. Kleinod, ibid. [10].
98. R. Becker, Int. Rep. 94- 9 (1994), Institut für Angewandte Physik der Universität Frankfurt/ M, Germany.
99. W. Krauss-Vogt et al., *Nucl. Instrum. Methods*, A268, 5, 1988.

Chapter 2/Section 12

CHARACTERIZATION
OF ION SOURCES

12   VACUUM ARC ION SOURCES

Advantages: High current of metal ions, very high pulse currents, multiply charged
  ions, large beams possible.
Disadvantages: Pulsed operation, noisy discharge, insulating and low-conducting
  material not usable, less suited for gas ions.

12.1   Ion Source History

Vacuum arcs have been studied intensively by many people since the 1920s with
focus on high-power switches and vacuum valves (Hg arcs). The production of metal
ions in vacuum arcs has also received attention during the last 20 years. A list of
publications is given in Reference 1.

The first reliable metal vapor vacuum arc (MEVVA) ion source was developed at
Berkeley by Brown in the early 1980s,[2-5] Humphries, Jr.,[6] and Picraux.[7] A multicathode
ion source was introduced by Brown in 1987[8] that extended lifetime of the ion source
and its flexibility in operation. Similar MEVVA sources have been constructed in
other laboratories.[9] Zhang designed a MEVVA with a movable rotating cathode to
extend cathode lifetime[10] and E. Oks uses an additional gas discharge to trigger his
vacuum arc ion source[11] and Hirshfield[12] suggested a laser-triggered vacuum arc ion
source. A dc operated MEVVA was constructed by Brown in 1991.[13]

12.2   Working Principle of the Vacuum Arc Ion Source and Description
of the Discharge

The metal-vapor vacuum arc occurs between hot cathode spots and a cold anode
in vacuum. The principal arrangement of the vacuum arc ion source electrodes and
the electric circuitry is shown in Figure 12.1. After ignition by a high-voltage spark
the vacuum arc plasma is maintained between cathode spots and the anode. Material
is vaporized from the cathode spots and feeds the discharge. The ionization behavior
in the region of the cathode spots mainly defines the arc plasma parameters. Ion
production from vacuum arcs was investigated in detail by Kimblin.[14]

The behavior of cathode spots is rather complicated. An extensive description
can be found in the book by Lafferty[15] and an application of the theory of vacuum arcs

0-8493-2502-1/95/$0.00+$.50
© 1995 by CRC Press Inc.

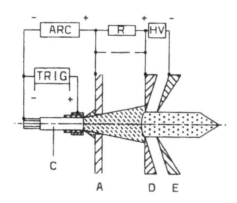

FIGURE 12.1
Vacuum arc ion source principle and circuitry.

to the discharge behavior of vacuum arc ion sources in Reference 16. The cathode spots occur randomly and have a lifetime of only microseconds. One cathode spot carries several amperes of arc current, which means several spots are necessary to carry the arc current of typically 100 to 300 A. The high current density in the cathode spots (10^6 A/cm^2) creates a high pressure gradient, which accelerates the plum-shaped plasma toward the anode and partly through the anode hole into the expansion area. The dimensions of the expansion area allow the adjustment of the plasma density and surface to the needs of the extraction system.

Usually the arc has to be pulsed to limit the power dissipation at the cathode. Duty cycles of a few percent can be reached in most designs. With a large and well-cooled cathode, however, the dc operation of a vacuum arc ion source is also possible.[13]

About 10% of the arc current is supposed to be ionic and about half of it can be utilized for extraction, which means about 10 A of metal ions can be extracted under optimized conditions.[4,16] Vacuum arc ion sources can deliver small emittance beams of metal ions for particle accelerators or large-area beams for ion implantation.

12.2.1 Vacuum Arc Ion Source Operation

The vacuum arc is triggered by an electrode ring insulated from the cathode usually by an alumina pipe. The trigger electrode is connected to a supply that delivers high-voltage pulses (about 10 kV, 10 μs long) with a current of 10 to 30 A.

The main arc is connected to a power supply of about 250 V with a buffer, a serial resistor, and a transistor switch (Figure 12.1) or to a transmission line. The dimensions of the power supply and the pulse-forming electronics depend on the pulse length and the duty cycle of the vacuum arc ion source (see Chapter 6). The arc is usually pulsed with 0.1- to 5-ms and up to 100-Hz pulses depending on ion source design and the cathode material. (See Figure 12.2 for typical pulse shapes.) With 1 Hz and a pulse length of 200 μs, typical cathode lifetimes are more than a day for one cathode.

The arc plasma expands through a hole in the anode into an expansion area that is usually connected to the anode via a resistor. The expansion area is terminated by an extractor grid for ion beam extraction. The size of the expansion area and the grid depends on the ion source application. The ion current yield during the lifetime of a cathode is nearly constant and also varies little from one cathode to the next.

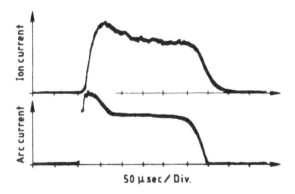

50 μ sec / Div.

FIGURE 12.2
Typical pulse shapes of arc current and ion current.

The total ion current increases linearly with increasing arc current within the range of operation of the ion source, limited by the saturation current for a given extraction voltage (Figure 12.3).

12.2.2 Charge State Distribution

Vacuum arc ion sources produce a large amount of ions with charges higher than 1 because they originate from the dense near-cathode spot plasma. Charge state distributions for most metals have been measured by the time-of-flight method[2] (see Table 12.1) or by using a magnetic analyzer to detect the different charge states.[17] Spectra for titanium and platinum are shown in Figures 12.4 and 12.5, respectively.

The charge state distribution depends mainly on the ionization energy of a charge state and on the evaporation temperature T_v of the cathode material. The higher T_v, the higher the potential drop at the cathode spot necessary to evaporate enough material. For heavier elements, higher charges can carry additional arc current so less particles are necessary. Lower particle density reduces recombination processes and thus increases the higher charge states in the extracted ion beam.

The linear dependency on the arc current (Figure 12.3) was also observed for the different charge states (Figure 12.6) but with varying gradients, i.e., the charge state distribution varies slightly with arc current towards higher-charged ions.

Investigations of different anode geometries and positions show little influence on the charge state distribution and on the total ion current. A shorter distance of the anode from the cathode seems to favor higher charge states and the anode hole diameter should be around one to two times the cathode-anode distance.

12.2.3 Influence of Magnetic Fields

An axial magnetic field changes little with the discharge behavior of a vacuum arc ion source, but influences the plasma density in the extraction area by compression of the expanding plasma plum. In this way the current density of the extracted ion beam can be intensified by increasing the magnetic field until the plasma is compressed to the size of the extraction area (see Figure 12.7).

If the magnetic field is increased to values of 0.1 T or more, the charge state distribution of the extracted ion beam is shifted toward higher charges depending on the element used. Figure 12.8 shows this effect for uranium ions.[18]

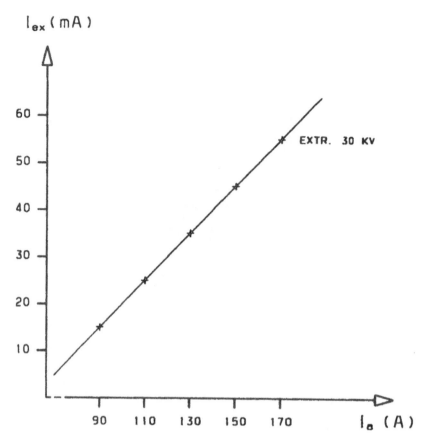

FIGURE 12.3

Total ion current as a function of the arc current.

TABLE 12.1

Charge State Fractions for Various Elements[4]

Element	Electrical current (%)				Element	Electrical current (%)				
	1+	2+	3+	4+		1+	2+	3+	4+	5+
C	100	—	—	—	Rh	28	52	18	2	—
Mg	23	77	—	—	Pd	24	69	7	—	—
Al	38	52	10	—	Ag	18	66	16	—	—
Si	38	58	4	—	In	79	21	—	—	—
Ti	3	80	17	—	Sn	36	64	—	—	—
Cr	14	73	13	—	Gd	3	78	19	—	—
Fe	18	74	8	—	Ho	8	79	13	—	—
Co	30	62	8	—	Ta	5	30	33	28	4
Ni	35	58	7	—	W	3	25	39	27	6
Cu	26	49	25	—	Pt	21	63	17	—	—
Zn	76	24	—	—	Au	28	69	3	—	—
Zr	4	47	38	11	Pb	47	53	—	—	—
Nb	2	36	43	19	Th	1	10	72	17	—
Mo	6	40	36	18	U	1	29	62	8	—

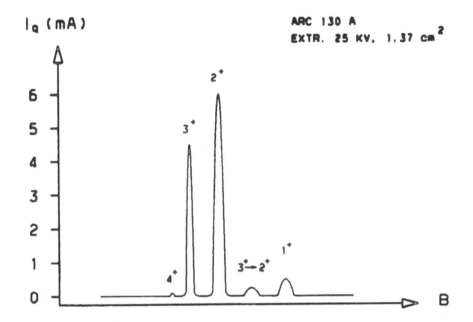

FIGURE 12.4
Charge state spectrum for titanium.

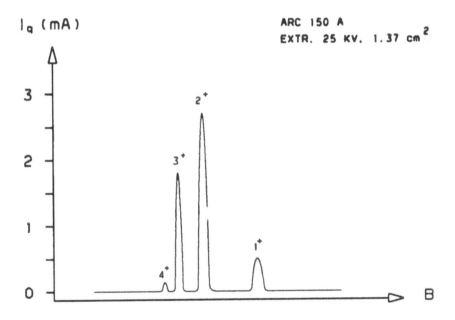

FIGURE 12.5
Charge state spectrum for platinum.

12.2.4 Plasma Instabilities and Ion Beam Modulations

The vacuum arc produces strong plasma oscillations due to its origin in the short-living cathode spots. Naturally, the ion current is modulated at higher frequencies (especially some 10 kHz) reaching about 30% of the total current level. This modulation influences the space charge compensation of the ion beam and increases losses

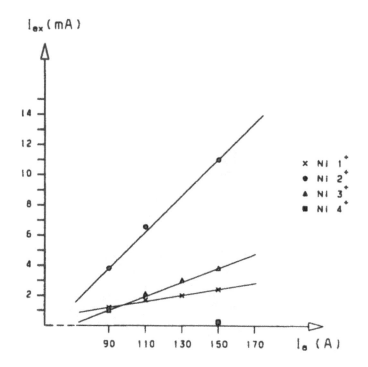

FIGURE 12.6
Ion current of nickel charge states as a function of the arc current.

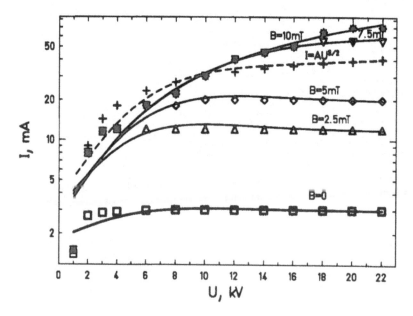

FIGURE 12.7
Variation of ion beam current with magnetic field.

in the low-energy beam transport systems. With special grid arrangements in the expansion area (see Figure 12.9), however, it is possible to reduce these modulations significantly.[19,20]

The modulation of the ion beam is much stronger for higher charge states, as can be seen from Figure 12.10, reaching nearly 100% for charge 3 and higher. The same

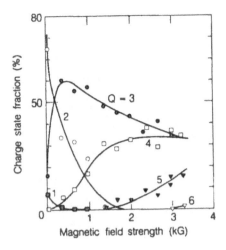

FIGURE 12.8
Variation of different charge states with magnetic field.[18]

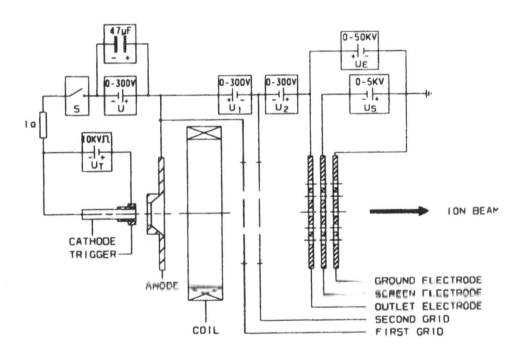

FIGURE 12.9
Grid arrangement for noise reduction.

charge states are shown when the ion source is operated with proper adjustment of the grids.[20]

12.2.5 Influence of Additional Gas

Feeding additional gas into the vacuum arc ion source can improve beam formation and space charge compensation of the extracted ion beam.[21] Additional gas also influences the charge state distribution by shifting it to lower charges in most cases. Figure 12.11 shows, for example, the combination Mo with Ar dependent on the Ar

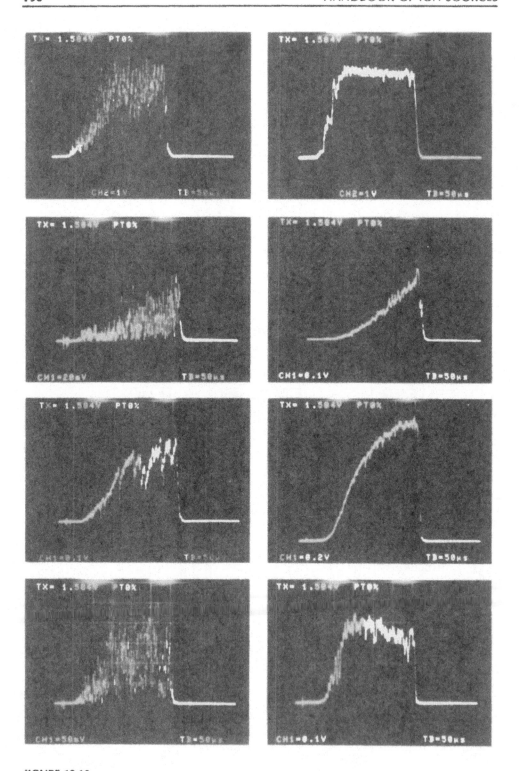

FIGURE 12.10

Modulation of ion beam pulse for different charge states with (right) and without (left) grids. The first row shows the total current (20 mA/div). The second shows the singly charged Fe (left, 0.4 mA/div; right, 2 mA/div). The third shows the doubly charged Fe (left, 2.0 mA/div; right, 4 mA/div). The last row shows the Fe^{3+} (left, 1.0 mA/div; right, 2 mA/div).

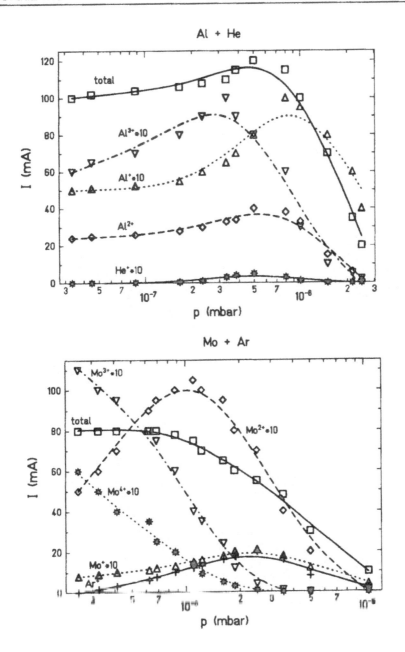

FIGURE 12.11
Influence of additional gas on the charge state distribution: (a) Al plus He, (b) Mo plus Ar.

gas pressure (measured in the beam line after extraction; the pressure inside the source is two to three orders of magnitude higher). Charge-exchange processes in the expanding plasma with the gas atoms seem to be the main reason.[21]

A small amount of gas ions is also produced if a gas is added to the vacuum arc, depending on arc current and the type of gas used. Up to 10% gas ions have been found in ion source operation without a magnetic field. A much higher percentage of gas ions can be produced if a magnetic field of about 10 mT is applied to the vacuum arc ion source (see Figure 12.12).[21] A Penning discharge is expected to occur in the arc region, ionizing the gas atoms.

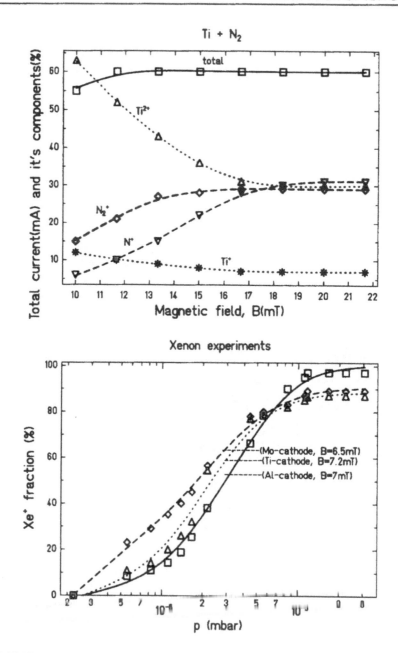

FIGURE 12.12
Yield of gas ions depending on (a) the magnetic field for Ti plus N_2 and (b) the gas pressure for Xe.

By proper selection of the gas, the metal, and the ion source parameters a simultaneous implantation of the correct amount of two elements is possible with a vacuum arc ion source, as shown in Figure 12.13.

12.3 Specific Ion Source Design

12.3.1 The LBL-MEVVA Generation

13.3.1.1   MEVVA-2 (Figure 12.14)

- Special design and construction details of the source

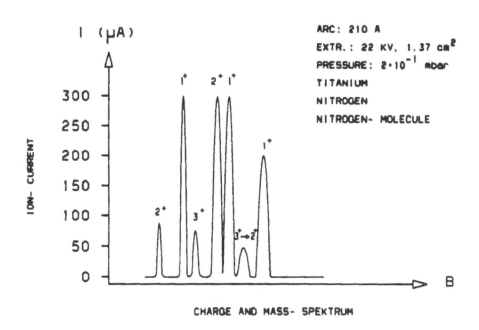

FIGURE 12.13
Spectrum for Ti and N₂.

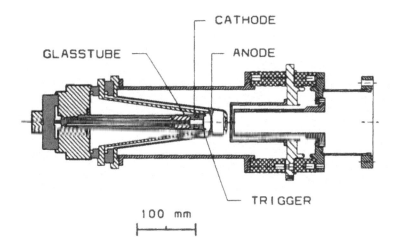

FIGURE 12.14
MEVVA-2.[13]

The MEVVA 2 is the basis of all other MEVVA ion sources. It has one cathode of 6- to 12-mm diameter surrounded by an alumina insulator and a stainless-steel trigger ring, both fixed to the cathode by a cement glue. The contact to the trigger ring is made by a cable guided inside the hollow cathode support. Anode and plasma expansion end electrodes are mounted on insulated conical cylinders. The insulation to the extractor ground electrode was originally done by a Pyrex glass cylinder

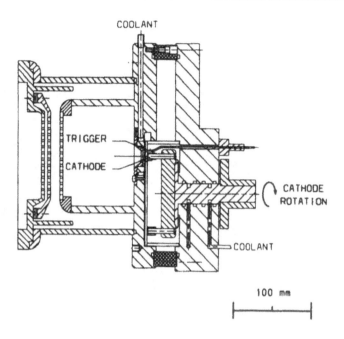

FIGURE 12.15
MEVVA-5. (Courtesy of I. G. Brown, Lawrence Berkeley Laboratory, Berkeley, CA.)

that allowed one to see the plasma inside the source. The source could be put inside a solenoid coil to investigate or use the effects of an axial magnetic field. The source can be operated with duty cycles up to 2%.

- Ion source material and vacuum conditions

  Anode, plasma expansion, and trigger ring, stainless steel; cathode support, Cu; cathode end flange, Al; trigger ring insulator, alumina; other insulators, polymer material. The base pressure is ≤10⁻⁴ Pa.

- Application area of the source

  Linear accelerators, synchrotrons, ion implantation, ion beam modification of materials.

- Deliverer or user

  Plasma and Ion Source Technology Group, Lawrence Berkeley Lab., Berkeley, CA 94720, U.S.A.

  Nippon Steel Corporation, 10-1 Fuchinobe 5-chrome, Sagamihara shi, Kanagawa 229, Japan

  Australian Nuclear and Technology Organization, Private Mail Bag 1, Menai, NSW, Australia

12.3.1.2 MEVVA-5 (Figure 12.15)

- Special design and construction details of the source

  This MEVVA uses the same cathode and anode geometry as MEVVA 2 but uses a support with 18 cathodes that can be used one after the other without breaking the vacuum. The expansion area is enlarged to allow extraction with a 10-cm-diameter grid with 5-mm holes and an extraction area of ~50 cm². A 100-kV supply can be applied to the extraction gap of 10 mm and the extraction alumina insulator of 12-cm length. Ion currents up to 3.5 A can be extracted with a duty cycle of 4%.

- Ion source material and vacuum conditions

  Cathode wheel, water-cooled copper; cathode insulator, alumina or BN; trigger-ring source body and extraction grid, stainless steel; cathode end flange, Al.

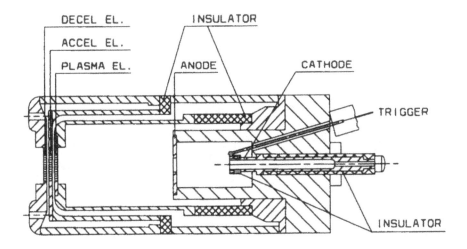

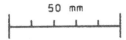

FIGURE 12.16
Mini-MEVVA. (Courtesy of I. G. Brown, Lawrence Berkeley Laboratory, Berkeley, CA.)

- Application area of the source

  Broad-beam ion implantation and ion beam modification of materials.

- Deliverer or user

  Plasma and Ion Source Technology Group, Lawrence Berkeley Lab., Berkeley, CA 94720, U.S.A.

12.3.1.3   Mini-MEVVA (Figure 12.16)

- Special design and construction details of the source

  The Mini-MEVVA is a compact version of MEVVA 2; cathode, anode, and expansion area have the same size but the housing is optimized to a smaller size. The overall length is 22 cm and the diameter 8.5 cm. The extraction grid is 20 mm in diameter and the three-electrode system can be used to slow ions down to a few kilovolts.

- Ion source material and vacuum conditions

  Stainless steel and alumina.

- Application area of the source

  Ion implantation and surface treatment.

- Deliverer or user

  Plasma and Ion Source Technology Group, Lawrence Berkeley Lab., Berkeley, CA 94720, U.S.A.

12.3.1.4   Micro-MEVVA

- Special design and construction details of the source

  The Micro-MEVVA is just 2 cm in diameter, 10 cm long, and weighs only about 100 g. It is constructed from a set of coaxial stainless-steel and alumina tubes. The cathode diameter is 2 mm and the single aperture extraction has a 3-mm diameter. Ion current of 10 to 20 mA can be extracted with up to 20 kV. The entire source is

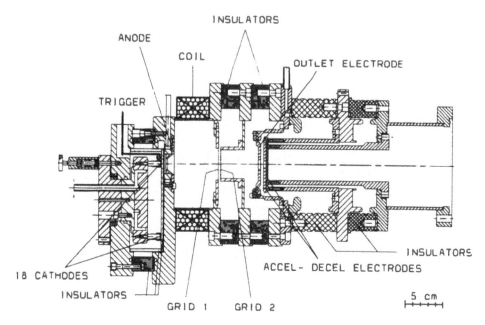

FIGURE 12.17
Sectional view of the multicathode GSI-VA ion source.

placed inside a vacuum tank and has no special cooling, so the repetition rate has
to stay low (~1 Hz).

- Ion source material and vacuum conditions

  Stainless steel and alumina.

- Application area of the source

  Metal ion injection into EBIS sources, surface analysis.

- Deliverer or user

  Plasma and Ion Source Technology Group, Lawrence Berkeley Lab., Berkeley, CA
  94720, U.S.A.

12.3.2 The GSI Multicathode Vacuum Arc Ion Source (Figure 12.17)

- Special design and construction details of the source

  This multicathode ion source is of a design similar to MEVVA 4.[4] The source has 18
  cathodes that are made either from the same material for long-term operation or
  from different materials if a change in ion species is needed. The cathodes are
  screwed into a water-cooled copper cathode block and can be moved in front of the
  anode by a telecommand motor.[22,23] A solenoid coil creates a magnetic field of up to
  20 mT in the expansion area and a double grid between anode and extraction can
  be used for noise reduction of the extracted ion beam. The source fits to the GSI
  standard accel-decel extraction system also used for the CHORDIS ion sources.

- Ion source material and vacuum conditions

  Cathode support, water-cooled copper; source body, stainless steel; mesh, Mo or W;
  mesh width, 0.2 to 0.5 mm; wire diameter, 0.02 mm.

- Application area of the source.

  Injection into synchrotron accelerator, injection into RFQ accelerators, ion implantation, ion beam modification of materials.

- Deliver or user

  GSI, Planckstr. 1, 64291 Darmstadt, Germany

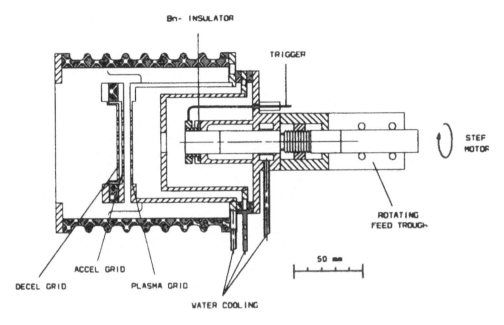

FIGURE 12.18
Sectional view of the rotating cathode VA ion source.[10]

12.3.3 The Rotating Cathode Vacuum Arc Ion Source (Figure 12.18)

* Special design and construction details of the source

  This vacuum arc ion source follows the Berkeley geometry but instead of burning a cathode down and replacing it here the cathode (16-mm Ø) rotates slowly and is moved in the cathode insulator about 0.1 mm every 10 min, which gives a cathode life time of about 20 h (4 h for a nonmovable one) at a high-duty cycle of 6% and an average metal ion beam of 50 mA.

  There are several versions of this source for 30-, 50-, and 100-kV extraction voltage and extraction areas between 3- and 18-cm Ø.

* Ion source material and vacuum conditions

  Cathode insulator, BN; extractor grids, stainless steel; insulators, quartz glass or alumina.

* Application area of the source

  Ion implantation, ion beam modification of materials.

* Deliverer or user

  Institute of Low Energy Nuclear Physics, Beijing Normal University, Beijing 100875, China

  Beijing General Research Institute of Non-Ferrous Metals, Beijing 100088, China

12.3.4 The Titan Vacuum Arc Ion Source (Figure 12.19)

* Special design and construction details of the source

  This VA ion source uses a separate gas discharge (Penning type) to ignite the vacuum arc and, thus, metal and gas ions can be produced simultaneously or separately. The ignition with the gas discharge improves the cathode lifetime by about tenfold compared to a MEVVA cathode. The source parts are surrounded by cooled oil. The cathode spot is stabilized by a small magnetic field generated by a permanent magnet behind the cathode and a solenoid around the expansion area. An accel-decel mesh system serves for extraction of the ions.

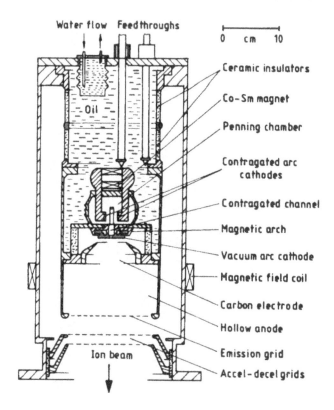

FIGURE 12.19
Sectional view of the Titan VA ion source. (From Oks, E. M., *Rev. Sci. Instrum.*, 63, 2422, 1992. With permission.)

- Ion source material and vacuum conditions

  Anode, carbon; other electrodes, stainless steel; extraction grid, 0.3-mm Ø W wire.

- Application area of the source

  Ion implantation into metals.

- Deliverer or user

  High Current Electronics Institute, Academy of Science, Tomsk 634055, Russia

12.3.5 The ITEP Low-Duty Cycle Vacuum Arc Ion Source (Figure 12.20)

- Special design and construction details of the source

  This source is specially designed for low-duty cycle operation of 0.02% and for low-duty cycle accelerator injection. The cathode is not cooled and is pressed by a spring to the conical insulator, giving a long lifetime of a cathode. Extraction is done directly from an expansion cup of 20- to 40-mm Ø. With a composite cathode, more than one element can be produced at the same time and also low-melting-point materials can be used with high arc current if contained in a matrix of holes in a high-melting-point material (such as Pb in Mo). A modified version of this source uses an 3-kV electron beam injected through a hole in the cathode to generate higher charge states than usually found in vacuum arc ion sources.

- Ion source material and vacuum conditions

  Insulator, alumina; anode, copper.

- Application area of the source

  Accelerators, ion implantation, high charge states.

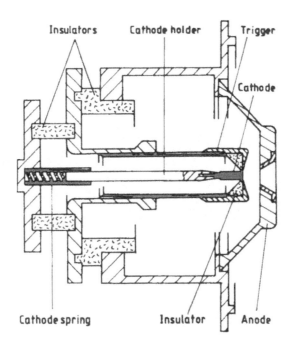

Insulators   Cathode holder   Trigger

Cathode

Cathode spring   Insulator   Anode

FIGURE 12.20
Sectional view of the ITEP VA ion source. (Courtesy of V. A. Batalin, ITEP, Moskow.)

- Deliverer or user

  Institute of Theoretical and Experimental Physics, B.25, Moscow 117259, Russia

12.3.6 Operating Data of Vacuum Arc Ion Sources

Arc voltage	10–40 V	
Arc current	50–500 A	
Ignition voltage	10–20 kV	For TITAN, 3 kV
Ignition pulse	10–20 μs	
Duty cycle	0.02–6%	Depending on design, also dc possible
Pulse length	0.1–3 ms	
Ion current	10 mA–10 A	Depending on extraction area
Average current	0.1 100 mA	
Charge states	Up to 6	Depending on element and magnetic field
Extraction area	0.2–1000 cm²	
Extraction voltage	10–100 kV	With deceleration from 1 kV

REFERENCES

1. H. C. Miller, A Bibliography and Author Index for Electrical Discharges in Vacuum (1897–1986), General Electric Co. Rep. GEPP-TIS-366e, UC-13, 1988.
2. I. G. Brown et al., Report LBL-21999 Rev. LBL Berkeley, 1987.
3. I. G. Brown et al., *Rev. Sci. Instrum.*, 57, 1069, 1987.
4. I. G. Brown, in *The Physics and Technology of Ion Sources*, I. G. Brown, Ed., John Wiley, New York, 1989, 331.
5. R. A. MacGill et al., *Rev. Sci. Instrum.*, 61 II, 580, 1990.
6. S. Humphries, Jr. et al., *Appl. Phys. Lett.*, 47, 468, 1985.

7. R. J. Adler and S. T. Picraux, *Nucl. Instrum.*, 6, 123, 1985.
8. I. G. Brown et al., *J. Appl. Phys.*, 63, 4889, 1988.
9. P. Spädtke et al., *Rev. Sci. Instrum., Nucl. Instrum. Meth.*, A278, 643, 1989.
10. X. Zhang et al., *Rev. Sci. Instrum.*, 63, 2431, 1992.
11. S. P. Bugaev et al., *Rev. Sci. Instrum.*, 63, 2422, 1992.
12. J. L. Hirshfield, *IEEE Trans. Nucl. Sci.*, 23, 1006, 1976.
13. I. G. Brown et al., *Rev. Sci. Instrum.*, 63, 2417, 1992.
14. J. Kimblin, *Appl. Phys.*, 44, 3074, 1973.
15. J. M. Lafferty, *Vacuum Arcs-Theory and Application*, John Wiley, New York, 1980.
16. S. Andre et al., *Proc. Mevva Workshop*, Beijing, 1993.
17. B. H. Wolf et al., Preprint GSI-92–71, GSI, Darmstadt, 1992; *Rev. Sci. Instrum.*, 65, 3091, 1994.
18. I. G. Brown et al., *Nucl. Instrum. Methods*, B43, 455, 1989.
19. S. Humphries, Jr. et al., *J. Appl. Phys.*, 59, 1790, 1986.
20. E. Oks et al., *Rev. Sci. Instrum.*, 65, 3109, 1994, in print.
21. P. Spädtke et al., *Rev. Sci. Instrum.*, 65, 3113, 1994.
22. P. Spädtke et al., Report GSI-88–20, GSI, Darmstadt, 1988.
23. B. H. Wolf et al., *Rev. Sci. Instrum.*, 61 II, 408, 1990.
24. I. G. Brown, P. Spädtke, et al., *Nucl. Instrum. Methods*, A295, 12, 1990.

CHARACTERIZATION OF ION SOURCES

K. N. Leung

13 LARGE AREA ION SOURCES

13.1 Ion Source History

Large-area, multiple-aperture ion sources were first developed as ion thrusters for space exploration.[1,2] Today, they find important applications in fusion research and in ion implantation of large flat panel displays. The development of large-area ion sources for fusion application stems from the success of neutral beam plasma heating. All large tokamak experiments, JET in Europe, JT-60 in Japan, TFTR at Princeton, and DIII-D at General Atomic in La Jolla, use high-power neutral beam injection of hydrogen or deuterium for auxiliary heating of the reactor plasma. For this purpose, the ion source for the injector should be capable of producing large volumes of dense, uniform, and quiescent plasma with a high percentage of atomic species. The permanent magnet plasma generator (also known as the multicusp or bucket source)[3] is able to fulfill most of these requirements, and it has been developed by various laboratories throughout the world to be used as ion sources for neutral beam injection systems.[4-6] These large-area multicusp sources have much in common. Their operation is based on the same principles. They have similar dimensions, and their performance is quite comparable.

13.1.1 Working Principle of the Large-Area Multicusp Ion Source and Description of the Discharge

The basic principle of the multicusp ion source was described in Chapter 6. In this chapter, we describe the production of high-current H^+/D^+ and H^-/D^- ion beams by large-area multicusp sources. In order to form large ion beams, a sizable ion source chamber is needed. To achieve long cathode lifetime, the big multicusp sources are normally equipped with large numbers of tungsten filaments. Electrons are emitted from these filament cathodes thermally. They can ionize the background gas particles and form a discharge plasma. The loss rate for the energetic primary electrons on the chamber walls (the anode) are drastically reduced by installing rows of permanent magnets on the external surface of the chamber as illustrated in Figure 13.1. The magnetic cusp-fields are localized near the chamber wall leaving the interior volume almost free of magnetic field. With this arrangement, the plasma uniformity, gas, and discharge efficiencies of the ion source are much enhanced.

0-8493-2502-1/95/$0.00+$.50
© 1995 by CRC Press Inc.

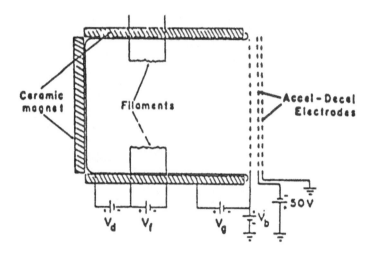

FIGURE 13.1
Schematic diagram of a large-area multicusp ion source.

It has been demonstrated that if a magnetic filter field is present in the source,[7] the primary electrons are prevented from reaching the extraction region. This arrangement can enhance the atomic hydrogen or deuterium ions percentage in the extracted beam. Atomic ion fractions in excess of 80% are routinely achieved in this way. Both the surrounding multipoles and the magnetic filters can take on different configurations. In the U.S., most multipoles are aligned parallel to the ion beam direction. In JAERI (Japan), the orthogonal orientation is chosen, i.e., the multipoles are arranged in hoops that are perpendicular to the beam axis. In this way, the interior fields tend to be higher and the magnetic filter action occurs naturally.[6] In the U.S. common long-pulse source (CLPS), the magnetic filter field must be superimposed to help the species mix.[8,9] The permanent magnets in the JET PINI source are arranged in a so-called checker-board fashion,[5] which results in the best plasma uniformity.

TABLE 13.1

Large Ion Sources for Fusion

Tokamak ion source	JET PINI	JT-60 JAERI source	TFTR CLPS	DIII-D CLPS
Pulse length (s)	≤10[a]	≤10	≤2[b]	≤5[b]
Current (A)	≤60 H⁺ ≤30 D⁺	≤35 H⁺	≤74 D⁺	≤83 H⁺ ≤68 D⁺
Current density (A/cm²)	≤0.20 H⁺	≤0.28 H⁺	≤0.23 D⁺	≤0.23 H⁺
Accel. size (cm)	16×45	12×27	12×43	12×48
Transparency (%)	40	40	60	60
Apertures (type & number)	262 holes	1020 holes	45 slots	55 slots
Aperture size	1.2-cm diam	0.4-cm diam	0.6×12 cm	0.6×12 cm
Bucket type	Checkerboard	Azim. line cusps	Axial line cusps	Axial line cusps
Atomic ion fraction (%)	≤87	≤91	≤85	≤74

[a] Limit of system: source has been designed and tested at 15-s pulse length.
[b] Limit of system: source has been designed for and tested at 30-s pulse length.

The superimposed filter in this case is produced by a special "magnetic tent" arrangement where a complete hoop of north poles is mounted near the accelerator end and the corresponding south poles all lie along a line in the center of the back plate. All these filter arrangements have been very effective, and atomic ion fractions in excess of 80% have been obtained in these large multicusp ion sources. Table 13.1 summarizes the performance of the large ion sources now being used in JET, JT-60, TFTR, and DIII-D.[10]

13.2 Specific Ion Source Design (the U.S. CLPS)

In order to be usable for neutral beam injectors, the ion source must satisfy the following criteria: (1) be capable of long-pulse operation, and have (2) uniform current density profile, (3) low fluctuation level, (4) high atomic ion species, (5) low impurity level, (6) high gas and electrical efficiencies, and (7) high reliability and durability. The U.S. CLPS provides a typical example of the large-area, multi-aperture ion source now being employed in the neutral beam injector system of all major tokamaks. A schematic diagram of the CLPS is shown in Figure 13.2. The plasma generator is a rectangular magnetic multipole bucket with an extraction area of 12 × 48 cm.[8-12] The chamber walls form a box 24 cm wide, 57 cm long, 30 cm deep, and with 40 rows of samarium-cobalt magnets around the outside perimeter (Figure 13.3). Because of its superior thermal characteristics, copper was selected as the housing wall despite its structural limitations. A relatively thin wall cross section, 3 mm, in the area of the magnet pole faces is required to keep cusp-field levels high on the plasma side of the wall. Stainless-steel bars brazed to the housing wall, at intervals consistent with magnet and water-cooling subassembly locations, provide the necessary reinforcement to withstand the vacuum loading. The filament sandwich consists of two large adjoining plates on which the 32 (or 34) filaments are mounted. The filament sandwich is made of gun-drilled, water-cooled copper. Typical filament currents were 4000 A during preheat, and 3100 A during arc conditions. Arc currents were typically 900 A for 70-V, 65-kW operation. The back copper plate is actively cooled to dissipate power loading by backstreaming electrons from the accelerator. On the rear side of the back plate a central row of magnets is mounted, and combined with a loop around the border of the 10 × 40 cm region these form the cusp fields for the back plate. The chamber walls and the back plate normally serve as anode for the source.

13.3 Large-Area H⁻ or D⁻ Sources

In order to heat plasmas in future fusion reactors and to drive current in tokamaks, multiamperes of very-high-energy neutral beams will be required.[13] The high neutralization efficiency (>60%) of H⁻ or D⁻ ions enables them to form atomic beams with energies in excess of 150 keV.[14]

There are different approaches for producing H⁻ or D⁻ ions. A large self-extraction H⁻ source based on surface conversion of positive ions has already been operated successfully to generate a steady-state H⁻ ion beam current greater than 1 A.[15,16] In this H⁻ source, a water-cooled cesiated molybdenum converter is inserted into the multicusp plasma generator. By biasing the converter negatively with respect to the plasma, positive ions are accelerated across the sheath and they impinge on the converter surface. H⁻ ions that are formed at the converter are then accelerated back through the sheath by the same potential. The bias voltage on the converter thus

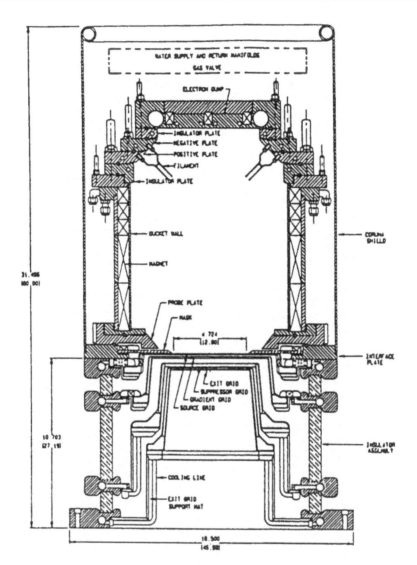

FIGURE 13.2
Schematic of the U.S. common long-pulse source (CLPS).

determines the energy of the negative ion leaving the source (self-extraction). The converter surface is normally curved to geometrically focus the H⁻ ions through the plasma to the exit aperture. In this type of source, the H⁻ output current is directly proportional to the area of the converter.

The advantages of the surface conversion H⁻ source are low operating gas pressure (~1 mtorr) and very small electron content in the H⁻ beam. The source can be operated in steady-state or long-pulsed mode. However, source operation always requires the presence of cesium, which can cause voltage breakdown in the accelerator column.

Barium is a metal that has a reasonably low work function and it has a much lower vapor pressure than cesium. Experimental work at the FOM Institute, Amsterdam, demonstrated that a pure barium metal surface can provide reasonable probabilities for H⁻ ion formation.[17] Steady-state beams of D⁻ ions with currents greater than 100 mA have been achieved by employing a large multicusp source with

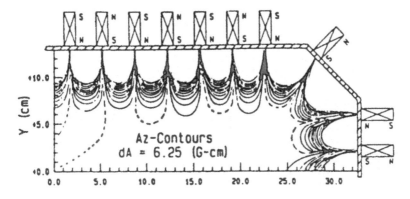

FIGURE 13.3
Magnetic field distribution of the U.S. common long-pulse source.

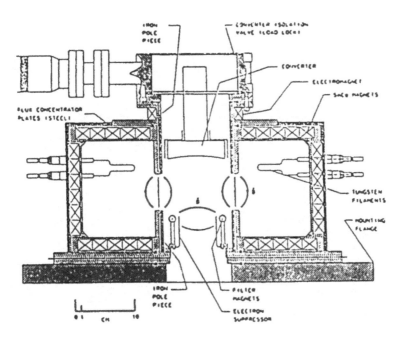

FIGURE 13.4
Schematic diagram of a barium surface conversion H⁻/D⁻ source.

a 6-cm diameter barium converter (Figure 13.4).[18] Experimental investigations have been performed to develop large surface conversion D⁻ sources with the use of barium metal for the production of multiampere D⁻ beams for fusion applications.[19]

H⁻ ions can also be extracted directly from a hydrogen discharge plasma. This technique of generating H⁻ ions has the advantage over surface production processes in that it requires no cesium, and it uses presently developed multicusp positive ion sources. Experimental results demonstrate that low-emittance H⁻ beams with current densities higher than 250 mA/cm² can be obtained from a magnetically filtered multicusp volume H⁻ source.[20] Surprisingly, it has been found that the H⁻ output current is greatly enhanced (more than fivefold) if cesium vapor is added to a hydrogen discharge, resulting in an extractable current density exceeding 1 A/cm².[21] The improvement in H⁻ yield is accompanied by a large reduction in the extracted electron current as well as the optimum source operating pressure. Based on these

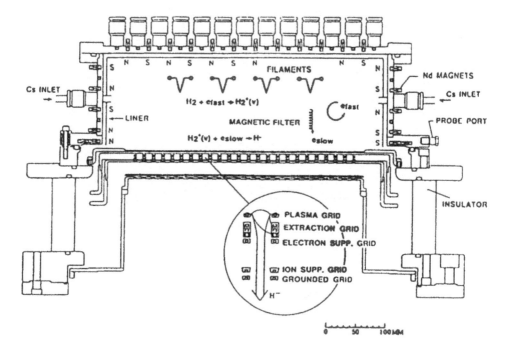

FIGURE 13.5

Cross-sectional view of the JAERI large-volume H⁻ source together with the multiaperture extraction system.

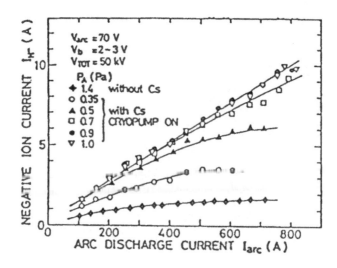

FIGURE 13.6

The dependence of H⁻ current on the arc current for source operation with and without cesium.

findings, large cesiated multicusp volume H⁻ sources are now developed by various fusion laboratories to generate multiampere H⁻ beams.[22] Figures 13.5 and 13.6 show the large, multiaperture, multicusp source developed in JAERI, Japan, which produces H⁻ current higher than 10 A.[22] In fact, one could now operate these large-area sources to provide H⁻ ions in much the same manner as is now done to provide H⁺ ions for neutral beam systems.

13.3.1 Operating Data for the U.S. CLPS (Other Sources Have Similar Data)

- Arc voltage: 60 to 90 V
- Arc current: 500 to 1000 A
- Arc power: tens of kilowatts
- Cathode data: ~10 V/100 A (for a 1.5-mm-diam tungsten filament)
- Duty cycle: depends on the tokamak operation
- Time structure of the discharge: pulse length in the order of several seconds
- Magnetic field data: typically 4 kG for SmCo magnets
- Gas pressure: several millitorr

13.3.2 Ion Source Performance

See Table 13.1.

13.3.3 Ion Source Application

- Research: in neutral beam systems for fusion research
- Industry: same sources has been used to produce large nitrogen ion beams for surface modification

13.3.4 Deliver or User

The large-area multicusp sources are developed by Lawrence Berkeley Laboratory (U.S.A.), Culham Laboratory (U.K.), NIFS (Japan), and JAERI (Japan).

Users include Princeton Plasma Physics Laboratory (U.S.A.), General Atomic (U.S.A.), JET (U.K.), NIFS (Japan) and JT-60 (Japan), Harwell Laboratory (U.K.), Nissin Co. (Japan), and Hitachi Co. (Japan).

REFERENCES

1 G. R. Brewer, Ion Propulsion, Gordon and Breach, New York, 1970.
2 H. R. Kaufman, Adv. Electron Electron Phys., 36, 265, 1974
3. R. Limpaecher and K. R. MacKenzie, Rev. Sci. Instrum., 44, 726, 1973.
4. K. W. Ehlers and K. N. Leung, Rev. Sci. Instrum., 50, 1353, 1979.
5. A. J. T. Holmes, T. S. Green, and A. F. Newman, Rev. Sci. Instrum., 58, 6119, 1983.
6. Y. Okumura, H. Horiike, and K. Mizuhashi, Rev. Sci. Instrum., 55, 1, 1984; Y. Ohara, M. Akiba, H. Horiike, H. Inami, Y. Okumura, and S. Tanaka, J. Appl. Phys., 61, 1323, 1987.
7. K. W. Ehlers and K. N. Leung, Rev. Sci. Instrum., 52, 1452, 1981.
8. P. A. Pincosy et al., Rev. Sci. Instrum., 57, 2705, 1986.
9. M. C. Vella, W. S. Cooper, P. A. Pincosy, R. V. Pyle, P. D. Weber, and R. P. Wells, Rev. Sci. Instrum., 59, 2357, 1988.
10. Wulf B. Kunkel, Rev. Sci. Instrum., 61, 354, 1990.
11. L. A. Biagi et al., J. Vac. Sci. Technol., A2, 666, 1984.
12. J. A. Paterson et al., Proc. 11th Symp. Fusion Engineering, Austin, TX, 1985, 153; R. P. Wells, ibid., p. 160.
13. L. D. Stewart et al., in Proc. 2nd Int. Symp. Production and Neutralization of Negative Hydrogen Ion and Beams, Brookhaven National Lab., Upton, NY, 1980, 321.
14. K. H. Berkner, R. V. Pyle, and J. W. Stearns, Nucl. Fusion, 15, 249, 1975.
15. K. N. Leung and K. W. Ehlers, Rev. Sci. Instrum., 53, 803, 1982.

16. J. W. Kwan et al., *Rev. Sci. Instrum.*, 57, 831, 1986.
17. C. F. A. van Os et al., *SPIE Proc.*, 1061, 568, 1989.
18. C. F. A. van Os et al., *Bull. Am. Phys. Soc.*, 35, 2099, 1990.
19. J. W. Kwan et al., *Rev. Sci. Instrum.*, 63, 2705, 1992.
20. K. N. Leung et al., *Rev. Sci. Instrum.*, 59, 453, 1988.
21. K. N. Leung et al., *Rev. Sci. Instrum.*, 60, 531, 1989.
22. In *Proc. 6th Int. Symp. Production and Neutralization of Negative Ions and Beams*, Brookhaven, Upton, NY, 1992.

CHARACTERIZATION OF ION SOURCES

T. Jolly

14 INDUSTRIAL ION SOURCES AND ION SOURCE APPLICATIONS

14.1 Introduction

Ion sources have been used for a variety of industrial applications over the last 20 years. During this time, although applications have come and gone, the general range of applications has steadily increased. For example, the etching of bubble memory components, one of the first applications in the early 1970s, has ceased because of the disappearance of this product, and through-hole vias in gallium arsenide devices are now etched using reactive ion etching in a plasma discharge. Such losses are more than balanced by the arrival of new applications such as the deposition of laser gyroscope mirrors, and the etching of sensors and other micromechanical components.

In this chapter we restrict ourselves to the types of ion sources that are available commercially, and to the applications of these sources. We thus consider only broadbeam gridded ion sources, and gridless assist ion sources. The chapter starts by looking at current applications. A review of the requirements for practical industrial ion sources is followed by a description of current ion source types. The chapter concludes with a section on ion beam neutralization.

Important applications for which industrial sources are purchased fall into three categories. In the first category is the milling or cleaning of surfaces by means of inert or reactive ion beams. The process is known as ion beam milling (or etching) or reactive ion beam etching (RIBE). The second category concerns the deposition of coatings, by using the flux sputtered off a target by the ion beam. This is known as ion beam sputter deposition (IBSD). (There is also a direct deposition technique, as is described below.) The third category involves the concurrent use of an ion beam to improve the properties of a coating that is being deposited using a different technique. This is ion-assisted deposition (IAD). It is most often used to assist electron beam evaporation of oxide and metallic coatings.

Until recently, any one type of ion source was expected to be capable of performing all of the above applications, and this was generally the Kaufman type of source, as described below. A recent trend is the adoption of more specialized sources for certain applications. The first example to appear was the end-Hall source, designed specifically for IAD. Another example would be the electromagnet ECR ion source, which is only suitable for milling applications where conventional neutralization of the ion beam is not necessary.

There are a number of ion sources in industrial use that are not covered in this chapter. The Freeman source[1-3] (see Chapter 2.4) and recent rf equivalents[4] (Chapter 2.7) used in ion implanters, and the liquid metal ion source (LMIS)[5] (Chapter 2.15) used mainly in lithography mask repair systems, are not covered because they are only met as a component of dedicated systems. Likewise, sources encountered in analysis techniques such as secondary ion mass spectrometry (SIMS),[6] and in charge neutralization devices for scanning electron microscopes[7] are not described either.

14.2 Applications

14.2.1 Ion Beam Milling

One of the first industrial applications of ion beam milling was in the etching of patterns in metal coatings on bubble memory devices. The surface of the device was coated with metal before applying a thick photoresist pattern. The device was then exposed to the ion beam. Photoresist and exposed metal were sputtered away together until all of the exposed metal was removed. Excess resist was then removed to leave the pattern of metal pole pieces. Many magnetic recording heads are currently made in a similar way.

Ion beam milling is of particular value where one wishes to etch through a layered structure with the least possible change in etch rate, for example, gold on gallium arsenide, both of which etch at about 150 nm/min. For a more complete list of etch rates, see Reference 8. Industrial processes may demand large ion sources and cassette-to-cassette operation to achieve the required throughput, and Figure 14.1 shows such a machine with a 12-in. wafer capability, and a 40-cm diameter ion source giving a beam of up to 4 A at 2000 eV, and 1.5 A at 100 eV.

14.2.2 Reactive Etch Processes

A number of important processes employ reactive gases, either introduced into the ion source (RIBE) or aimed directly at the substrate, as in chemically assisted ion beam etching (CAIBE).* This requires considerable extra complexity in the system, and in an industrial context usually makes a loadlock necessary to allow the throughput in safety.

The reactive process may be used to allow extra throughput, and etch rates of up to 200 nm/min have been reported in diamond by using CAIBE with NO_2,[9] and several micrometers per minute in GaAs with chlorine, running the ion source on argon concurrently.[10] It may also be used for selectivity of the substrate etch rate over the mask, and it allows the etching of deep vertical walls in many III-V compounds.[11] (See Figure 14.2.) One use is in the manufacture of solid-state laser diodes.[12]

An advantage over similar plasma processes such as reactive ion etching (RIE) is that the ion flux can be inclined from normal incidence, and several proprietary processes use this capability. Many other RIBE and CAIBE processes have been investigated over the last decade,[13] but the useful ones are generally considered proprietary. It is probable that no major silicon processes are performed at present with reactive ion beam processes, a state of affairs that is discussed in Reference 14.

---

* Also known as ion beam-assisted etching (IBAE).

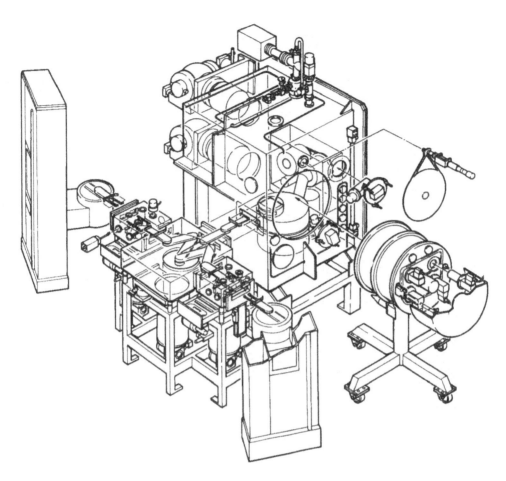

FIGURE 14.1
Cassette-to-cassette ion beam milling system with 40-cm ion source, for substrates of up to 12-in. diameter. (Oxford Instruments, England. With permission.)

14.2.3 Ion Beam Sputter Deposition

Continued development of multilayer dielectric mirrors for laser gyroscopes shows that the highest quality of optical coatings is still produced by using ion beam sputter deposition (IBSD). In this technique, a broad beam ion source is used to sputter coating material off a metal or dielectric target onto the substrates. At present, mirrors can be produced with total losses of 1.6 ppm at 850 nm.[15]

These coatings are also used for high-power laser mirrors, both visible light and infrared. However, they are only just starting to be used more generally for commercial-grade optical coatings, as well as in other specialized applications such as X-ray mirrors and graded-index optical coatings. A summary of recent work in all these fields is given in Reference 16.

Figure 14.3 shows a typical layout for a deposition system. The geometry is based on the prime requirements of ensuring (1) that the absolute minimum of beam current hits any component other than the sputter target, and (2) that the maximum amount of the sputtered flux is used to coat the substrates, and the minimum amount is back-sputtered into the ion source and onto its grids.

This results in the use of a large chamber, with the major components grouped as closely as possible together in the geometry shown. Target and substrates are both mounted vertically to minimize contamination, as may be seen in Figure 14.4.

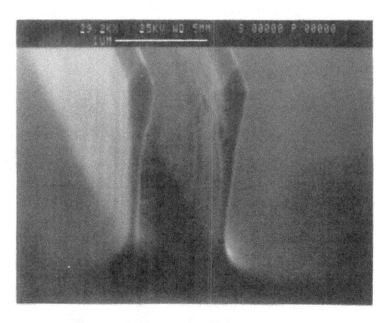

FIGURE 14.2
Chlorine reactive ion beam etched GaAs. Sample was masked by e'-beam direct-write using SAL603 negative resist (visible on top of the features.) Etching was with 300-eV chlorine ions at 0.6 mA/cm². The undercut wall angle is largely due to the divergence of the ion beam. (Vawter, G. A. and Wendt, J. R., Sandia National Laboratory, U.S.A. With permission.)

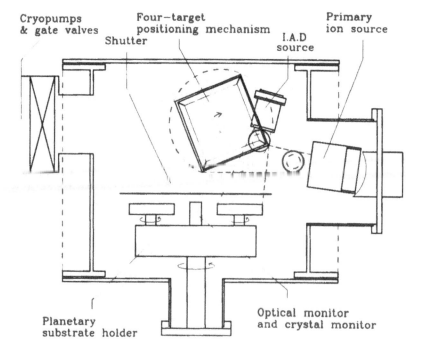

FIGURE 14.3
Optical coating system with 15-cm deposition source and four targets for making multilayer optical coatings by ion beam sputter deposition.

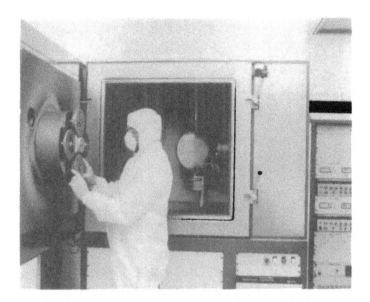

FIGURE 14.4
Optical coating system installation of the type shown in Figure 14.3. (Howe, A. T. and Phillips, D. D., Amoco Laser Company, Naperville, IL, U.S.A. Reprinted from the July 1993 issue of *Photonics Spectra*, © Laurin Publishing Co., Inc. With permission.)

14.2.4   Direct Ion Beam Deposition

Several techniques have been developed for depositing diamondlike coatings (DLC) with ion sources. These generally involve the sputtering of graphite targets.[18] However, most of these applications are not commercially viable, and have been superseded by techniques that are plasma based. One exception is the direct deposition of DLC onto plastics. A hydrocarbon such as methane is fed into the ion source, and the ion source flux coats substrates put in its way. As long as the beam energy is less than a few hundred electron volts, the coating is deposited more quickly than it is sputtered away by the same flux.[19,20]

It is clearly possible to deposit coatings of other materials directly if the ion beam energy is reduced further. Promising work has been published on the deposition of X-ray mirrors by decelerated beams of ions,[21] which suggests that this technique may become important despite its complexity.

14.2.5   Ion-Assisted Deposition

Many coating systems fitted with electron beam evaporation hearths have had ion assist guns fitted. A typical installation is shown in Figure 14.5. The simultaneous bombardment of a growing film can improve many properties of the film. In the case of metal coatings one is often aiming for better adhesion,[22] and in the case of optical coatings, for lower stresses and improved adhesion as well as higher density and lower porosity.[23] For summaries of the benefits of ion-assisted deposition, see References 22 and 24.

Ion assistance is also used with ion beam sputter deposition, but less often because the improvements are much less marked. However, in work at Laserdot,[25] 10.6-μm mirrors have been fabricated with assisted ion beam sputter deposition. The mirrors were made of Y_2O_3 overlaid on silver, deposited on a silicon substrate. These mirrors showed no microdamage in tests that extended up to 15 h at 100 kW/cm^2.

FIGURE 14.5
Production vacuum coater configured for ion-assisted electron beam deposition using an end-Hall ion source. (From Yates, D., Omitec Thin Films Ltd, England. With permission.)

IAD can also prove useful in the manufacture of X-ray mirrors by means of IBSD. It is found that the surface roughness and degree of amorphousness of SiO_2 layers is strongly affected by the energy of the assist ions.[26] The roughness is minimized by the use of 100-eV argon ions, while the amorphousness is maximized at about 50 eV. Platinum and carbon behave similarly, though with different optimum energies. However, ion assistance shows no effect on the roughness of nickel.

14.3   Ion Source Requirements

14.3.1   Source Parameters

In the typical ion beam milling application one wishes to maximize the yield of sputtered atoms per unit of heat flow into the substrate. The sputtering yield S, measured in atoms sputtered per incident ion, increases with ion energy, and reaches a maximum at an ion energy E of a few thousand electron volts. (At higher energies, ion implantation occurs instead.) However, if the yield per unit of heat S/E is plotted against E instead, the maximum occurs at a much lower energy. For many materials, this yield optimizes at an ion beam energy in the region of 500 eV.[8] Some applications do require higher energy, and sputter deposition in particular is often performed at 1000 to 1200 eV. This is because the higher energy leads to a higher deposition rate, and can also improve the coating quality (due to the energy put into the growing film by elastically scattered ions) in both cases at the expense of higher target temperatures.

Reactive etching and direct deposition is performed at energies between 50 and 300 eV. Higher energies are not necessary because much of the process is chemically driven. Indeed, one of the benefits of the process is that low energies may be used, resulting in the creation of fewer defects, which is especially important in III-V compounds. Similar energies are used for direct deposition. Energies below 50 eV are not of general industrial interest at present, which is fortunate because most gridded sources have greatly reduced output at these low energies. (See below.)

Typical beam current densities for many industrial processes are about 1 mA/cm^2 measured at the target of the ion beam, and this is the current density used in standard ion beam milling rate data.[8] Current densities of up to 5 mA/cm^2 are used in high-rate milling and deposition processes, but above these densities it can become very difficult to cool the substrate adequately: a problem that is found in many other vacuum processes.[27]

Direct deposition is a very severe test of grid design, because the current density needs maximizing at low energy, but other low-energy processes generally do not need such high currents, since other factors limit the rate of the process. In CAIBE, for example, one may be limited by the maximum feasible partial pressure of the process gas in the chamber, or by the diffusion of reaction products out of a deep trench.

Ion assistance also uses energies of about 50 to 300 eV. Where the end-Hall source is used, the ions are actually generated with a wide range of energies up to the maximum cited, but the ion energy is not critical anyway. The ion energy is high compared with the energy of the evaporated flux from the electron beam hearth, since 1 eV = 11,600 K. Thus, current densities need only to be relatively low.[24]

14.3.2 Process Gases

The source needs to be able to withstand the process gases in use. Although argon is by far the most common gas, chlorine and oxygen occur sufficiently often that most sources are designed to be resistant to them. Preferred materials are thus stainless steel (316 or 304), alumina, quartz, molybdenum, tantalum, and high-density graphite. Copper wire is also regularly used, as well as 400-series stainless steel for magnetic pole pieces, and tungsten or tantalum for filaments.

Industrial ion sources are fabricated generally of stainless steel, and so must be protected from wet chlorine, especially if they contain magnetic components made of the low-chromium 400-series stainless steel. This is one of the reasons for the advisability of loadlocks on production systems. In theory, one should suffer problems from such reactions as copper with chlorine, and graphite with oxygen. The effects are relatively minor in practice due to the low partial pressure of the process gases during use. Even graphite grids on oxygen ion sources do not erode as fast as simple ion arrival-rate calculations would suggest. Weight-loss studies on a 15-cm graphite grid set in an oxygen rf ion source predict a screen grid lifetime of about 500 h at 1 mA/cm^2.[28]

Certain problematic materials such as bromine and iodine require either the elimination of stainless steel, or the heavy plating of stainless-steel components with a metal such as nickel.

Gridded sources need to be operated at relatively low pressures, or else there will be too much direct impingement of the grids by scattered ions, shortening their life. As is the case with most ion source parameters, there are no rigid rules, but pressures generally need to be below 5.0×10^{-2} Pa measured in the process chamber, and preferably below 2.0×10^{-2} Pa. This means that the maximum gas flow that can be fed into the ion source is often set by the pumping system fitted to the process chamber.

The optimum gas flow depends on a number of factors, including the efficiency of the ion source and the required beam current. A significant amount of gas reenters the ion source from the process chamber, and so a poor pumping system will actually make an ion source seem more efficient! Typical flow rates, however, are about 2 to 5 sccm for a 3-cm ion source and 30 to 50 sccm for a 40-cm ion source.

Lower gas flows always start to reduce the beam current available. Minimum gas flows for a Kaufman source often correspond to a chamber pressure of about 1.0 × 10^{-2} Pa, but ECR sources in particular can usually operate at pressures below 10^{-3} Pa.

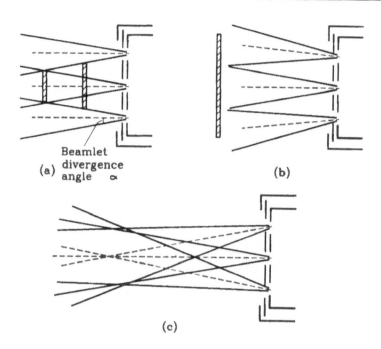

FIGURE 14.6
Microdivergence. (a) An ion beam with parallel beamlets. The area of uniform etching decreases with distance as shown. (b) With divergent beamlets. The area of uniform etch rate is increased, but the microdivergence is not uniform. (c) With focused beamlets. Note the sudden expansion of the overall diameter downstream of the ion source.

14.3.3 Uniformity

For an assist source, a wide beam divergence and a beam profile of a broadly Gaussian shape is often appropriate, because the flux can be aimed to cover a planetary substrate holder with a similar spread to the evaporation hearth. For etching applications, however, many processes require uniformities of ±3%.

When an ion source is specified, an important issue here is often overlooked. As may be seen in Figure 14.6a, an ion beam where all of the apertures in the grids are arranged parallel, produces parallel beamlets. In an ion source of, say, 150-mm beam diameter, it is not possible for such a beam to provide a uniform flux profile over a target diameter as large as 150 mm, because points at the edge of the target will clearly be struck by flux from fewer beamlets than points at the center of the target. The further the beam travels, the smaller the uniform area becomes. It is possible to arrange for uniform etching over a substrate of 150-mm diameter, or even larger, but only if one carefully aims the beamlets from the apertures, or blocks off some of the central beamlets. (See Figure 14.6b.) This first action in particular has severe effects on the microdivergence of the beam, and this is discussed further in the next section.

14.3.4 Microdivergence

The divergence angle of a beamlet, α, is usually defined in ion beam processing technology as the half-angle that contains 95% of the ion beam flux. (See Figure 14.6a.) It is also known as the microdivergence. The concept of microdivergence was introduced[29] to emphasize the independence of the external divergence of an ion beam (its macroscopic divergence) from the divergence at any point inside the beam. In practice, in simple two- and three-grid designs, the nature of the grid as a structure that accelerates the ions and then decelerates them ensures that all of the beamlets are divergent to some degree. Parallel or convergent beamlets are not possible. On the other hand, the external divergence of an ion beam can be made convergent or divergent by the expedient of deflecting the beamlets as shown in Figures 14.6b and 14.6c.

The nature of the microdivergence becomes critical in situations where all aspects of an etch process, including the wall angle and the etch rate at the bottom of trenches, must be constant all the way across a substrate. In this case, it is not sufficient for the current density to be the same at all points across the substrate; the angular distribution of the arriving flux must be the same at all points also (Figure 14.6a). An example of the effects of a nonuniform microdivergence can be seen in Figure 14.7.

14.3.5 Grid Design

(See also Chapter 4.)

Broad-beam gridded ion sources are in use with ion optics consisting of one, two, or three grids. All of the gridded ion sources described below can use two grids as shown in Figure 14.8. This, as the most common configuration, will be described first, using the following nomenclature: the inner (screen) grid has a thickness t_s, and apertures of diameter d_s; the front (accelerator) grid has a thickness t_a, and apertures of diameter d_a; and the effective gap between the grids is l_e. The potentials on the accelerator and screen grids are V_a and, approximately, V_b. The ratio of beam energy to total energy, R, is defined as

$$R = V_b / (V_b - V_a) \tag{1}$$

In two-grid ion optics, the inner (screen) grid is attached to the discharge chamber and so is at approximately the same potential as the plasma inside: typically +50 to +1500 V. The energy of the extracted ions is set by the voltage of the beam power supply, which is connected to the anode in the discharge chamber, and is unaffected by the voltage on the accelerator grid.

The accelerator grid is held negative to ground for a number of reasons. First, the voltage must be below ground by more than 100 V approximately, or electrons backstreaming into the ion source will be recorded as ion beam current. The minimum acceptable value for V_a to prevent backstreaming is given by the semiempirical equation[30]

$$(1 - R)l_e/d_a \sim 0.2 \exp(-t_a/d_a) \tag{2}$$

Second, the potential difference between the grids must be adjusted to vary the focusing effect of the field between the grids. Thus, the potential difference can be set for minimum impingement of the ion beam onto the accelerator grid, or for minimum divergence of the ion beam, or for any suitable compromise between the two. In some applications, of course, one wants to adjust the voltages for a greater divergence anyway. An impingement current of <5 to 10% of beam current will generally ensure an acceptable grid life. Third, large magnitudes of accelerator voltage can cause erosion by returning ions of the front face of the accelerator grid. This will not usually affect grid life, but may well cause unacceptable process contamination. The effect appears to be caused by low-energy charge-exchange ions generated by collisions in front of the ion source between the ion beam and chamber gas. Below a certain

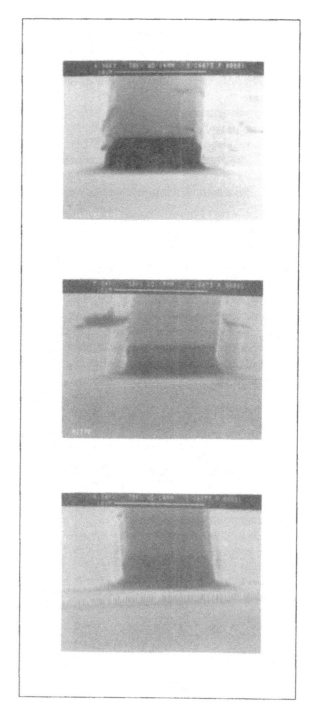

FIGURE 14.7
Micrographs showing the effect of uniform etch rate, but nonuniform microdivergence on a 3-in. silicon wafer, in an inert-gas ECR ion source etch process. (Fuchs, J., University of Jena, Germany. With permission.)

chamber pressure and ion current the effect is not noticed. Fourth, and obviously, the potential difference between the grids must not exceed the breakdown value for the grid design in use: damage will not occur immediately, but repeated breakdowns are not advisable.

The extracted current depends on the potential difference V between the grids, and solving Poisson's equation for one-dimensional flow leads to the well-known result for this application. For ions of charge q and mass m, the current in each beamlet is given by:

$$I = \left(\pi\varepsilon_0/9\right)\left(2q/m\right)^{1/2}\left(V^{3/2}d_s^2/l_e^2\right) \tag{3}$$

From this it will be noted that the extracted current depends on $V^{3/2}$. The current per beamlet is a function of d_s/l_e, and so is unchanged if the grid design is scaled down. However, the current out of a given size of source is then increased, since the total number of beamlets is increased. Practical engineering and maintenance concerns limit the grid gap to about 1/200 of the source beam diameter for pyrolytic graphite grids, and about 1/500 of the diameter for dished molybdenum grids. Likewise, a screen grid thinner than 0.25 mm is generally impractical, and this implies a minimum screen grid hole size of about 1.0 to 1.5 mm.

As the optimum grid gap varies from about $0.5d_s$ for maximum beam current to about $1.0d_s$ for low-divergence operation, it will be seen that the above argument suggests that the maximum current density out of the grids is limited in small ion sources by the screen grid thickness, and in larger sources by grid gap stability. This is basically true, but one must also note that Equation 3 may not be obeyed so well in smaller grids. The deterioration is often stated to start at about $d_s = 2.0$ mm (see Reference 33, for example), but more detailed work suggests that holes as small as 1.0 mm can be used with full advantage if the plasma density in the discharge chamber is high enough.[31]

Small ion sources may have grids of any reasonable material, even silicon can be used to advantage with oxygen and in silicon etching,[32] but larger sources use pyrolytic graphite or dished molybdenum, and the largest sources, over 20-cm diameter, almost invariably use dished molybdenum.

Under normal operating conditions, as can be seen from Figure 14.8, the plasma boundary retreats over the apertures in the screen grid. This gives a focusing effect to the ions as they pass through the grid structure, and so the size of aperture in the accelerator grid can be made smaller with a relatively modest effect on the extracted beam current. This is usually done in larger ion sources in order to minimize the gas consumption of the ion source. Diameters as small as $d_a \sim 0.67d_s$ are used. On smaller ion sources the apertures are commonly the same size on each other to allow the maximum extracted beam current.

Three-grid ion optics can be used on smaller ion sources, but the extra complexity is seldom justified in industrial applications. The normal implementation is to add a third decelerator grid, biased to ground potential, downstream of the accelerator grid. The chief benefit of adding this grid is that it results in a lower beamlet divergence angle α under typical operating conditions. For example, where a two-grid source running at 500 eV, with $V_a = 100$ to 500 V, has a divergence of $\alpha = 7.7$ to 18.3°, depending on V_a and the grid gap, a three-grid source will have a divergence of 6.6 to 13.3°. In both cases, the higher voltage corresponds to higher current, as is shown by Equation 3. In practice, though, as two-grid ion optics can be made smaller than three-grid optics, they have more reserve beam current capacity, and this makes the focusing advantage less significant.

A good summary of two- and three-grid ion optics design is given in References 33 to 35. For a broader study of grid design, the reader is recommended to Chapter 4 of this volume and to Chapter 5 of Reference 36.

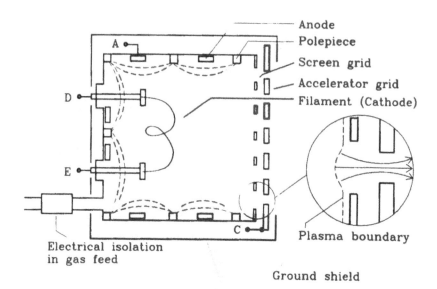

FIGURE 14.8
Kaufman ion source.

Single-grid ion sources have occasionally been used in industrial applications. In grid design, the apertures are made very small, 100 to 300 μm in diameter. The ions are accelerated between the plasma sheath and the single (accelerator) grid. Due to the great number of holes, the ion beam current can be very high at low energies, even though Equation 3 is not obeyed at all well for apertures below about 1 to 2 mm in diameter. At high energies, the grids glow red hot due to bombardment by ions that are not extracted, but at low energies, 10 to 200 eV, they produce very high beam currents, more than 10 mA/cm². Grid lifetime is very limited by erosion, if not by accidental breakage; beam uniformity is strongly affected by warping of the grid material. These factors have limited the acceptance of single-grid optics. However, they do have certain advantages in that they generate a precisely measurable beam current of very-low-energy ions, with a relatively well-defined beam energy. Grids are manufactured by photolithography, in nickel, silicon, molybdenum, or other metals. For more details see References 17, 32, and 37.

14.3.6 Ion Beam Shaping

The dominant factors on the shape of the ion beam from a gridded ion source are the radial uniformity of the plasma inside the ion source, the radial uniformity of the perveance of the grids, and the microdivergence of the beamlets. In all ion sources the plasma density drops towards the perimeter of the discharge chamber. In addition, ECR sources and helical antenna rf ion sources tend to have a drop in density at the middle.[38] This can actually be an advantage, because a modest drop in beam intensity at the middle can improve the uniformity of the beam at the target. The microdivergence α, as defined above, depends principally on the beam energy V_b, the accelerator voltage V_a, the grid dimensions l_e and d_a, and the beam current I_b. It is also affected to a smaller extent by the screen grid thickness t_s and the potential difference between the plasma and the screen grid (equivalent to the discharge voltage V_d in a Kaufman ion source).

It is strongly recommended that one has the means to calculate beam divergence. This is best done by computer simulation, and several packages are available for desktop computers.[39] This information is important in process development for two main reasons: (1) the beam divergence is often a process variable in its own right, as can be seen from Figure 14.2, and (2) it may well be necessary to change a parameter, such as the beam current, and to maintain the same divergence, to avoid disturbing the etch profile.

Assuming a uniform plasma and parallel beamlets, then the area of uniform etching decreases quite quickly with working distance, and at a distance d from the ion source, it is obvious from Figure 14.6a that the area of optimum uniform etching from a source of diameter D_s is only D_e, where

$$D_e \sim D_s - 2d \tan \alpha \qquad (4)$$

To make the ion source diverge and obtain greater areas of etch or assist uniformity (noting the comments above about uniform microdivergence), one needs to deflect the beamlets outward as in Figure 14.6b. There are two ways used to achieve this. One way is to deliberately fabricate the grids so that the holes do not align.[40] As shown in Figures 14.6b and c, the beamlets are deflected in the direction opposite to the offset; one can regard it as being due to attraction of the ions towards the protruding edge of the accelerator aperture as they pass particularly near to it. The effect is limited to modest beam currents and to modest deflections of 10 to 20°, or else direct impingement of the accelerator grid becomes excessive.

The other way to deflect beamlets is to use convex-dished grids, of the type originally developed for ion thruster applications in space,[30] and arrange the beamlets to emerge at normal incidence to the grid surface. This allows the beamlets to operate through aligned apertures at their maximum perveance, giving higher currents. (See also Chapter 4.)

Offset grids and concave-dished grids[41] can also be used to focus the ion beam, and this can be of value for sputter deposition off a target, and, just occasionally, for milling small areas at high rates. As can be seen from Figure 14.6c, it is not easy to arrange for the beam as a whole to converge significantly, and the uniformity of the beam at the point of crossover will have degenerated to an approximately Gaussian profile. Beyond the crossover point, the beam diverges very rapidly.

14.3.7 Maintenance

Ion source maintenance can be a significant factor in the downtime of a production system, especially if the downtime is unscheduled. Many of the improvements to ion sources over the last decade have been introduced to reduce the maintenance time of the ion source, or to make the source intrinsically more reliable. Examples of the first type would be self-aligning grids and plug-in exchange neutralization guns; examples of the second type would include spare filaments in Kaufman sources, the wider use of gridless ion sources in assist applications, and the general adoption of switch-mode power supplies.

There is a strong move towards the maintenance of ion sources by using "drop-in" replacements, so that parts can be maintained off-line. Probably most industrial ion sources can now take prealigned grid sets, and items such as quickly replaced quartz liners and filament assemblies are also becoming more common. The reader will have to approach the individual ion source vendors for more information.

Reduced maintenance is one of the major factors cited by users intending to change over to rf and ECR ion sources in production systems. Nonetheless, when choosing an ion source one should consider each application on its individual merits.

For example, in many optical deposition processes, the front of the grids on the ion source need cleaning every 100 h or so because of the amount of back-sputtered material.[42] This interval is not very much longer than the filament lifetime in a multi-filament Kaufman ion source. As another example, in one 40-cm ion source production etch process known to the author, the filaments are changed weekly, in 4 min, as part of the routine system maintenance. Also, of course, rf and ECR sources may have process disadvantages over filament sources, as is discussed below.

14.4 Ion Source Types

14.4.1 Kaufman Ion Sources

(See also Chapters 2.2, 2.6, and 2.13.)

The first broad-beam electron-bombardment ion sources were developed in 1960[43] as part of the NASA's development of space propulsion thrusters. Development work in the U.S. and Europe led to the production of highly reliable ion sources, generally running on mercury, and using hollow cathodes both to generate the plasma and for use as beam neutralizers.[30]

In the 1970s, these sources were adapted for use in industrial ion beam milling and deposition applications, using argon instead of mercury, and refractory metal filaments instead of hollow cathodes.[44] Hollow-cathode sources have indeed been used industrially, but the extra complexity has limited their popularity, despite their greatly superior lifetime over filaments. Hollow-cathode neutralizers are more common.

The essential ingredients of a Kaufman type ion source are shown in Figure 14.8. Electrons are emitted into the discharge chamber, usually by a hot filament. Tungsten and tantalum are the common materials, although others have occasionally been used, notably thoriated iridium for use with oxygen.[45] (See also Chapter 2.1.) The electrons are confined by a weak magnetic field, which causes them to oscillate and prevents them from reaching the anode except by diffusion. The magnetic field may be a divergent solenoidal field, or a multipole field, as shown. The anode can be in rings as shown, or, alternatively, can be in larger panels, depending on the shape of the magnetic field used.

The plasma will consist of neutrals with a density of about 10^{19} m^{-3}, and of ions and electrons with a density about 10% of this. The Bohm criterion requires that ions arrive at the boundary with a minimum energy of

$$V_0 > kTe/2 \tag{5}$$

This condition is sufficient to ensure that the arrival rate of ions at the boundary of the ion source is substantially greater than that of neutrals, allowing an extraction rate for ions greater than for neutrals. This velocity also means that the ions travel equally in all directions to the boundary of the discharge chamber.

The ion beam is extracted from the front of the source, through grids that are either in a field-free region, or are oriented normal to the divergent magnetic field. Two-grid optics are used, or the three-grid or one-grid alternatives described above. Sources are currently available in sizes from about 1 cm in diameter to greater than 40 cm in diameter. (See Figure 14.9.)

The circuit for this ion source is shown in Figure 14.10. The discharge voltage is an important variable in the control of the sources, because it directly affects the energy of the ionizing electrons. If the voltage is too high, significant fractions of

FIGURE 14.9
Forty-centimeter ion source with beam current of 1.5 A at 100 eV, and 4 A at 1500 eV. (Oxford Instruments, England. With permission.)

doubly charged ions will be produced. Doubly charged ions are not welcome, not just because they confuse the measurement of beam current, but also because they can do excessive damage in the etching of III-V compounds.[46,47] In small ion sources, the most likely mechanism for their production is direct excitation to the doubly charged state. For argon, the appearance potential for this process is just over 40 eV. In large ion sources, the chances of a singly charged ion being ionized again become significant, and this will occur at a lower voltage.[48]

The hallmarks of a Kaufman source would in fact seem to be (1) the confinement of electrons in a large discharge chamber by means of a weak magnetic field, and (2) the use of a grid extraction system, either in a field-free region or at right angles to the divergent magnetic field. Nonetheless, similar ion sources have been developed in other technologies, notably the duopigatron,[49] developed as part of the fusion plasma neutral beam work. (See Chapter 2.3.) This source, as Forrester pointed out in his taxonomy of ion sources in Reference 36, is functionally equivalent to a hollow-cathode Kaufman ion source.

One main benefit of Kaufman type ion sources is that they can readily be made in a wide variety of shapes. One shape of industrial importance is the rectangular source shown in Figure 14.11.[50] This figure actually shows three ion sources arranged to treat the surface of a 1-m wide sheet of material passing in front. Note that each source has its own neutralizer. Similar arrangements have also been produced to perform ion beam sputter deposition onto passing sheets of material.

Kaufman ion sources have given excellent service for many years. In RIBE their main drawback is, of course, filament lifetime, especially in oxygen, and many users of RIBE are moving over to filamentless ion sources. However, other concerns, such as contamination from the filaments in particular, have been greatly exaggerated. Indeed, all commercial ion beam sputter deposition of high-quality optical coatings

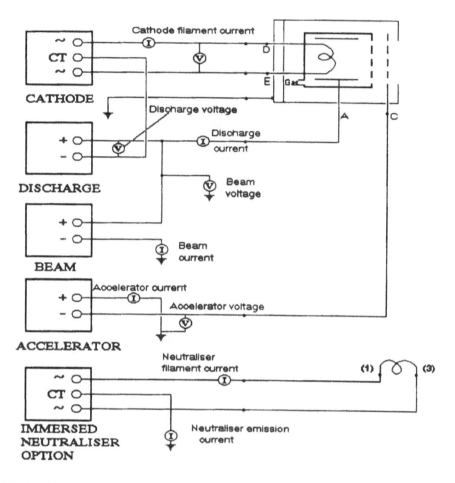

FIGURE 14.10

Circuit of power supply for Kaufman ion source.

is currently undertaken with Kaufman sources. There is not much evidence that material from the filament is deposited in the coatings. Experiment suggests that the prevention of contamination by the chamber wall and target holder, for example, are dominant factors.[51]

Filament-driven ion sources are used in ion beam milling and IBSD partly, of course, because these were the first type of high-power ion source to become available. However, they do have certain advantages for these types of application. In particular, they are stable when coated with large amounts of flux from sputter targets, either metals or oxides. They also are very efficient, and use less gas than rf ion sources, especially when the grids are optimized for minimum gas conductance with high perveance. The mirrors described in Reference 15 were made with this type of ion source.

14.4.2 ECR Ion Sources

Industrial ECR sources seem to have developed out of the various ECR plasma products that have been produced for the semiconductor etching market. The first, and simpler, type of wafer-processing plasma reactor consists of a plasma trapped

FIGURE 14.11

Linear ion source configuration. (From Ion Tech Inc., Fort Collins, CO, U.S. With permission.)

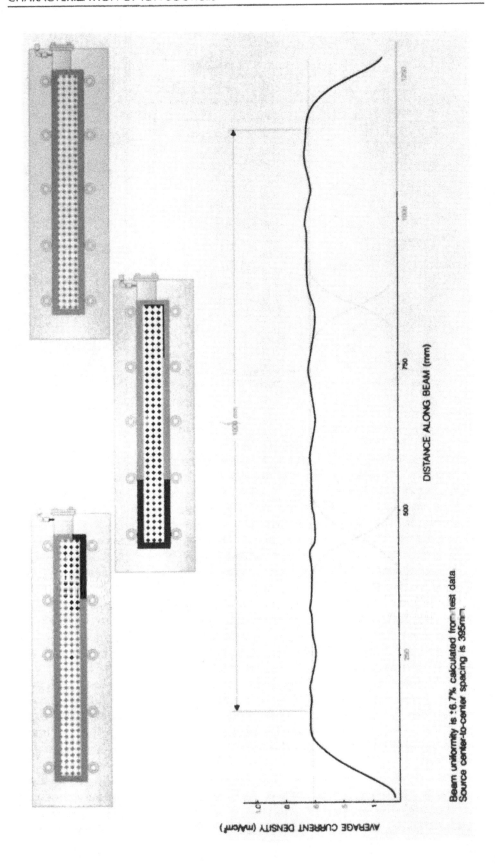

AVERAGE CURRENT DENSITY (mA/cm²)

DISTANCE ALONG BEAM (mm)

Beam uniformity is ±6.7% calculated from test data
Source center-to-center spacing is 395mm

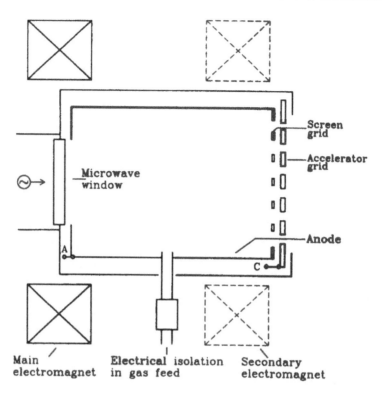

FIGURE 14.12
ECR ion source with electromagnets.

between two electromagnets, forming magnetic mirrors.[52,53] The microwave energy is generated by a magnetron, passed through a TE_{10} to TM_{01} mode converter, and injected into the chamber through an axial quartz window. To obtain true electron cyclotron resonance at 2.45 GHz, a convenient frequency for industrial use, requires a magnetic field of 0.0875 T. The magnets are usually arranged so that the field strength drops from a higher strength to this value a few centimeters into the discharge chamber, and at this point very strong absorption of the energy occurs. The layout looks similar to the ECR ion source layout of Figure 14.12, but without the ion optics, of course. (See Chapter 2.8.)

The second plasma reactor variant involves the injection of the energy at the periphery of the chamber, perhaps by using an array of antennae, as in distributed ECR. By means of permanent magnets attached to the outside of the discharge chamber, the ECR condition can be met just in the locality of the antennae.[54] All of these concepts are encountered in ECR ion sources.

The simplest ECR ion source consists of grids attached to a liner inside a chamber of the first type above (see Figure 14.12).[38] The liner is used to raise anode potential above ground, and the grids extract the ion beam as normal. The theory of the energy absorption is well understood,[53] but the detailed behavior of the plasma is still being worked on, and the reader is referred to the plasma reactor literature for more details.[53,55,56]

There are differences, however. First, the screen grid serves as one mirror for the electrons in the plasma, and so, as shown in Figure 14.12, the second electromagnet is not essential, though useful for optimizing beam current or uniformity.[38] Second, because of the solenoidal magnets, the magnetic field strength is sufficient to prevent any effective neutralization of the ion beam: even at a typical working distance of the ion source of one to two times the source diameter D_s, the magnetic field strength will still be 0.01 T, or more. The magnetic field may suppress visible sparking, but when etching sensitive devices, one must beware of damage to them due to the neutralization current that must necessarily try to flow through them.

Note, however, that unlike ECR plasma etching, the trajectories of the ions emerging from the source, assuming they have an energy of a few hundred electron volts, are relatively unaffected by the field lines. This is in contrast to the plasma case, where the ions naturally follow the diverging field.

It should also be mentioned that these ion sources are radically different from the ECR ion sources that have been developed to produce multiply charged ions for accelerators and atomic physics experiments.[57] In order to produce large fractions of multiply charged ions, these sources have specific features: they run at low pressures of 10^{-3} to 10^{-4} Pa to prevent recombination of the ions, and they need to confine the plasma with multipole magnet fields to suppress MHD oscillations. (See Chapter 2.9.)

Another type of ECR ion source that has been developed in recent years is the multipolar ECR ion source.[58] This is more akin to the second type of plasma reactor described above, in that it uses permanent magnets of neodymium-iron to generate fields of 0.0875 T along the wall of the ion source as shown in Figure 14.13. By using a multipolar structure, the field is less than 5 mT at the substrate, and normal neutralization systems can be used. (See also Chapter 2.8.)

14.4.3 RF Ion Sources

At frequencies much below the electron plasma frequency, ω_{pe}, electromagnetic waves cannot propagate into the plasma. For practical plasma densities, this frequency is in the microwave region. Thus, rf ion sources can only couple into the plasma capacitively or inductively, using the electric or magnetic fields to excite the plasma. In practice, it is usual for both effects to be occurring to some extent. (See Chapter 2.7.)

Rf ion sources have a much longer history than is sometimes recognized, and a short review of some work of the previous 20 years on inductively coupled ion sources can be found in Reference 59 of 1969. The most extensive published work is without doubt the three decades of development work on rf ion sources at Giessen University.[60] In this work, inductively coupled rf ion sources designed for use as space thrusters were created with diameters from 5 to 35cm, with maximum currents up to 4 A, and thrusts to 0.4 N. (Note, however, that these sources have not, of course, been optimized for industrial use; the 4-A current was achieved with an extraction voltage of V = 5.75 kV.) Neutralization is by hollow cathode.

An industrial version of this type of source has a layout of the type shown in Figure 14.14.[61] In this design, the helical antenna is wrapped around a quartz discharge chamber, and the rf magnetic field couples through to the plasma. In practice, there will be also electrostatic coupling from the antenna, unless electrostatic shields of the type employed in inductively coupled plasma (ICP) sources are used.[62] ICP sources, similarly to ECR plasma sources, are products of the semiconductor plasma etching industry.

The advantage of magnetic coupling is that it results in a very low temperature of plasma, with essentially zero sputtering of discharge chamber components. It also makes starting much more difficult, because the energy is basically transferred to free electrons in the plasma. Thus, a source with a modest amount of electrostatic coupling is easier to start.

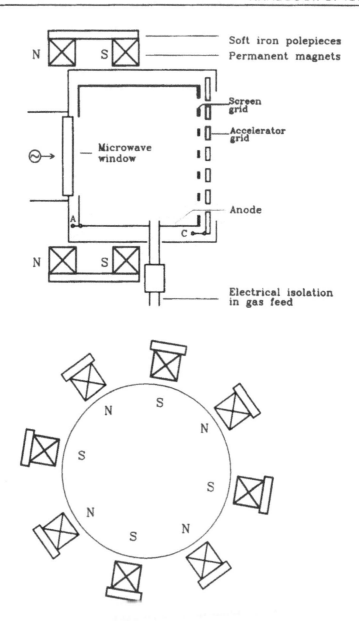

FIGURE 14.13
ECR ion source with permanent magnets.

Another factor to be considered very carefully is whether one really wants zero sputtering in the discharge chamber in any case. A modest amount of resputtering of the quartz discharge chamber in the region of the antennae clearly delays the need for cleaning the source when backstreaming conducting material is entering the source from the process; too much resputtering will contaminate the process with the materials of which the source is made. One opinion is that a good compromise is for there to be enough resputtering for the quartz nearest to the antenna to stay clean almost indefinitely. In this case, the rest of the source will slowly become coated in backstreaming materials from the process, but will only need cleaning every few hundred hours. The author is currently evaluating such a source in IBSD of optical coatings. Preliminary results suggest that run in this way the emission of iron and

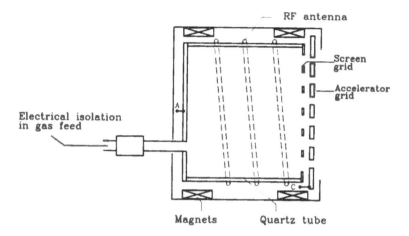

FIGURE 14.14
Rf ion source with helical antenna.

silicon is quite small compared to published data from Kaufman sources.[63] Certainly, one would need to beware of a source that ended up being continuously self-cleaning!

A number of designs of capacitively coupled ion sources have been made. At their simplest they take the layout of Figure 14.14 and instead of having a helical antenna, they apply a rf potential between the grids and the back plate of the discharge chamber. Various examples of this concept have been used.[64,65] However, most of them suffer to a greater or lesser extent from sputtering of the electrodes by direct ion impingement from the plasma in the ion source. One ingenious way of controlling this problem is the ion source geometry shown in Figure 14.15.[35,66] In this source, the cathode is a ceramic-coated axial rod, and the anode is the cylindrical wall of the discharge chamber. There is an axial magnetic field. Electrons can then move in the cathode plasma sheath with cycloidal paths in the magnetic field. The field also minimizes electron losses to the anode. Sputtering of the cathode is minimized by the high efficiency and its small size.

14.4.4 Assist Sources

Most of the ion sources described above have been used in assist applications, but there are certain sources that are particularly well adapted for this rôle. The end-Hall source is without doubt becoming the best known. The concept for this source comes from work undertaken in the former Soviet Union on space thrusters.[67] In its most usual industrial form[68–70] it has a layout similar to Figure 14.16. A strong divergent axial magnetic field emerges from the base of the source (as shown) and goes through the aperture in the anode. A single filament (or a hollow-cathode electron gun) is used, and this produces electrons both for ion generation and for neutralization.

In operation, gas is injected into the body of the source, and electrons from the hot filament are attracted toward the anode. The magnetic field hinders their reaching the anode, and so increases their likelihood of ionizing the gas. Most of the plasma is formed in the region of the aperture in the anode, as the density of neutrals is much higher here. The ions formed are accelerated toward the cathode and also toward the axis of the source. Ions thus have a spread of energies and trajectories depending on where they were formed in the ion source, and the result is a divergent ion beam with a wide spread of energies.

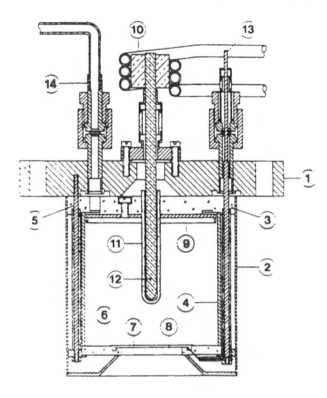

FIGURE 14.15
Rf ion source, capacitively coupled. (From Plasma Consult GmbH, Wuppertal, Germany. With permission.)

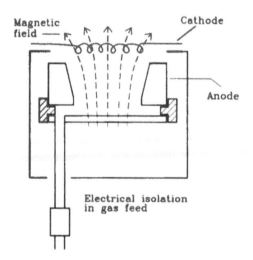

FIGURE 14.16
End-Hall ion source.

In very many assist applications these properties are no drawback at all: the ion flux profile is similar to the flux profile of the electron beam evaporation hearth (Figure 14.5) and the same planetary tooling can produce a uniform process for both fluxes at the same time. Neither the ion current nor the energy distribution can be measured directly, but most assist processes are tolerant of variations in these parameters, and in an industrial context the parameters are adjusted to obtain the correct results.

Gridless sources can also be used for surface treatments such as adhesion promotion for plastics, and in such applications the facts that one can make rectangular end-Hall sources and can operate them by means of an electron gun rather than a hot filament can both be of value.

Where the application is research, or the results are strongly affected by energy (see Reference 26, for example), or by contamination from the filament, then a gridded Kaufman or rf source may be more appropriate.

In the context of assist sources, mention should be made of free-radical sources. These sources generate a beam of dissociated atoms using an rf plasma. A number of important uses for them are currently developing in surface cleaning by hydrogen atoms,[71] the oxidation of high-temperature superconductor (HT_c) materials,[72] and doping with nitrogen atoms to produce blue solid-state lasers.[73]

14.5   Neutralization and Beam Transport

14.5.1   Neutralization

In all of the above discussions, the use of a neutralized beam has been assumed. The most obvious effect of using a relatively unneutralized ion beam is that if the ion beam strikes an insulating surface, that surface will charge up in potential until it loses its charge in clearly visible microdischarges. Such discharges will usually destroy a component being etched, and can even damage a nearby substrate in the process of sputter deposition. The excessive divergence of the beam due to the electrostatic repulsion of the ions will also be clearly seen. (Magnetic pinch does not apply at these energies.) In most applications apart from IBSD using oxide targets, at least part of the beam will be directed at a conducting surface, and so secondary electrons emitted by the impact of the ions will produce at least partial neutralization of the ion beam.

For most processes, full neutralization is the preferred mode of operation; that is to say, with the beam containing the same number density of ions and electrons. The electrons are injected into the beam by the neutralizer; the beam can be pictured as a potential well needing to be filled. One reason for the grid structures discussed above is to ensure that electrons are not lost into the source. However, if the target is insulating, then electrons are lost to the target at the same rate as the ion current. Thus, in cases of doubt, neutralizer systems are normally run at the same current as the ion source. This will ensure that the ion beam potential is within a few volts positive of ground. An ion beam is run at sufficient volts above ground to ensure that it does not lose its electrons.

It should be pointed out that an ion beam will always diverge by more that the grid design suggests.[74] This is not due to inadequate neutralization and electrostatic repulsion, but to the potential differences across the beam, including the one described above: the beam is a plasma, retaining electrons and losing ions by sideways diffusion. It can be shown that the amount of extra divergence will depend on the potential at the center of the beam, and, hence, should be reduced by neutralizing the beam by lower-energy electrons. Thus, increasing the chamber pressure should cool the electrons and reduce beam divergence as well. In practice, these effects are usually only dominant at pressures below 1×10^{-2} Pa. As pressures are increased to

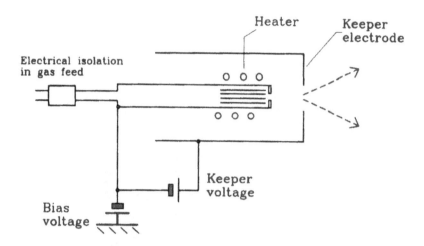

FIGURE 14.17
Hollow-cathode neutralizer.

higher values typical of industrial processes, ion-neutral scattering is liable to spread the beam again. These effects have not disturbed any industrial processes to the author's knowledge, but they are indicative of the need for a stable and well-controlled process if one wishes to ensure reproducible results. Neutralization by background gas is discussed further in Chapter 4.

14.5.2 Immersed Wire Neutralizer

On early industrial ion sources, and in many current research applications, the usual neutralizer is an immersed wire in the beam, using the circuit of Figure 14.10. This is heated until it emits electrons into the beam, although ion impact into the hot filament is also a significant factor. It has to be physically in the beam, or else the electron emission currents will be space charge limited, and the beam potential will rise to very high values to try to extract the electrons. With the wire in the beam, the beam will be typically 5 to 10 V above ground.[33]

14.5.3 Hollow-Cathode Neutralizer

The hollow-cathode neutralizer, of the type shown in Figure 14.17, was developed for use with Kaufman ion thrusters on spacecraft.[75] In its original form it ran on cesium, but for use with inert gases, it has often been fitted with an insert coated with carbonates of barium, strontium, and calcium,[76] or impregnated with barium-calcium-aluminate.[77] The rest of the structure is refractory metal. In operation, the neutralizer is heated to about 1000°C, and the carbonates dispense an atomically thin layer of oxides to the region of the nozzle. Because of the low work function of these oxides, a plasma may readily be struck between the inside of the bore and the keeper. The plasma inside the bore generates enough heat for the heater itself to be turned down, or even switched off. When neutralization is needed, the plasma is struck between the nozzle and the ion beam, forming a plasma bridge between them.

Such a neutralizer can operate for thousands of hours, and is still the basis of the neutralizers being used on current space-thruster programs.[60] It is also the basis of the

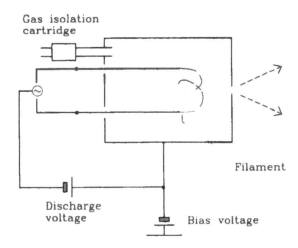

Gas isolation
cartridge

Filament

Discharge
voltage

Bias voltage

FIGURE 14.18
Plasma bridge neutralizer.

hollow cathodes used to drive ion thrusters recently developed by Hughes Research.[78] However, the control system is clearly fairly complex, and in some circumstances there may be concerns over contamination by the carbonates. Severe damage is done to the neutralizer if it is exposed while hot to air or gases that are not inert.

Essentially identical neutralizers can easily be run without the oxides, but they will then run much hotter, and have a shorter life. It is also possible to run a neutralizer in a Penning mode, using a Penning discharge in the hollow cathode to maintain the discharge. There appears to be no published data, though Reference 79 shows a similar device.

14.5.4 Filament Plasma Bridge Neutralizer

Several designs have been developed for filament-driven neutralizers.[80,81] These are another form of hollow-cathode neutralizers. The plasma is generated by means of a hot filament, resulting in easy starting and easy maintenance. The simplest layout is shown in Figure 14.18. In operation, the filament is able to establish an intense plasma inside the neutralizer, even on a gas flow of 2 to 4 sccm, because the pressure inside is high: the aperture is less than 1 mm in diameter. A plasma bridge is struck through the aperture between the plasma inside the neutralizer and the ion beam. The emission current is then controlled by means of the bias voltage. Because the filament is shielded from the process, filament lifetimes in excess of 24 h are attained.

14.5.5 Rf Neutralizer

The rf neutralizer is another form of hollow-cathode neutralizer, in which the plasma is maintained by an inductive rf antenna. (See Figure 14.19.) In operation it is very similar to the filament plasma bridge neutralizer, with the plasma being sustained by the rf power, and the coupling controlled by the bias voltage supply. Eliminating the filament gives a substantial increase in the complexity of the neutralizer. However, in the context of a loadlock processing system kept under vacuum except for maintenance, this may make economic sense.

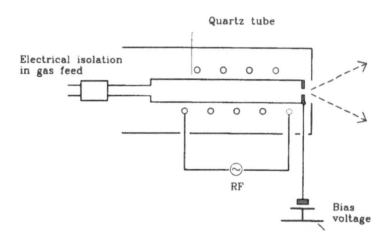

FIGURE 14.19
Rf neutralizer.

REFERENCES

1. Freeman, J. H., *Ion Implantation*, North-Holland, Amsterdam, 1973.
2. Aitken, D., The Freeman ion source, in *The Physics and Technology of Ion Sources*, Brown, I. G., Ed., Wiley-Interscience, New York, 1989.
3. Rose, P. H., The evolution of ion sources for implanters, *Rev. Sci. Instrum.*, 61, 342, 1990.
4. Sakudo, N., Tokiguchi, K., and Seki, T., Beam qualities of a microwave ion source, *Rev. Sci. Instrum.*, 61, 309, 1990.
5. Melngailis, J., Focused ion beam technology and applications, *J. Vac. Sci. Technol.*, B5, 469, 1987.
6. Storms, H. A., Brown, K. F., and Stein, J. D., *Anal. Chem.*, 49, 2023, 1977.
7. Kimball Physics, Kimball Hill Road, Wilton, NH 03086–9742.
8. Collections of data on etch rates for common materials are available from the ion source vendors.
9. Efremow, N. N., Geis, M. W., Flanders, D. C., Lincoln, G. A., and Economou, N. P., Ion-beam-assisted etching of diamond, *J. Vac. Sci. Technol.*, B3, 416, 1985.
10. Geis, M. W., Lincoln, G. A., Efremow, N. N., and Piacentini, W. J., A novel anisotropic dry etching technique, *J. Vac. Sci. Technol.*, 19, 1390, 1981.
11. Hagberg, L., Jonsson, B., and Larsson, A., Fabrication of ultrahigh quality vertical facets in GaAs using pattern corrected electron beam lithography, *J. Vac. Sci. Technol.*, B10, 2243, 1992.
12. Vettiger, P., Benedict, M. K., Bona, G.-L., Buchman, P., Cahoon, E. C., Datwyler, K., Dietrich, H.-P., Mosler, A., Seitz, H. K., Voegli, O., Webb, D. J., and Wolf, P., Full-wafer technology — a new approach to large-scale laser fabrication and integration, *IEEE J. Quantum Electron.*, 27, 1319, 1991.
13. Harper, J. M. E., Cuomo, J. J., and Kaufman, H. R., Technology and applications of broad-beam ion sources used in sputtering. Part II. Applications, *J. Vac. Sci. Technol.*, 21, 737, 1982.
14. Scheer, H.-C., Ion sources for dry etching: aspects of reactive ion beam etching for Si technology, *Rev. Sci. Instrum.*, 63, 3050, 1992.
15. Rempe, G., Thompson, R. J., Kimble, H. J., and Lalezari, R., Measurement of ultralow losses in an optical interferometer, *Opt. Lett.*, 17, 5, 363, 1992.
16. Jolly, T. W. and Lalezari, R., Ion beam sputter deposition techniques for the production of optical coatings of the highest quality, Paper 1782–41, *Int. Symp. Optical System Design*, Berlin, 1992.
17. Harper, J. M. E., Cuomo, J. J., Leary, P. A., Summa, G. M., Kaufman, H. R., and Bresnock, F. J., Low energy ion beam etching, *Proc. 9th Int. Conf. Electron and Ion Beam Science and Technology*, 80–6, 518, 1980.
18. Kitabatake, M. and Wassa, K., Growth of diamond at room temperature by an ion-beam sputter deposition under hydrogen-ion bombardment, *J. Appl. Phys.*, 58, 1693, 1985.

19. Mirtich, M. J., Kussmaul, M. T., Banks, B. A., and Sovey, J. S., Dual ion beam processed diamondlike films for industrial applications, *Technology 2000*, sponsored by NASA, the Technology Utilization Foundation, and NASA Tech Briefs Magazine, Washington, D.C., November 27 to 28, 1990.

20. Kimock, F. M., Knapp, B. J., and Finke, S. J., Abrasion wear resistant polymeric substrate product, U.S. Patent 5 190 807, March 2, 1993.

21. Kataoka, I., Ito, K., Hoshi, N., Yonemitsu, T., Etoh, K., Yamada, I., and Delaunay, J.-J., The effects of ion energy on carbon and tungsten films fabricated by direct ion beam deposition and ion beam sputtering deposition, *Mater. Res. Soc. Symp. Proc.*, Vol. 223, 1991.

22. Smidt, F. A., Use of ion beam assisted deposition to modify the microstructure and properties of thin films, *Int. Mater. Rev.*, 35, 61, 1990.

23. Gibson, U. J., Ion beam processing of optical thin films, *Phys. Thin Films*, 13, 109, 1987.

24. Martin, P. J. and Netterfield, R. P., Ion-assisted dielectric and optical coatings, in *Handbook of Ion Beam Processing*, Cuomo, J. J., Rossnagel, S. M., and Kaufman, H. R., Eds., Noyes Publications, Park Ridge, NJ, 1989.

25. Daugy, E., Pointu, B., Villela, G., and Vincent, B., Dual ion beam sputtering: a new coating technology for the fabrication of high power CO_2 laser mirrors, *SPIE*, 1502, 203, 1991.

26. Kataoka, I., Yonemitsu, T., Sekine, K., Yamada, I., Itoh, K., and Hoshi, N., Investigation of oxide/metal multi-layers for soft X-ray optics fabricated by ion beam sputtering, *Mater. Res. Soc. Symp.*, 157, 1990.

27. For example, see King, M. and Rose, P., Experiments on gas cooling of wafers, *Nucl. Instrum. Methods*, 189, 169, 1981.

28. Jolly, T. W., unpublished data, 1993.

29. Huth, Ch., Scheer, H.-C., Schneemann, B., and Stoll, H.-P., Divergence measurements for characterization of the micropatterning quality of broad ion beams, *J. Vac. Sci. Technol.*, A 8, 4001, 1990.

30. Kaufman, H. R., Technology of electron-bombardment ion thrusters, in *Advances in Electronics and Electron Physics*, Vol. 36, Marton, L., Ed., Academic Press, New York, 1974, 265.

31. Rovang, D. C. and Wilbur, P. J., Ion extraction capabilities of closely spaced grids, AIAA paper AIAA-82–1894, 1982.

32. Korzec, D., Engemann, J., Bansky, J., and Keller, H. M., Integrated silicon grid ion extraction system for O_2 processes, *J. Vac. Sci. Technol.*, B8, 1716, 1990.

33. Kaufman, H. R. and Robinson, R. S., Ion source design for industrial applications, *AIAA J.*, 20, 6, 1982.

34. Kaufman, H. R., Cuomo, J. J., and Harper, J. M. E., Technology and applications of broad-beam ion sources used in sputtering. Part 1. Ion source technology, *J. Vac. Sci. Technol.*, 21, 1982.

35. Korzec, D. and Engemann, J., Optimized eight-inch extraction system for reactive ion beam etching, *J. Vac. Sci. Technol.*, B7, 1448, 1989.

36. Forrester, A. T., *Large Ion Beams*, Wiley-Interscience, New York, 1988.

37. Forrester, A. T., Crow, J. T., Massie, N. A., and Goebel, D. M., A multipole containment single grid extraction ion source, UCLA report PPG-224, 1975.

38. Jolly, T W and Blackburrow, P., Microwave ion beam sources for reactive etching and sputter deposition applications, *Rev. Sci. Instrum.*, 61, 747, 1990.

39. Spadtke, P., Computer modelling, in *The Physics and Technology of Ion Sources*, Brown, I. G., Ed., Wiley-Interscience, New York, 1989.

40. Homa, J. M. and Wilbur, P. J., Ion beamlet vectoring by grid translation, AIAA paper 82–1895, 1985.

41. Harper, J. M. and Gambino, *J. Vac. Sci. Technol.*, 16, 1901, 1979.

42. Barnes, S. E., Grindrod, D. C., Jolly, T. W., and Shaw, C. J., The design of industrial systems for ion beam sputter deposition of optical coatings, *Ion and Plasma Assisted Techniques 6th Int. Conf.*, 1987, 441.

43. Kaufman, H. R. and Reader, P. D., *Am. Rocket Soc. Pap.*, 1374–60. See also references cited in Reference 30.

44. Kaufman, H. R., Technology of ion beam sources used in sputtering, *J. Vac. Sci. Technol.*, 15, 272, 1978.

45. Guarnieri, C. R., Ramanathan, K. V., Yee, D. S., and Cuomo, J. J., Improved ion source for use with oxygen, *J. Vac. Sci. Technol.*, A6, 1988, 2582.

46. Skidmore, J. A., Green, D. L., Young, D. B., Olsen, J. A., Hu, E. L., and Coldren, L. A., Investigation of radical-beam ion-beam etching-induced damage in GaAs/AlGaAs quantum-well structures, *J. Vac. Sci. Technol.*, B9, 3516, 1991.

47. Sugata, S. and Asakawa, K., Characterization of damage on GaAs in a reactive ion beam etching system using Schottky diodes, *J. Vac. Sci. Technol.*, B6, 876, 1988.
48. Kaufman, H. R. and Robinson, R. S., Plasma processes in inert gas thrusters, AIAA paper 79–2055, 1979.
49. Davis, R. C., Morgan, O. B., Stewart, L. D., and Stirling, W. I., A multi-ampere duoPIGatron ion source, *Rev. Sci. Instrum.*, 43, 278, 1972.
50. Ion Tech Inc., 2330 E. Prospect, Ft. Collins, CO 80525.
51. Daugy, E., Pointu, B., Audry, C., and Hervo, C., Dependence between optical and chemical properties of Y_2O_3 and ZrO_2 thin films produced by ion beam sputtering, *J. Opt.*, 21, 99, 1990.
52. Chen, F. F., *Introduction to Plasma Physics and Controlled Fusion, Vol. 1: Plasma Physics*, Plenum Press, New York, 1984, 30.
53. Lieberman, M. and Gottscho, R. A., *Physics of Thin Films*, Francombe, M. and Vossen, J., Eds., Academic Press, New York, 1992; and references therein.
54. Cooke, M., Magnetically confined ECR increases uniformity of high-density plasmas, *Res. Dev.*, 97, 1988.
55. Gorbatkin, S. M., Berry, L. A., and Roberto, J. B., Behavior of Ar plasmas formed in a mirror field ECR ion source, *J. Vac. Sci. Technol.*, in press.
56. Carl, D. A., Williamson, M. C., Lieberman, M. A., and Lichtenberg, A. J., Axial radio frequency field intensity and ion density during low to high mode transition in argon electron cyclotron resonance discharges, *J. Vac. Sci. Technol.*, B9, 339, 1991.
57. Jongen, Y. and Lyneis, C. M., Electron cyclotron resonance ion sources, in *The Physics and Technology of Ion Sources*, Brown, I. G., Ed., Wiley-Interscience, New York, 1989, 207.
58. Heard, P. J. and Jolly, T. W., Oxford Instruments, Plasma Technology, unpublished data, 1989.
59. Carter, R. G. and Newton, R. H. C., The extraction of ions from a radio-frequency discharge, *J. Phys. D*, 241, 1969.
60. Groh, K. H. and Loebt, H. W., State-of-the-art radio-frequency ion thrusters, *J. Propulsion*, 7, 573, 1991.
61. Oxford Instruments, Plasma Technology, North End, Yatton, Bristol, England, BS19 4AP.
62. Hopwood, J., Review of inductively coupled plasmas for plasma processing, *Plasma Sources Sci. Technol.*, 1, 109, 1992.
63. Vitkavage, D. J. and Mayer, T. M., Target contamination by cathode sputtering in broad beam ion sources, *J. Vac. Sci. Technol.*, A6, 154, 1988.
64. Lossy, R. and Engemann, J., RF broad-beam ion source for reactive sputtering, *Vacuum*, 36, 973, 1986.
65. Lejeune, C., Grandchamp, J. P., and Kessi, O., RF multipolar plasma for broad and reactive ion beams, *Low Energy Ion Beams-4*, Brighton, England, April 1986.
66. Plasma Consult GmbH, Müngsterner Str. 10, 42285 Wuppertal, Germany.
67. Morosov, A., *Physical Principles of Cosmic Electro-Jet Engines*, 1, 13, Atomizdat, Moscow, 1978.
68. Kaufman, H. R., Robinson, R. S., and Seddon, R. I., End-Hall ion sources, *J. Vac. Sci. Technol.*, A5, 2081, 1987.
69. Kaufman, H. R. and Robinson, R. S., *Operation of Broad-Beam Sources*, Commonwealth Scientific Corporation, Alexandria, VA, 1987, 57.
70. Commonwealth Scientific Corporation, 500 Pendleton St., Alexandria, VA 22314.
71. Rouleau, C. M. and Park, R. M., GaAs substrate cleaning for epitaxy using a remotely generated atomic hydrogen beam, *J. Appl. Phys.*, 73, 4610, 1993.
72. Park, R. M., Troffer, M. B., Rouleau, C. M., DePuydt, J. M., and Haase, M. A., *Appl. Phys. Lett.*, 57, 2127, 1990.
73. Locquet, J.-P. and Machler, E., Characterization of a radio frequency plasma source for molecular beam epitaxy growth of high-T_c superconductor films, *J. Vac. Sci. Technol.*, A10, 3100, 1992.
74. Crow, J. T., Space Charge Effects in Ion Beams, Doctoral dissertation, University of California, Los Angeles (1977); as detailed in Forrester, A. T., *Large Ion Beams*, Wiley-Interscience, New York, 1988, 139.
75. Ernstene, M. P., James, E. L., Purmal, G. W., Worlock, R. M., and Forrester, A. T., Surface ionization engine development, *J. Spacecr. Rockets*, 3, 744, 1966.
76. Rawlin, V. K. and Pawlik, E. V., A mercury plasma-bridge neutralizer, *J. Spacecr. Rockets*, 5, 814, 1968.
77. Walther, S. E., Groh, K. H., and Loebt, H. W., Experimental and theoretical investigations of the Giessen neutralizer system, AIAA/DGLR 13th International Electric Propulsion Conference.

78. Beattie, J. R. and Mattosian, J. N., Xenon ion sources for space applications, *Rev. Sci. Instrum.*, 61, 348, 1990.
79. Gruzdev, V. A., Kreindel, Yu. E., Rempe, N. G., and Troyan, O. E., An electron gun with a plasma emitter, *Instrum. Exp. Tech.*, 28, 151, 1985.
80. Reader, P. D., White, D. P., and Isaacson, G. C., Argon plasma bridge neutralizer operation with a 10 cm beam diameter ion etching source, *J. Vac. Sci. Technol.*, 15, 1093, 1978.
81. Lejeune, C., Grandchamp, J. P., and Kessi, O., Electrostatic reflex plasma source as a plasma bridge neutralizer, *Vacuum*, 36, 857, 1986.

Chapter 2/Section 15

CHARACTERIZATION OF ION SOURCES

G. Alton

15 HIGH BRIGHTNESS FIELD IONIZATION AND FIELD EVAPORATION (LIQUID-METAL) ION SOURCES

Applications
Ultrahigh-brightness microfocused ion beams for use in ion implantation, SIMS, RBS, hydrogen profiling, etc.

LMIS
Advantages: High brightness, wide range of species, high intensity, long lifetime, low power consumption.
Disadvantages: High angular divergence, energy spread.

GFIS
Advantages: High brightness, low power consumption.
Disadvantages: Species limited to gaseous feed materials, highly divergent, low intensity, short needle lifetime.

15.1   Introduction

Ultrabright ion beams are in demand for a diverse variety and growing number of focused ion beam applications, including ion microprobe surface analysis, ion microscopy, secondary ion mass spectrometry (SIMS), microfabrication, microetching, micromachining, lithography, ion implantation with submicron imaged beams, material characterization, ion beam deposition, mask repair, and ion propulsion. Ion beams that meet these application criteria have been made possible, primarily, by the development of field ionization and field evaporation ion sources. Both source types utilize needle type geometry emitters and rely on the phenomena of either field ionization or field evaporation and, therefore, are essentially point sources. The gas field ionization source (GFIS) requires the use of gaseous feed materials, while the field evaporation ion source utilizes molten materials for generation of high-brightness ion beams; the latter source is commonly referred to as a liquid-metal ion source (LMIS). Sources based on these ionization principles have the capability of generating ion beams with central core brightness of 10^5 times those of conventional gaseous discharge sources[1,2] with useful on-target beam intensities suitable for ultrabright focused beam applications. The utilization of beams from these sources have, in turn,

0-8493-2502-1/95/$0.00+$.50
© 1995 by CRC Press Inc.

been made possible by the development of ion optical systems capable of transporting and focusing beams onto sample surfaces with high resolution; these systems have evolved from laboratory instruments, with limited current densities, to ultrahigh current densities. For microfocused beam applications, the sources are used, in combination, with high-resolution, low-aberration optical systems to provide beams with on-target diameters less than 1 μm at current densities as high as ~1 A/cm² and ion beam intensities up to ~5 nA.[3]

This chapter is devoted to the description of the principles of operation, design aspects, performances, and ion beam characteristics of the GFIS and LMIS. Less attention is given to the ion optical systems required for the extraction, mass separation, transport, and focusing of the ion beams. Because of length limitations and the vast number of publications that have been devoted to these sources and their applications, a comprehensive review of all aspects of the subjects will not be possible or practical. Therefore, the reader is referred to the citations made within this manuscript for specific details on individual source types and their applications or the beam transport optical systems that have been developed for their use. The rather expansive list of references that have been published on these source types and their applications includes a book by Prewett and Mair that deals with the physics and technology of the LMIS with emphasis on the use of microfocused beams for microscopy and analysis, micromachining and deposition, microcircuit lithography, and ion implantation[4] and a recent review article by Orloff that emphasizes the use of the LMIS for microfocused beam applications.[3] Comprehensive bibliographies on the subject of field ionization and evaporation sources and their applications can be found in Reference 5 and in the article by MacKenzie and Smith;[6] the latter reference contains ~1100 references to both field ionization and liquid-metal ion sources and their applications, as well as several review articles on these source types.

15.2   Factors that Affect Transport and Focusing of Intrinsically Ultrabright Ion Beams

15.2.1   Liouville's Theorem

Liouville's theorem states that the motion of a group of particles under the action of conservative force fields is such that the local number density in the six-dimensional phase space volume (hypervolume) $xyzp_xp_yp_z$ everywhere remains constant. The theorem applies to an ion beam subjected to conservative force fields. The quality of an ion beam is usually expressed in terms of emittance ε and brightness B. Both are related to Liouville's theorem.

15.2.2   Emittance

For ion beam transport, the components of phase space transverse to the direction of beam motion are usually the most important. If the transverse components of motion of a group of particles are independent in configuration space, the motion of the particles in the orthogonal planes (x,p_x), (y,p_y), and (z,p_z) will be uncoupled and, therefore, the phase spaces associated with each of these planes will be separately conserved.

These conserved areas of phase space are referred to the emittance ε of the ion beam in the respective directions.

The following definitions dependent on energy E and velocity v have been adopted, historically, for the definitions of normalized emittance ε_{nx} and ε_{ny}:

$$\varepsilon_{nx} \approx \pi \left(\iint \frac{dxdx'}{\pi} \right) \beta\gamma \text{ and } \varepsilon_{ny} \approx \pi \left(\iint \frac{dydy'}{\pi} \right) \beta\gamma \tag{1}$$

or

$$\varepsilon_{nx} \approx \pi \left(\iint \frac{dxdx'}{\pi} \right) \sqrt{E} \text{ and } \varepsilon_{ny} \approx \pi \left(\iint \frac{dydy'}{\pi} \right) \sqrt{E} \tag{2}$$

where the integrations are performed over the emittance contour that contains a specified fraction of the beam (e.g., 10%, 20%, etc.); the units of normalized emittance defined by Equations 1 are expressed in units of $\pi \cdot mm \cdot mrad$, while those defined by Equations 2 are usually expressed in units of $\pi \cdot mm \cdot mrad \ (MeV)^{1/2}$.

15.2.3 Brightness

Another figure of merit often used for evaluating ion beams is the brightness B. Brightness is defined in terms of the ion current d^2I per unit area dS per unit solid angle $d\Omega$ or

$$B = \frac{d^2I}{dSd\Omega} \tag{3}$$

In terms of normalized brightness, Equation 3 can be shown to be equivalent to[7]

$$B_n = \frac{2d^2I}{\varepsilon_{nx}\varepsilon_{ny}} \tag{4}$$

15.2.4 Contributions to Emittance Growth

Active ion optical elements can increase the phase space of an ion beam through three dominant types of aberrations: geometrical, chromatic, and parasitic or mechanical. Geometrical aberrations arise from third-order contributions to the image size; chromatic aberrations reflect the fact that lenses are strongly dependent on the energy, angular divergence, and energy spread of the ion beam; parasitic aberrations arise because of imperfect mechanical alignment. Of the several classifications of geometrical aberrations, spherical aberrational effects are of principal concern in the ion beam transport systems. The control of parasitic aberrations is well understood and, therefore, in principle, can be rendered negligibly small. Space charge effects during extraction from the source and transport through the beam transport system can also affect the emittance and brightness of ion beams.

15.2.4.1 Spherical Aberration

An object imaged by a lens system will be increased by an amount Δr_s given by[8]

$$\Delta r_s = \mathcal{M} C_{so} \theta_0^3 \tag{5}$$

where $\mathcal{M}$ is the magnification, C_{so} is the coefficient of spherical aberration referred to the object side of the lens, and θ_0 is the half-angle of arrival at the lens. The spherical aberration coefficient depends on the focal length of the lens and is, therefore, lens specific. The image grows directly with the coefficient and, therefore, a low value is desired. For a given spherical aberration coefficient, image growth can be controlled by limiting θ_o.

15.2.4.2 Chromatic Aberration

The Einzel lens potential difference and ion energy occur in the expression for the focal length of an electrostatic lens. Distortion will occur in the image due to (1) changes in the accelerating potential during ion transit through the lens, (2) energy spread in the ion beam, (3) inelastic collisions between the ion and solid or gas scatterers, and (4) large angular divergence in the ion beam.

For the case of chromatic aberrations, an object imaged by a lens system will be increased in size by an amount

$$\Delta r_c = \mathcal{M} C_{co} \theta_o \frac{\Delta E}{E_1} \tag{6}$$

where $\mathcal{M}$ is the magnification, C_{co} is the coefficient of chromatic aberration referred to the object plane, θ_o is the angle of departure from the object position, and ΔE is the energy spread of the beam of mean energy E_1 at the object side of the lens. From this expression, we see that it is important to limit the lens entrance angle, and the energy spread, if possible, and to accelerate the beam to high energies prior to focusing with the lens. Unfortunately, very few data are available for chromatic aberration coefficients of electrostatic lenses. However, analytical expressions for upper limits for the chromatic aberrations of Einzel and immersion lenses have been derived by El Kareh and El Kareh.[8] The aberrational properties of quadrupole lenses are tabulated in a text by Hawkes.[9]

For example, the upper limit for the growth in size Δr_c due to energy spread ΔE in an ion beam passing through an Einzel lens operated in the acceleration/deceleration entrance/exit mode, referred to the image plane, is given by

$$\Delta r_c = \mathcal{M} \frac{\Delta E r_m}{E^{1/4} \phi_m^{3/4}} \tag{7}$$

where ΔE is the energy spread in the ion beam of energy E_1, $\mathcal{M}$ is the magnification, ϕ_m is the variable potential difference impressed on the Einzel lens, and r_m is the radius of the beam as it passes through the lens.

15.2.4.3 Space Charge Effects

As a result of space charge, an ion beam of current I, diameter d, energy E, and particle mass M traveling a distance z will grow in size by an amount Δr_{sc} given by[10]

$$\Delta r_{sc} = r_0 \left\{ \cosh \left[2 \left[\frac{2e}{M} \right]^{-1/4} I^{1/2} \left[\frac{E}{e} \right]^{-3/4} \frac{z}{d} \right] - 1 \right\} \tag{8}$$

where r_0 is the original size of the beam at $z = 0$. From this expression, we again see the merit of accelerating beams to high energies.

15.2.4.4   Accumulative Effects

The performance of an optical system is determined by the size d of the focused beam on target. Contributions to the radius r of the beam due to different effects in a given element (lens or magnet) are usually assumed to add in quadrature according to:

$$r^2 = r_o^2 + \Delta r_c^2 + \Delta r_s^2 + \Delta_{sc}^2 \qquad (9)$$

where r_o is the apparent radius of the source at the point of generation of the beam. Contributions to the increase in r due to parasitic aberrational effects are not included in Equation 9; for a multicomponent beam transport system, the corresponding terms appropriate for each optical component must be added.

15.3   Field Ionization Type Ion Sources

15.3.1   General Principles of the GFIS

Field electron emission refers to the transfer of electrons from the surface of a metal into the vacuum as a result of the action of very high electric fields at the surface, which lowers the potential barrier so that electrons can leak out of the metal. Ionization of atoms or molecules adsorbed at or near a surface at high potential can also occur whenever the field polarity is reversed, and thus the technique can be used to form positive ions. The field ionization process usually requires fields of the order of 10^8 V/cm. To achieve surface fields of such high values at moderate potential differences, small-diameter, spherically tipped solid needle, or thin, hollow, capillary electrodes are usually employed. Solid needle ionizers typically have radii of ~100 nm.

Field ionization is the reverse of field emission and takes place whenever an atom or molecule is in an extremely high electric field; the presence of the strong electric field polarizes the atom or molecule and lowers the potential barrier, which makes it possible for an electron to tunnel from the atom or molecule through the barrier into the metal. The first observation of the field ionization phenomenon was made by Müller,[11] which subsequently led to the development of field emission ion sources for focused ion beam applications, first used by Levi-Setti[12,13] and by Orloff and Swanson.[14–16] The brightness of the GFIS depends on the virtual source size δ and, therefore, this characteristic of the source is fundamentally important for ultrabright focused beam applications.

The GFIS source is the principal component of the field ion microscope. Field ionization microscopes can be used to image structural features of materials with subnanometer spatial resolution and are, therefore, capable of delineating features of the sample on an atomic scale. For this application, the object or specimen is a very finely etched needle having an apex end radius ~100 nm, so that high electric fields can be achieved by impressing a very modest potential difference (typically, 1 to 10 kV) between the needle and an extraction electrode. The image of the specimen is viewed on a fluorescent screen and is created by the ion beam that is generated by field ionization of atoms in close proximity to the specimen surface; by projecting the image back to the source of origin, the virtual source site δ from which the ions appear to emanate is estimated to be ≤1 nm.

Two source concepts that produce cylindrically symmetric ion beams are illustrated in Figure 15.1a and b. The hemispherically tipped ionizer (Figure 15.1a) produces ions from atoms that pass within the critical distance x_c of the tip. At high

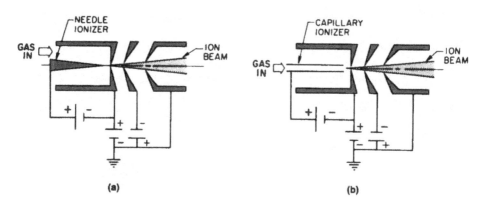

FIGURE 15.1

Illustration of two principal types of field ionization sources: (a) needle ionizer; (b) capillary ionizer.

pressures, the unionized particles in the path of the accelerated ion beam may act as scattering centers and thus degrade the beam quality. An attractive method shown in Figure 15.1b reduces this effect and thus may be more desirable. It consists of a thin capillary tube through which the gaseous or vaporous material to be ionized can be fed. The $1/r^2$ relationship between the electric field $\bar{E}$ and distance r near the tip of the needle results in a field strength $\bar{E} \simeq 10^8$ V/cm at relatively low voltages. Efficient ionization of atoms or molecules sufficiently close to the surface end of the apex of the needle typically requires voltages in the range 1 to 10 kV applied to a field emitter. Operation of the source is effected by introducing a gas at low pressure (typically, 0.01 to 1 mtorr) in the region of the field emitter to supply the atoms or molecules to be ionized. The needles in the GFIS may be operated at ambient temperature or chilled to low temperatures.

15.3.2 Theory of Field Ionization

The basic mechanisms of the field emission and field ionization processes are essentially the same and involve the tunneling of electrons from the metal into the vacuum (field emission) or electrons from the atom or molecule into the surface (field ionization). The tunneling mechanism can only be explained by quantum mechanical theory and has no classical analog. Analytical expressions for the probability of field emission have been derived by Fowler and Nordheim[17] for an abrupt potential step at the metal surface that neglects the image potential term $-e^2/4x$, where x is the distance of the electron from the surface. More rigorous theoretical calculations that include the image term have been made by Nordheim.[18] Summaries of work in field emission have been given by Good and Müller[19] and texts have been written on the subject of field emission and field ionization by Gomer[20] and on the principles of field ionization microscopy by Müller and Tsong.[21] Computational simulation studies of field emission have also been recently made by Herrmannsfeldt et al.[22]

The problem of field ionization (FI) can be treated in an analogous manner to that of field emission, but is slightly more difficult because the image potential term now must include both nuclear and electronic terms, as well as polarization effects. Theoretical treatments of the problem have been described by Gomer,[20] Haydock and Kingham,[23] and by Forbes,[24] the latter reference reviews several FI models. The following analysis closely follows that made by Gomer in Reference 20.

An atom near a surface in a strong electric field is strongly polarized as a result of the superposition of the field and image effects. The effect is illustrated in Figure 15.2,

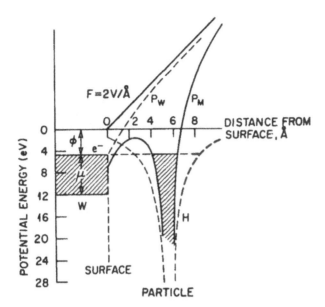

FIGURE 15.2
Potential energy diagram of a hydrogen atom in an electric field of 2 V/Å near a tungsten surface: μ, Fermi level; ϕ, work function of surface; P_m, atomic potential; P_w, superposition of applied field and image potentials.

which displays the potential energy diagram of an H atom in its ground state under the influence of a strong electric field of $\bar{E} = 2$ V/Å. The potential barrier is lowered by the superposition of the strong electric field $\bar{E}$ and the induced image charges of the electron and nucleus V_{im}. The one-dimensional potential energy for the atom is, therefore, the sum of the applied and induced image potentials or

$$V(x) = -e\bar{E}x + V_{im} \tag{10}$$

where e is the electronic charge and x is the distance of the atom from the surface. For the transfer process to occur, the electric field must raise the potential energy of the atomic electron at least to the Fermi level of the metal. The critical distance x_c (in centimeters) for which tunneling can occur is given approximately by

$$x_c = \left(\frac{I_i - \phi - V_{im}}{|\bar{E}|} \right) \cong \frac{(I_i - \phi)}{|\bar{E}|} \tag{11}$$

where I_i is the first ionization potential of the atom and ϕ the work function of the metal.

15.3.2.1 The Probability of Ionization

The probability of barrier penetration P_t or tunneling and, hence, ionization can be estimated by applying the Wentzel-Kramers-Brillouin (WKB) quantum theory approximation, provided realistic potential energy curves V(x) are available. For the case where the image potential in Equation 10 is assumed to be a Coulomb potential of the form

$$V_{im} = \frac{Ze^2}{x} \tag{12}$$

where Z is the effective nuclear charge, then P_t becomes

$$P_t \cong \exp\left\{-6.8\times10^7 \frac{(I_i)^{3/2}}{|E|}\left(1 - 7.6\times10^{-4}Z^{1/2}\frac{|E|^{1/2}}{I_i}\right)^{1/2}\right\} \tag{13}$$

The transition rate of barrier penetration is equal to the product of P_i and the frequency of arrival of electrons at the barrier, given by

$$P_i = \upsilon_e P_t \tag{14}$$

where

$$\upsilon_e = 1/2\, v_e/r_e \tag{15}$$

so that the probability of ionization per unit time P_i is approximately given by $P_i = \upsilon_e P_t$. In Equation 15, v_e is the velocity of the electron and r_e the effective orbital radius of the electron in an s-state; for example, the frequency of arrival of the electron at the barrier is of the order of 10^{15} to 10^{16}/s.

The time τ (lifetime) during which the transition takes place is given by

$$\tau = \left(\upsilon_e P_t\right)^{-1} \tag{16}$$

The field ionization current can be calculated in principle by multiplying the arrival rate of atoms at the surface of the ionizer, dn_0/dt, and the probability of ionization P_i. However, the mechanism of generation depends in a complex way on the ambient temperature, field strength, and polarizability of the atom or molecule. These factors affect the voltage-current characteristics of the GFIS, which depend in a complex way on the temperature of the ionizer needle, the field strength, and the gas particle density in the vicinity of the tip of the needle. A typical current-voltage curve for a GFIS is shown in Figure 15.3, which displays the behavior for the low-field region (I) and the high-field region (II).

The ion current field emitter is a complex function of the electric field and temperature of the ionizer needle and can best be approximated by consideration of certain limiting cases for the conditions that exist during operation of the sources. We shall consider the low-field and high-field cases for purposes of illustration.

15.3.2.2 Ion Current

Low-Field Ionization

At low fields, where the total rate of ionization is small compared to the arrival rate of atoms or molecules the current is determined by the equilibrium number of particles near the ionizing tip multiplied by P_i the probability of ionization factor (Equation 14). For the case when $T \neq T_t$, the steady-state or equilibrium population

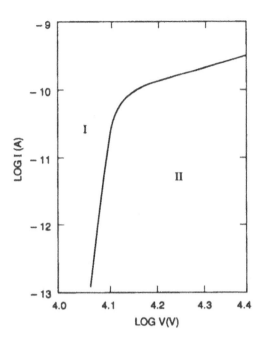

FIGURE 15.3
Typical V-I relationship for a GFIS; the low-field regime is designated in the region labeled I and the high-field regime is designated by the region labeled II.

density of fully thermally accommodated atoms or molecules in the immediate vicinity of the needle n_t can be expressed as

$$n_t = n_o \left(\frac{T}{T_t}\right)^{1/2} \exp\left[-V(\bar{E})/kT\right] \tag{17}$$

where n_o is the concentration of atoms or molecules far from the needle tip and T and T_t are, respectively, the temperatures of the gas far from and at the tip of the ionizer. Under the assumption that the field ionization process does not appreciably affect the number density of particles n_t, the total ion current can be estimated by integrating over the volume that extends between the emitter tip of radius r_t to the transition region that defines the boundary between n_t and n_o, or

$$i = 2\pi n_o \int_{r_t+x_c}^{\infty} Pi(T/T_t)^{1/2} \exp\left[-V(r)/kT\right] r^2 dr \tag{18}$$

For cases in which $-V(\bar{E}) \geq 3/2\ kT$ and the polarization energy $-V(r) = -V(\bar{E}) = {}^1\!/_2\alpha\ \bar{E}^2$ is dominant and where α is the polarizability of the atom or molecule, the current can be approximated from Equation 18 by the following relation:

$$i \simeq 2\pi n_o r_t^2 x_c (T/T_t)^{1/2} P_i \exp\left[-1/2\alpha \bar{E}^2/kT\right] \tag{19}$$

Equation 19 is valid for cases when the ionization rate is low compared to the arrival rate of particles striking the tip of the needle.

Supply-limited (high-field) ionization At very high electric fields $\bar{E}$, all particles approaching the tip are assumed to be ionized before reaching the tip. For this case, the current is determined by the supply function. Molecules passing near the ionizing tip are attracted by polarization forces.[19] For the case where the apex of the ionizer is spherical with radius r_t, the field at a distance r can be expressed by

$$\bar{E}(r) = \bar{E}(r_t)\left(\frac{r_t}{r}\right)^2 \tag{20}$$

The potential energy of a polarizable particle with a permanent dipole moment in this central force field is then the sum of the potential energies given by

$$-V(\bar{E}) = p\bar{E} + \frac{1}{2}\alpha\bar{E}^2 \tag{21}$$

where α is the polarizability of the atom or molecule and p the dipole moment of the particle. In the one-dimensional centrifugal force field approximation, the potential energy can be expressed by

$$V'(\bar{E}) = V(\bar{E}) + \frac{M}{2}v^2\frac{\rho^2}{r^2} = V(\bar{E}) + \frac{3}{2}kT\frac{\rho^2}{r^2} \tag{22}$$

where ρ is the distance of closest approach to the apex of the ionizer if the field $\bar{E}$ were zero and v is the velocity of the particle of mass M far from the tip of the needle. When the particle strikes the needle tip at grazing incidence, then from conservation of energy, the total energy of the particle E_T taken as 3/2 kT, can be equated to the potential energy of the particle as expressed in Equation 22 so that E_T becomes

$$E_T = 3/2\,kT = V(\bar{E}_o) + 3/2\,kT\left(\frac{\rho}{r_t}\right)^2 \tag{23}$$

From Equation 23, we can derive an expression for the effective tip surface area

$$\sigma = \left(\frac{\rho}{r_t}\right)^2 = 1 - \frac{2}{3}V(\bar{E}_o)/kT \tag{24}$$

where $-V(\bar{E}_o)$ is a positive quantity. The effective area A_{eff} of the ionizer is increased by a factor of σ over the geometric area A_t or

$$A_{eff} < A_t\sigma = 2\pi r_t^2\,\sigma = 2\pi r_t^2\left(1 - \frac{2}{3}\frac{V(\bar{E}_o)}{kT}\right) \tag{25}$$

The arrival rate of particles at the tip of the needle is enhanced by polarization forces by a factor of $\sigma(\Phi)$, where Φ is defined as $\Phi = \frac{1}{2}\alpha|\bar{E}|^2/kT$. For simple geometries, the enhancement factor $\sigma(\Phi)$ may be calculated analytically. Southon[25] derived an expression for a sphere, given by

$$\sigma(\Phi) \approx \sqrt{\pi\Phi} \text{ for } \Phi > 2 \tag{26}$$

while van Eekelen[26] derived an expression for the enhancement factor $\sigma(\Phi)$ for a hyperboloid, which can be expressed according to

$$\sigma(\Phi) = 1/4\left(\Phi + 2.7\Phi^{2/3} + 2.7\Phi^{1/3} + 1\right) \tag{27}$$

At very high fields, all particles arriving at the ionizer tip are ionized so the current is limited only by the rate of arrival. Since polarization forces attract atoms in the vicinity of the ionizer, the rate of arrival is faster than that associated with thermal motion. At very high fields, where the ionization probability tends to be unity, the ion current is determined by the number of particles that strike the ionization zone that surrounds the emitter.

The high-field ion current i from the tip of a hemispherical ionizer of radius r_t and area $A = 2\pi r_t^2$ in the central field approximation for polarization forces is

$$i = \frac{n_o}{4}q\bar{v}\sigma A_t = \frac{n_o}{4}q\bar{v}2\pi r_t^2\sigma \tag{28}$$

By substituting $n_0 = P/kT$ and $\bar{v} = \sqrt{\frac{8}{\pi}\frac{kT}{M}}$ into Equation 28 where $\bar{v}$ is the velocity of the particle of mass M and charge q and P is the pressure, Equation 28 becomes

$$i = 2\pi r_t^2 \frac{qP}{(2\pi MkT)^{1/2}}\sigma \tag{29}$$

Since $-V(\bar{E}_0)$ is a positive quantity, the rate of arrival due to polarization forces is increased by the factor σ over that associated with thermal motion alone. The quantity σ may have values from 10 to 100 whenever the particles do not have permanent dipole moments.

15.3.3 Practical Field Ionization Sources and their Technologies

GFISs have been developed that operate either at ambient or cryogenic temperatures. However, it is advantageous to cool the emitter to low temperatures in order to enhance the density of atoms in the high-field region of the source. The atoms or molecules are accommodated to lower thermal temperatures due to transfer of kinetic energy from the atoms and molecules to the cold needle; the low-energy particles, therefore, spend more time in the high-field region, where they may be condensed on the apex of the ionizer. Field ionization then occurs at a faster rate because more particles reside in close proximity to the emitter surface where the local radius of curvature is small. Polarization forces play an important role in trapping particles near the surface of the apex end of the field emitter; ionization then takes place by the tunneling of electrons through the distorted potential well of the atom into the metal.[20] The tip temperature also affects the threshold field strength required to ionize the gas, i.e., the threshold voltage required to ionize a gaseous feed material decreases with decreasing tip temperature (see, e.g., References 27 to 30). The fields required for ionization are approximately one half those required for operating room temperature sources. This is a direct effect of the concentration of the species very close to the surface due to the pumping effect of the low-temperature surface. The effective distance between the atom or molecule and ionizer surface is low in the region where the electric field strength is extremely high.

The method of field ionization has been applied in mass spectrometry ion sources, in field ion microscopy (FIM), and as sources of intense ion beams for general use. Several GFISs have been developed for specific focused beam applications, including those described in References 2, 14 to 16, and 31 to 37 for H; References 27 to 30 and 38 to 39 for He; Reference 28 for Ne; References 28, 31, and 40 for Ar; and Reference 28 for Xe.

15.3.3.1 Needle Fabrication and Surface Morphology for Angular Confinement

Needle Fabrication

Needles for the GFIS can be formed by electrochemical etching (see, e.g., References 2, 20, and 41) followed by field evaporation to achieve an atomically smooth end form (see, e.g., Reference 32). W is favored as an emitter material by field ion microscopists because it is easily formed into a sharply pointed tip by standard NaOH etching techniques (see, e.g., Section 15.4.3.1, Figure 12, and References 20 and 41). However, W is particularly susceptible to the effects of reactive gas etching due to the presence of contaminant gases in the vacuum system. For the source of Reference 2, W emitters were found unsuitable for long-term use because of rapid degradation at vacuum levels of 10^{-7} torr. Because of the chemical inertness properties of Ir, it has been utilized as needle material because of its resistive to reactive gas etching.[2,14-16,31,32] However, it is more difficult to form into sharp, fine-tapered needles and dilute NaClO, for example, is required for etching the needles to tip radii of 0.1 to 0.3 μm.[2]

Techniques have been developed for creation of surface contours on microscopic (100 to 1000 nm) and near-atomic scales (10 to 100 nm) that utilize the morphological changes resulting from ion beam sputtering from high-field buildup and from high-temperature, high-field evaporation of single crystalline emitters to alter the emission or ionization sites on the apex end of emitters. A number of researchers, including Swanson and Crouser, have used thermal field buildup and high-field evaporation of single-crystal W (100) emitters to redistribute or reshape the emission surface and to thereby increase the angular intensity of electron and ion beams extracted from the emitters.[42] By structuring on a near-atomic scale, the angular divergence can be confined and the stability of the ion beam increased by local enhancement of the surface electrostatic field.[29,42-45] Control of the emitter contours on a microscopic scale also allows the manipulation of the supply of neutral molecules to the ionization sites on the apex end of the needle. These contours also affect beam stability. For example, the local field strength of the surface of an emitter can be enhanced by sputtering the apex end of single-crystal needles with a heavy ion beam such as He⁺ with the field reversed (field emission mode). This procedure has been used to create what is referred to as a supertip or W (111) that has an emission area of ~10 nm in diameter with a protruding height of a few atomic layers.[29,46,47] Enhanced field emission can also be effected by building up single crystals at the apex end of other needle materials such as Ir[31,32] that confine the angular divergence of beams when operated

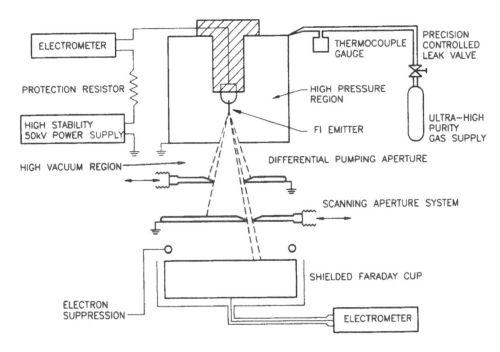

FIGURE 15.4
Schematic representation of the room temperature H_2 GFIS, equipped with an Ir needle type ionizer, described in Reference 2; the experimental apparatus for measuring the angular divergence of ion beams generated in the source is also displayed.

in the FI mode. The field buildup technique is accomplished by heating the field emitter between 1300 and 1600°C with a positive voltage field strength of $\sim 5 \times 10^7$ V/ cm applied to the needle, which initiates migration of atoms on the surface of the needle, resulting in a reshaping of the needle tip.[45] All of these techniques are effective in reducing the virtual sizes of the source of emissions and, therefore, improve the brightness of the source for focused beam applications. Structuring of field emitter surfaces to manipulate the surface electrostatic field is also of interest for scanning tunneling microscopy (STM) for stable, and reproducible high-resolution images of surface atomic structures.

15.3.3.2 Types of Gas Field Ion Sources

A schematic representation of the GFIS ion source developed by Allan et al.[2] is displayed in Figure 15.4. This source is operated at room temperature; no provisions are made to improve the intensity capabilities of the source by operating the needle at cryogenic temperatures. In the high-field regime, ion beam intensities up to 21 nA, comprised of 90% H^+, have been achieved at an applied voltage of 20 kV and a pressure of $\sim 5 \times 10^{-3}$ torr. The source has an apparent size of 100 nm and can be effectively operated over a range of pressures between 1×10^{-3} to $\sim 10^{-2}$ torr. W needles were found to begin to degrade after ~ 5 min of operation, while the Ir needles lasted in excess of 60 h. Ir, although more difficult to chemical etch, is essentially impervious to chemical erosional effects that occur at high pressures and was first used by Orloff and Swanson[14-16,31,32] to extend the operational lifetime of this source type. By cooling the emitter tip to liquid-N_2 temperatures, it is estimated that ion currents at low fields could be increased by approximately a factor of four. The measured brightness B of the source is B $\simeq 7.65 \times 10^6$ A/cm²·sr at 17 kV.

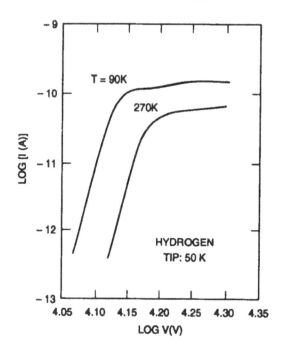

FIGURE 15.5

V-I relationship for an H_2 GFIS operated at two different gas temperatures. The needle temperature for both cases was 50 K. (From Reference 28.)

The hydrogen GFIS of Sato et al. utilizes a W needle with end radius of ~0.01 μm operated at liquid-N_2 temperature.[35] The source generates total ion beam currents of 200 nA at a pressure of 3.5×10^{-3} torr at 17 kV. The angular intensity of the source is ~1.2 μA/sr. This source type can also be operated with other gases, including He and Ar. The Ar source described in Reference 40 also utilizes a W needle operated at liquid-N_2 temperature. The source generates Ar^+ ion beams of ~1 nA at ~2.2×10^{-3} torr and 12.3 kV.

An H_2 and rare GFIS that utilizes field evaporated W [110] needles chilled to liquid-He temperatures has been described by Hanson and Siegel.[36] This source generates angular current densities $dI/d\Omega$ of H_2^+ and H^+ up to 10 μA/sr at 6 kV; angular intensities up to 60 μA/sr have been observed on occasions. The source has also been utilized for He^+ and Ar^+ generation.

An H_2 GFIS equipped with a thermally field built-up, single-crystal Ir [110] emitter with typical tip radii ranging between 0.08 and 0.1 μm, operated at liquid N_2 temperature, is described in Reference 31. The source generates highly angularly confined beam intensities, consisting principally of H^+ and H_2^+, up to ~70 nA at ~3×10^{-2} torr and extraction voltage of 14 kV.

15.3.4 Operational Characteristics

15.3.4.1 Current-Voltage Characteristics

The current-voltage characteristics of a GFIS depend on the field strength $\bar{E}$, the radius of the tip of the needle r_t, the character of the emission region of the tip of the needle, the temperature of the gas at the needle tip, the temperature of the gas far from the needle tip, and the pressure. The GFIS source described in Reference 36 is

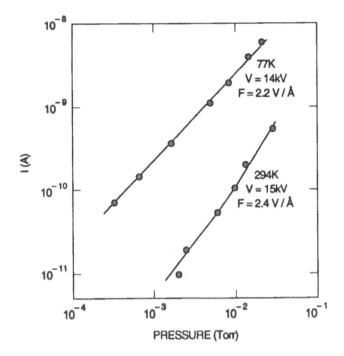

FIGURE 15.6

Ion beam intensity vs. H_2 pressure for a GFIS equipped with a polycrystalline Ir needle type ionizer maintained at 77 K and 294 K. (From Reference 14.)

equipped with a thermally annealed, field evaporated W (110) emitter estimated to be capable of angular intensities exceeding 10 µA/sr when operated at liquid-He temperatures. Figure 15.5 displays the current-voltage characteristics for a GFIS operated with H_2 at two different temperatures.[28] As noted, the low- and high-field regions are clearly delineated and the advantages of lowering the temperature of the needle are obvious.

15.3.4.2 *Dependence on Gas Pressure*

According to Equation 29, the ion beam intensity from a GFIS increases linearly with pressure in the high-field regime. This behavior is usually observed at lower pressures as illustrated in References 2 and 28. For example, Jousten et al. observed a linear pressure dependence on intensity with pressure within the pressure range of 10^{-6} to 10^{-3} torr.[28] Figure 15.6 illustrates the ion beam intensity of an H_2 GFIS equipped with an Ir ionizer operated at 77 and 294 K.[14] The benefit of operating at lower temperatures is clearly illustrated. Nonlinear dependencies are usually observed at higher pressures (see, e.g., Reference 40).

15.3.5 **Characteristics of Beams Generated in the GFIS**

Spherical and chromatic aberrations of the ion optical system, the angular intensity, and the energy spread combine to limit beam size and current density on target for very-high-resolution applications. The angular intensity $dI/d\Omega$, current density dI/dA, and energy spread are fundamental figures of merit of the GFIS for high-resolution applications.

15.3.5.1   Ion Beam Intensities, Angular Distributions, Angular Intensities,
and Current Densities

The use of the GFIS as a tool for focused beam applications depends critically upon the ion beam characteristics of the sources, i.e., the angular intensity and the energy spread within the beam. Space charge effects will result in increases in the emittances and reduction of brightness, as well as increases in the energy spread within the ion beam. However, space charge effects are usually not important because of the low ion currents typical of the GFIS. The effective source size δ, defined as the area from which the ions appear to originate, is less than ~1 nm for the GFIS. The source exhibits angular intensities of ~1 μA/sr, which correlates to a brightness of ~10^9 A/cm^2. Since the virtual size of the GFIS is very small, the properties of the beam that is transported to the target are determined by aberrations in the GFIS during extraction and induced in the beam by the beam transport system. Spherical and chromatic aberrations of the ion optical system and the energy spread combine to set the limit of spot size and current density for very-high-resolution probes with sources of energy spreads as low as 1 eV. The GFIS can be characterized in terms of the angular intensity $dI/d\Omega$ (A/sr) for which the focused beam intensity is the product of $dI/d\Omega$ and the solid angle $\Delta\Omega$ subtended by the beam-limiting aperture used in the focusing system; the physical size of the image is determined by the current density distribution within the image plane, which is a function of the beam-limiting aperture size and the aberrational aspects of the GFIS induced into the beam during extraction by the optical components used to transport the beam through the system. The solid angle $\Delta\Omega$ is given by $\Delta\Omega = \pi r^2/R^2 = \pi \tan^2\theta_0$, where r is the radius of the beam-limiting aperture and R is the distance from the point of origin to the aperture where $r/R = \tan\theta_0$; in the small-angle approximation $\Delta\Omega = \pi\theta_0^2$. The merits of the focusing system can be measured in terms of the image size and the beam intensity I on target where I is the current passing through the beam-limiting aperture, or $I = (dI/d\Omega)\,\Delta\Omega$.

The angular intensity $dI/d\Omega$ of a GFIS depends on the electric field strength and the gas density at the apex of the emitter; the latter is determined by the gas pressure and the temperature of the tip of the ionizer. The highest angular intensity that has been observed from a conventional GFIS operated at room temperature is $dI/d\Omega \cong$ 1 μA/sr.[32] The angular intensity $dI/d\Omega$ can be increased by cooling the tip of the ionizer to cryogenic temperatures; for example, cryogenic techniques have been used by Hanson and Siegel to cool the tip of the needle of a GFIS to condense the gas on the apex of the emitter, thereby providing a higher density of atoms in the proximity of the ionizer.[36] This technique has resulted in increases of the angular intensities from the GFIS $dI/d\Omega$ up to 10 μA/sr. However, because of the ease at which atoms can leave the surface when energy is transferred to the surface of the ionizer during operation, stable operation is often difficult to achieve when this technique is utilized. A GFIS with a specially prepared field emitter referred to as a supertip that increases the effective angular intensity from ~1 to ~10 μA/sr has been developed.[29,46,47]

The beam angular distribution influences the size of the image of the focused ion beam. The angular distribution of H$^+$ ion beams from a source equipped with a polycrystalline Ir emitter is shown in Figure 15.7.[2] The shape of this distribution is controlled by the crystallographic structure of the tip, which produces localized regions of high electric field that promote FI. As noted, the full width of the beam subtends a half-angle up to 500 mrad; the limits of the distribution are determined by the location of the emission sites on the tip of the emitter. The roughly Gaussian distribution for the angular distribution shown in Figure 15.7 is typical for

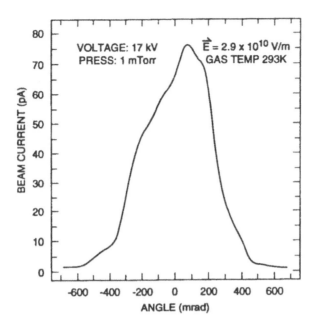

FIGURE 15.7

Ion beam intensity vs. angular divergence for the room temperature GFIS, equipped with a poly-crystalline Ir needle type ionizer, described in Reference 2.

polycrystalline emitters; approximately 85% of the beam is found within a cone of half-angle of 300 mrad or ~17°.

As indicated in Section 15.3.3.1, heating a field emitter between 1600 to 1900 K in the presence of a positive applied field of 5 to 6×10^7 V/cm can effect a thermody-namic reshaping (buildup) of the emitter surface through surface migration.[29] This process is accomplished by operating the source in the electron emission mode and subjecting the needle tip to heavy-ion bombardment.[45] Earlier work by Crewe et al.[44] has shown that TF buildup of a ⟨130⟩ oriented W emitter increased the angular confinement of electron beams. Schwoebel and Hanson developed a procedure in-volving field emission through H_2 films whereby ion emission structures can be formed *in situ* on ⟨110⟩ oriented W field emitters.[43] Through this process, sites are built up that are capable of producing 10 to 15 nA beams of H_2^+ with half angles of 16 mrad. These sites are protuberances of 200 Å in diameter and a few atoms in height, which results in field enhancement factors of ~1.5. Figure 15.0 displays the angular distri-bution from a thermally built-up ⟨110⟩ Ir ionizer scanned along the (1̄11) ⟺ (111) directions.[16]

15.3.5.2   Mass Spectra

The type and numbers of species that can be generated in a GFIS depends on the electric field strength, the ionization potential of the species in question, and whether the feed gas is atomic or molecular. When ultrahigh-purity H_2 is used as the source feed gas, the field ion current consists of H^+, H_2^+, and $_3^+$ ions in relative amounts that depend upon the strength of the electric field $\vec{E}$ at the emitter tip and the protrusion of atomic sites on the emitter surface. In the high-field regime ($\vec{E} > 2.3 \times 10^8$ V/cm²), H^+ is the major component in the mass spectrum while at lower fields, H_2^+ is the dominant species.[2] At a field close to the field evaporation limit for Ir ($\vec{E} \cong 5 \times 10^8$ V/cm²), 90% of the beam is H^+.[48] Operation in the high-field regime yields

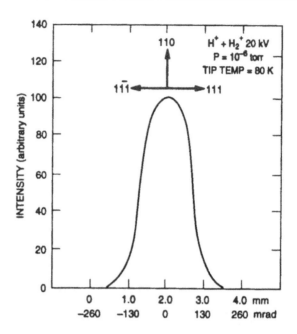

FIGURE 15.8
Relative Ion beam density vs. angular divergence for an H_2 GFIS equipped with an evaporated and thermally annealed W (100) ionizer illustrating the effect of crystalline orientation on angular confinement. (From Reference 16.)

maximum current, but significantly increases the energy spread within the ion beam.[49,50] The high-field ion current depends directly on feed gas species and the number of gas molecules present in the high-field region surrounding the tip. The mass spectra observed for noble gas feed materials are predominantly singly charged atomic species (see, e.g., References 27 to 29 and 37 to 39).

15.3.5.3 Energy Distributions

The energy spread in the GFIS depends on the field strength utilized during generation; in the transition region between the low- and high-field current-voltage regimes, approximately 50% of the beam is H^+, which exhibits an energy spread of $\Delta E \sim 2$ eV.[49] The energy spread increases significantly as the field strength increases and spreads up to and exceeding $\Delta E \cong 20$eV for the H^+ component have been observed.[51] Energy spreads within the GFIS are observed to increase with increasing field strength. This fact comes as no surprise and is, in part, an artifact of the tunneling mechanism responsible for ionization of atoms or molecules as they pass within some variable distance (less than or equal to the critical distance above the apex of the needle) where the potential is a strongly varying function of position. Because ions are born at different equipotentials, they are accelerated through different potential differences and, consequently, to different final energies; thus, the tunneling mechanism is responsible, in part, for the observed energy distributions in beams generated in the GFIS. Of course, collisional processes, space charge effects, and Coulomb interactions also contribute to the observed energy spreads.

Measurements of the energy distributions of beams extracted from a H_2 GFIS have been performed by Tsong and Müller[49] for H_2, He, Ne, and Ar sources and by Hanson and Siegel[50] for H_2. For angular intensities of less than $10\mu A/sr$, Hanson and Siegel found that 90% of the H_2^+ field ions from bright site emission of fields of

1.5 V/Å occur within an energy width of 1.2 to 1.9 eV. Within this narrow energy spread, only H_2^+ ions are produced. The beam intensities and angular intensity from a single bright site can be increased by raising the source gas pressure and the applied voltage to the needle. Energy broadening occurs at high intensities as a demonstrated by the fact that at 3.5 μA/sr, 90% of ions have an energy spread of 1.9 eV (0.94 eV FWHM), while at 18 μA/sr, 90% of ions have an energy spread of 2.2 eV (0.98 eV FWHM).

Because of the relatively low beam intensities from the GFIS and the inconveniences and problems associated with operation of the source in the cryogenic mode, the GFIS has largely been abandoned in favor of LMIS, where angular intensities of $dI/d\Omega \sim 20$ μA/sr are obtainable.

15.4 Field Evaporation (Liquid-Metal) Ion Sources

15.4.1 General Principles of the LMIS

The liquid-metal ion source (LMIS) has the advantage of higher beam intensities and longer lifetimes, and avoids the gas loading and scattering problems associated with the field ionization source (GFIS). In addition, the LMIS is simple in structure, easy to operate, has a very long lifetime, and also consumes very little power. The LMIS differs from the GFIS in that the GFIS emission geometry is determined by the radius of curvature of the solid or capillary needle used to effect ionization, while the molten liquid forms the emitting surface in the LMIS. The LMIS utilizes either a needle or capillary that serves as the conduit for a liquid metal having a high surface tension and a low vapor pressure at its melting point. For stable beam generation, the center of the liquid metal needle emitter must be precisely located in relation to the optical axis of the ion extraction system. LMISs utilize thin (~100-μm-diameter) needles with end radii ~10 μm or thin capillary tubes through which the liquid metal flows. The needle or capillary is heated to the melting point of the metal during operation. The apex end of the needle or capillary tube, which must be wetted with liquid metal, is placed in close proximity to a ground electrode so that the required high electric field (10^7 to 10^8 V/m) can be generated by application of a moderate voltage (~10 kV, typically). Two basic types of liquid-metal sources are used: (1) capillary type and (2) needle type; these source geometries are illustrated in Figure 15.9. However, the needle type source is usually preferred over the capillary type because they are generally more mechanically stable and easier to operate. Most ion beam focusing systems that are designed for high spatial resolution (<1 μm) utilize needle type LMISs. The radius of the apex of the needle in the needle geometry source, prior to wetting with the liquid metal, is approximately 10 μm. A strong electric field, applied between a capillary or needle wetted with molten metal at anode potential and an extraction aperture, pulls the metal up into a modified Taylor cone (apex angle, ~98.6°) with a rounded apex.[52] The cone apex end of the molten metal emitter has a radius of about 5 nm. The conical shape of the end of the needle is determined by a balance between electrostatic and surface tension forces. Stable Taylor cone formation requires an adequate flow of liquid metal to the tip of the emitter. For best performances of needle type LMISs, the apex of the needle, prior to wetting, is usually blunt (typically with a radius of 5 μm or more). If the field is sufficiently strong (~10^8 V/cm) at the surface of the molten cone and in a direction so as to accelerate positive ions, ion desorption will occur; the process whereby ions are removed from the tip region of the molten surface is often referred to as field evaporation; neutral atoms and clusters of atoms may also accompany the field

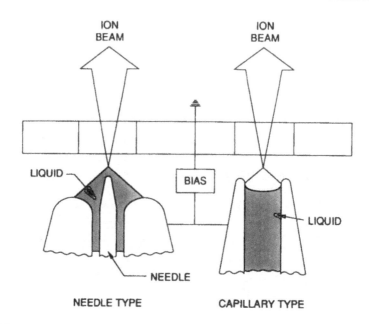

FIGURE 15.9
Illustration of two principal types of liquid-metal ion sources: needle ionizer, and capillary ionizer.

evaporation process. Extremely sharp needles are more difficult to operate, presumably due to the fact that the tip does not allow the formation and sustenance of a stable liquid cone. However, if the radius of curvature of the apex of the needle is made too large, multiple emission sites can form along the periphery of the apex. It is highly likely that field ionization and field evaporation processes both contribute to the ionization process. In this way, beam intensities of a few to several tens of microamps are readily initially formed. The beam current rises steadily and can reach values up to 1 mA for a single needle or several tens of milliamps for multiple arrays. As with the GFIS, the remarkable feature of this source type is that the area of ion emission is extremely small — having a radius of curvature of a few nanometers.[53-55] Thus, the LMIS is, in principle, the brightest of existing positive-ion sources. The liquid-metal ion source offers a means for generating beams with central brightnesses of $>1 \times 10^6$ A/sr·cm^2 and, consequently, ultrabright beams from the LMIS can be focused into submicron images with current densities of a few amps per centimeter squared.

The practical application of the LMIS for focused beam applications was first demonstrated by Seliger et al.[1] In more recent years, the applications for focused ion beams based on LMIS technology have grown rapidly. The LMIS has made possible the transport and focusing of beams on target with sizes less than 100 nm in diameter and current densities in excess of 1 A/cm^2. Such high-current-density, small-diameter beams have been used to great advantage in scanning ion microscopy, surface analysis, micromachining, mask repair, direct ion implantation, high-resolution ion lithography, and are particularly well suited for direct, high-resolution writing with substitutional dopants during ion implantation doping of semiconducting materials.

15.4.2 Theory of Field Evaporation

Although the mechanism underlying ionization in the LMIS is not fully understood, several attempts have been made to explain the formation mechanism in terms

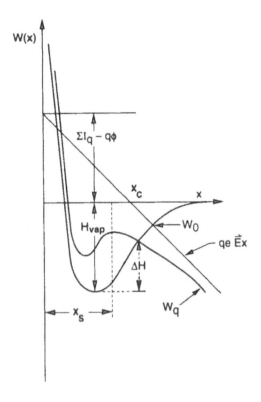

FIGURE 15.10
Potential energy $W(x)$ as a function of position x from a metal surface for an atom W_0 and ion W_q in an external, uniform electric field $\vec{E}(x)$.

of an evaporation model. For example, the "image-hump" model has been used to explain the field evaporation-ionization process for the high-temperature case by several groups. For further details of models used to describe the process, the reader is referred to the review article by Marriot and Riviere[56] and the more comprehensive field evaporation model described by Forbes.[57] The model views the field evaporation process as the thermal escape of an ion over a one-dimensional barrier. The process whereby a neutral atom is evaporated from a liquid-metal surface under the influence of a strong electric field is referred to as field evaporation. The potential well of a surface atom is strongly modified by the external electric field, and the atom has an increased probability of overcoming the energy barrier. This barrier height, ΔH in the simple "image-hump" formalism, is made up of the ionization potential I of the atom, the heat of vaporization H_{vap} of the atom, the work function ϕ of the surface, and a term involving the "image" attraction and the applied electric field referred to as the Schottky term and the polarization energy. For this emission mechanism, field strengths of around 1 to 5 V/Å are needed for the microamp current emission produced by the liquid metal ion source. Figure 15.10 displays the potential energy diagram of an atom or ion in a strong electric field during evaporation from a metal surface.

The ion current density j that can be extracted through a barrier height of ΔH can be expressed as

$$j = qNA(\omega)\exp\left\{-\frac{\Delta H}{kT}\right\} \tag{30}$$

where q is the charge on the ion, N is the number of active sites per unit area, A(ω) is a factor that depends on the vibration frequency of the atoms in the potential well, k is Boltzmann's constant, and T is the absolute temperature of the surface. ΔH can be expressed according to the following simple relationship:

$$\Delta H = H_{vap} + 1/2\alpha |\bar{E}|^2 + \Sigma_q I_q - q\Phi - \left(\frac{e^3 q^3 |\bar{E}|}{4\pi\varepsilon_0}\right)^{1/2} \tag{31}$$

where H_{vap} is the heat of vaporization, I_q is the ionization energy required to ionize the atoms to charge state q, Φ is the work function of the emission surface, $|\bar{E}|$ is the electric field strength at the surface, ε_0 is the permitivity of free space, and α is the polarizability of the atom. The last term in Equation 31 is attributable to the Schottky effect, which lowers the escape barrier.[58] In this way, beam intensities of a few to several tens of microamps can be formed.

Energy Deficit. The peak position of the energy distribution, measured by the use of a retarding field analyzer, is always observed to be offset with respect to the potential required to bring the particle to zero energy. This energy difference is referred to as the energy deficit and can be explained by reference to the potential energy diagram shown in Figure 15.10. For an ion at the critical distance x_c above the liquid-metal surface the energy required to escape the liquid-metal surface can be expressed by

$$E_i = \Delta H - H_{vap} \tag{32}$$

Following extraction and transporting the ion beam to the retarding field analyzer, which is set at an identical potential, an unpolarized particle will arrive at the electrode surface with a final energy E_f just equal to the difference between the first ionization potential I_i and the work function Φ_e of the electrode

$$E_f = I_i = \Phi_e \tag{33}$$

The energy at which the ion will strike the electrode will be identically equal to the difference between Equations 32 and 33. The resulting expression is referred to as the energy deficit ΔE given by

$$\Delta E = E_f - E_i = H_{vap} + I_i - \Phi_e - \Delta H \tag{34}$$

15.4.3 Practical Liquid-Metal Ion Sources

15.4.3.1 Needle Fabrication, Morphology, and Wetting Procedures

Needle etching. Needle preparation is usually by electrolytic etching of a thin W or other metal wires (for example, wires of 250 to 400 μm in diameter) in NaOH solution to form a needle with a truncated, parabolic end with end radius of curvature of between 1 and ~10 μm. The choice of the electrolytic etchant depends on the chemical properties of the needle in relation to those of the etchant. For example, W needles are usually etched in NaOH[41] while more chemically inert Ir needles require

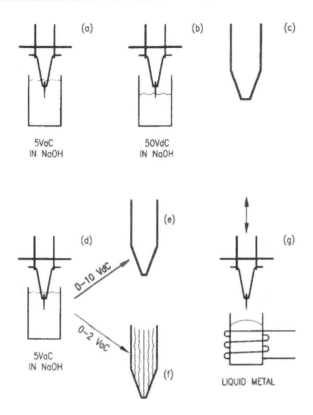

FIGURE 15.11
Illustration of procedure utilized to chemically etch needles for the LMIS as described in Reference 41.

special electrolytes such as $NaClO$.[2] The etching procedure is usually effected by the use of both dc and ac power. The use of ac power results in longitudinal grooves along the axis of the needle; the size of these grooves affect the flow of metal to the apex of the needle. The cleaning, etching, and needle wetting procedures utilized at Bell Laboratories to prepare a needle type LMIS are illustrated schematically in Figure 15.11.[41]

Needle wetting. The vacuum wetting stage is the most important part of the fabrication process. The liquid metal must wet the reservoir and needle assembly in order for the source to operate properly. To effect wetting, the assembly must be clean and free of oxides or other chemical compounds. The wetting cycles and temperatures required for wetting the LMIS assembly depend on the needle and reservoir materials. A scheme for wetting a Ga LMIS is illustrated in Figure 15.12; the needle assembly and Ga reservoir are simultaneously heated by electron bombardment to ~600°C.[4] The wetting temperatures for other metals are metal specific. For example, In requires approximately 800°C, while Au wets at 1200 to 1500°C.[59]

15.4.3.2 Types of Liquid-Metal Ion Sources

Liquid-metal ion sources rely on the wicking of a molten elemental or alloy metal along the outer surface of a metal needle (e.g., References to 59 to 65), the inner surface of a metal capillary tube (e.g., References 66 and 67), or through a sintered matrix (e.g., References 68 to 71). Stability of the conical emission area in the needle type LMIS is affected by the strong surface tension forces that are present between the thin layer of molten metal and the wetted needle. To ensure liquid flow along the needle during operation, the needle and reservoir must be heated to the melting point

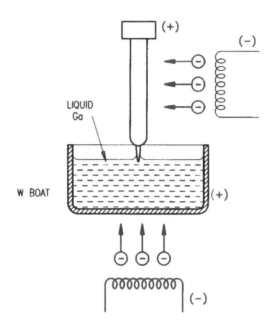

FIGURE 15.12
Illustration of a method for wetting a needle type LMIS using electron bombardment heating.[4]

of the metal. These components may be heated directly or indirectly. Directly heated sources are heated by passing a current through the reservoir and needle assembly; indirectly heated sources are equipped with a heater that indirectly transfers heat to the wetted assembly. The source may be composed of single or multiple emitters. The ion currents achievable from the LMIS increase linearly with the number of emitters.[70,71] Many examples of each of these source types have been developed as documented in the bibliography on sources of this type in Reference 6.

Three of the most often utilized needle geometry LMISs and the methods utilized for heating the assembly are illustrated in Figure 15.13. An example of a Ga LMIS developed by Prewitt and Jefferies is shown, schematically, in Figure 15.14;[65] the Ga source assembly used to make first measurements of the emittances of beams extracted from a LMIS at the Oak Ridge National Laboratory is illustrated, schematically, in Figure 15.15.[62,63] High-melting-temperature metals, such as Au, require provisions for heating the metal to temperatures exceeding 1000°C; a number of high-temperature sources have been developed for high-temperature metals including those described in References 72 and 73. For chemically active metals such as the group IA elements (Li, Na, K, Rb, and Cs), provisions must be made for *in situ* wetting the needle. As examples of this technology, Li and Cs sources equipped with provisions for retracting the needle back into molten Cs for wetting are described, respectively, in References 74 and 75. Figure 15.16 shows a schematic representation of the Li needle LMIS equipped with provisions for inserting or withdrawing the needle in the molten Li metal for wetting/rewetting, which was developed for use in the high-voltage terminal of a Van de Graaf accelerator for microprobe applications.[74]

Porous type LMISs have also been developed by Ishikawa et al.[68–71] In this source type, the emitter includes a porous region, prepared by shaping selected tungsten powders into a cone followed by sintering to form a porous tip through which the liquid metal flows. The porous emitter can also be used in combination with a solid needle. By mixing tungsten powders of different granule size, the flow impedance of the source can be varied. This type of ion source is claimed to be able to operate at

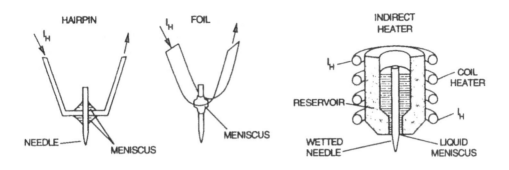

FIGURE 15.13
Schematic illustration of directly heated filament type, directly heated foil type, and indirectly heated LMISs.

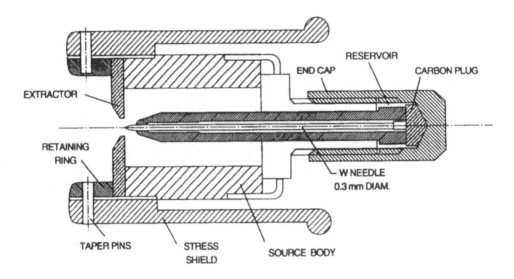

FIGURE 15.14
Schematic drawing of the gallium LMIS assembly designed and developed by Prewitt and Jefferies.[65]

higher temperatures than possible in ordinary needle type LMISs due to the reduced surface area from which thermal evaporation can occur and, thus, would be particularly well suited for high-vapor-pressure liquid metals. Variations of this type of source are shown in Figure 15.17 [68].

15.4.4 Operational Characteristics

15.4.4.1 Metal Flow Balance Conditions

Structurally, an LMIS consists of a liquid metal that flows along the surface of a needle or through a capillary into the region of a strong electric field that causes the liquid metal to be pulled into a conical geometry with a rounded apex. The operating characteristics of liquid-metal sources, therefore, are strong functions of flow characteristics of the particular metal or alloy along the surface of the metal (tungsten) needle or capillary tube. Wagner has explained the flow characterization of the LMIS by use of a hydrodynamic model balancing the pressures due to surface tension and

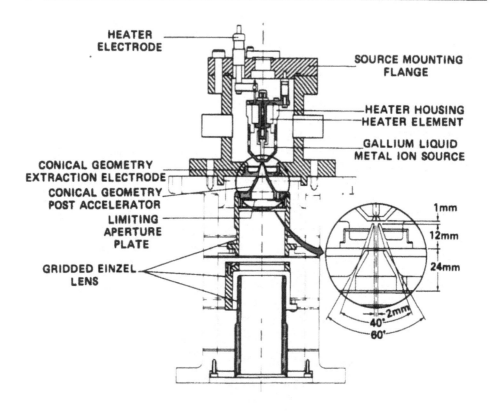

FIGURE 15.15
Schematic drawing of the gallium liquid metal ion source equipped with a conical geometry ion
extraction/postacceleration stage.[62,63]

electrostatic forces on the end of the liquid cone of the source and the viscous
capillary flow along the tip.[76] In the static case, a fluid can be in stable equilibrium by
the opposing and precisely balanced forces of surface tension and electrostatic stress,
provided that the liquid assumes the shape of a cone with a half-angle of 49.3°,
commonly referred to as a Taylor cone;[52] when the forces are in balance, the condition
for stability can be expressed by

$$\frac{2\gamma}{r} = \frac{1}{2} \varepsilon_0 \bar{E}^2 \qquad (35)$$

where γ is the surface tension, r is the principal radius of the cone, $\bar{E}$ is the electrostatic field strength, and ε_0 is the permittivity of free space. In order to satisfy this
equation, the electric field must be proportional to $r^{-1/2}$ and the potential must have
the form

$$V(r,\theta) = Ar^{1/2}P_{1/2}(\cos\theta) \qquad (36)$$

where $P_{1/2}(\theta)$ is a Legendre polynomial of order $\frac{1}{2}$ and A is a constant. The potential
is constant along the surface of the cone, which has a half angle of 49.3°. For an
infinitely sharp cone apex, the electric field would be infinitely high and, therefore,
during operation, the conical end does not form because of evaporation of material.
This action causes a spherical tip to form on the apex of the cone; the radius of the tip

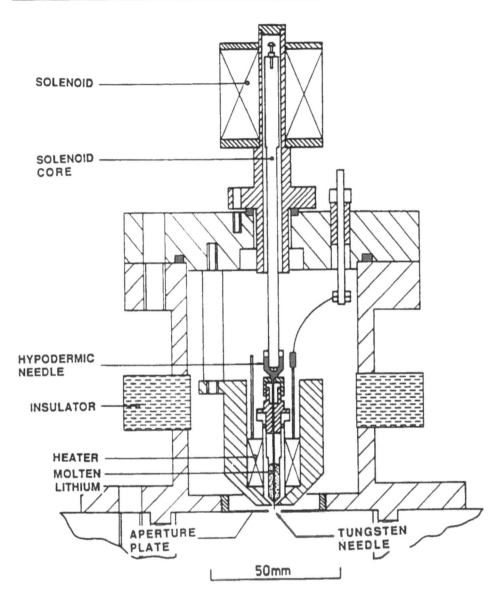

SOLENOID

SOLENOID
CORE

HYPODERMIC
NEEDLE

INSULATOR

HEATER
MOLTEN
LITHIUM

APERTURE
PLATE

TUNGSTEN
NEEDLE

50mm

FIGURE 15.16
Schematic drawing of a needle type Li$^+$ LMIS equipped with provisions for remote wetting or rewetting the needle.[734]

of the cone is ~5 nm.[53-55] Due to surface tension forces, the liquid material, lost through field or conventional evaporation processes, is replaced by new material flowing up the needle. This condition can be expressed in the following form:[77]

$$\frac{2\gamma}{r} = p + \frac{1}{2}\varepsilon_0 \vec{E}^2 \tag{37}$$

where $p = (1/2)\rho v^2$ is the pressure due to the flow of a liquid of density ρ moving along the surface of the cone at velocity $\bar{v}$. In order to achieve stable operation through a balance of surface tension and electric field stress forces in the presence of space charge, protrusions, in the form of a cusp, on the end of the Taylor cone are required

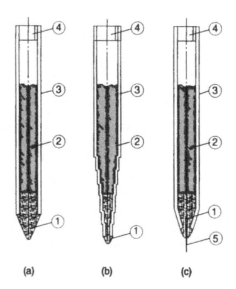

(a)   (b)   (c)

FIGURE 15.17
Schematic illustrations of impregnated-electrode type LMISs: (a) porous W tip, (b) porous W tip with small porous surface, and (c) porous W tip with a solid W needle; (1) sintered porous W tip, (2) liquid metal, (3) cylindrical W reservoir, (4) BN cap, and (5) W needle.[68]

to form. Their presence during operation of a number of LMISs has been experimentally observed (see, e.g., References 78 to 82) and predicted through the modeling studies described by Kingham and Swanson.[77]

15.4.4.2 Voltage-Current (V-I) Relationships

The critical voltage V_{oc}. Whenever a positive potential is maintained between a needle wetted with a molten metal and an extraction electrode, ion emission will occur at some critical voltage V_{oc} whenever the electric field reaches sufficient strength so as to pull the liquid metal into a cone at the apex end of the needle. For this condition to occur, the electric field stress must be equal to or exceed the surface tension of the metal at the apex of the needle, as previously discussed. Assuming the liquid-metal film conforms to the radius of the needle, an expression can be derived that relates the critical starting voltage V_{oc} to the surface tension γ and the radius of the tip of the needle r_t and the gap between the apex of the needle and the extraction electrode h.[79,83,84] The relation is given by

$$V_{oc} = \ln\left(2h/r_t\right)\left(\gamma r_t/\varepsilon_o\right)^{1/2} \tag{38}$$

Despite the simplicity of the expression and the variability of the gap h due to protrusions at the apex of the cone during operation, the expression is found to agree reasonably well with experiment. Bell and Swanson have made a detailed study of the dependence of the critical voltage on source operating parameters.[85]

The extinction voltage V_{ox}. The minimum value of V_{ox}, strong enough to hold the liquid-metal cone in place, is lower than the critical starting voltage V_{oc} due to the fact that a higher voltage is required to pull the liquid up into a cone. The minimum voltage V_{ox} is referred to as the extinction voltage. An expression for the extinction voltage V_{ox} is found to hold for capillary type LMISs that have low-impedance flow characteristics.[86] The expression appropriate for low-impedance flow is

$$V_{ox} = \left[2fR\gamma \cos\theta / \varepsilon_o \right]^{1/2} \tag{39}$$

where f is a logarithmic function of the ratio of the gap h between the capillary and the extraction electrode and the radius of the capillary R, and θ is the complement of the base angle of the cone, equal to $\pi/2 - \theta_T$, where θ_T is the Taylor angle, $\theta_T = 49.3°$.[87] This formula is also found to predict extraction voltage values for the capillary in good agreement with experiment.[86]

Estimation of the V-I characteristics. Mair has developed a simple formula for estimating the current-voltage relationship for capillary or smooth-needle type LMISs.[86] The model ignores flow effects but considers space charge effects at the apex end of the liquid-metal emitter. The relation is

$$i = 3\pi R\gamma \cos\theta (2e/M)^{1/2} \left[V_{oc}/V_{ox} \, 01 \right] / \left(V_{ox} \right)^{1/2} \tag{40}$$

where the extinction voltage V_{ox} is determined by extrapolating the current-voltage relation to the i = 0 condition.

V-I characteristics. The LMIS exhibits nonlinear current-voltage characteristics, as expected from space charge-dominated ion beams; the energy distribution and composition of ion beams are sensitive functions of the total ion beam intensity extracted from the source. Because the energy spread depend sensitively on the total ion beam intensity, the highest performances are achieved at low total ion beam intensities.

The field strength required to pull the metal into a geometry commensurate for field emission to occur depends on several factors including the following: the voltage applied to the needle or capillary, the radius of the apex end of the needle and the roughness of the needle surface, the surface tension of the metal, and the temperature of the liquid metal. At the critical voltage, ions begin to be field evaporated from the apex of the cone. For example, the critical and extinction voltages are illustrated in Figure 15.18, which displays the V-I characteristics of an In needle type LMIS for increasing and decreasing voltages.[59] The effect of temperature on ion emission for a needle type Cs LMIS is illustrated in Figure 15.19;[88] the V-I characteristics for Au LMISs with various needle radii are displayed in Figure 15.20;[83] the solid circles represent the threshold for ion emission. Surface quality effects for needles and capillary Ga and In LMISs are displayed in Figure 15.21.[59,77,89]

15.4.4.3   Ion Beam Intensities and Species Capabilities

The currents that can be achieved with this source type range from a few to several hundred microamps. The current that can be achieved can be enhanced by use of multiple emitters such as described in References 70 and 71. Clampitt et al. have reported high-brightness beams of Li^+, Cs^+, Sn^+, Ga^+, and Hg^+ from single emitters with beam intensities ranging from 500 to 700 μA.[67] Current densities from these sources reach densities in excess of 10^4 A/cm^2 at extraction voltages of 3 to 10 kV.

A number of sources have been developed that are predicated on ion beam generation from elemental metals. The versatility of the technique has been significantly extended by using the low-melting (eutectic) property of alloys. By using low-melting-point alloys, the LMIS can be used to generate beams of most of the metallic elements. The liquid metal may consist of single or multiple elements as in an alloy LMIS; several ionic species will be present in the beam when multiple element alloys are utilized. Typically, small percentages of multiply charged ions will also be present

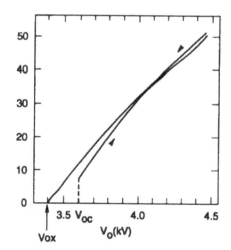

FIGURE 15.18
V-I characteristics of an In needle type LMIS for both increasing and decreasing voltages.[59]

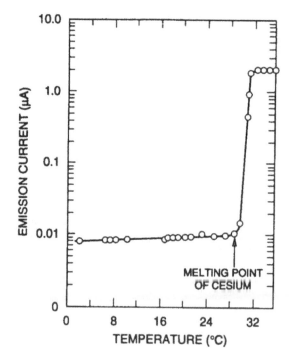

FIGURE 15.19
Temperature dependence of ion emission from a Cs needle type LMIS.[88]

in the mass spectrum of beams extracted from a LMIS. The liquid-metal ion source has been adapted for many focused beam applications in more recent years. Ion beams have been generated with liquid-metal ion sources from many elemental metals or alloys, including Al, As, Au, B, Be, Bi, Cs, Cu, Fe, Ga, Ge, In, Li, P, Pb, Pd, Pr, Rb, Si, Sb, Sn, U, and Zn. In the past decade, focused ion beam systems with liquid metal ion sources have had a significant impact on the semiconductor industry; they have been applied to new and greatly improved methods of failure analysis, as well

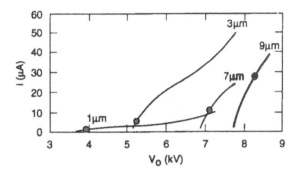

FIGURE 15.20
V-I characteristics for Au ion sources with varying needle radii. The dots represent the threshold voltage for ion emission in each case.[823]

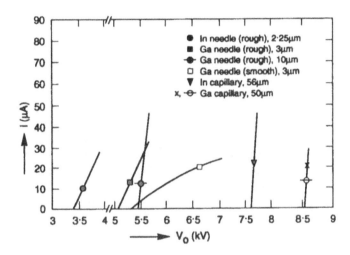

FIGURE 15.21
V-I characteristics for Ga and In needle and capillary LMISs illustrating the effect of needle radii and outer radii and surface character on the threshold voltage required for emission.[59,77,89]

as circuit repair and modification, *in situ* processing, and lithographic mask repair. Elements such as As, B, Be, and Si, which are particularly interesting for implantation into semiconductor materials such as Si or GaAs, can only be produced from alloy sources; the last four species have been obtained using liquid-metal alloys because of the reactivity or volatility of the pure metal species. The versatility of this source type in terms of species is illustrated in Table 15.1. The table provides examples of some of the elements and alloys that have been successfully utilized to form ion beams from otherwise difficult elements and references a number of liquid-metal ion sources that have been developed to generate a variety of ion beams from both elemental and eutectic alloy materials.[41,64,73,88,90–101]

15.4.5 Characteristics of Ion Beams Extracted from the LMIS

The form of the species (monomer, dimer, trimer, etc.) and energy widths of beams extracted from the LMIS depend on the operational regimes. At very low total beam intensities, <10 μA, the beam consists primarily of singly charged monomer

TABLE 15.1

A Partial Listing of Metals and Alloys for Use
in Liquid-Metal Ion Sources

Species	Liquid element or alloy	Ref.
Be	Au-Be, Au-Be, Au-Si-Be, Ga-Si-Be	91, 92, 93, 94
B	Pt-B, Pd-Ni-B, Pt-Ni-B	91, 91, 95
Al	Al	64
Si	Au-Si, Ga-Si-Be	91, 95, 94
Ni	Pd-Ni-B	91
Cu	Cu	76
Zn	Zn	96
Ga	Ga	88
Ge	B-Pt-Au-Ge	97
As	Pd-Ni-B-As, As-Sn-Pb, As-Pt	91, 97, 98
Rb	Rb	99
Pd	Pd-Ni-B	91
In	In	88
Sn	Sn, As-Sn-Pb	73, 97
Sb	Sb-Pb-Au	95, 98
Cs	Cs	73
Pt	Pt-B	91, 73
Au	Au	43, 90
Pb	Pb, Sb-Pb-Au	73, 97
Bi	Bi	100
U	U	101

ions with a low percentage of doubly charged ions present. Neutral particles are also present in the beam as it leaves the solid surface. The neutral particles are atoms evaporated from the liquid metal that are not ionized; their intensity depends on the total beam intensity. As the total beam is increased, a proportion of the total beam current is carried by singly ionized dimers, trimers, and higher-order clusters. At total currents I > 10 to 100 μA, depending on the nature of the metal, charged droplets may also begin to appear. The ion production mechanism is thought to be through field evaporation at low ion beam intensities. The beam is very unstable for this operational regime. For higher beam intensities, the characteristic frequencies measured for this mode of operation imply that the presence of a cusp at the apex of the cone, which is elongated by the electric field stress, results in droplet formation. The angular distribution of the droplets is confined to a few degrees about the optical axis of the emission tip of the LMIS, while the angular distribution of the ion beam has a half-angle ≈25 to 30°.

15.4.5.1 Mass Spectra of Ion Beams

During operation of a LMIS electrons, positive ions, neutral atoms, and clusters as well as optical emission occurs. The heavy particles include singly and multiply charged monoatomic ions, singly charged cluster ions, neutral atoms, and cluster droplets. The relative magnitudes of the species in question depend on the mode of source operation and specific liquid metal. For example, the ion species extracted from a Ga or an Al LMIS are predominantly singly charged while in sources that utilize heavier metals such as Au or Sn, the doubly charged species may dominate.[88,102] An example of the composition of ion species from an Au LMIS is displayed in Figure 15.22.[103] Cluster ions with large numbers of atoms per cluster may be emitted, particularly during operation in the high-current mode. For example, Ga ions with as many as 26 atoms have been observed[104] while clusters of ions with up to 13 atoms have been measured for In and Pr LMISs.[59]

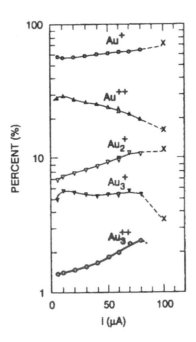

FIGURE 15.22

Distributions of Au^+, Au^{++}, Au_2^+, Au_3^+, and Au_3^{++} ions extracted from an Au LMIS as a function of ion beam intensity.[103]

15.4.5.2 Total Particle Emission

Heavier particles are usually observed, especially during high-current operation; these particles are usually described as droplets or macroparticles.[105–107] Several attempts have been made to quantify the mass loss during LMIS operation (see, e.g., References 105 to 109). Not all particles emitted during LMIS operation are ionized as evidenced by the data represented in Figure 15.23, which displays the loss of particles in atoms per second vs. ion beam intensity for Ga and In LMISs.[105,106,108,109] Droplet or macroparticle emission constitute a significant fraction of total mass loss, even in the low-ion-current mode of operation.

The highly divergent ion beams emitted from the conical emission region of the liquid metal contains neutral atoms, singly and multiply charged ions, neutral and charged clusters of atoms, and droplets. The relative proportions of these compo nents vary with the metal used and the operating conditions of the source and are not, in general, well known, even for the gallium source, which has been widely used. The beam current fluctuations are remarkably small, and can be attributed mostly to shot noise. Most of the ions have energies slightly less than that which corresponds to the applied accelerating voltage; the peak in the energy distribution is observed to be a few electron volts less than the level determined by the applied voltage. This difference is called the energy deficit. The spread in energy of the beam (measured at the half-maximum of the peak) is a few electron volts. Optical emission from a very small volume at the needle apex is seen during source operation, and the apex region is found to be heated.

15.4.5.3 Current Densities, Angular Distributions, and Chromatic Angular Intensity

Current densities. Perhaps the most important attribute of a LMIS is that it is essentially a "point" source capable of a high angular intensities. As discussed previously, the physical diameter of the ion emitting region at the apex of the liquid

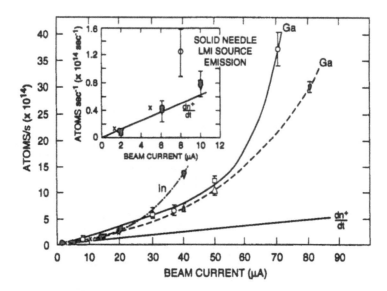

FIGURE 15.23

Total mass emission rate and singly charged, monomer ion emission rate vs. emission current (i) for Ga and In needle LMIS.[108] ○, Mair and von Engel;[105] □, Thompson no electron suppression;[106] △, Thompson with electron suppression;[106] •■▼, Papadopoulos;[108] ×, Wagner et al.[109]

is of the order of 1 to 10 nm but, unfortunately, its effective or apparent size is of the order of 50 nm because of perturbations caused by spherical and chromatic aberrations, space charge effects, and Coulomb interactions between individual charged particles in the generation region of the source. The current density $J(r)$ in an ultrabright beam transport system is usually aberration limited. As the beam-limiting aperture angle is increased, $J(r)$ is observed to have increasingly long tails, which are attributable to spherical and chromatic aberrations and/or Coulomb interactions that take place at the point of origin of the ion beam. The resolution of a focused ion beam system depends principally on the FWHM of the beam current distribution at the target. For ion beam deposition applications, the current in the tail of the distribution may constitute a significant part of the total ion beam intensity on target. While $J(r)$ may be of the order of 1 A/cm² at the peak of the distribution, it may rapidly decrease by a few orders of magnitude in the tail of the distribution. Because the area covered by the tail of the distribution is usually large with respect to the central peak area, a significant fraction of the total beam current is contained within the annular region surrounding the peak. Kubena and Ward have measured the current density profiles $J(r)$ of a Ga beam and found beam tails due to these effects extending to several times the FWHM.[110] Based on measurements of the angular intensity, a brightness of 10⁶ A/cm²·sr can be estimated, where brightness as defined by Equation 4. Brightness values of 3.3×10^6 A/cm²·sr have been by estimated from experimental resolution and current measurements of by Seliger et al.[111]

Angular distributions and angular intensities. In order to utilize the source for submicron imaging applications (e.g., microprobe surface analysis, ion implantation, microfabrication, and microetching), the angular divergence and, consequently, ion beam intensity must be severely limited. This is a consequence of the highly divergent characteristic of ion beams from this source type and the difficulty of accelerating and focusing beams without severely distorting their phase spaces through spherical aberrational effects (such effects increase as θ_0^3 where θ_0 is the half-angular divergence into the optical device; see Equation 7). The half-angular divergence of

LMIS beams range from a few degrees at low beam intensities to a few tens of degrees at higher intensities (see, e.g., References 65, 100, and 112 to 117).

Although the angular intensities are an order of magnitude greater than those that can be generated in a conventional GFIS, the brightness is usually less because of the larger virtual source size. The virtual source size δ of the LMIS is approximately two orders of magnitude greater than for the GFIS.[124] However, this is compensated for because the beam intensities from the LMIS are orders of magnitude higher than those for a GFIS. In practice, the beams from the LMIS are usually aperture limited for high-resolution applications. Angular distributions of ion beams have been measured for a number LMISs including those for Al,[112,113] Ga,[65,100,109,114-117] In,[118] Pr,[119] Au,[120,121] and Bi.[100] In general, the angular distribution for a given LMIS depends, among other factors, on beam intensity, the particular element in question, needle diameter, extraction electrode geometry, and impedance to flow along the needle surface. Although some variations occur in the shapes of measured angular distributions for an elements such as Ga, the distributions are of two general shapes: (1) a Gaussian profile with higher on axis angular intensity and (2) a hollow-beam profile that characteristically has higher angular intensities symmetrically displaced about the axis of symmetry beyond which the angular intensity rapidly decreases. The hollow-beam effect is characteristic of space charge-dominated ion beams; however, the causes for these differences in angular distributions have not been clearly delineated to date. Angular intensities typically vary between 10 to ~30 μA/sr for the Ga LMIS operated at 10 kV; these features are evident in the angular intensity distributions displayed in Figure 15.24 for a Ga^+ LMIS operated at several intensities.[117] Figure 15.25 displays angular intensity vs. half-angle for Ga and Bi sources at two different intensities.[100,122] Because space charge effects vary inversely with particle velocity, heavier particles exhibit larger angular spreads due to the greater radial forces within the ion beam at the same beam intensity and energy. However, as pointed out by Dixon and Von Engle, space charge effects cannot fully account for the experimentally observed increases in angular divergence with beam intensity for a particular species.[123] In a more comprehensive analysis, Kingham and Swanson were able to explain the increase on the basis of the changes in length of the field-induced protrusions that occur on the apex of the cone during ion beam generation and that effectively increase the angular divergence of the beam when the ion beam intensity is increased;[77] this effect, then, is an optical effect brought about by the dynamic changes in the emission geometry during extraction. The angular intensity is found to vary with the condition of the needle surface or impedance to liquid-metal flow along the surface of the needle; smooth needles exhibit greater angular intensities. The mass effect is illustrated by the comparisons made in Figure 15.26 of the theoretical results of Kingham and Swanson[77] with the experimental measurements of the angular intensity as made by Mair.[86] The physical size and taper of the needle can also affect the angular intensity as clearly demonstrated by Bell and Swanson.[124] Changes in the shapes of angular intensity vs. ion beam current are frequently, but not always, observed when differently tapered needles are used; this effect is illustrated in Figure 15.27, which displays the measurements of Torii and Yamada for an Al LMIS source operated with differently tapered needles.[125]

Angular distribution of multiply charged and cluster ion beams. The angular spreads for doubly charged and singly charged cluster ion beams are often observed to be sharper than the corresponding singly charged species as evidenced by the measurements of Papadopoulos for Ga[59] and Papadopoulos et al. for Au.[121]

Chromatic angular intensity. For ultrabright beam applications, it is useful to consider what is termed the chromatic angular intensity, which is defined by the following expression

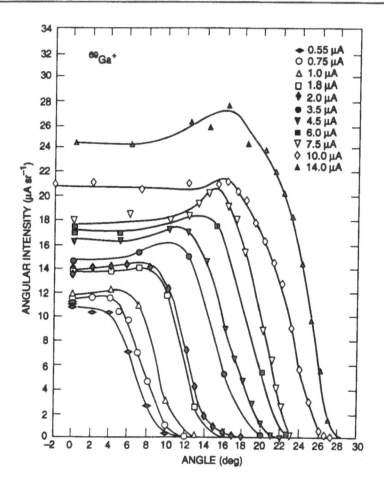

Angular intensity vs. beam half-angle for a Ga⁺ LMIS operated at several ion beam intensities.[117]

$$\left(dI/d\Omega\right)_c = dI/d\Omega(\Delta E)^2 \tag{41}$$

where $dI/d\Omega$ is the angular intensity and ΔE the energy spread of the ion beam within the aperture-limited beam on target. Several groups have provided chromatic angular intensity information for needle type Ga LMISs including those reported in References 126 to 128.

15.4.5.4 Energy Distributions

The quality of the beam that can be transported through a system is affected by a loss of brightness in the LMIS due to space charge effects at the source that result in radial broadening and to energy spreads within the ion beam. The enlarged apparent source size and energy distribution are caused primarily by space charge effects and spherical aberrations near the emitter.

Because of the small dimensions of the end radius of the emitter (~5 nm in diameter), very high source current densities of ~10^6 A/cm^2 at the source can be realized by extracting a few microamperes of current. Space charge effects are present in the LMIS that affect both the transport and focusing of LMIS beams generated in the source. The energy spread of the ion beam is an important parameter for microfocused ion beam applications in that it affects the achievable image size

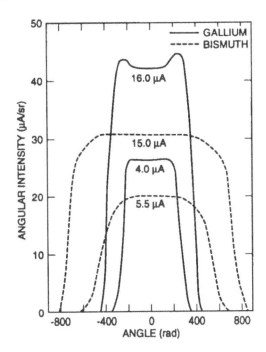

FIGURE 15.25
Angular intensity (di/dΩ) vs. half-angle θ_0 for various ion beam intensity levels (aperture settings: 108 and 146 μsr for Ga and Bi, respectively).[100,122]

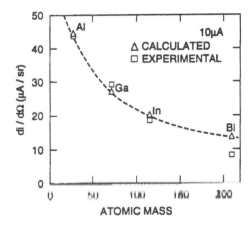

FIGURE 15.26
On-axis experimental angular intensity (di/Ω) vs. atomic mass compared with theory for a current of 10 μA.[86]

because of chromatic aberrations and dispersive effects if magnetic analysis is utilized. The energy distribution in the LMIS is much broader than for a GFIS, because the energy distribution depends on the current density (see, e.g., Reference 129) and mass of the particle.[54] For a LMIS, the energy spread at FWHM is typically >5 eV for ion beam intensities $1 \geq 1$ μA; the energy spread is observed to increase rapidly with total beam intensity.[60,121,129–141]

Dependence of energy spread on ion beam intensity and mass. Inherent in LMIS beams are space charge effects that cause increases in the angular and energy spreads; these, in turn, lead to increases of the phase space of the ion beam through spherical and

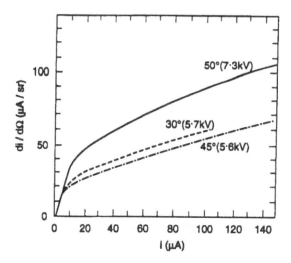

FIGURE 15.27
On-axis angular intensity ($di/d\Omega$) vs. source ion beam intensity for an Al LMIS for various needle taper angles. The operating voltage in each case is indicated.[125]

chromatic aberrations. Energy spread attributable to space charge influences near the region of ion emission may be caused by collective and individual particle Coulomb interactions (Boersch effect).[142] Theoretical treatment of the Coulomb-collision-dominated problem in a high-density, laminar-electron beam predicts an energy spread ΔE proportional to the square root of the ion current I or $\Delta E \propto I^{1/2}$.[143] For the case of a high-density, highly divergent beam dominated by Coulomb interactions, an energy spread $\Delta E \propto I^{2/3}$ is theoretically predicted.[144] Energy spreads ΔE in the LMIS are found to follow, approximately, the $I^{1/2}$ to $I^{2/3}$ relation[145] and have widths of several electron volts (see, e.g., References 60, 134, 145, 146). The magnitude of the energy spread resulting from pairwise collisions between particles has been calculated by Gesley and Swanson,[130] Gesley et al.,[147] and by Knauer.[144] The method used by Knauer predicts that ΔE increases according to $I^{2/3}M^{1/3}$,[144] while the calculations of Gesley and Swanson[130] and Gesley et al.[147] predict that ΔE increases according to $I^{1/2}M^{1/4}$. The energy spread ΔE associated with collective space charge effects increases linearly with ion current or $\Delta E \propto I$. The data displayed in Figure 15.28 indicates that the FWHM of the distribution increases linearly according to $\Delta E \propto IM^{1/2}$ for both singly and doubly charged species.[130] The resultant energy distribution associated with these effects is superposed on other distributions intrinsic to the source that are independent of ion current (e.g., energy spreads attributable to thermal and ion formation processes). All additively affect the emittance through chromatic aberrational and dispersive effects in the extraction postacceleration, lensing, and momentum analysis systems used in the experiments.

The shape of the energy distributions depend on particular species. The distributions for singly ionized atomic or monomer ion beams tend to be slightly asymmetric, Gaussian distributions regardless of the element in question. Many measurements have been reported in the literature including those for Au+,[121,148–150] Bi+,[151] Cs+,[152] Pr+,[119] and Sn+[153] from elemental feed material sources while energy distributions from alloy type sources include the measurements reported for Au+,[154–156] B+,[157] Be+,[154] Cu+,[158] Ga+,[156] In+,[158,159] Ni+,[160] P+,[158] Pb+,[161] Sb+,[161] and Si+.[54,154,156,160] In general, the asymmetric character is found to be independent of beam intensity and the FWHM of the distribution increases in width while the peak of the distribution shifts with energy. The FWHM widths of the energy distributions for Au+ taken from the

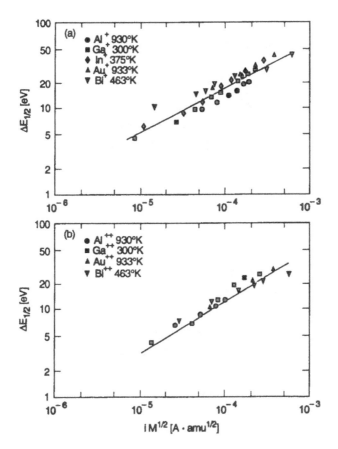

FIGURE 15.28
FWHM energy spread $\Delta E_{1/2}$ vs. $IM^{1/2}$ for (a) singly, and (b) doubly charged ions.[130]

literature and compiled by Marriott[117] are displayed in Figure 15.29. As noted, the FWHM tend to increase in slope within the ion beam intensity range between 2 and ~30 μA. Within this range of beam intensities, the FWHM increases with ion beam intensity approximately according to $\Delta E \cong i^{0.7}$; this has been widely interpreted as an indication of the validity of the presence of the multiparticle potential energy transfer process as proposed by Knauer[144] in order to explain the energy broadening observed in electron beams (Boersch effect).[142]

The energy distributions vary in shape and position with temperature as noted in Figure 15.30, which displays the results of measurements made by Swanson, et al.[100] As noted, the shape of the distribution becomes more distorted with increasing temperature. The shift in peak may be attributable to an increase in the population of excited state atoms brought about by an increase in the electron bombarding current during extraction, which would lower the energy required to ionize the atoms. If true, this deficit in energy between the low-temperature and high-temperature cases should be identically equal to the excitation energy for the first excited state of the atom in question.

In general, the energy spreads increase with increasing mass of the particle in question for singly charged monomer ions for both elemental and alloy type sources. This is again in agreement with energy broadening theories because particles with higher mass-to-charge ratios spend a longer time in the high particle density region during extraction where more influence can be imparted to the particles by Coulomb, Boersch, and space charge effects.

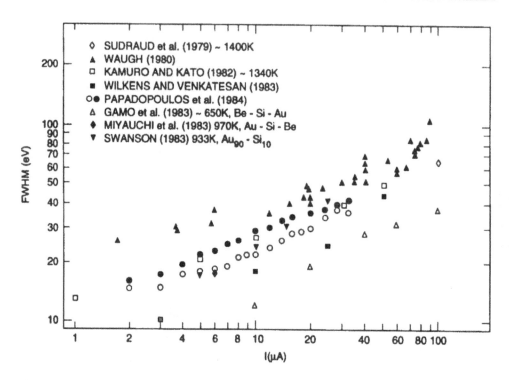

FIGURE 15.29
FWHM energy spread vs. ion beam intensity I for Au[+] ions.[117]

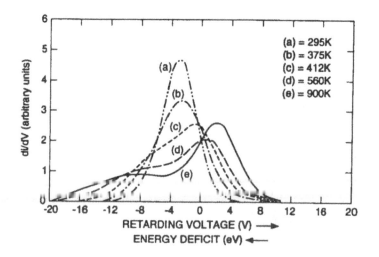

FIGURE 15.30
Retarding field analyzer energy distribution measurements for total ion beams of I = 2 μA extracted from a Ga LMIS at various operational temperatures.[100]

The energy spreads for multiply charged components are in general smaller than their singly charged counterparts, ostensibly due to the decrease in time spent in the high-density region during extraction. The shapes of the energy distributions for singly charged molecular ions vary from species to species and with emission current. The energy distributions for Ga^{2+} and Ga^{3+} as measured by Culbertson et al. were found to be single peaked with energy deficits of several tens of electron volts that rapidly increased with increasing ion beam intensity.[162,163]

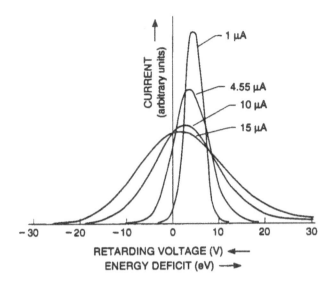

FIGURE 15.31
Retarding field analyzer energy distribution measurements for various intensity Ga+ ion beams.[134]

Energy deficit measurements. Several groups have measured energy deficits for Ga+, including Mair et al.,[134] Swanson et al.,[54] Sakurai et al.,[164] Culbertson et al.,[162] and Marriot.[117] Figure 15.31 displays retarding field analyzer energy distribution measurements made by Mair et al. for Ga+ ion beams at low ion beam intensities.[134] Comparisons between measured and calculated energy deficits are shown in Table 15.2 for several ion species;[54] reasonable agreement is typically found with the values calculated from Equation 34. Mair et al. obtained very close results between measured (~4.6 eV) and calculated values (~4.3 eV) for Ga+ ion beams.[134] The agreements found between measured and calculated energy deficits are often used as arguments in support of an evaporation mechanism, particularly at low ion beam intensities. These arguments have been further enforced by the energy deficit measurements made for several ion species, including Al,[54,64] Au+,[116] Bi+,[54,165] In+,[53,165] Pb+,[116] and Sn+.[153]

15.4.5.5 Modeling and Computational Simulation Studies

Several groups have presented models related to the ion formation mechanisms in attempts to explain energy and angular distribution characteristics of the LMIS. These include the studies reported in References 144 and 166 to 179. The effect of space charge and Coulomb interactions on the brightness of beams extracted from a LMIS have been studied by several groups, among which were the works of Ward and Seliger[175] and Ward et al.[176-178] In the initial study, Ward and Seliger used a sphere-on-orthogonal-cone model (SOC) without a protrusion to analyze beam spreading due to space charge effects.[176] In a later study, Ward et al. used Monte Carlo techniques to model an SOC LMIS emitter.[176] Increases in the emittances of the beam were predicted. Based on the results of these studies predictions were made of the dependence of beam size on target current density. In a separate study, Ward et al. computed the effects of thermal broadening on ion beams extracted from LMISs.[177] Monte Carlo techniques have been utilized by Ward to estimate virtual source sizes to be of the order of 50 and 100 nm.[179] The dynamics of the LMIS have been recently simulated by Whealton et al.[180] Figures 15.32 and 15.33 display ion trajectories and emittance, respectively, for a Ga ion beam of 50 μA obtained by solving Poisson-Vlasov equations[181] by use of the computer code described in Reference 182. The

TABLE 15.2

Comparison of Energy Deficits ΔE Calculated from
Equation 34 Assuming ΔH = 0 and Measured by
Extrapolating Experimental Results to i = 0

Ion	ΔE (calc.) (eV)	ΔE (exp.) (eV)
Al⁺	5.0	3.6±1
Ga⁺	4.9	3.4±1
Ga⁺⁺	21.5	19±1
In⁺	5.8	3.2±1
Bi⁺	6.0	4.7±1

From Reference 54.

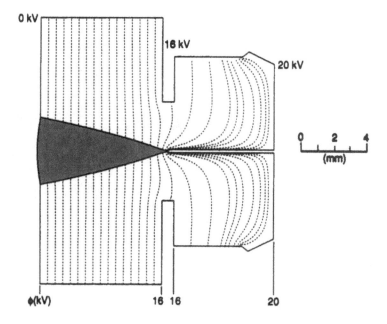

FIGURE 15.32
Simulation of extraction from a Ga LMIS.[182] Ion beam intensity: 50 μA; extraction voltage: 4 kV.

aberrational effects induced during extraction of the highly divergent beam are
evident in the emittance diagram.

15.4.5.6 Emittances and Brightnesses

Emittances of beams extracted from LMISs are of fundamental importance. The
emittances, defined by Equations 1 and 2, can be related to the brightness of the LMIS
through Equation 4. Figure 15.34 displays an emittance device designed specifically
for measuring the emittances of high-brightness, highly divergent ion beams such as
those extracted from a GFIS or an LMIS. Figures 15.35 and 15.36 show x and y
emittance contour data for a 26-keV Ga⁺ ion beam extracted from the LMIS source
displayed in Figure 15.15, as measured by Alton and Read.[184,185] As noted, the x-
direction emittance contours are larger due to dispersion effects attributable to energy spread within the ion beam and chromatic and second-order aberrations by the
magnet. The dispersive effects are not present in the y-direction emittance contours
and, therefore, typical y-direction emittances are taken as representative of the intrinsic

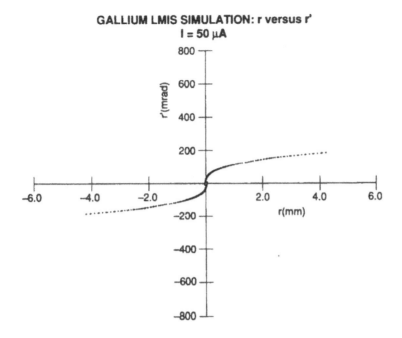

FIGURE 15.33
Computed r′ vs. r emittance diagram for Ga ion beam shown in Figure 15.32. The aberrational effects associated with space charge and spherical aberrations increased during extraction.[182]

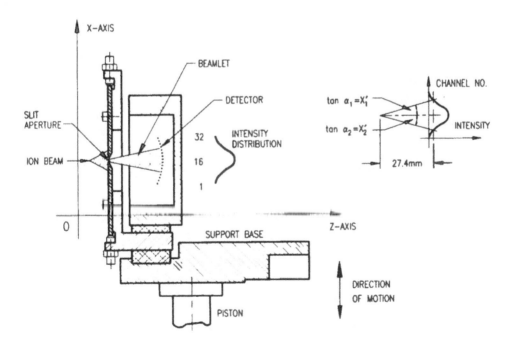

FIGURE 15.34
Stepping motor-controlled, high-resolution emittance measuring device for measuring the emittances of high-brightness, highly divergent ion beams. (From Reference 183.)

emittance of the source and averages made at a given percent of ion beam intensity for all data taken during the experiments. Fits to the average y-direction emittance data results in the emittance vs. percentage total ion beam relation shown in Figure 15.37.[186]

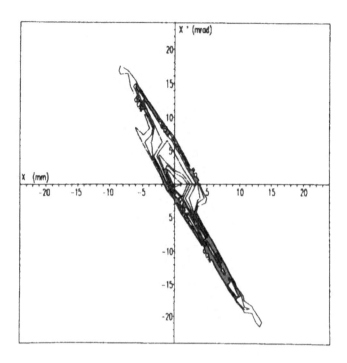

FIGURE 15.35

X-direction emittance contours for a 1-μA Ga⁺ ion beam extracted from the LMIS shown in Figure
15.15. Aperture angle: α = 7°; extraction voltage: 6 kV.

15.4.5.7   Factors that Affect the Emittances and Beam Sizes on Target

For high-resolution applications, the energy spread in the beam is of fundamental importance due to the intrinsically small size of the virtual source from which the
ions appear to emanate; for such applications, the achievable beam size on target is
limited by aberrations induced by lenses, steerers, magnets, and other beam transport
components (see, e.g., Equations 7 to 10).

The beam diameter d for Gaussian profile, low-intensity, low angular divergence
after transport through a system with magnification $\mathcal{M}$ can be approximated by
substituting ion beams at low angular divergences. By substituting

$$\theta_v = \frac{I}{\pi dI/d\Omega} \tag{42}$$

into Equation 10, the final on-target beam diameter d can be estimated from the
following expression[4]

$$d^2 = \frac{1}{4}\mathcal{M}^2 d_o^2 + \frac{1}{4}\mathcal{M}^2 \left(\frac{\Delta E}{E_1}\right)^2 \left(\frac{I}{\pi dI/d\Omega}\right) C_{oc}^2 + \frac{1}{4}\mathcal{M}^2 \left(\frac{I}{\pi dI/d\Omega}\right)^3 C_{os}^2 \tag{43}$$

From Equation 43, the effects of three distinct aperture settings on the beam size on
target can be delineated: (1) at large angular aperture settings, spherical aberration
dominates and $d \propto I^{3/2}$; (2) at intermediate aperture settings, chromatic aberration is
the dominant effect and $d \propto I$; (3) for the smallest aperture settings, the physical size
of the source dominates and d is independent of I.

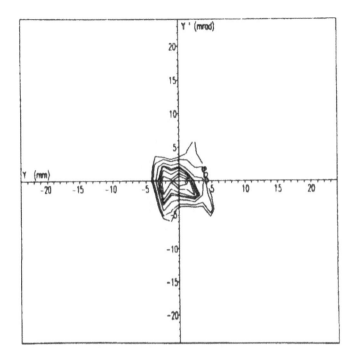

FIGURE 15.36
Y-direction emittance contours for a 1-μA Ga⁺ ion beam extracted from the LMIS shown in Figure
15.15. Aperture angle: $\alpha = 7°$; extraction voltage: 6 kV.

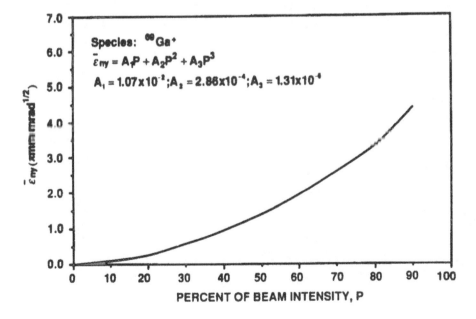

FIGURE 15.37
Average y-direction ε_{ny} emittance as a function of percentage P of a Ga⁺ ion beam contained within
a particular emittance contour. (Data derived from experiments described in References 62 and 63.)

ACKNOWLEDGMENT

The author is particularly grateful to Ms. Jeanette McBride for her enduring patience and skill in typing several iterations of the manuscript before arriving at the final version.

REFERENCES

1. R. L. Seliger, J. W. Ward, V. Wang, and R. L. Kubena, *Appl. Phys. Lett.*, 34, 310, 1979.
2. G. L. Allan, J. Zhu, and G. J. F. Legge, in *Proc. Fourth Australian Conf. Nuclear Techniques of Analysis*, R. J. McDonald, Ed., Lucas Heights, NSW, Australia, 1985, 49; G. L. Allan, G. J. F. Legge, and J. Zhu, *Nucl. Instrum. Methods*, B34, 122, 1988.
3. J. Orloff, *Rev. Sci. Instrum.*, 64, 1105, 1993.
4. P. D. Prewett and G. L. R. Mair, *Focused Ion Beams from Liquid Metal Ion Sources*, Wiley, Chichester, 1991.
5. M. K. Miller and A. R. McDonald, *Atom Probe Field-Ion Microscopy and Related Topics: A Bibliography 1978–87*, ORNL/TM-11157, Oak Ridge National Laboratory, Oak Ridge, TN; M. K. Miller and A. R. Hawkins, *Atom Probe Field-Ion Microscopy and Related Topics: A Bibliography 1988*, ORNL/TM-11370, Oak Ridge National Laboratory, Oak Ridge, TN; M. K. Miller, A. R. Hawkins, and K. F. Russell, *Atom Probe Field-Ion Microscopy and Related Topics: A Bibliography 1989*, ORNL/TM-11696, Oak Ridge National Laboratory, Oak Ridge, TN; K. F. Russell and M. K. Miller, *Atom Probe Field-Ion Microscopy and Related Topics: A Bibliography 1991*, ORNL/TM-12223, Oak Ridge National Laboratory, Oak Ridge, TN; *Atom Probe Field-Ion Microscopy and Related Topics: A Bibliography 1992*, ORNL/TM-12625, Oak Ridge National Laboratory, Oak Ridge, TN.
6. R. A. D. MacKenzie and G. D. W. Smith, *Nanotechnology*, 1, 163, 1990.
7. T. R. Walsh, *J. Nucl. Energy*, C4, 53, 1962; C5, 17, 1963.
8. A. B. El Kareh and J. C. J. El Kareh, *Electron Beams, Lenses, and Optics*, Vol. 2, Academic Press, New York, 1970.
9. P. W. Hawkes, *Quadrupoles in Electric Lens Design*, Suppl. 7, Academic Press, New York, 1970.
10. V. K. Zworykin, G. A. Morton, E. G. Ramberg, J. Hiller, and A. W. Vance, *Electron Optics and the Electron Microscope*.
11. E. W. Müller, *Phys. Rev.*, 102, 87, 1956; E. W. Mueller, *J. Appl. Phys.*, 27, 474, 1956.
12. R. Levi-Setti, *Scan. Electron Microsc.*, 125, 1974.
13. R. Levi-Setti, *Nucl. Instrum. Methods*, 168, 139, 1980.
14. J. Orloff and L. W. Swanson, *J. Vac. Sci. Technol.*, 12, 1209, 1975.
15. J. Orloff and L. W. Swanson, *Scan. Electron Microsc.*, 10, 57, 1977.
16. J. Orloff and L. W. Swanson, *J. Vac. Sci. Technol.*, 15, 845, 1978.
17. R. H. Fowler and L. W. Nordheim, *Proc. R. Soc. London Ser. A*, 119, 173, 1928.
18. L. W. Nordheim, *Proc. R. Soc. London Ser. A*, 121, 628, 1928.
19. R. H. Good and E. W. Müller, *Handbook der Physik*, Vol. 21, Springer, Berlin, 1956, 176.
20. R. A. Gomer, *Field Emission and Field Ionization*, Harvard University Press, Cambridge, MA, 1961.
21. E. W. Müller and T. T. Tsong, *Field Ion Microscopy Principles and Applications*, American Elsevier, New York, 1969.
22. W. B. Herrmannsfeldt, R. Becker, I. Brodie, A. Rosengreen, and C. A. Spindt, *Nucl. Instrum. Methods* A298, 39, 1990.
23. R. Haydock and D. Kingham, *Surf. Sci.*, 239, 1981.
24. R. Forbes, *J. Phys. D*, 18, 973, 1985.
25. M. J. Southon, Thesis, Cambridge University, 1963.
26. H. A. M. van Eekelen, *Surf. Sci.*, 21, 21, 1970.
27. M. J. Southon and D. G. Brandon, *Philos. Mag.*, 579, 1963.
28. K. Jousten, K. Böhringer, and S. Kalbitzer, *Appl. Phys.*, B46, 313, 1988.
29. K. Böhringer, K. Jousten, and S. Kalbitzer, *Nucl. Instrum. Methods*, B30, 289, 1988.
30. K. Jousten, K. Böhringer, R. Börrett, and K. Kalbitzer, *Ultramicroscopy*, 26, 301, 1988.
31. S. Orloff and L. W. Swanson, *J. Appl. Phys.*, 50, 2494, 1970.

32. J. Orloff and L. W. Swanson, *J. Appl. Phys.*, 50, 6026, 1979.
33. G. N. Lewis, H. Paik, J. Mioduszewski, and B. M. Siegel, *J. Vac. Sci. Technol.*, B4, 116, 1986.
34. R. J. Blackwell, J. A. Kubby, G. N. Lewis, and B. M. Siegel, *J. Vac. Sci. Technol.*, B3, 82, 1985.
35. S. Sato, T. Kato, and N. Igata, *Proc. 11th Symp. Ion Sources and Ion-Assisted Technol. (ISIAT)*, 1987, 91.
36. G. R. Hanson and B. J. Siegel, *J. Vac. Sci. Technol.*, 16, 1875, 1979.
37. A. E. Bell, K. Jousten, and L. W. Swanson, *Rev. Sci. Instrum.*, 61, 363, 1990.
38. K. Horiuchi, T. Itakura, and H. Ishikawa, *J. Vac. Sci. Technol.*, B6, 937, 1988.
39. M. Konishi, M. Takizawa, and T. Tsumori, *J. Vac. Sci. Technol.*, B6, 498, 1988.
40. M. Sato, *J. Phys.*, 48-C6, 183, 1987.
41. A. Wagner and T. M. Hall, *J. Vac. Sci. Technol.*, 16, 1871, 1979.
42. L. W. Swanson and L. C. Crouser, *J. Appl. Phys.*, 40, 4741, 1969.
43. P. R. Schwoebel and G. R. Hanson, *J. Vac. Sci. Technol.*, B3, 214, 1985.
44. A. V. Crewe, D. N. Eggenberger, J. Wall, and L. M. Welter, *Rev. Sci. Instrum.*, 39, 576, 1968.
45. P. C. Bettler and F. M. Charbonnier, *Phys. Rev.*, 119, 85, 1960.
46. Ch. Wilbartz, Th. Maisch, D. Hüttner, K. Böhringer, K. Jousten, and S. Kalbitzer, *Nucl. Instrum. Methods*, B63, 120, 1992.
47. Th. Maish, Ch. Wilbartz, Th. Miller, and S. Kalbitzer, *Nucl. Instrum. Methods*, B80/81, 1288, 1993.
48. T. C. Clements and E. W. Müller, *J. Chem. Phys.*, 37, 2684, 1962.
49. T. S. Tsong and E. W. Müller, *J. Chem. Phys.*, 3279, 1964.
50. G. R. Hanson and B. M. Seigel, *J. Vac. Sci. Technol.*, 19, 1176, 1981.
51. H. D. Beckey, *Principles of Field Ionization and Field Desorption Mass Spectrometry*, Pergamon Press, Oxford, 1977, 12.
52. G. I. Taylor, *Proc. R. Soc. A*, 280, 313, 1964.
53. P. D. Prewett, G. L. R. Mair, and S. P. Thompson, *J. Phys. D*, 15, 1339, 1982.
54. L. W. Swanson, *Nucl. Instrum. Methods*, 218, 347, 1983.
55. L. W. Swanson, G. A. Schwind, and A. E. Bell, *J. Appl. Phys.*, 51, 3453, 1980.
56. P. Marriot and J. C. Riviere, *Emission Mechanisms in the Liquid Metal Ion Source: A Review*, Rep. No. AERE-R 11294, Harwell Laboratory, Oxfordshire, 1984.
57. R. G. Forbes, *J. Phys. D*, 15, 1301, 1982.
58. W. Schottky, *Phys. Z.*, 15, 872, 1914.
59. S. Papadopoulos, Ph.D. thesis, Oxford University, 1986.
60. L. W. Swanson, G. A. Schwind, and A. E. Bell, *J. Appl. Phys.*, 51, 3453, 1980.
61. G. L. R. Mair, D. C. Grindod, M. S. Mousa, and R. V. Latham, *J. Phys. D*, 16, L209, 1983.
62. G. D. Alton and P. M. Read, *J. Appl. Phys.*, 66, 1018, 1989.
63. G. D. Alton and P. M. Read, *J. Phys. D*, 22, 1029, 1989.
64. A. E. Bell, G. A. Schwind, and L. W. Swanson, *J. Appl. Phys.*, 53, 4602, 1982.
65. P. D. Prewett and D. K. Jefferies, *J. Phys. D*, 13, 1747, 1980.
66. R. Clampitt, *Nucl. Instrum. Methods*, 189, 111, 1981.
67. R. Clampitt, K. L. Aitken, and D. K. Jefferies, *J. Vac. Sci. Technol.*, 12, 1208, 1975.
68. J. Ishikawa and T. Takagi, *J. Appl. Phys.*, 56, 3050, 1984.
69. J. Ishikawa and T. Takagi, *Vacuum*, 36, 825, 1986.
70. J. Ishikawa, H. Tsuji, and T. Takagi, *Rev. Sci. Instrum.*, 61, 502, 1990.
71. Y. Gotoh and J. Ishikawa, *Rev. Sci. Instrum.*, 63, 2438, 1992.
72. P. D. Prewett, D. K. Jefferies, and T. D. Cockhill, *Rev. Sci. Instrum.*, 52, 562, 1981.
73. R. Clampitt, *Nucl. Instrum. Methods*, 189, 111, 1981.
74. P. M. Read, J. T. Maskrey, and G. D. Alton, *Rev. Sci. Instrum.*, 61, 502, 1990.
75. P. D. Prewett and E. M. Kellog, *Nucl. Instrum. Methods*, B6, 135, 1985.
76. A. Wagner, *Appl. Phys. Lett.*, 40, 440, 1982.
77. D. R. Kingham and L. W. Swanson, *Appl. Phys.*, A34, 123, 1984.
78. H. Gaubi, P. Sudraud, N. Tence, and J. Van de Walle, *Proc. 29th Int. Field Emission Symp.*, H.-O. Andren and H. Norden, Eds., Almqvist and Wiksell, Stockholm, 1982, 357.
79. K. L. Aitken, D. K. Jefferies, and R. Clampitt, Culham Laboratory Report CLM/RR/E1/21, 1975.
80. K. L. Aitken, *Proc. Field Emission Day, ESA, Noordwijk (Holland)*, ESA, Noordwijk, 1976, 23.
81. G. Benassayag and P. Sudraud, *J. Phys. Colloq.*, C9, Suppl. to No. 12, 45, C9–223, 1984.
82. R. Clampitt, K. L. Aitken, and D. K. Jefferies, *J. Vac. Sci. Technol.*, 12, 1208, 1975.
83. A. Wagner and T. M. Hall, *J. Vac. Sci. Technol.*, 10, 1871, 1977.
84. G. L. R. Mair, *Nucl. Instrum. Methods*, 172, 567, 1980.
85. A. E. Bell and L. W. Swanson, *Appl. Phys.*, A41, 335, 1986.

86. G. L. R. Mair, *J. Phys. D*, 17, 2323, 1984.
87. G. I. Taylor, *Proc. R. Soc. A*, 313, 453, 1969.
88. R. Clampitt and D. K. Jefferies, *Nucl. Instrum. Methods*, 149, 739, 1978.
89. G. L. R. Mair, *Vacuum*, 36, 847, 1986.
90. T. Venkatesan, A. Wagner, and D. L. Barr, *J. Appl. Phys.*, 53, 787, 1982.
91. V. Wang, J. W. Ward, and R. L. Selinger, *J. Vac. Sci. Technol.*, 19, 1158, 1981.
92. R. L. Kubena, C. L. Anderson, R. L. Selinger, R. A. Jullens, and E. H. Stevens, *J. Vac. Sci. Technol.*, 19, 916, 1981.
93. E. Miyauchi, H. Hashimoto, and T. Utsumi, *J. Appl. Phys.*, 22, L225, 1983.
94. E. Miyauchi, H. Arimoto, H. Hashimoto, T. Furuya, and T. Utsumi, *Jpn. J. Appl. Phys.*, 22, L287, 1983.
95. K. Gamo, T. Ukegawa, Y. Inomoto, Y. Ochiai, and S. Namba, *J. Vac. Sci. Technol.*, 19, 1182, 1981.
96. T. Okutani, M. Fukuda, T. Noda, H. Tamura, H. Watanabe, and C. Sheperd, *17th Symp. Electron, Ion and Photon Beam Technology*, Los Angeles, CA, May 1983.
97. K. Gamo, T. Ukegawa, Y. Inomoto, K. K. Ka, and S. Namba, *Jpn. J. Appl. Phys.*, 19, L595, 1980.
98. K. Gamo, Y. Inomoto, Y. Ochiai, and S. Namba, *Proc. 10th Conf. Electron and Ion Beam Science and Technology*, 83–2, The Electrochemical Society, Pennington, NJ, 1982.
99. N. D. Bhaskar, C. M. Klimcak, and R. P. Fraukolz, *Rev. Sci. Instrum.*, 61, 366, 1990.
100. L. W. Swanson, G. A. Schwind, A. E. Bell, and J. E. Brady, *J. Vac. Sci. Technol.*, 16, 1864, 1979.
101. J. Van de Walle and P. Sudrand, *Proc. 29th Int. Field Emission Symp. (IFES)*, H.-O. Andren and H. Norden, Eds., Almqvist and Wiksell, Stockholm, 1982, 341.
102. R. Clampitt and D. K. Jefferies, *Inst. Phys. Conf. Series*, No. 38, Institute of Physics, Bristol, 1978, chap. 1, p. 12.
103. A. R. Waugh, *Proc. 28th Int. Field Emission Symp.*, L. Swanson and A. Bell, Eds., Oregon Graduate Center, Beaverton, OR, 1981, 80.
104. D. L. Barr, *J. Vac. Sci. Technol.*, B5, 184, 1987.
105. G. L. R. Mair and A. von Engel, *J. Phys. D*, 14, 1721, 1981.
106. S. P. Thompson, *Vacuum*, 34, 223, 1984.
107. S. P. Thompson and A. Von Engel, *J. Phys. D*, 15, 925, 1982.
108. S. Papadopoulos, *J. Phys. D*, 20, 530, 1987.
109. A. Wagner, T. Venkatesan, P. M. Petroff, and D. Barr, *J. Vac. Sci. Technol.*, 19, 1186, 1981.
110. R. L. Kubena and J. W. Ward, *Appl. Phys. Lett.*, 51, 1960, 1987.
111. R. L. Seliger, J. W. Ward V. Wang, and R. L. Kubena, *Appl. Phys. Lett.*, 34, 310, 1979.
112. Y. Torii and H. Yamada, *Jpn. J. Appl. Phys.*, 21(3), L132, 1982; Y. Torii and H. Yamada, *Jpn. J. Appl. Phys.*, 22(7), L444, 1983.
113. H. Yamada and Y. Torii, *Rev. Sci. Instrum.*, 57(7), 1282, 1986.
114. F. Zhou, M. Ma, W. Wang, and M. Hua, *Nucl. Inst. Mech. Phys. Res.*, B6, 143, 1985.
115. B. Wilkens and T. Vankatesan, *J. Vac. Sci. Technol.*, B1(4), 1132, 1983.
116. M. Komuro, *Thin Solid Films*, 92, 155, 1982.
117. P. Marriott, *Angular and Mass Resolved Energy Distribution Measurements with a Gallium Liquid-Metal Ion Source*, Rep. No. AERE-R-12674, Harwell Laboratory, Oxfordshire, 1987.
118. M. J. Higatsberger, P. Pollinger, H. Studnicka, and F. G. Rüdenauer, *Secondary Ion Mass Spectrometry SIMS III*, Springer Verlag, Berlin, 1982, 38.
119. S. Papadopoulos, *J. Phys. D*, 20, 1302, 1987; S. Papadopoulos, *J. Phys.*, 48-C6, 200, 1987.
120. S. Papadopoulos, D. L. Barr, W. L. Brown, and A. Wagner, *J. Phys.*, 45, C9, 217, 1984.
121. S. Papadopoulos, D. L. Barr, and W. L. Brown, *J. Phys.*, 47, C2, 101, 1986.
122. L. W. Swanson, G. A. Schwind, and A. E. Bell, *Scan. Electron Microsc.*, I, 45, 1979.
123. A. J. Dixon and A. von Engel, *Inst. Phys. Conf. Series*, No. 54, Institute of Physics, Bristol, 1980, chap. 7, p. 292.
124. A. E. Bell and L. W. Swanson, *Appl. Phys.*, A41, 335, 1986.
125. Y. Torii and H. Yamada, *Proc. Int. Ion Eng. Congr. ISIAT and IPAT ('83)*, T. Tagagi, Ed., Institute of Electrical Engineers of Japan, Tokyo, 1983, 363.
126. T. R. Fox, R. Levi-Setti, and K. Lam, *Proc. 28th Int. Field Emission Symp.*, Portland, OR, 1981.
127. G. L. R. Mair and T. Mulvey, *J. Microsc.*, 142, 191, 1986; G. L. R. Mair and T. Mulvey, *Scan. Electron Microsc.*, III, 959, 1985.
128. H. P. Mayer and K.-H. Gaulkler, *Proc. 33rd Int. Field Emission Symp.*, Berlin, 1980.
129. A. E. Bell, K. Rao, G. A. Schwind, and L. W. Swanson, *J. Vac. Sci. Technol.*, B6, 927, 1988.
130. M. A. Gesley and L. W. Swanson, *J. Phys.*, C9, 45, 167, 1984.
131. G. L. R. Mair, *J. Phys. D*, 15, 2523, 1982.
132. G. L. R. Mair, *J. Phys. D*, 20, 1657, 1987.

133. P. Marriott, *Appl. Phys.*, A44, 329, 1987.
134. G. L. R. Mair, D. C. Grindrod, M. S. Mousa, and R. V. Latham, *J. Phys. D*, 16, L209, 1938.
135. G. L. R. Mair, T. Mulvey, and R. G. Forbes, *J. Phys.*, C9, 45, 179, 1984.
136. M. Komuro, *Proc. 27th Int. Field Emission Symp.*, Tokyo, 1980.
137. T. Ishitani Y. Kawanami, T. Ohnishi, and K. Umemura, *Appl. Phys.*, A44, 233, 1987.
138. G. Benassayag and P. Sudraud, *J. Phys.*, C9, 45, 223, 1984.
139. A. J. Dixon, *J. Phys. D*, 12, L77, 1979.
140. G. L. R. Mair, T. Mulvey, and R. G. Forbes, *J. Phys.*, C9, 45, 179, 1984.
141. G. L. R. Mair, R. G. Forbes, R. V. Latham, and T. Mulvey, *Microcircuit Engineering 83*, H. Ahmed, J. R. A. Cleaver, and G. A. C. Jones, Eds., Academic, 1983, 171.
142. H. Boersch, *Z. Phys.*, 139, 115, 1954.
143. K. H. Loeffler, *Z. Agnew Phys.*, 27, 145, 1969.
144. W. Knauer, *Optik*, 59, 335, 1981.
145. J. W. Ward and R. L. Seliger, *J. Vac. Sci. Technol.*, 19, 1082, 1981.
146. P. Marriott, *J. Phys.*, 48-C6, 189, 1987.
147. M. A. Gesley, D. L. Larson, L. W. Swanson, and C. H. Hinrichs, *Proc. SPIE*, 471, 66, 1984.
148. P. Sudraud, C. Colliex, and J. Van de Walle, *J. Phys. Lett.*, 40, L207, 1979.
149. A. R. Waugh, *J. Phys. D*, 13, L203, 1980.
150. M. Komuro and T. Kato, *Proc. 6th Symp. ISIAT '82*, T. Takagi, Ed., 1982, 63.
151. M. Komuro, *Proc. Int. Ion Eng. Conf.*, T. Takagi, Ed., IEE of Japan, Tokyo, 1983, 337.
152. H. Helm and R. Möller, *Rev. Sci. Instrum.*, 54(7), 837, 1983.
153. A. Dixon, C. Colliex, R. Ohana, P. Sudraud, and J. Van de Walle, *Phys. Rev. Lett.*, 46, 865, 1981.
154. K. Gamo, T. Matsui, and S. Namba, *Jpn. J. Appl. Phys.*, 22(11), L692, 1983.
155. E. Miyauchi, H. Hashimoto, and T. Utsumi, *Jpn. J. Appl. Phys.*, 22(4), L225, 1983.
156. M. Okunuki, R. Aihara, and N. Anazawa, *Proc. 6th Symp. ISIAT '82*, T. Takagi, Ed., 1982, 71.
157. T. Ishitani, K. Umemura, S. Hosoki, S. Takayama, and H. Tamura, *J. Vac. Sci. Technol.*, A2, 1365, 1984.
158. T. Ishitani, K. Umemura, and H. Tamura, *Jpn. J. Appl. Phys.*, 23(5), L330, 1984.
159. T. Ishitani, K. Umemura, and H. Tamura, *Jpn. J. Appl. Phys.*, 24(6), L451, 1985.
160. T. Ishitani, K. Umemura, S. Hosoki, S. Takayama, and H. Tamura, *J. Vac. Sci. Technol.*, A2(3), 1365, 1984.
161. K. Gamo, Y. Ochiai, T. Matsui, and S. Namba, *Proc. 29th Int. Field Emission Symp.*, H.-O. Andren and H. Norden, Eds., 1982, 330.
162. R. J. Culbertson, G. H. Robertson, and T. Sakurai, *J. Vac. Sci. Technol.*, 16, 1868, 1979.
163. R. J. Culbertson, G. H. Robertson, and Y. Kuk, *J. Vac. Sci. Technol.*, 17(1), 203, 1980.
164. T. Sakurai, R. J. Culbertson, and G. H. Robertson, *Appl. Phys. Lett.*, 34(1), 11, 1979.
165. L. W. Swanson, A. E. Bell, G. A. Schwind, and D. Larson, *Proc. 28th Int. Field Emission Symp.*, Portland, Oregon, 1981.
166. J. W. Ward, *J. Vac. Sci. Technol.*, B3, 207, 1985.
167. G. A. Evans, M. D. MacGregor, and R. Smith, *Vacuum*, 34, 47, 1984.
168. R. G. Forbes, *J. Phys.*, 45-C9, 161, 1984.
169. R. G. Forbes and N. N. Ljepojevic, *J. Phys.*, 50-C8, 3, 1989.
170. R. G. Forbes and N. N. Ljepojevic, *Vacuum*, 39, 1153, 1989.
171. N. K. Kang and L. W. Swanson, *Appl. Phys.*, A30, 95, 1983.
172. D. R. Kingham and L. W. Swanson, *J. Phys.*, 45-C9, 133, 1984.
173. D. R. Kingham and L. W. Swanson, *Vacuum*, 34, 941, 1984.
174. D. R. Kingham and L. W. Swanson, *Appl. Phys.*, A41, 157, 1986.
175. J. W. Ward and R. L. Seliger, *J. Vac. Sci. Technol.*, 19, 1082, 1981.
176. J. W. Ward, M. W. Utlaut, and R. L. Kubena, *J. Vac. Sci. Technol.*, B5, 169, 1987.
177. J. W. Ward, R. L. Kubena, and M. W. Utlaut, *J. Vac. Sci. Technol.*, B6, 2090, 1988.
178. J. W. Ward, R. L. Kubena, and M. W. Utlaut, *J. Vac. Sci. Technol.*, B6, 2090, 1988.
179. J. W. Ward, *J. Vac. Sci. Technol.*, B3, 207, 1985.

CHARACTERIZATION OF ION SOURCES

J. Ishikawa

16 NEGATIVE-ION SOURCES

16.1 Brief History of Negative-Ion Source Development

Negative-ion sources have been developed according to requirements from their application fields such as accelerators, fusion research, and material science. Light negative ions, especially hydrogen negative ions, were required for high-energy accelerators and fusion research. They stimulated the progress of surface effect type hydrogen negative-ion sources. During the development of hydrogen negative-ion sources, a new mechanism in hydrogen negative-ion production, i.e., volume production, was found, and then the development of hydrogen negative-ion sources based on the volume production started. On the other hand, heavy negative ions were required for tandem accelerators. Sputter type heavy negative-ion sources were successfully developed and have recently been able to deliver sufficient negative-ion currents for material science applications. Negative ions could also be produced by means of charge transfer from a positive-ion beam.

16.1.1 Surface Effect Type Hydrogen Negative-Ion Sources

The original negative-ion source was the off-axis duoplasmatron in which the center axis of the anode and ion extraction electrodes was shifted by about 1 mm from the center axis of the intermediate electrode.[1] Several ten microamperes of hydrogen negative-ion current could be extracted. From this result it was recognized that denser negative ions were produced just near an electrode surface, i.e., at the edge of the ion source plasma. The hollow-discharge duoplasmatron (HDD), in which an isolated metal pole was put into the generated plasma on the center axis, could deliver a pulsed hydrogen negative-ion beam current of 9 mA (0.25 A/cm^2: no cesium) or 18 mA (0.57 A/cm^2: cesium supply).[2,3] Thus, the fact that on a cesiated electrode surface negative ions were very effectively produced was clarified. Thereafter, several kinds of high-current hydrogen negative-ion sources, such as the magnetron,[3-8] PIG,[9-11] multicusp,[12-14] and duopigatron[15] types, were developed. In the course of the development of these ion sources some detailed investigations for maintaining the optimal condition of a cesiated surface for negative-ion production were vigorously performed. In this way the mechanism of negative-ion production was considered as a result of the reflection of positive hydrogen ions on the cesiated surface. These ion sources could deliver amperes of hydrogen ion beams. Some typical ion sources are described in Section 16.3.

16.1.2 Volume Production Type Hydrogen Negative-Ion Sources

Bacal et al. found that the dissociative electron attachment cross section of highly excited molecular hydrogen was extremely large and hydrogen negative ions could be effectively produced in the plasma (volume production).[16–18] In order to realize the optimum condition for this negative-ion production, an ion source plasma was separated into two regions with different electron temperatures by a magnetic filter: one region for the generation of highly excited molecular hydrogen with high-temperature electrons and the other region for the production of negative hydrogen ions by dissociative electron attachment with low-temperature electrons.[19,20] Several ten milliamperes per centimeter squared of negative ions could be extracted. It was recently found that the current density was increased by an order of magnitude by supplying cesium vapor to the plasma chamber.[21,22] Some typical ion sources are described in Section 16.3.

16.1.3 Sputter Type Heavy Negative-Ion Sources

Various kinds of heavy negative ions were strongly required for tandem accelerators. For this purpose, Mueller and Hortig originally developed a sputter type heavy negative-ion source as shown in Figure 16.1.[23] They used a rotated ring target (180 to 400 turns per minute): on one side neutral cesium particles were sprayed to the target surface and on the other, Kr^+ ions with an energy of 10 to 30 keV bombarded the target for sputtering. Several microamperes of C^-, Cr^-, Cu^-, Ag^-, Au^-, etc. were obtained as described in Table 16.1. Mueller and Hortig were the first to show that heavy negative ions were effectively produced on a cesiated metal surface by sputtering.

A few years later two potential sputter type negative-ion sources were developed: one was the universal negative-ion source (UNIS) by Middleton[24–26] as shown in Figure 16.2 and the other was the Aarhus negative-ion source (ANIS) by Andersen[27,28] as shown in Figure 16.3. In the UNIS, a high-energy cesium ion beam with an energy of 20 to 30 keV that was extracted from a surface ionization source was used for both sputtering a cone-shaped sputter target and supplying cesium particles to it. Since the sputtering yield by the high-energy ion beam is quite high, this source was especially suited for negative-ion production of low sputtering yield materials. Cesium particle density, however, on the target was usually insufficient in order to minimize the work function of the sputter target surface. Forty six kinds of heavy negative ions with currents ranging from submicroamperes to tens of microamperes were produced as described in Table 16.1. Just after the UNIS, several types of modified UNIS sources were developed such as an inverted type by Chapman[29] and a reflected-beam mode type by Brand.[30]

On the other hand, the ANIS was a compact source in which a sputter target was placed in a PIG-discharged Xe plasma and was biased by about –1 kV. Xe^+ ions uniformly bombarded the sputter target and at the same time neutral cesium particles were sprayed onto the target surface. Thus, this source is considered as the original plasma-sputter type heavy negative-ion source. Negative ions produced on the sputter target surface were focused to a small extraction aperture because of a spherical target surface. This source was suited for negative-ion production of high sputter yield materials because sufficient cesium particles were present on the sputter target surface, and could deliver various elements of negative-ion currents with

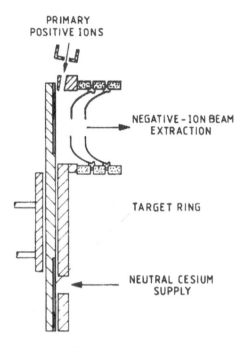

PRIMARY
POSITIVE IONS

NEGATIVE-ION BEAM
EXTRACTION

TARGET RING

NEUTRAL CESIUM
SUPPLY

FIGURE 16.1
Sputter type heavy negative-ion source developed by Mueller and Hortig.[23]

several to several ten microamperes as described in Table 16.1. The gas pressure of
the PIG-discharged plasma region, however, was considered relatively high, and,
therefore, most of the negative ions produced on the sputter target surface might
suffer an electron detachment collision with gas particles during passage through the
plasma region. After the ANIS, some modified ANIS sources were developed such
as that by Alton[31] and a simple negative-ion sputter source (SNICS) by Smith, Jr.[32,33]

TABLE 16.1

Negative-Ion Currents Obtained by Historically Representative
Heavy Negative-Ion Sources

Negative-ion source	Mueller and Hortig	UNIS	ANIS
	C^- 2.2	B^- 2.3	Li^- 1.0
	C_2^- 3.4	C^- 50	B^- 0.04
	Cr^- 0.1	Al^- 0.2	BO^- 2.5
	Cu^- 6.0	Si^- 27	C^- 20
	Ag^- 14	P^- 0.6	C_2^- 15
Negative-ion	Au^- 12	Ge^- 1.1	Al^- 2.3
current		As^- 1.0	Ti^- 0.9
(μA)		In^- 0.03	Ni^- 55
		Sb^- 0.32	Cu^- 51
		Te^- 1.8	Nb^- 1.0
		Ni^- 6.6	Ag^- 36
		Cu^- 7	Ta^- 3
		Nb^- 0.027	Au^- 80
Positive-ion energy	Kr^+ or Ar^+	Cs^+	Cs^+ + (Xe^+ or Ar^+)
and current for	10–30 keV	20–30 keV	1–2 keV
bombardment		1–1.5 mA	~10 mA

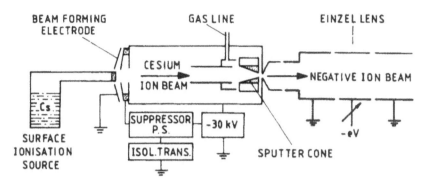

FIGURE 16.2
Universal negative-ion source (UNIS) developed by Middleton et al.[24]

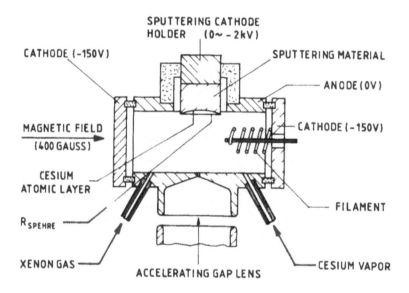

FIGURE 16.3
Aarhus negative-ion source (ANIS) developed by Andersen et al.[27]

In order to overcome some disadvantages of these sources, i.e., insufficient cesium particle supply to the sputter target surface in the UNIS and high gas pressure of the negative-ion transit region in the ANIS, two different types of negative-ion sources were developed: a versatile negative-ion source (VNIS) by Middleton et al.[34,35] and a NIABNIS by Ishikawa et al.[36,37] These ion sources are described in Section 16.3.

The power needed to generate one sputtered atom becomes lower as the ion kinetic energy is lowered. A quite high target current, however, is then needed to produce a high flux of sputtered atoms. A large sputter target placed in a dense plasma could collect a high current of positive ions with low kinetic energy. If a cesiated target surface is optimized for negative-ion production and a dense plasma is generated at a quite low gas pressure, a high current of negative ions could be extracted. This kind of negative-ion sources is called plasma-sputter type heavy negative-ion source, and can deliver milliamperes of heavy negative-ion current. Recently several kinds of plasma-sputter type sources have been developed, such as a multicusp type,[38] a microwave discharge type,[39] and an rf discharge type.[40,41] These ion sources are also described in Section 16.3.

16.1.4 Negative-Ion Production by Charge Transfer

Conventionally, a small amount of negative ions are produced by charge transfer when a positive-ion beam with an energy of 20 to 50 keV passes through a donor gas cell. If the donor gas is the same element as the positive ions, the charge transfer efficiency is as low as about 1%. If alkaline metal or alkali earth metal is used as a charge transfer cell gas, the charge transfer efficiency is increased to 10% or more. Because lithium is less active and lightest in mass among the alkaline metals, it is frequently used as a donor. Freeman et al. obtained 200 microamperes of Te⁻ current using a lithium donor.[42]

16.2 Mechanism of Negative-Ion Production and Fundamental Properties of Negative Ions

In an ordinary ion source plasma, an ionization cross section for negative ions by particle collision is several orders of magnitude lower than that for positive ions. Moreover, an electron detachment of negative ions due to a collision with plasma electrons frequently takes place because the absolute value of electron affinity is as low as about 1 eV. Therefore, the percentage of the negative ions is quite low, and high-current negative-ion extraction from the plasma is considered extremely difficult.

In order to increase the negative-ion yield, the following methods are utilized: volume production, charge transfer, and secondary negative-ion emission.

16.2.1 Volume Production

A hydrogen negative ion is produced through collisions of a hydrogen atom, a molecule, a molecule ion, or an excited molecule with an electron. In these collisions, the highly vibrationally excited molecule (excitation levels from 5 to 10) has a large cross section (about 10^{-15} cm²) for negative-ion production through a dissociative attachment reaction at an electron energy of around 1 eV.[43,44] Its value is by several orders of magnitude larger than other cross sections. If the highly vibrationally excited H_2 molecule is numerously generated in a hydrogen discharge plasma, hydrogen negative ions can be effectively produced. A method to produce negative ions through this process is called volume production. In this process, in order to effectively generate highly vibrationally excited H_2 molecules, high-energy electrons are needed. Therefore, the plasma region of an ion source should be divided into two regions by a magnetic filter: a high electron temperature region (or electron bombardment region) for generation of high vibrationally excited states of H_2 molecules and a low electron temperature region for the production of negative ions through dissociative attachment.

16.2.2 Charge Transfer

A negative ion is produced through a collision between a positive ion and a neutral particle where two electrons are shifted from the neutral particle to the positive ion, i.e., charge transfer. Alkaline metals or alkaline earths metals are selected as the neutral particles because they easily release their electrons. In this negative-ion production process, a positive-ion beam produced in advance is passed through a charge-transfer cell where alkaline metal or alkaline earths metal vapor is filled in. The efficiency of the charge transfer depends on the parameters of length of the cell and ion energy. The maximum negative-ion production efficiency by charge transfer can be obtained at a certain ion energy when the length of the charge-transfer

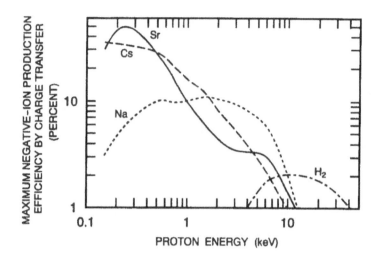

FIGURE 16.4

Maximum hydrogen negative-ion production efficiency by charge transfer as a function of hydrogen positive-ion energy.[45]

cell is tuned to an optimum condition. For example, the maximum negative-ion production efficiency by charge transfer for protons is shown in Figure 16.4.[45] The efficiency has a maximum value of 10 to 50% at a relatively low ion energy of a few hundred electron volts. On the other hand, the efficiencies for heavy positive ions have maximum values of the negative-ion production by charge transfer at a relatively high ion energy of a few ten kiloelectron volts as shown in Figure 16.5.[46] In elements having a high electron affinity, such as Cl, I, Te, and Au, the maximum value of the efficiency shows a quite high value of over 40%.

16.2.3 Secondary Negative-Ion Emission

16.2.3.1 Negative-Ion Production Probability

The negative-ion production efficiency is considerably high when an atom leaves a low-work-function metal surface at a velocity quite a bit higher than the thermal velocity. This ion production mechanism is called secondary negative-ion emission. In this mechanism, a negative ion is produced by a shift of an electron between the Fermi level of the metal surface and the electron affinity level of an atom to be negatively ionized during atom ejection processes such as sputtering or reflection.

Consider, for example, an extractable ion current produced through secondary negative-ion emission by sputtering or reflection as illustrated in Figure 16.6.[47] The negative-ion current extracted after passing a low gas pressure region is given by

$$I^- = I^+ A \, \eta^- \exp(-n_o L \sigma_d) \tag{1}$$

where I^+ is the positive-ion current for sputtering or reflection, A is the sputtering yield or the reflection yield, η^- is the negative-ion production efficiency on a metal target surface, σ_d is the electron detachment cross section for the negative ion, and n_0 is the gas density in the path of the negative-ion beam of length L. Since the sputtered or reflected particle has a velocity distribution with a velocity distribution function $f(v)$, the relation between the negative-ion production probability $P^-(v)$ and the negative-ion production efficiency η^- is given by

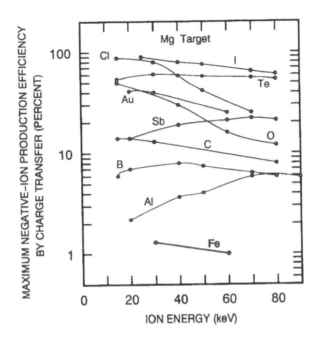

FIGURE 16.5

Maximum negative-ion production efficiency by charge transfer in a Mg target for various heavy elements as a function of positive-ion energy.[46]

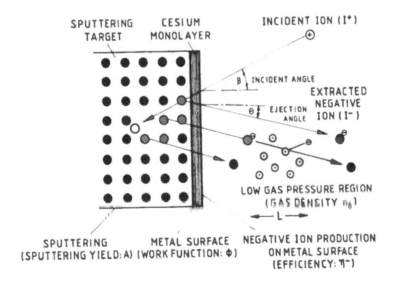

FIGURE 16.6

Illustration of the secondary negative-ion emission process by reflection or sputtering.

$$\eta^- = \int P^-(v)\, f(v)\, dv \qquad (2)$$

where v is the emitting particle velocity.

Rasser et al. have introduced a fairly simple equation giving the negative-ion production probability when a particle is emitted from a metal surface with a normal velocity of more than 10^4 m/s.[48] The equation is expressed as follows:

$$P^- = (2/\pi) \exp[-\pi(\phi-E_a)/2av\cos\theta] \qquad (3)$$

where ϕ is the work function of the metal surface, E_a is the electron affinity, θ is the ejection angle of an emitting particle ($\cos\theta$ means the normal velocity), and a is the decay factor for hydrogen, i.e., 2×10^{-5} eV/m from the experimental datum of van Wunnik et al.[49] In the case of heavy-particle sputtering, however, it often happens that the normal velocity of the emitting particle is much lower than 10^4 m/s, and then Equation 3 is not valid any more. Therefore, Ishikawa et al. proposed a modified semiempirical equation in which an effective local temperature T_{eff} proportional to the mass number of the emitting particle was introduced.[50]

$$P^- = (2/\pi) \exp[-(\phi-E_a)/(2av\cos\theta/\pi + kT_{eff})] \qquad (4)$$

where $kT_{eff} = 0.073 + 2.0 \times 10^{-3}M$.

The negative-ion production probability and efficiency are strongly dependent on the work function of the metal surface, and in order to obtain a maximum yield of negative-ion production by secondary negative-ion emission the minimum work function should be maintained on the metal surface.

The electron detachment cross section σ_d means the sum of the single- and double-electron detachment cross sections. The single-electron detachment cross section weakly depends on the following physical parameters of a negative ion and a target gas: the orbital radius of the electrons in the outermost shell of the negative ion, the number of these electrons, the electron affinity of the negative ion, negative-ion velocity, and the atomic radius of the target gas particle. Its value is between 0.6 $\times 10^{-15}$ and 2.5×10^{-15} cm^2 in an ion energy range of 5 to 50 keV in case of B, C, O, and Si negative ions on He, Xe, H$_2$, and N$_2$ gas targets.[51,52] The value of the double-electron detachment cross section is smaller by more than one order of magnitude than that of the single-electron detachment cross section.

16.2.3.2   Work Function of Cesium-Covered Metal Surface

Although cesium has the lowest work function (1.81 eV) of all elements, the work function of the cesium-covered surface is generally lower than that of bulk cesium. The work function of the cesium-covered surface for various metals was semiempirically calculated by Alton[53] (the minimum work function $\phi_{min} = 0.62(V_i + E_a) - 0.24\phi_0$ eV, where ϕ_0 is the inherent work function of the metal surface, V_i and E_a are the first ionization potential and electron affinity of an atom, respectively; in case of cesium, the sum of V_i and E_a is 4.35 eV, which is the smallest of all elements) and was measured by Graham.[54] The general behavior of the surface work function as a function of cesium coverage is illustrated in Figure 16.7. The inherent work function of usual metal surfaces ϕ_0 ranges from 4.5 to 6 eV. The work function of cesium-covered metal surfaces decreases with an increase in the cesium coverage, reaching a minimum value ϕ_{min} (1.37 to 1.76 eV) at a coverage of θ_{min} (about 0.6 atomic layers). It then increases with increasing cesium coverage up to a value of ϕ ($\theta = 1$) of 1.56 to 2.15 eV. Because no further cesium atoms adhere to the surface above room temperature due to weak Cs–Cs bonding, no further change in the surface work function is usually observed. When the metal surface is cooled too much, a cesium multilayer will be formed. In this case the work function changes to the bulk cesium work function ϕ_{Cs} (1.81 eV). In general, a margin of metal surface temperature where a cesium monolayer is formed is quite wide and ranges from room temperature to several hundred degrees Celsius, at which the bonding between metal atom and cesium is broken.

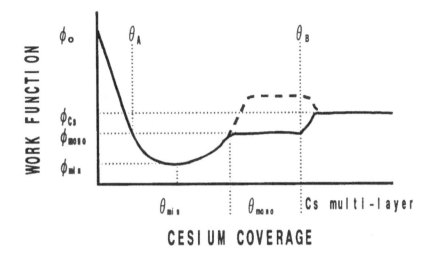

FIGURE 16.7
General behavior of the surface work function as a function of cesium coverage.

In the case of reflection of light ions on the cesiated surface, the cesium atom on the surface does not suffer sputtering due to the light ions. Then, once the minimum work function is realized on the cesiated surface, the optimal surface condition is maintained during operation of negative-ion production. On the other hand, in the case of the sputtering method, the cesium atom is sputtered together with metal atoms. Then, neutral cesium atoms must be continuously supplied to keep the optimal surface condition.

16.2.3.3 Experiments of Negative-Ion Production Efficiency

(a) Reflection Type
Granneman et al. measured the lithium negative-ion production efficiency as a function of the ion velocity normal to the cesiated W(110) surface with a minimum work function, as shown in Figure 16.8.[55] The efficiency clearly depends on the ion velocity normal to the surface, and it increases with it as indicated in Equation 3.

In reflection type hydrogen ion sources a cesiated metal surface (convertor) is placed in the generated plasma. Then, H^+, H_2^+, and H_3^+ ions bombard the surface, and at the same time, neutral hydrogen particles H^0 are absorbed on it. When these hydrogen particles are changed to negative ions on the surface, they have different negative-ion energies of 2, 1.5, 1.33, and 1 eV_{pc}, respectively, where V_{pc} is the potential difference between the plasma and the metal surface. Figure 16.9 shows the hydrogen negative-ion energy distribution as a parameter of the backscattered polar angle where an angle normal to the surface is regarded as $0°$.[56] In the direction normal to the surface, the H^0 particles mainly contribute to negative-ion formation since the main peak of the distribution exists around 1 eV_{pc}. In the high-polar-angle direction, however, the components of negative ions from H_3^+ and H_2^+ become main because the main peaks of these distributions shift to 1.3 to 1.5 eV_{pc}.

(b) Sputter Type
Ishikawa et al. measured the maximum negative-ion production efficiency in secondary negative-ion emission by sputtering.[50] The particles sputtered by 10- to 15-keV xenon ion beam bombardment on a cesiated surface were analyzed by use of a E × B mass-separator (for counting negative ions) and RBS measurement (for counting

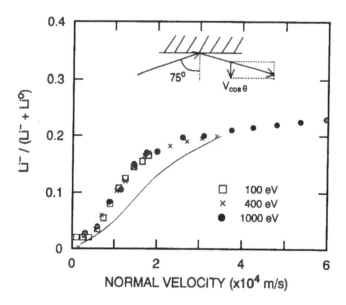

FIGURE 16.8
Lithium negative-ion production efficiency by secondary negative-ion emission by reflection as a function of normal velocity.[55]

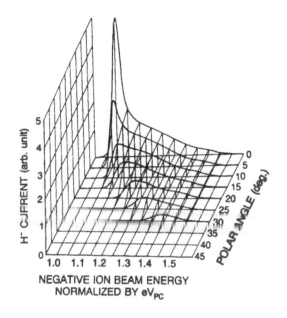

FIGURE 16.9
Hydrogen negative-ion energy distribution as a parameter of the backscattered polar angle.[56]

deposited sputtered particles) to obtain the ratio of negative ions to total sputtered particles. Table 16.2 shows an example of the measurement of maximum negative-ion production efficiencies for various heavy elements. For most of the metallic elements, the efficiencies are 10% or more.

TABLE 16.2

Maximum Negative-Ion Production Efficiencies

Element	C	Si	Cu	Ge	W
Efficiency(%)	18.3	15.6	12.1	13.6	9.4

16.3 Recent Negative-Ion Sources

16.3.1 Reflection Type Hydrogen Negative-Ion Sources

In the reflection type negative-ion source, positive ions bombard an electrode surface with a low work function. In most cases a convertor electrode is placed in a generated plasma and is negatively biased. The bombarding ion energy can be controlled by this bias voltage. The reflection type negative-ion sources can be classified according to whether the convertor electrode is one of the electrodes for the plasma generation (magnetron and PIG without a convertor) or is separated from it (PIG with a convertor, multicusp, and duopigatron types). In the latter sources, the important parameters for negative-ion production, such as the convertor voltage and current, can be independently controlled.

16.3.1.1 Magnetron Type

The magnetron type negative-ion source was historically the first ion source in which the negative-ion current was increased by a large margin in a cesiated mode of operation.[3-8] This type of ion source was also called surface plasma negative-ion source by Bel'chenko et al.[4-6] who first developed it. The fundamental structure of the ion source is illustrated in Figure 16.10.[3] In this source, a spool-shaped cathode is surrounded by an anode together with a perpendicular magnetic field of 1 to 2 kG and a cold electrode discharge is maintained between these two electrodes. Negative ions produced on the cathode surface are accelerated by the potential difference of the cathode fall and suffer resonant charge exchange collisions with hydrogen gas (about 1 torr) during their passage through the groove region. The slow negative ions that are produced in the groove region are extracted from a slit aperture together with accelerated negative ions. An H⁻ current of 880 mA (1 ms, 1 Hz) was obtained when the discharge current was 450 A in a cesiated mode of operation. Many versions of magnetron type negative-ion source were developed such as a multigroove type shown in Figure 16.11.[57]

16.3.1.2 PIG Type

(i) PIG Source Without a Convertor

In this source the negative-ion production mechanism is very similar to the magnetron type except its discharge mode, i.e., PIG discharge. The structure of the ion source is illustrated in Figure 16.12[10]. Since the cathode surface is not faced to an ion extraction aperture, only slow negative ions produced through resonant charge exchange are extracted. Several hundred milliamperes of pulsed H⁻ current were obtained from this type ion source.

(ii) PIG Source With a Convertor

Because direct extraction of negative ions produced on the metal surface is considered more efficient, a PIG type source with a convertor was developed. A dense plasma is generated by a cold or hot PIG discharge at a relatively low gas pressure of 0.1 or 4×10^{-3} torr, respectively. A 500- to 600-mA pulsed H⁻ current was obtained from a hot PIG discharge source as shown in Figure 16.13[11].

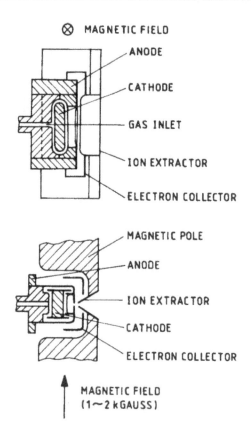

FIGURE 16.10
Typical structure of magnetron type negative-ion source.[3]

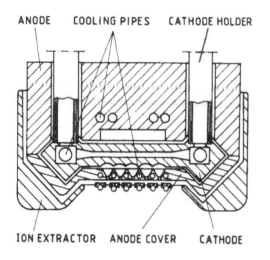

FIGURE 16.11
Magnetron type negative-ion source with multigrooves.[57]

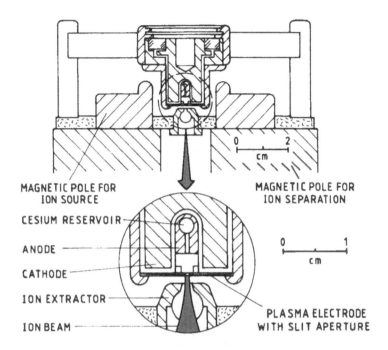

MAGNETIC POLE FOR
ION SOURCE

MAGNETIC POLE FOR
ION SEPARATION

CESIUM RESERVOIR

ANODE

CATHODE

ION EXTRACTOR

ION BEAM

PLASMA ELECTRODE
WITH SLIT APERTURE

FIGURE 16.12
PIG type negative-ion source without a convertor.[10]

16.3.1.3 *Multicusp Type*

Ehlers et al. developed a multicusp type negative-ion source with a convertor for
cw operation that could deliver more than 1 A of H$^-$ current.[12-14] The source structure
is illustrated in Figure 16.14. The plasma chamber is an elliptic cylinder, which is
surrounded with Sm-Co permanent magnets to form cusp magnetic fields. Electrons
for discharge are emitted from eight tungsten filaments placed at zero magnetic field.
Neutral cesium particles are supplied to the surface of the molybdenum convertor
(8 × 25 cm). A dense plasma (about 10^{12} cm^{-2}) was generated by an arc discharge
(90 V, 100 A) at a gas pressure of less than 1 × 10^{-3} torr, and a current of 14 A flowed
to the convertor in the cesiated mode. More than 1 A of negative-ion current could
be extracted.

16.3.2 *Volume Production Type Hydrogen Negative-Ion Sources*

Several kinds of high-current volume production type negative-ion sources were
developed using a multicusp plasma source. Multiampere H$^-$ current from a
multiaperture extraction electrode was obtained for neutral beam injection in fusion
research,[21,58] and several ten milliamperes H$^-$ current was extracted from a single
aperture.[20,22]

16.3.2.1 *Multiaperture Volume H$^-$ Source*

Okumura et al. developed a large hydrogen negative-ion source for fusion research that delivered several to ten amperes of H$^-$ beam with an energy of 50 to 75
kV (0.1 s).[21,59] The source structure is illustrated in Figure 13.5 in Section 13. The
plasma generator was a rectangular water-cooled copper chamber 24 × 48 cm and
15 cm in depth. The chamber is surrounded by Nd magnets to form line cusps, and

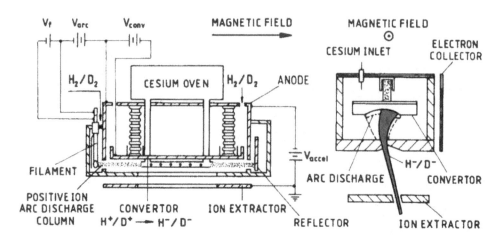

FIGURE 16.13
PIG type negative-ion source with a convertor.[11]

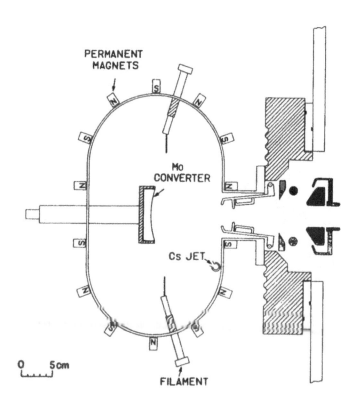

FIGURE 16.14
Multicusp type negative-ion source.[14]

the polarity is different on each side wall resulting in a transverse magnetic field, which acts as a magnetic filter. The arc discharge (70 V, 600 to 1200 A) is generated by eight tungsten filaments. Two kind of extractors are used: one has 434 apertures of 9-mm diameter within an area of 15 × 40 cm and the other has 253 apertures of 11.3-mm diameter within an area of 14 × 36 cm. The optimum xenon gas pressure for cesiated source operation (in the case of 10-A ion beam production) is 1.3×10^{-3} torr,

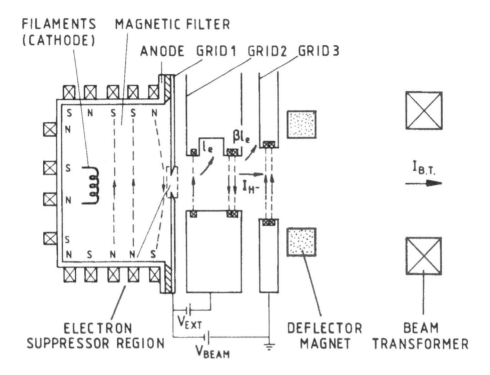

FIGURE 16.15
Single-aperture volume H⁻ source with a multicusp plasma chamber.[20,60]

while that for pure volume production (the extracted ion current then was about half of the current in the case of cesiated source operation) is 1.8×10^{-3} torr.

Kuroda et al. also obtained 8.5 A of H⁻ current from a multicusp plasma source in cesiated mode with a multiaperture electrode having 800 extraction holes of 9-mm diameter in an area of 25×50 cm.[58]

16.3.2.2 Single-Aperture Volume H⁻ Source

Holmes et al. investigated the performance of a cw volume source with a single extraction aperture as shown in Figure 16.15.[20,60] The ion source is a $19 \times 14 \times 10$ cm multipole bucket source, which has a source filter (dipole filter or tent filter) to divide two electron temperature regions. The extraction aperture on the source is 16 mm in diameter. In order to reduce the extracted flux of electrons an electron suppressor is placed near the extraction aperture. The triode (G1, G2, and G3) accelerator is used to extract the beam and collimate it with dumping the extracted electrons. The beam was extracted by a voltage, V_{ext}, of up to 20 kV and was then accelerated to its full beam energy, V_{beam}, of up to 100 kV. In the uncesiated mode, an ion current of 44 mA was extracted at an arc current of 210 A with a source gas pressure of 8 to 9×10^{-3} torr. Through the use of cesium in the ion source the H⁻ current was enhanced by about a factor of two (86 mA of H⁻).

Leung et al. developed an rf-driven multicusp source for particle accelerators that delivered an H⁻ current of about 40 mA without cesium and about 90 mA with cesium vapor from a single aperture.[22] The source structure is illustrated in Figure 16.16. The source chamber is cylindrical (10 cm in diameter and 10 cm long) and surrounded by Sm-Co permanent magnets to form longitudinal line-cusp magnetic fields. In order to prevent energetic electrons from reaching an extraction region, a magnetic filter is

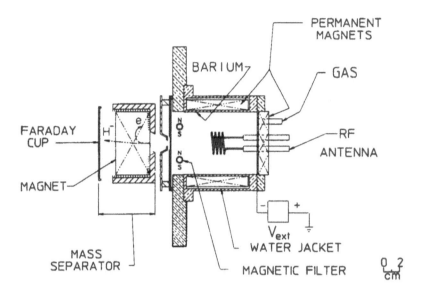

FIGURE 16.16
Rf-driven volume source with a multicusp plasma chamber.[22]

formed near the extraction region with a pair of permanent magnet rods. The plasma was generated by an rf power (about 2 MHz, 10 to 50 kW) at a pressure of about 1.2 × 10⁻² torr through inductive coupling of the plasma via a two-turn antenna. An H⁻ current of about 40 mA was extracted from a 5.6-mm-diameter aperture without cesium. When cesium was introduced by mounting a cylindrical collar in the extraction region, the H⁻ current was increased to about 90 mA.

16.3.3 Sputter Type Heavy Negative-Ion Sources

16.3.3.1 Beam Sputter Type

(a) Versatile Negative-Ion Source

Middleton et al. developed a versatile negative-ion source (VNIS) as shown in Figure 16.17.[34,35] Since a cooled sputter target is placed just in front of a cylindrical tantalum filament surface ionizer for cesium, sufficient cesium particles are supplied to the target surface and the gas pressure of the negative-ion transit region is decreased by a large margin. Thus, a few hundred microamperes of various kinds of heavy negative-ions could be successfully obtained from this source as described in Table 16.3. This source is compact and can deliver high output currents. Difficulties of biasing high voltages more than several kilovolts to the sputter target is a slight disadvantage of this design.

(b) NIABNIS

Ishikawa et al. developed a neutral and ionized alkaline metal bombardment type heavy negative-ion source (NIABNIS) as shown in Figure 16.18.[36,37] Because a high-energy cesium ion beam of 10 to 25 keV is extracted from a cesium plasma source, such as a compact microwave ion source, and a cooled sputter target is directly attached to the cesium ion extractor, both high-current cesium ion beam and sufficient neutral cesium particles are supplied on the target surface. Several hundred microamperes of carbon negative ions have been obtained as described in Table 16.3.

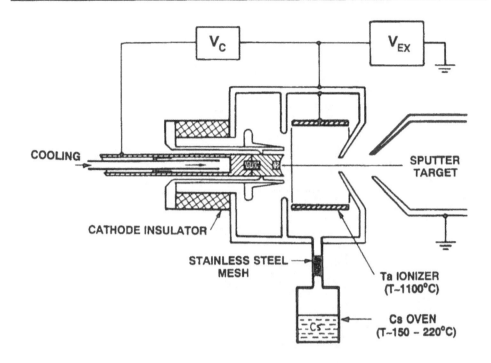

FIGURE 16.17
Versatile negative-ion source (VNIS) developed by Middleton et al.[34]

TABLE 16.3

Negative-Ion Currents Obtained by Recent
Representative Heavy Negative-Ion Sources

Negative-ion source	VNIS	NIABNIS
	Be⁻ 0.1	C⁻ 735
	B⁻ 35	C₂⁻ ~200
	C⁻ 300	B⁻ 15
	F⁻ 250	B₂⁻ 20
	Al⁻ 6	Cu⁻ 320
Negative-	Si⁻ 250	Al⁻ 6.0
ion	P⁻ 25	Al₂⁻ 7.7
current	O⁻ 100	Mo⁻ 0.6
(μA)	Fe⁻ 4	Sb₂⁻ 2.1
	Cu⁻ 150	As⁻ 2.4
	Pt⁻ 150	Ni⁻ 59
	Au⁻ 200	Nb⁻ 1.6
		Mo⁻ 1.0
		Ta⁻ 0.4
Positive-ion energy and current for bombardment	Cs⁺ 5–10 keV 5–15 mA	Cs⁺ 10–25 keV 1–10 mA

16.3.3.2   *Plasma-Sputter Type*

(a) Multicusp Plasma-Sputter Type

Mori and Alton developed a multicusp type heavy negative-ion source (BLAKE-II) as shown in Figure 16.19.[38] This source is the first negative-ion source where a large

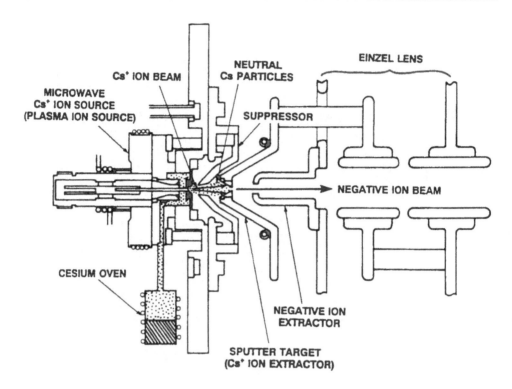

FIGURE 16.18
Neutral and ionized alkaline metal bombardment type heavy negative-ion source developed by
Ishikawa et al.[36]

sputter target is placed in a plasma generated at a very low gas pressure, and could
deliver several milliamperes of various heavy negative ions in a pulsed mode. The
plasma is effectively generated by electron bombardment in a cylindrical chamber
with cusp magnetic fields. Two sets of hot filaments made of lanthanum hexaboride
(LaB_6) are used for electron emission at a relatively low temperature of 1400 to
1500°C. At the center of the chamber a sputter target is set whose shape is a part of
a sphere, and neutral cesium particles are supplied to the target surface from a cesium
oven. Although the mechanism of negative-ion production of this source is similar to
that of the ANIS, the xenon gas pressure of the plasma region of this source (2.3×10^{-4}
torr) is lower by about two orders of magnitude than that of the ANIS. Then, most
of the negative ions produced on the sputter target surface do not suffer an electron
detachment collision during the passage through the plasma region. The pulsed arc
current and voltage during the normal operation were 10 to 20 A and 30 to 40 V (500
μs, 20 Hz), respectively. The sputter target voltage and current were −500 to −1000 V
and about 300 mA, respectively. The typical extracted beam currents were 10 mA for
Au^-, 10 mA for Cu^-, 6.4 mA for Pt^-, 6 mA for Ni^-, 5.4 mA for Ag^-, and 8 mA for C^-
beams. The 80% normalized beam emittance of the 6-mA Ni^- was about 0.3 π·mm·mrad.

(b) Microwave-Discharged Plasma-Sputter Type

Mori et al. developed a microwave-discharged plasma-sputter type heavy negative-ion source (BLAKE-IV) as shown in Figure 16.20.[39] In this source a xenon plasma
was generated by 2.45-GHz microwaves. A cylindrical plasma chamber is surrounded
by eight pieces of Sm-Co permanent magnets that form a magnetic cusp field to
confine the plasma and the magnetic field for the ECR condition at a position of about
7 mm inside the chamber wall. In order to remove the electrons extracted from the

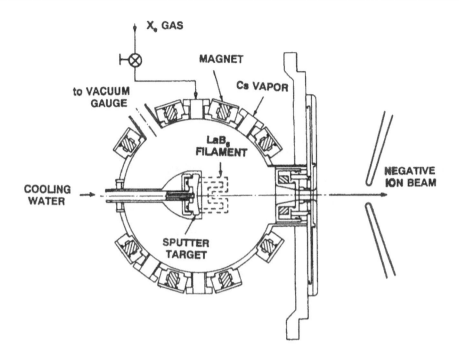

FIGURE 16.19
Multicusp plasma-sputter type heavy negative-ion source (BLAKE-II) developed by Mori and Alton.[38]

source, a small Sm-Co dipole magnet is placed at the ion extraction aperture. The microwave is introduced to the plasma through a sputter target that works as an antenna for the microwaves. Pulsed microwave power of about 3 kW (0.5 ms, 20 to 50 Hz) was supplied for discharge. The xenon gas flow rate was about 0.1 sccm. The water-cooled sputter target was isolated electrically and was biased by –0.2 to –1.5 kV. A Cu⁻ ion beam current of 6 mA was obtained at the optimum cesium supply condition.

(c) RF Plasma-Sputter Type

Ishikawa et al. developed an rf plasma-sputter type heavy negative-ion source[40,41,61-63] that can deliver milliamperes of boron, carbon, silicon, phosphor, and copper negative-ion currents in dc mode, and long operation is possible.

Figure 16.21 and Table 16.4 show a schematic diagram and performances, respectively, of the rf plasma-sputter type heavy negative-ion source. A dense xenon plasma is generated by introducing a few hundred watts of an rf (13.56 MHz) power to an rf coil. A sputter target is negatively biased to several hundred volts. Then, xenon ions bombard uniformly on a cesiated sputter target surface. A part of the sputtered particles are emitted as negative ions through a secondary negative-ion emission process. The negative ions surviving electron detachment during passage through the plasma region are extracted as a beam from an emission aperture. In the negative-ion extraction region a weak magnetic field perpendicular to the beam axis with a peak strength of about 500 G is applied in order to eliminate electrons from the beam.

For a high-current heavy negative-ion source, a dense plasma should be generated in a very low gas pressure condition in order to obtain a large amount of sputtered particles and to reduce neutralization of negative ions passing through the plasma region. At the same time, the work function of the cesiated target surface

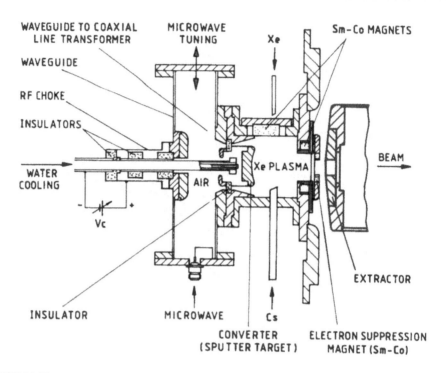

FIGURE 16.20
Microwave-discharged plasma-sputter type heavy negative-ion source (BLAKE-IV) developed by Mori et al.[39]

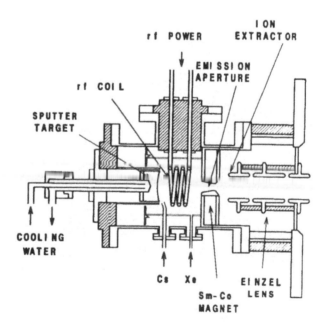

FIGURE 16.21
Rf plasma-sputter type heavy negative-ion source developed by Ishikawa et al.[40]

TABLE 16.4

Performance of RF Plasma-Sputter Type Heavy Negative-Ion Source

Constructive Dimensions

Sputtering target	42 mm in diameter, spherical surface, water cooling
Rf coil	50 mm in diameter, 3 turns, compressed air cooling
Plasma region	50 mm in diameter and 70 mm in length
Emission aperture	10 mm in diameter
Electron remover	480 G at peak by Sm-Co permanent magnets
Source chamber	180 mm in diameter and 240 mm in length

Typical Operating Conditions

Discharge gas pressure	$1–4 \times 10^{-4}$ torr of xenon gas
Input rf power	300 W of 13.56 MHz rf power
Sputter voltage	600 V
Target current	300 mA
Cesium oven temperature	200–250°C
Ion extraction voltage	15–20 kV

Maximum Extracted Negative-Ion Currents

B^-: 0.03 mA,	B_2^-: 1.0 mA	LaB_6 target
C^-: 1.6 mA,	C_2^-: 2.3 mA	High-purity graphite target
P^-: 0.8 mA		Polycrystalline InP target
Si^-: 3.8 mA,	Si_2^-: 0.27 mA	Polycrystalline silicon target
Cu^-: 12.1 mA,	Cu_2^-: 0.15 mA	Oxygen-free copper target

should be minimized to achieve maximum negative-ion production efficiencies in secondary negative-ion emission by sputtering. From the measurements of maximum negative-ion production efficiencies, which are around 10% for C, Si, Cu, Ge, and W, and their sputtering yields, which are between 0.2 and 2.5 by 600-eV xenon ion bombardment, the generated plasma density should be over 1×10^{11} cm^{-3} to obtain several milliamperes of negative ions. From the data of electron detachment cross sections, the xenon gas pressure of the plasma region (7-cm length) should be lower than 4×10^{-4} torr in order that more than 90% of generated negative ions survive. An rf discharge using an inner rf coil is suited for efficient ionization at a very low gas pressure as described in Table 16.4, and the ion source lifetime is expected to be very long because of the absence of filaments.

In Table 16.4 the maximum extracted negative-ion currents from the source are shown. Negative ion current components were calculated from the analysis of mass distribution as a percentage of the total extracted negative ion current. More than 10 mA of copper negative-ion current were obtained, and also milliamperes of boron, carbon, phosphor, and silicon negative-ion currents, which would be useful for implantation into semiconductors.

REFERENCES

1. Lawrence, G. P., Beauchamp, R. K., and McKibben, J. L., Direct extraction of negative ion beams of good intensity from a duoplasmatron, *Nucl. Instrum. Methods*, 32, 357–359, 1965.
2. Sluyters, Th. and Prelec, K., A hollow discharge duoplasmatron as a negative hydrogen ion source, *Nucl. Instrum. Methods*, 113, 299–301, 1973.

3. Sluyters, Th., Negative hydrogen sources for beam currents between one milliampere and one ampere, *Proc. 2nd Symp. Ion Sources and Formation of Ion Beams*, LBL Report-3399, VIII-2-1-VIII-2-12, 1974.

4. Bel'chenko, Yu. I., Dimov, G. I., and Dudnikov, V. G., A powerful injector of neutrals with a surface-plasma source of negative ions, *Nucl. Fusion*, 14, 113–114, 1974.

5. Bel'chenko, Yu. I., Dimov, G. I., and Dudnikov, V. G., H⁻ ions from a crossed-field discharge, *Sov. Phys. Tech. Phys.*, 18(8), 1083–1086, 1974.

6. Bel'chenko, Yu. I., Dimov, G. I., and Dudnikov, V. G., Surface-plasma source of negative ions, *Proc. 2nd Symp. Ion Sources and Formation of Ion Beams*, LBL Report-3399, VIII-1-1-VIII-2-6, 1974.

7. Green, T. S., Scaling laws for high-current H⁻ magnetron ion sources, *Nucl. Instrum. Methods*, 125, 345–348, 1975.

8. Prelec, K., Progress in the development of high current, steady state H⁻/D⁻ sources at BNL, *Proc. 2nd Int. Symp. Production and Neutralization of Negative Hydrogen Ions and Beams*, BNL Report-51304, 145–152, 1980.

9. Ehlers, K. W., Design considerations for high-intensity negative ion sources, *Nucl. Instrum. Methods*, 32, 309–316, 1965.

10. Allison, P., Smith, Jr., H. V., and Sherman, J. D., H⁻ ion source research at Los Alamos, *Proc. 2nd Int. Symp. Production and Neutralization of Negative Hydrogen Ions and Beams*, BNL Report-51304, 171–177, 1980.

11. Dagenhart, W. K., Stirling, W. L., Banic, G. M., Barber, G. C., Ponte, N. S., and Whealton, J. H., Short-pulse operation with the SITEX negative ion source, *Production and Neutralization of Negative Ions and Beams, 3rd Int. Symp.*, AIP Conference Proceedings No. 111, 353–360, 1984.

12. Ehlers, K. W. and Leung, K. N., Characteristics of a self-extraction negative ion source, *Proc. 2nd Int. Symp. Production and Neutralization of Negative Hydrogen Ions and Beams*, BNL Report-51304, 198–205, 1980.

13. Ehlers, K. W., Negative ion sources for neutral beam system, *Proc. Int. Ion Engineering Congr.*, Institute of Electrical Engineers of Japan, 59–69, 1983.

14. Lietzke, A. F., Ehlers, K. W., and Leung, K. N., The status of ≥1 ampere H⁻ ion source development at the Lawrence Berkeley Laboratory, *Production and Neutralization of Negative Ions and Beams, 3rd Int. Symp.*, AIP Conference Proceedings No. 111, 344–351, 1984.

15. Tsai, C. C., Feezell, R. R., Haselton, H. H., Ryan, P. M., Schechter, D. E., Stirling, W. L., and Whealton, J. H., Characteristics of a modified duoplasmatron negative ion source, *Proc. 2nd Int. Symp. Production and Neutralization of Negative Hydrogen Ions and Beams*, BNL Report-51304, 225–231, 1980.

16. Bacal, M., Bruneteau, A. M., Graham, W. G., Hamilton, G. W., and Nachman, M., H⁻ and D⁻ production in plasmas, *Low Energy Ion Beams 1980*, Inst. Phys. Conf. Ser. No. 54, 139–142, 1980.

17. Bacal, M., Bruneteau, A. M., Doucet, H. J., Graham, W. G., and Hamilton, G. W., H⁻ and D⁻ production in plasma, *Proc. 2nd Int. Symp. Production and Neutralization of Negative Hydrogen Ions and Beams*, BNL Report-51304, 95–102, 1980.

18. Bacal, M. and Bruneteau, A. M., H⁻ production and destruction mechanisms in hydrogen low pressure discharge, *Production and Neutralization of Negative Ions and Beams, 3rd Int. Symp.*, AIP Conference Proceedings No. 111, 31–42, 1984.

19. Leung, K. N. and Ehlers, K. W., Volume H⁻ ion production experiments at LBL, *Production and Neutralization of Negative Ions and Beams, 3rd Int. Symp.*, AIP Conference Proceedings No. 111, 67–78, 1984.

20. Holmes, A. J. T. and Green, T. S., Extraction and acceleration of H⁻ ions from a magnetic multipole source, *Production and Neutralization of Negative Ions and Beams, 3rd Int. Symp.*, AIP Conference Proceedings No. 111, 429–437, 1984.

21. Okumura, Y., Hanada, M., Inoue, T., Kojima, H., Matsuda, Y., Ohara, Y., Seki, M., and Watanabe, K., Cesium mixing in the multi-ampere volume H⁻ ion source, *Production and Neutralization of Negative Ions and Beams, 5th Int. Symp.*, AIP Conference Proceedings No. 210, 169–180, 1990.

22. Leung, K. N., Bachman, D. A., and McDonald, D. S., Production of H⁻ ions by an rf driven multicusp source, *Particles and Fields Series 53: Production and Neutralization of Negative Ions and Beams, 6th Int. Symp.*, AIP Conference Proceedings No. 287, 368–372, 1994.

23. Mueller, M. and Hortig, G., An ion source for negative heavy ions, *IEEE Nucl. Sci.*, 16(3), 38–40, 1969.

24. Middleton, R. and Adams, C. T., A close to universal negative ion source, *Nucl. Instrum. Methods*, 118, 329–336, 1974.

25. Middleton, R., A survey of negative ion sources for tandem accelerators, *Nucl. Instrum. Methods*, 122, 35–43, 1974.

26. Middleton, R., A survey of negative ions from a cesium sputter source, *Nucl. Instrum. Methods*, 144, 373–399, 1977.
27. Andersen, H. H. and Tykesson, P., A PIG sputter source for negative ions, *IEEE Nucl. Sci.*, 22(3), 1632–1636, 1975.
28. Tykesson, P., Andersen, H. H., and Heinemeier, J., Further investigations of ANIS, *IEEE Nucl. Sci.*, 23(2), 1104–1108, 1976.
29. Chapman, K. R., The inverted sputter source, *Nucl. Instrum. Methods*, 124, 299–300, 1975.
30. Brand, K., Improvement of the reflected beam sputter source, *Nucl. Instrum. Methods*, 154, 595–596, 1978.
31. Alton, G. D. and Blazey, G. C., Studies associated with the development of a modified university of Aarhus negative ion source, *Nucl. Instrum. Methods*, 166, 105–116, 1979.
32. Caskey, G. T., Douglas, R. A., Richards, H. T., and Smith, Jr., H. V., A simple negative-ion sputter source, *Nucl. Instrum. Methods*, 157, 1–7, 1978.
33. Smith, Jr., H. V., A modified Aarhus negative-ion source, *Nucl. Instrum. Methods*, 164, 1–10, 1979.
34. Middleton, R., A versatile high intensity negative ion source, *Nucl. Instrum. Methods*, 214, 139–150, 1983.
35. Middleton, R., A versatile high intensity negative ion source, *Nucl. Instrum. Methods*, 220, 105–106, 1984.
36. Ishikawa, J., Takeiri, Y., Tsuji, H., Taya, T., and Takagi, T., Neutral and ionized alkaline metal bombardment type heavy negative-ion source (NIABNIS), *Nucl. Instrum. Methods*, 232B(1), 186–195, 1984.
37. Ishikawa, J., Takeiri, Y., and Takagi, T., Mass-separated negative-ion-beam deposition system, *Rev. Sci. Instrum.*, 57(8), 1512–1518, 1986.
38. Alton, G. D., Mori, Y., Takagi, A., Ueno, A., and Fukumoto, S., High-intensity plasma-sputter heavy negative-ion source, *Rev. Sci. Instrum.*, 61(1), 372–377, 1990.
39. Mori, Y., Negative metal ion sources, *Rev. Sci. Instrum.*, 63(4), 2357–2362, 1992.
40. Ishikawa, J., Negative ion beam technology for materials science, *Rev. Sci. Instrum.*, 63(4), 2368–2373, 1992.
41. Ishikawa, J., Tsuji, H., Okada, Y., Shinoda, M., and Gotoh, Y., Radio frequency plasma sputter type heavy negative ion source, *Vacuum*, 44(3/4), 203–207, 1993.
42. Freeman, J. H., Temple, W., and Chivers, D., A note on the production of intense beams of negative heavy ions, *Nucl. Instrum. Methods*, 94, 581–582, 1971.
43. Mündel, C., Berman, M., and Domeke, W., Nuclear dynamics in resonant electron-molecule scattering beyond the local approximation: vibrational excitation and dissociative attachment in H_2 and D_2, *Phys. Rev. A*, 32(1), 181–193, 1985.
44. Wadehra, J. M. and Bardsley, J. N., Vibrational- and rotational-state dependence of dissociative attachment in e-H_2 collisions, *Phys. Rev. Lett.*, 41(26), 1795–1798, 1978.
45. Schlachter, A. S. and Morgan, T. J., Formation of H⁻ by charge transfer in alkaline-earth vapors, *Production and Neutralization of Negative Ions and Beams, 3rd Int. Symp.*, AIP Conference Proceedings No. 111, 149–160, 1984.
46. Schlachter, A. S., Formation of negative ions by charge transfer: He⁻ to Cl⁻, *Production and Neutralization of Negative Ions and Beams, 3rd Int. Symp.*, AIP Conference Proceedings No. 111, 300–310, 1984.
47. Ishikawa, J., A heavy negative ion sputter source: production mechanism of negative ions and their applications, *Nucl. Instrum. Methods*, B37/38, 38–44, 1989.
48. Rasser, B., van Wunnik, J. N. M., and Los, J., Theoretical models of the negative ionization of hydrogen on clean tungsten, cesiated tungsten and cesium surfaces at low energies, *Surf. Sci.*, 118, 697–710, 1982.
49. van Wunnik, J. N. M., Geerings, J. J. C., Granneman, E. H. A., and Los, J., The scattering of hydrogen from a cesiated tungsten surface, *Surf. Sci.*, 131, 17–33, 1983.
50. Ishikawa, J., Tsuji, H., Gotoh, Y., and Azegami, S., Measurement of heavy negative ion production probabilities by sputtering, *Particles and Fields Series 53: Production and Neutralization of Negative Ions and Beams, 6th Int. Symp.*, AIP Conference Proceedings No. 287, 66–74, 1994.
51. Ishikawa, J., Tsuji, H., and Maekawa, T., Electron detachment cross-sections in low energy heavy negative ion beam apparatus, *Vacuum*, 39(11/12), 1127–1130, 1989.
52. Tsuji, H., Ishikawa, J., Maekawa, T., and Takagi, T., Electron detachment cross-sections for heavy negative-ion beam, *Nucl. Instrum. Methods*, B37/38, 231–234, 1989.
53. Alton, G. D., Semi-empirical mathematical relationships for electropositive adsorbate induced work function changes, *Surf. Sci.*, 175, 226–240, 1986.

54. Graham, W. G., Properties of alkali metals adsorbed onto metal surfaces, *Proc. 2nd Int. Symp. Production and Neutralization of Negative Hydrogen Ions and Beams*, BNL Report-51304, 126–132, 1980.

55. Granneman, E. H. A., Geerlings, J. J. C., van Wunnik, J. N. M., van Bommel, P. J., Hopman, H. J., and Los, J., H- and Li- formation by scattering H^+, H_2^+ and Li^+ from cesiated tungsten surfaces, *Production and Neutralization of Negative Ions and Beams, 3rd Int. Symp.*, AIP Conference Proceedings No. 111, 206–217, 1984.

56. Wada, M., Pyle, R. V., and Stearns, J. W., Work function dependence of surface produced H- in the presence of a plasma, *Production and Neutralization of Negative Ions and Beams, 3rd Int. Symp.*, AIP Conference Proceedings No. 111, 247–253, 1984.

57. Prelec, K., Report on the BNL H- ion source development, *Production and Neutralization of Negative Ions and Beams, 3rd Int. Symp.*, AIP Conference Proceedings No. 111, 333–342, 1984.

58. Ando, A., Tsumori, K., Takeiri, Y., Kaneko, O., Oka, Y., Okuyama, T., Kojima, H., Yamashita, Y., Akiyama, R., Kawamoto, T., Mineo, K., Kurata, T., and Kuroda, T., Cesium-seeded experiments on the 1/3-scaled H- ion source for large helical device, *Particles and Fields Series 53: Production and Neutralization of Negative Ions and Beams, 6th Int. Symp.*, AIP Conference Proceedings No. 287, 339–350, 1994.

59. Inoue, T., Hanada, M., Mizuno, M., Ohara, Y., Okumura, Y., Suzuki, Y., Tanaka, M., and Watanabe, K., Development of a multi-ampere H- ion source at JAERI, *Fields Series 53: Production and Neutralization of Negative Ions and Beams, 6th Int. Symp.*, AIP Conference Proceedings No. 287, 316–323, 1994.

60. McAdams, R., King, R. F., and Newman, A. F., Physics tests of an electron suppressor with variable electric and magnetic fields, *Production and Neutralization of Negative Ions and Beams, 5th Int. Symp.*, AIP Conference Proceedings No. 210, 255–265, 1990.

61. Tsuji, H., Ishikawa, J., Gotoh, Y., and Okada, Y., Rf plasma sputter-type dc-mode heavy negative ion source, *Particles and Fields Series 53: Production and Neutralization of Negative Ions and Beams, 6th Int. Symp.*, AIP Conference Proceedings No. 287, 530–539, 1994.

62. Ishikawa, J., Negative-ion source for implantation and surface interaction of negative-ion beams, *Rev. Sci. Instrum.*, 65(4), 1290–1294, 1994.

63. Ishikawa, J., Applications of negative-ion beams, *Surf. Coat. Technol.*, 65, 64–70, 1994.

CHARACTERIZATION
OF ION SOURCES

B. H. Wolf

17   SURFACE AND THERMAL IONIZATION ION SOURCES

17.1   Surface Ionization Ion Source

Advantages: Stable surface for ion extraction, suited for usually difficult elements such as halogens and alkalines, highly element selective.
Disadvantages: Extensive heat dissipation to the environment, only suited for low-work-function or high-electron-affinity elements.

17.1.1   Ion Source History

Ionization at hot surfaces was first investigated by Langmuir and others in 1923.[1] Later the beneficial effect of oxygen or cesium coverage of the surface improved the ionization efficiency for positive or negative ion formation, respectively.[2,3] It became a high-current ion source with the development of cesium ion thrusters (late 1950s)[4] and the cesium negative-ion sputter ion source of Alton and Mori (see Section 16).[5]

17.1.2   Working Principle of Surface Ionization Ion Sources

Surface ionization can be a very efficient way of ionizing elements with low ionization potentials such as the alkalines (≤ 5 eV) or high electron affinities to form negative ions such as the halogens (≥ 1.8 eV).

Surface ionization ion sources (see Figure 17.1) consist of a high-temperature ionizer made of high-work-function material, such as tungsten, rhenium, iridium, or zeolite for positive-ion production, or low-work-function material, such as tungsten with a Cs monolayer, platinum coated with C, or lanthanum hexaboride for the generation of negative ions.

The ionization efficiency can be calculated using a formula given by the Langmuir-Saha equation for thermal equilibrium conditions

$$\frac{n_i}{n_o + n_i} = \left(1 + \frac{g_o}{g_i} e^{q(\phi_i - \Phi_s)/kT}\right)^{-1}$$

with n_i and n_0 the number of ions or atoms evaporated from the surface, g_i and g_0 statistical weights for ions or atoms, ϕ_i the ionization potential of the atom, and Φ_s the

0-8493-2502-1/95/$0.00+$.50
© 1995 by CRC Press Inc.

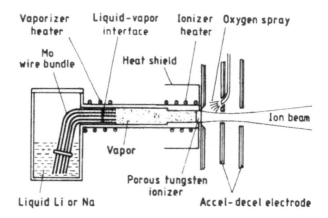

FIGURE 17.1
Surface ionization ion source of Daley.[6]

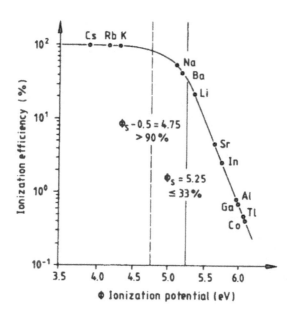

FIGURE 17.2
Ionization efficiency vs. ionization potential for a surface with $\Phi_s = 5.25$.

work function of the surface. Figure 17.2 shows the ionization efficiency of various elements on a surface with $\Phi_s = 5.25$. If the difference $\Phi_s - \phi_i$ is ≥ 0.5, the ionization efficiency is above 90% and for $\Phi_s = \phi_i$ the efficiency is 33%, in this example, or lower, depending on the statistic weights g, which are 2J + 1 (J = total angular momentum of the ion or atom).

For negative-ion production a similar formula is valid with the difference between the work function of the surface and the electron affinity of the atom or molecule in the exponential ($\Phi_s - A_e$) (see Section 16).[11]

Table 17.1 gives the value of the work function for various materials and also the ionization potential or electron affinity of various elements and molecules suitable for surface ionization.

TABLE 17.1

Work Function Φ_s, Ionization Potential ϕ_i, and Electron Affinity A_e of Various Materials or Elements.[7-11]

Material	Φ_s (eV)	ϕ_i(eV)	A_e(eV)
Ni	4.61	7.6	1.1
Mo	4.15	7.2	1.3
Ta	4.12	7.8	0.6
W	4.54	8.0	0.6
W + O	6	—	—
Ir	5.40	9.0	1.9
Pt	5.32	9.0	2.5
Re	4.85	7.9	0.2
Li	2.46	5.4	0.6
Na	2.28	5.1	0.55
Al	4.2	6.0	0.45
K	2.25	4.3	0.5
Ca	3.2	6.1	−1.5
Ga	4.16	6.0	0.3
Rb	2.13	4.2	0.49
Sr	2.74	5.7	−1
In	—	5.8	0.3
Cs	1.81	3.9	0.4
Ba	2.11	5.2	−0.5
La	3.3	5.6	0.5
Re. E.	~3.5	5.6–6.9	0.5–0.6
Th	3.38	~4	—

Re. E: rear earth metals.

Material	Φ_s (eV)	A_e (eV)
Ba on W	1.56	—
Cs on W	1.36	—
Th on W	2.63	—
BaO	1.5	—
SrO	2.0	—
BaO + SrO	0.95	—
Cs-oxide	0.75	—
LaB_6	2.70	—
ThO_2	2.54	—
TaC	3.14	—
ZrO_2	4.2	—
MgO	4.4	—
BeO	4.7	2.8
Al_2O_3	4.7	—
SiO_2	5.0	—
CuO	5.34	—
W-oxide	6.24	—
Ni-oxide	6.34	—
Pt-oxide	6.55	—

Despite the fact that the ionization efficiency decreases with increasing temperature, the ionizer has to be hot enough to evaporate the respective element. On the other hand, the diffusion or the surface coverage has to stay low enough ($\leq$10% of a monolayer) to preserve the ionization conditions (Φ_s) of the ionizer material. The control of this surface coverage is the main task when designing surface ionization

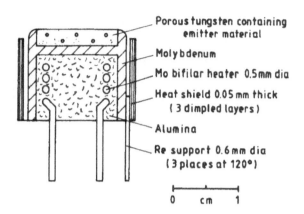

FIGURE 17.3
Surface ionizer with the element contained in a W-sponge.[15]

ion sources. Ion source lifetime in continuous operation can be more than 2000 h, strongly depending on the ion source design and the reservoir for atoms to be ionized.

17.1.3 Examples of Surface Ionization Ion Sources

The source of Daley shown in Figure 17.1 keeps the reservoir with the liquid material separate from the porous tungsten ionizer and, thus, only vapor can reach the ionizer, which is protected from clogging liquid. A similar method of differential heating was used in an indium source by the same author.[12] The preparation technique of surface ionizers is very similar to that of dispenser and oxide cathodes (see Section 1). The ionizer can be made of porous material where the element to be ionized can diffuse through from the rear side (Figure 17.1) or can be contained inside a W-sponge (Figure 17.3).[13] The ionizer can be coated with material containing the element to be ionized or vapor of the respective element is directed to the hot surface where it may be ionized and reflected, adsorbed at the surface, and reevaporated as ion or atom or just reflected as atom.

The preparation of aluminosilicate (zeolite) layers doped with the desired alkali element was described by R. K. Feeney.[14] Figure 17.3 shows a commercially available surface ionizer where the element to be ionized is contained in a highly porous tungsten sponge (porosity 70%).[15] Figure 17.4 shows a surface ion source where a pellet made from Cs-aluminosilicate material is used. The pellet is coated with a tungsten layer and heated sufficiently to evaporate the material. The ion current can be controlled by a potential difference through the pellet.[16]

In contrast to plasma ion sources where the extraction optics depends on the plasma meniscus, the extraction is defined by the ionizer geometry independent of the ion current density. The shape of the ionizer surface can be planar or concave. The emittance of surface ionization ion sources is mainly determined by the surface shape and the surface temperature, which gives an energy spread of ~0.2 eV.[17] Due to the stable emission surface it is possible to operate large-area, single-aperture acceleration systems (see Section 4). Intensive studies of high-current, large-area surface ionization ion sources for inertial confinement accelerators have been made by the group at LBL in Berkeley.[17,18] Their source, a potassium-doped zeolite, operated at a temperature of about 1000°C and delivered more than 10 mA/cm² of K⁺ ions. From a 2.54-cm-diameter source heated with 160 W they accelerated 75 mA with an

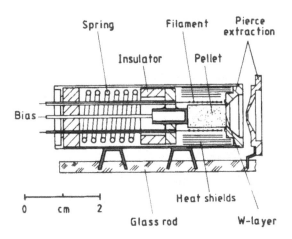

FIGURE 17.4
Surface ionization ion source with pellet.[16]

acceleration voltage of 130 kV. For long-term operation the reservoir in the zeolite is the limiting factor and other designs using porous tungsten may be advantageous. In the latter case pore clogging by the diffusing element to be ionized or poisoning of the surface by condensing background gases and electron bombardment may lead to inhomogenous ion emission with negative effect on beam stability and emittance. The lifetime of this ion source depends strongly on the duty cycle and was between a few days and one year (the total amount of charges is limited).

For the application at isotope separators a great advantage of surface ionization is that the neighboring elements mostly have quite different ionization potentials or electron affinities, a fact that makes this method highly element selective as described in Section 18.4.4. Figure 18.10 shows a negative surface ionization ion source used at CERN at ISOLDE. A LaB_6 pellet, which has a low work function of 2.6 eV, is positioned in the particle stream from the target, acts as ionizer, and negative ions can be extracted and accelerated. Electrons, which are generated as well, are bent by a magnetic field to an electron catcher.[19] More on negative ion production by surface ionization can be found in Section 16 and in Reference 20.

17.2 Thermoionization Ion Source

Advantage: High ionization efficiency.
Disadvantages: Low ion current, extensive heat dissipation to the environment.

17.2.1 Ion Source History

This type of ion source was developed around 1970 by R. G. Johnson at LBL[22] and by G. J. Beyer at JINR.[21]

17.2.2 Working Principle of Thermal Ionization Ion Sources

Thermoionization is, in principle, the same as surface ionization but is used in connection with a hot cavity as ionizer. In this way the ionization efficiency can be improved since the particles are trapped inside the cavity and undergo several collisions with the cavity walls before leaving the source. Elements with ionization potential up to ~8 can be ionized in these ion sources using temperatures up to 3000°C. Thermoionization is discussed in Section 18.4.5 in detail.

Hot cavity sources are highly selective too. At low plasma densities in the source, the degree of ionization in the volume exceeds by far the one at the surface, but stays below the theoretical limit given by the Saha equation. A comprehensive review on hot cavity ion sources is given by R. Kirchner.[23]

REFERENCES

1. K. H. Kingdom and I. Langmuir, *Phys. Rev.*, 21, 385, 1923.
2. J. A. Becker, *Advances in Catalysis VII*, Academic Press, New York, 1955.
3. M. Müller and G. Hortig, *IEEE Trans. Nucl. Sci.*, 16, 38, 1969.
4. A. T. Forrester and R. C. Speiser, *Astronautics*, 4, 34, 1959.
5. G. D. Alton, Y. Mori, et al., *Nucl. Instrum. Methods*, A270, 194, 1988; Nucl. Instrum. Methods, A273, 5, 1988; *Rev. Sci. Instrum.*, 63, 2362, 1992; Rev. Sci. Instrum., 65, 1141, 1994.
6. H. L. Daley and J. Perel, *Rev. Sci. Instrum.*, 42, 1324, 1971.
7. C. J. Smithells, *Metals Reference Book*, Vol. III, Butterworths, London, 1967, 737ff.
8. M. v. Ardenne, *Tabellen zur Angewandten Physik I*, VEB Deutscher Verlag der Wissenschaften, Berlin, 1956 and 1962.
9. E. W. McDaniel and M. R. C. McDowell, *Case Studies in Atomic Collision Physics II*, North Holland, Amsterdam, 1972.
10. G. D. Alton, *Nucl. Instrum. Methods*, B73, 221, 1993.
11. G. D. Alton, *Surf. Sci.*, 175, 226, 1986.
12. H. L. Daley et al., *Rev. Sci. Instrum.*, 37, 473, 1966.
13. O. Heinz and R. T. Reaves, *Rev. Sci. Instrum.*, 39, 1229, 1968.
14. R. K. Feeney et al., *Rev. Sci. Instrum.*, 47, 964, 1976.
15. Spectra-Mat, Inc., Highway 1, Watsonville, CA 95076, U.S.A.
16. A. E. Souzis et al., *Rev. Sci. Instrum.*, 61, 788, 1990.
17. H. L. Rutkowski et al., *Rev. Sci. Instrum.*, 65, 1728, 1994.
18. A. Warwick, Heavy Ion Fusion, Report LBL-18840, Lawrence Berkeley Lab., Berkeley, 1984, 27.
19. B. Vosiki et al., *Nucl. Instrum. Methods*, 186, 307, 1981.
20. G. D. Alton, *Rev. Sci. Instrum.*, 65, 1141, 1994.
21. G. J. Beyer et al., *Nucl. Instrum. Methods*, 96, 437, 1971.
22. R. G. Johnson et al., *Nucl. Instrum. Methods*, 106, 83, 1972; R. A. Meyer et al., *Bull. Am. Phys. Soc.*, 16, 539 and 1162, 1971.
23. R. Kirchner, *Nucl. Instrum. Methods*, A292, 203, 1990.

CHARACTERIZATION OF ION SOURCES

B. H. Wolf

18 HIGH EFFICIENCY ION SOURCES

Advantages: High gas efficiency, low energy spread, good beam quality.
Disadvantage: Moderate ion current.

18.1 Ion Source History

The development of high-efficiency ion sources is connected to the introduction of the on-line isotope separators (ISOL) to detect radioactive isotopes produced with ion beams at accelerator facilities or neutron beams at nuclear reactors. The development started in 1966 with TRISTAN at the reactor in Ames, Iowa[1] and at CERN with ISOLDE in 1967[2] where modified electron bombardment ion sources of the Nielson[3,4] and Nier-Bernas[5] types have been used (see Section 2). In 1976 the forced electron beam-induced arc discharge source (FEBIAD) was introduced by Kirchner (see Section 2).[6]

The first ECR ion source designed for high efficiency was designed by Bechtold in 1987[7] and started a series of similar designs.[8,9] Sortais built a compact ECR ion source, Nanogan, using permanent magnets only (see Section 9).[10]

Surface ionization ion sources are used for high-efficiency negative-ion production[11] and also for positive ions (see Section 17).[12,13]

With the availability of high-power lasers, the laser resonance photoionization ion sources became possible in the mid-1980s.[14-16] There was a series of international conferences on electromagnetic isotope separators with EMIS 12 held in 1991 in Sendai Japan[17] and one on radioactive nuclear beams started in 1989.[18] A comprehensive survey article was given by Van Duppen and Kirchner at ICIS 91.[19]

18.2 Working Principle of High-efficiency Ion Sources

Unlike the ion sources discussed in the previous sections, high-efficiency ion sources are not a separate design but are ion sources of various types optimized for the special needs of ISOL operation. Therefore, the whole system from radioactive particle production via particle transfer to the ion source and efficient and fast ion production and detection has to be optimized together (Figure 18.1).

0-8493-2502-1/95/$0.00+$.50
© 1995 by CRC Press Inc.

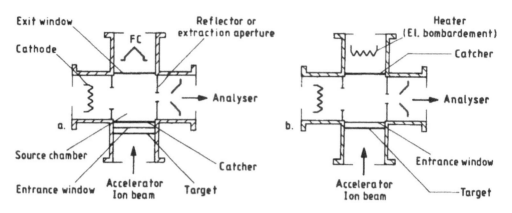

FIGURE 18.1
ISOL arrangement with target and ion source.

18.3 ISOL Criteria

Apart from more general criteria like reliability, good beam characteristics, and reproducibility, there are some very specific criteria that are of importance for the design and operation of high-efficiency ion sources.

- High ionization efficiency is essential to be able to detect the small number of isotopes produced ranging from less than one per second to about 10^{12} per second. The efficiency is defined as number of ions extracted from the ion source divided by the number of particles entering the ion source. Losses due to radioactive decay inside the source are not taken into account. Efficiencies above 10% and up to 80% can be reached depending on the element and the ion source used.

- A short confinement time inside the ion source is essential to reduce particle decay losses. The highest possible electron density is one way to realize a fast ionization of the incoming particles. To avoid condensation at the walls the transfer line and the ion source chamber have to be kept on a sufficiently high temperature. In general, target and ion source have to be as close together as possible. For the detection of short-living isotopes, the diffusion to the source has to be immediate and permanent. The diffusion can be forced by a gas stream of a noncontaminating element. In some cases the reaction products are accumulated in the target or catcher foil for some time and then released partly element selectively by heating the foil slowly that one after the other diffuses to the discharge chamber.

- High element selectivity of the ionization process is important to avoid not only mass contamination, which can be filtered by the analyzing system as long as the energy spread of the ions is small, but also isobaric contamination, which would drastically reduce the analyzer transmission if a high-resolution analyzer system had to be used. The best selectivity is achieved with laser resonance ionization.

18.4 Ion Sources for High Efficiency

Various ion sources are used for high-efficiency operation.

18.4.1 Nier-Bernas Ion Source

Figure 18.2 shows the ISOCELE ion source used at the Orsay synchrocyclotron. It is a Nier-Bernas type ion source (described in Section 2) with an integrated molten target and the possibility of exchanging the target and the cathode filament through

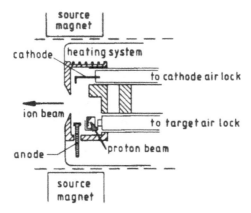

FIGURE 18.2
ISOCELE ion source with integrated target. (Raven, H. L., *Nucl. Instr. Meth.*, 139, 281, 1976. With permission.)

air locks. The target is hit by the proton beam from the accelerator and releases the reaction products that are immediately ionized and detected, giving a very short response time of the ISOL system.[20]

18.4.2 FEBIAD Ion Source

Figure 18.3 shows the FEBIAD ion source, already described in Section 2, which has an excellent efficiency of ~50% for most elements.

Figure 18.4 shows the efficiency of the FEBIAD source in comparison with the modified Nielsen source (MNIS) for Xe+ ions in dependence on the gas pressure. One can clearly see the excellent performance at low gas pressure due to the forced electron arc discharge. The ionization efficiency of different FEBIAD models and various elements is summarized in Figure 18.5, which shows efficiencies between 20 and 60% for most of them except neon (2%).

The separation efficiency of the FEBIAD source, depending on the half-life, is shown for various elements in Figure 18.6. For most elements with a half-life above 5 s, an efficiency of 50% is reached except for the noble gases neon, argon, and krypton. The separation efficiency of carbon, nitrogen, and oxygen was also low, due to the formation of refractory molecules with the hot tantalum discharge chamber.

The FEBIAD ion source delivers ion beams with low energy spread depending on the potential of the extraction aperture (Figure 18.7).

Since the efficiency of the FEBIAD source does not change much with the element there is no selectivity for one element in the ionization process. This can be compensated by collecting particles in a cooled target and selectively releasing them with increasing target temperature.[21] A review of experience with the FEBIAD ion source is given in Reference 22.

18.4.3 ECR Ion Source

There are two major applications for high-efficiency ECR ion sources (Figure 18.8).

- Production of singly charged light ions especially from C, N, O, and Ne, which have low efficiencies in electron bombardment ion sources.
- Production of multiply charged ions from radioactive isotopes for the efficient acceleration of radioactive beams for isotope separation and cleaning, ion implantation, and physics experiments.

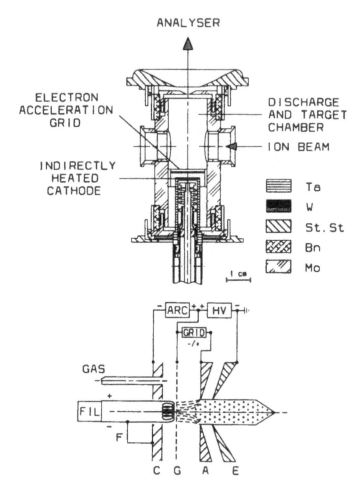

FIGURE 18.3
FEBIAD ion source scheme and circuitry. (Courtesy of R. Kirchner, GSI, Darmstadt, Germany.)

Electron cyclotron resonance ion sources developed for the production of multiply charged ions (see Section 9) can be used unchanged in the second case and can be easily modified for high-efficiency generation of low charge states.

Special design features are

- Single-stage arrangements give high efficiency for low charge states.
- Moderate microwave frequency of 5 to 10 GHz.
- Quartz discharge chambers reduce absorption of the rear reaction products at the chamber wall and avoids the formation of refractory compounds of C, N, and O with the wall materials. The high temperature of the plasma chamber also reduces condensation of the reaction products.
- The use of an iron yoke avoids interference of the source field with the primary ion beam.
- Compact sources with small discharge chambers but good magnetic confinement give high efficiencies of up to 15% for C, 25% for N, 55% for O, 50% for Ne, and 90% for Xe.[7,9]
- Helium or other light carrier gases are also good support gases for gas mixing (see Section 9), which can also improve the source efficiency.

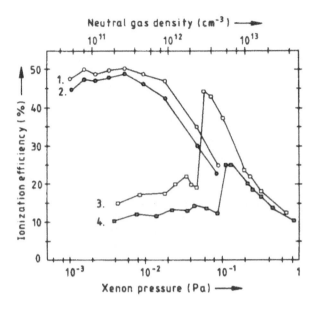

FIGURE 18.4
Ionization efficiency vs. gas pressure for FEBIAD and MNIS. (Courtesy of R. Kirchner, GSI, Darmstadt, Germany.)

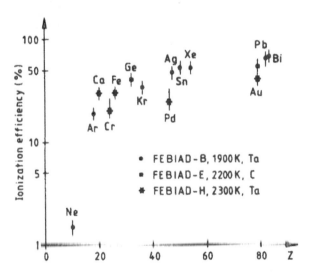

FIGURE 18.5
Ionization efficiency of different FEBIAD ion sources for various elements. (Courtesy of R. Kirchner, GSI, Darmstadt, Germany.)

- The best efficiencies are usually realized with low (≤100 W) microwave power and at low gas pressure. Figure 18.9 shows the ionization efficiency for ^{13}N vs. background pressure and for comparison the ^{15}N efficiency vs. N_2 or CO_2 pressure showing clearly the increase of the efficiency toward lower gas pressure.

18.4.4 Surface Ionization Ion Source

Surface ionization can be a very efficient way of ionizing elements with low ionization potentials such as the alkalines (≤5 eV) or high electron affinities to form negative ions such as the halogens (≥1.8 eV). The great advantage of surface ionization

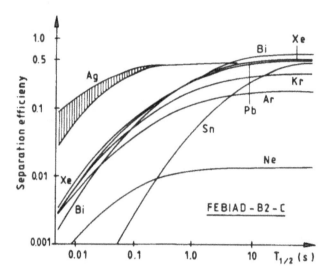

FIGURE 18.6
Separation efficiency vs. half-life of various elements. (Courtesy of R. Kirchner, GSI, Darmstadt, Germany.)

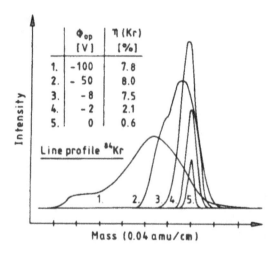

FIGURE 18.7
Energy spread for various potential differences between grid and extraction aperture. (Courtesy of R. Kirchner, GSI, Darmstadt, Germany.)

is that the neighboring elements mostly have quite different ionization potentials or electron affinities, a fact that makes this method highly element selective. Figure 18.10 shows a negative surface ionization ion source used at CERN at ISOLDE. A LaB$_6$ pellet, which has a low work function of 2.6 eV, is positioned in the particle stream from the target, acts as ionizer, and negative ions can be extracted and accelerated. Electrons, which are generated as well, are bent by a magnetic field to an electron catcher.

18.4.5 Thermoionization Ion Source

Thermoionization is, in principle, the same as surface ionization but is used in connection with a hot cavity as ionizer. In this way the ionization efficiency can be improved since the particles are trapped inside the cavity and undergo several

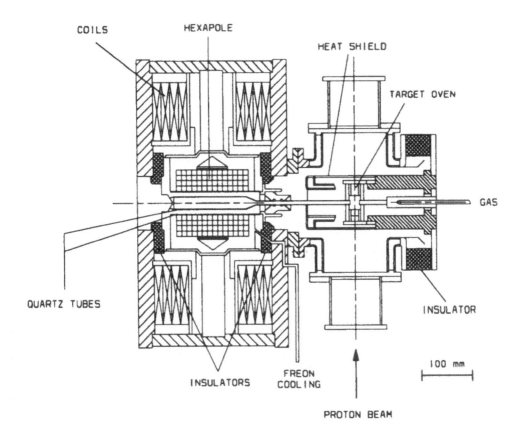

FIGURE 18.8
ECR high-efficiency ion source.[7]

collisions with the cavity walls before leaving the source. Elements with ionization potential up to ~7 can be ionized in these ion sources. Figure 18.11 shows a hot cavity source with the vaporizer and rhenium ionizer intensively heated by electron bombardment. Hot cavity sources are also highly selective, as can be seen in Figure 18.10 for a tungsten cavity of 2850 K. At low plasma densities in the source, the degree of ionization in the volume exceeds by far the one at the surface. The respective amplification factor N is calculated for thermal equilibrium to 7500 for the example in Figure 18.12 and it is shown that a factor N = 150 would fit the experimental ionization efficiency values. The lower value N is due to the imperfect cavity with entrance and extraction holes on both ends. A detailed discussion of thermal ion sources for on-line isotope separators can be found in Reference 23.

18.4.6 Laser Resonance Ion Source

In contrast to laser ion sources described in Section 10, where the ions are created in the plasma produced by the high power density of the laser beam focused onto a solid target, laser resonance ion sources use the selective photoionization transition of a specific element. Tunable high-power laser systems are necessary to reach a high ionization efficiency. To date, such high laser power was available just in pulsed lasers at very low repetition rate. To improve the overall efficiency the reaction products are stored in a moderately heated cavity where the particles are prevented from condensation at the walls and have a high probability of being ionized by the

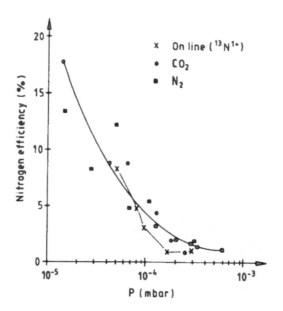

FIGURE 18.9
Nitrogen ionization efficiency vs. gas pressure.[9]

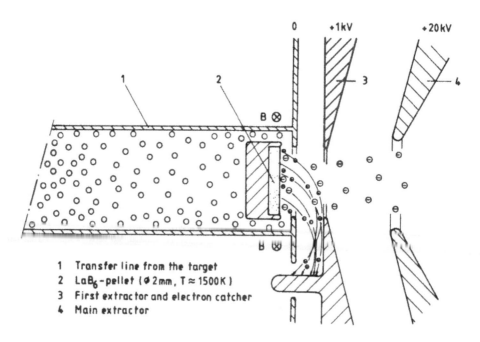

1 Transfer line from the target
2 LaB$_6$-pellet (ϕ 2mm, T $\approx$ 1500K)
3 First extractor and electron catcher
4 Main extractor

FIGURE 18.10
Negative surface ionization ion source from CERN-ISOLDE. (From Vosiki, B., *Nucl. Instr. Meth.*, 186, 307, 1981. With permission.)

pulsed laser beam sent into the cavity through the extraction hole or through a separate window. Depending on the element investigated, there is a reduction in selectivity by the competing process of thermoionization by the hot cavity. Figure 18.13 shows a laser resonance source developed for on-line operation.[24]

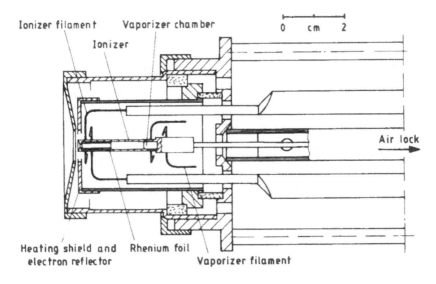

FIGURE 18.11
Hot cavity ion source.[11]

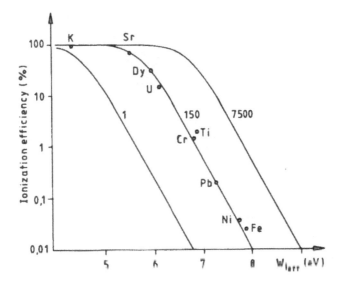

FIGURE 18.12
Ionization efficiency for various elements vs. ionization potential. The calculated curves give the limit for pure surface ionization (N = 1) and for thermal equilibrium (N = 7500). (Courtesy of R. Kirchner, GSI, Darmstadt, Germany.)

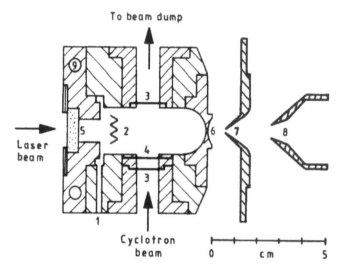

FIGURE 18.13
Laser resonance photoionization ion source. (Courtesy of P. Van Duppen, Univ. Löven, Belgium.)

REFERENCES

1. O. W. Talbert et al., *Nucl. Instrum. Methods*, 38, 306, 1965.
2. G. Anderson, *Ark. Fys.*, 36, 61, 1967; A. Kjelberg and G. Rudstam, The ISOLDE Collaboration Rep. CERN 70–3, 1970.
3. O. Almen and K. O. Nielsen, *Nucl. Instrum. Methods*, 1, 302, 1957.
4. S. Sundell, *Proc. 7th Int. Conf. Electromagnetic Isotope Separators and the Technique of their Application*, Marburg, 1970.
5. L. Chavet and R. Bernas, *Nucl. Instrum. Methods*, 51, 77, 1967.
6. R. Kirchner and E. Röckel, *Nucl. Instrum. Methods*, 133, 187, 1976; *Nucl. Instrum. Methods*, B26, 235, 1987.
7. V. Bechtold et al., *Proc. 7th Workshop ECR Ion Sources*, Jülich, 1986, 248.
8. M. Domsky et al., *Nucl. Instrum. Methods*, A295, 291, 1990.
9. P. Decrock et al., *Nucl. Instrum. Methods*, B58, 252, 1991.
10. A. Chabert et al., *Proc. 2nd Int. Conf. Radioactive Nuclear Beams*, Hilger, London, 1991; P. Sortais, *Proc. 11th Int. Workshop ECR Ion Sources*, Rep. KVI 996, 1993, 97,
11. B. Vosiki et al., *Nucl. Instrum. Methods*, 106, 007, 1081.
12. G. J. Beyer et al., *Nucl. Instrum. Methods*, 96, 437, 1971.
13. R. G. Johnson et al., *Nucl. Instrum. Methods*, 106, 83, 1972; R. A. Meyer et al., *Bull. Am. Phys. Soc.*, 16, 539 and 1162, 1971.
14. H. J. Kluge et al., *Proc. Accelerated Radioactive Beams Workshop*, L. Buchmann, Ed., TRIUMF Rep. TRI-85-1, 1985, 119.
15. S. V. Andreev et al., *Opt. Commun.*, 57, 317, 1986.
16. V. S. Letokhov, *Laser Photoionization Spectroscopy*, Academic Press, Orlando, 1987.
17. *Proc. 12th Int. Conf. Electromagnetic Isotope Separators and their Applications*, Sendai, Japan, 1991; *Nucl. Instrum. Methods*, B70, 1992.
18. *Proc. 1st Int. Conf. Radioactive Nuclear Beams*, World Scientific, Singapore, 1989; *Proc. 2nd Int. Conf. Radioactive Nuclear Beams*, Hilger, London, 1991.
19. P. Van Duppen et al., *Rev. Sci. Instrum.*, 63, 2381, 1992.
20. J. C. Puteaux et al., *Nucl. Instrum. Methods*, 121, 615, 1974.
21. R. Kirchner et al., *Nucl. Instrum. Methods*, A247, 265, 1986; *Nucl. Instrum. Methods*, B26, 204, 1987; *Nucl. Instrum. Methods*, B70, 56, 1992.
22. R. Kirchner, *Proc. EMIS 12*, *Nucl. Instrum. Methods*, B70, 186, 1992.
23. R. Kirchner, *Nucl. Instrum. Methods*, 186, 275, 1981; *Nucl. Instrum. Methods*, A292, 203, 1990.
24. P. Van Duppen et al., *Hyp. Interaction*, 74, 193, 1992.

CHARACTERIZATION OF ION SOURCES

B. H. Wolf

19 OTHER ION SOURCES

In the following section some highly specialized ion sources are briefly described and review articles are given as reference for more detailed information in the respective field.

19.1 Polarized Ion Sources

Polarized ion sources generate beams of nuclear spin-polarized ions and consist of a complex system of particle generator, polarizer, and ionozer. Polarized ion beams are used for a variety of nuclear physics experiments. There are several designs used mainly for polarized H^+, D^+, H^-, and D^- ions, but also some heavier ions such as Li^+ or Na^+ have successfully been generated in polarized states.

Usually, an atomic beam is polarized in Stern-Gerlach type magnetic sextupole fields and further cleaned of hyperfine substates in rf transition stages. The polarized beam is then ionized by an intensive electron beam (similar to an EBIS; see Section 11) or in an ECR discharge (see Section 9). Negative ions are produced in charge-exchange cells. In optically pumped polarized ion sources, intensive laser beams are used to polarize the electrons in alkaline vapors, which can transfer the polarization to a proton or deuteron beam. For more information see the articles by P. Schiemenz,[1] T. B. Clegg,[2] and T. O. Niinikoski.[3]

19.2 Cluster Ion Sources

Cluster ions are used for micromachining as slow but high-momentum beams. Cluster formation itself is a wide field of research. Clusters are generated in a high-density gas jet expanding into vacuum by condensation of atoms or molecules in the supersaturated gas flow. Clusters can also be produced by sputtering of materials with heavy ions. In this process cluster ions are also generated directly.[4]

The cluster particle beam is ionized by a transversal electron beam of low energy or by photoionization using laser light. Extraction and acceleration of cluster ion beams is done in the usual way by electric fields. For more information, see the articles by O. F. Hagana[5] and T. D. Märk,[6] and the special reports.[7] For the theory on cluster formation see, H. H. Andersen.[8]

0-8493-2502-1/95/$0.00+$.50

19.3   Ion Sources for Space Application

Ion thrusters for space propulsion are of similar design as large-area (see Section 13) and industrial ion sources (see Section 14); in reality, the latter have been developed from ion thrusters. Ion thrusters have to deliver a high momentum and therefore use heavy ions, mainly xenon or mercury. Neutralization of the generated ion beam is essential to avoid charge-up of the satellite or space probe. The main design problem is to find a minimum weight but extremely sturdy design to resist the gravitational forces and the vibration during the launching of the space object. On the other hand, the design can be an open structure as in a space vacuum that is delivered free. Besides traditional filament discharges, rf discharges are used and show some advantages such as simple construction, less power supplies (all at ground potential), and long lifetime limited only by the extraction grid erosion.

For more information see the articles by H. Bassner,[9] E. Stuhlinger,[10] R. C. Finke,[11] and K. H. Groh.[12]

19.4   Ion Diodes

Ion diodes, together with other special geometries, are designed to deliver short pulses (nanoseconds) of extremely high currents (kA/cm^2) for drivers of inertial confinement fusion reactors. Most of these ion sources are designed for lithium and other light ion beams. Some originated from the SDI (Star Wars) projects on ion guns. The wide variety of such ion sources is described by R. A. Gerber[13] in detail and also by S. Humphries.[14]

REFERENCES

1. P. Schiemenz, *Rev. Sci. Instrum.*, 63, 2519, 1992.
2. T. B. Clegg, Proc. IEEE Particle Accelerator Conf. San Francisco 1991, IEEE Conf. Record, Nov. 1991 2083, *Rev. Sci. Instrum.*, 61, 385, 1990.
3. T. O. Niinikoski, Proc. 7th Int. Conf. on Polarization Phenomenon in Nuclear Physics 1990, Colloq. Phys., C6, 22, 1990, 191.
4. M. Müller and G. Hortig, *Trans. Nucl. Sci.*, 16, 368, 1969.
5. O. F. Hagena, *Rev. Sci. Instrum.*, 63, 2374, 1992.
6. T. D. Märk and A. W. Castleman, Jr., *Adv. At. Mol. Phys.*, 20, 65, 1985; *Int. J. Mass Spectrom. Ion Processes*, 79, 1, 1987.
7. Z. Phys. D Atoms, *Molecules and Clusters* 12, 1989, and 19/20, 1991.
8. H. H. Andersen, *Nucl. Instrum. Methods*, B68, 1992.
9. H. Bassner and D. E. Koelle, IAF-Paper 87–496, Brighton, U.K., 1987.
10. E. Stuhlinger, *Ion Propulsion for Space Flight*, McGraw-Hill, New York, 1964.
11. R. C. Finke, Electric propulsion and its application to space missions, Prog. Astronaut. Aeronaut., 79, 1981.
12. K. H. Groh and H. W. Loeb, *Rev. Sci. Instrum.*, 63, 2513, 1992.
13. R. A. Gerber, Light ion sources for inertial confinement fusion, in E. G. Brown, Ed., *The Physics and Technology of Ion Sources*, John Wiley & Sons, New York, 1989, 371.
14. S. Humphries, Jr., *Nucl. Fusion*, 20, 1549, 1980.

Chapter 3

PRODUCTION OF IONS FROM NONGASEOUS MATERIALS

B. H. Wolf

CONTENTS

INTRODUCTION

Most ion sources are designed to produce a plasma from neutral gas particles (exceptions are liquid-metal, vacuum arc, and some surface ionization ion sources). Most materials, however, are not in the gas phase at room temperature and need special preparation for ionization. For the application of ion sources, metal ions are of greater interest compared to gases. There have been various methods developed for ion production from nongaseous materials.[1-5] In this chapter these methods,

0-8493-2502-1/95/$0.00+$.50

which can be applied to most types of ion sources, are discussed. The selection criteria for the right material combination for the design of high temperature vaporiser arrangements are given in various tables and are discussed for specific applications.

1   METHODS OF ION PRODUCTION FROM SOLIDS

The selection of the best method to feed solid material into an ion source depends on the operational mode of the ion source (dc or pulse), the variety of ion species needed (specialized ion source or universal), the temperature conditions inside the ion source, ion source reliability, necessary operation time, and ease of operation and maintenance of the ion source. The best solution will vary if one has to use an existing ion source, or if one can start with a new design.

The most important methods for feeding solids into ion sources are

1. Evaporation from an external or internal furnace
2. Use of volatile chemical compounds
3. On-line chemical synthesis[1,3]
4. Cathodic sputtering of the solid (including negative ion and RF sputtering)
5. Evaporation by vacuum arc[6] or laser beam[7,8]

In general, the highest ion current can be expected using the pure substance as in methods 1 and 5, whereas the amount of compound ions or support gas ions reduces the metal ion yield in methods 2 to 4. In some cases, however, the ion source needs a support gas for best results, such as ECR ion sources, and can run with a properly chosen compound gas under optimum conditions. The main characteristics of the different methods are as follows.

1.1   Evaporation

Evaporation of pure material gives similar ion yields as with gases if all parts of the ion source are at higher temperatures than the condensation temperature of the entire material. This is possible for materials with vapor pressure above 1 Pa at temperatures below ~1200°C and increasingly difficult, or even impossible, for higher temperatures.

Figure 1 shows vapor pressure curves for some typical elements.[9] Table 1 gives temperatures necessary for a given vapor pressure and the melting point of the elements.

Besides condensation inside the ion source and the extraction system, alloying of the used element with ion source parts quite often limits the evaporation method (see Section 2.4). An additional problem involves the precise temperature control of the oven (see Section 3.2), especially at lower temperatures, where the temperature response is very slow, depending on the mass of the oven and the heat contact to cooled parts of the ion source. Thus, the temperature can easily run too high, which means the oven charge is evaporated within a short time. The influence of the heat transfer from the ion source plasma, by radiation or particle (ion or electron) impact, to the oven can also be substantial especially at low temperatures.

Because of the T^4 dependency of radiation cooling (Stefan-Boltzmann law) the temperature control gets easier at higher temperatures, but the power necessary to heat the oven also increases with T^4.

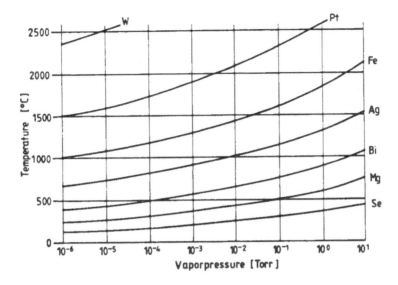

FIGURE 1

Vapor pressure dependence on temperature for selected metals.

TABLE 1

Temperatures Necessary to Reach Selected Vapor Pressures and Melting Point of Non-gaseous Elements[1,11-14]

Element	t_m °C	Vapor pressure in torr at t_v °C							
		10^{-6}	10^{-5}	10^{-4}	10^{-3}	10^{-2}	10^{-1}	1	760
Li	179	280	325	377	439	514	607	725	1331
Be	1284	820	942	1029	1130	1246	1395	1580	2477
B	2300	1550	1700	1850	1980	2170	2380	2700	3920
C	3550s	1850	1950	2100	2220	2400	2620	2900	
Na	98	128	158	195	238	291	356	437	880
Mg	657	240	287	331	383	443	515	605	1100
Al	659	810	890	985	1095	1220	1370	1537	2450
Si	1420	1080	1175	1290	1400	1570	1720	1900	2630
P_{red}	597	82	107	129	157	185	222	262	431
S	118	2	17	38	75	100	135	180	445
K	64	60	91	123	161	207	265	338	762
La	880	355	400	463	528	605	700	817	1482
Sc	1397	957	1058	1161	1282	1423	1595	1804	2727
Ti	1610	1220	1321	1431	1558	1703	1877	2083	3260
V	1857	1327	1432	1551	1687	1847	2037	2287	3350
Cr	1903	977	1062	1162	1267	1392	1557	1737	2660
Mn	1244	637	697	767	852	947	1067	1224	2150
Fe	1539	1017	1107	1207	1322	1467	1637	1847	2880
Co	1492	1072	1162	1262	1377	1517	1697	1907	3100
Ni	1453	1060	1140	1240	1321	1470	1669	1884	2840
Cu	1084	880	946	1035	1141	1273	1432	1628	2595
Zn	420	180	211	248	292	343	405	492	906
Ga	37	688	757	842	937	1057	1197	1372	2240
Ge	936	952	1037	1142	1262	1407	1582	1797	2830
As	817	159	174	205	240	275	305	362	616
Se	220	120	145	170	199	234	287	348	685
Br	-7	-145	-137	-128	-98	-85	-68	-49	59
Rb	39	39	64	95	133	176	228	300	690
Sr	770	298	342	394	456	531	623	742	1370
Y	1552	1100	1249	1362	1494	1649	1833	2056	3030
Zr	1852	1702	1850	2020	2200	2410	2680	2950	3600

TABLE 1 (continued)

Temperatures Necessary to Reach Selected Vapor Pressures and Melting Point of Non-gaseous Elements[1,11-14]

Element	t_m °C	Vapor pressure in torr at t_v °C							
		10^{-6}	10^{-5}	10^{-4}	10^{-3}	10^{-2}	10^{-1}	1	760
Nb	2468	2057	2194	2355	2539	2740	2980	3262	4930
Mo	2577	1850	1987	2167	2377	2627	2927	3297	5560
Ru	2427	1767	1931	2058	2230	2431	2666	2946	3900
Rh	1966	1460	1581	1715	1850	2030	2258	2507	4500
Pd	1550	1060	1156	1271	1387	1547	1727	1967	3980
Ag	961	720	767	832	922	1032	1167	1337	2210
Cd	321	120	149	182	221	267	321	392	767
In	156	602	667	747	837	947	1077	1242	2075
Sn	232	800	860	980	1080	1220	1373	1609	2680
Sb	630	345	390	425	480	570	620	760	1640
Te	452	230	260	295	330	385	445	520	990
J	114	-74	-55	-43	-30	-11	12	39	183
Cs	30	22	46	74	110	153	207	277	690
Ba	710	370	420	478	544	629	730	858	1640
La	920	1157	1262	1377	1527	1697	1897	2147	4230
Ce	804	927	1004	1091	1190	1305	1439	1599	2930
Pr	935	950	1070	1200	1320	1510	1730	2000	3030
Nd	1024	877	957	1062	1192	1342	1537	1777	3180
Sm	1052	460	510	580	653	742	850	960	1630
Eu	826	360	405	466	532	611	705	840	1500
Gd	1312	1030	1160	1290	1400	1560	1770	2000	2730
Tb	1356	1030	1110	1250	1370	1540	1720	1980	2530
Dy	1407	740	820	897	997	1117	1260	1490	2330
Ho	1461	770	850	947	1052	1117	1270	1500	2330
Er	1500	825	920	1000	1130	1250	1450	1630	2630
Tm	1720	554	600	680	757	847	960	1130	2130
Yb	824	270	310	360	405	470	560	660	1530
Lu	1652	1140	1240	1370	1500	1660	1880	2100	1930
Hf	2222	1730	1860	2010	2220	2450	2710	3030	5000
Ta	2996	2220	2407	2600	2820	3065	3350	3750	5425
W	3382	2380	2554	2767	3016	3309	3605	3990	5900
Re	3172	2200	2350	2530	2760	3050	3370	3914	5900
Os	2697	1947	2101	2264	2451	2667	2920	3221	5500
Ir	2454	1677	1797	1947	2107	2307	2527	2827	5000
Pt	1773	1465	1570	1700	1852	2070	2300	2582	4530
Au	1063	930	1000	1100	1220	1365	1500	1707	2970
Hg	-39	-42	-24	-5	18	48	82	126	357
Tl	304	365	412	468	535	615	713	837	1457
Pb	328	425	483	548	625	718	832	975	1750
Bi	271	405	450	508	578	661	762	892	1560
Th	1827	1560	1686	1831	1999	2196	2431	2715	4200
U	1132	1320	1442	1582	1737	1927	2157	2447	3818

$$P_{rad} = \sigma f (T - T_0)^4$$

with

$$\sigma = \frac{2\pi^5 k^4}{15c^2 h^3} = 5.67 \times 10^{-8} \, W \cdot m^{-2} \cdot K^4$$

where T is the oven temperature in kelvin, T_0 ambient temperature in kelvin, f, surface area in meters squared, k, Boltzmann constant, c, speed of light, and h, Planck constant. To heat an oven with a surface of 10 cm^2 to a temperature of 1500°C one needs about 275 W without losses through heat contact.

Extremely careful design of the oven and the discharge chamber with efficient head shields is essential for high-temperature evaporation (see Section 2.3). Four to five layers of reflecting material (with low thermal contact between the layers) cuts the power consumption to about a quarter of the above value.

Evaporation is a continuous process and thus best adjusted to dc ion source operation, which is commonly used for ion implantation at lower energies (<500 keV). For pulsed operation, the evaporation is less suitable especially for very low-duty cycles as used for synchrotron accelerators because of condensation during the time the discharge is switched off. Dc operation of the ion source (and the use of a chopper after beam extraction) would overcome this problem but would lead to a very poor material efficiency of the ion source.

1.2 Chemical Compounds*

The use of chemical compounds that are volatile at room temperature is as easy as the use of pure gas. For those that need higher temperatures to be evaporated, the remarks of Section 1 are valid. Table 2 gives the vapor pressure of some compounds used or usable for ion sources.

Besides possible chemical reactions of the compound elements with ion source parts, especially with hot filaments (see Section 2.4), the main disadvantage of this method is the amount of unwanted ions of compound elements. The yield of the different ion species is not only given by the stoichiometric proportion, but is also influenced by the extraction. Light ions are extracted more efficiently than heavy ones (the ion current decreases with square root of the mass; see Chapter 4). Usually, compound elements are light such as H, C, N, O, F. The extraction system may also be influenced by the additional ion species, since it is only matched by one ion mass.

1.3 On-line Chemical Synthesis

This is a combination of methods 1 and 2. The solid material is positioned in a furnace at a proper temperature. Chlorine- or fluorine-containing gas, usually CCl_4 or Cl_2 or ClF_3, is fed into the hot furnace and reacts with the sample, forming a gaseous chloride or flouride that easily decomposes again. Both Cl_2 and CCl_4 give similar results; Cl_2 is easier to handle, but more corrosive than CCl_4.[18] ClF_3 can be used for noble metals such as Pd, Pt, and Ir. The chloride or fluoride gas is fed into the ion source discharge chamber and its compounds are ionized and extracted. This method has been developed for isotope separation with calutrons but was successfully adapted to other ion sources.

Table 3 summarizes evaporation data of compounds used in on-line chemical synthesis.

The chlorination method has, in principle, the same characteristics as method 2, but the source operation is more delicate and the corrosion problems more severe. It can be used for many elements in a very similar way. As with methods 1 and 2, chlorination is better suited for dc operation of the ion source.

---

* Most compounds are hazardous or toxic and have to be handled with extreme care, respecting the security standards of your country.[15]

TABLE 2

Commercially Available Gaseous or Easy-to-Evaporate Compounds[1,16,17]

Element	Compound	State (25°C)	Boiling point (°C) at 760 torr	Comments[15] (Check safety regulations!!)
B	B_2H_6	Gas	−92	Toxic and inflammable
	BF_3	Gas	−99	Toxic, corrosive
	BCl_3	Gas	13	Toxic, corrosive
C	CO	Gas	−191	Toxic but highest current
C	CH_3	Gas	−164	
	CO_2	Gas	−78	Convenient, fits to rear gas connectors
Si	SiH_4	Gas	−112	Inflammable, decomposes above 400°C
	SiF_4	Gas	−95	Stable up to 2000°C
	$SiCl_4$	Gas	58	Very toxic, corrosive
P	PH_3	Gas	−88	Very toxic, inflammable
	PF_3	Gas	−101	Decomposes at 100°C
	PF_5	Gas	−84	Corrosive, stable
S	SH_2	Gas	−61	Toxic
	SO_2	Gas	−10	Corrosive
	SF_6	Gas	−64	Fits to rear gas connectors
Ge	GeH_4	Gas	−88	Very toxic, inflammable
As	AsH_3	Gas	−62	Very toxic decomposes slowly above 30°C
	AsF_5	Gas	−53	Corrosive, toxic
Se	SeH_2	Gas	−41	Very toxic
	SeF_6	Gas	−46	Stable, corrosive
Br	HBr	Gas	−67	Corrosive
I	HI	Gas	−51	Corrosive
Te	TeF_6	Gas	−39	Corrosive
C,S	CS_2	Liquid	46	Inflammable
Al	$Al(CH_3)_3$	Liquid	130	Inflammable
Ti	$TiCl_4$	Liquid	136	Stable
Cr	CrO_2Cl_2	Liquid	117	Decomposes at 360°C
Fe	$Fe(CO)_5$	Liquid	105	Ignites in air
Ni	$Ni(CO)_4$	Liquid	43	Extremely toxic!!!
Zn	$Zn(CH_3)_2$	Liquid	66	Toxic, inflammable
	$Zn(C_2H_5)_2$	Liquid	118	Toxic, inflammable
Ga	$Ga(CH_3)_3$	Liquid	56	Stable
	$Ga(C_2H_5)_3$	Liquid	—	Stable
Ge	$Ge(CH_3)_4$	Liquid	43	Stable
	$Ge(C_2H_5)_4$	Liquid	163	Unknown
Br	Br_2	Liquid	59	Corrosive
Mo	MoF_6	Liquid	36	Corrosive
In	$In(C_2H_5)_3$	Liquid	—	Unknown
Sn	$SnCl_4$	Liquid	113	Corrosive
	$Sn(CH_3)_4$	Liquid	78	Toxic
Sb	SbF_5	Liquid	150	Corrosive
Re	ReF_6	Liquid	48	Corrosive
Pb	$Pb(CH_3)_4$	Liquid	110	Ignites in air, toxic
Bi	$Bi(CH_3)_3$	Liquid	110	Ignites in air, toxic
V	VF_5	Solid	111	Stable
	$VOCl_3$	Solid	127	Decomposes under electron bombardment
Cr	$Cr(CO)_5$	Solid	420	Decomposes at 250°C
Ga	GaI_3	Solid	345	
Ge	GeI_4	Solid	377	Decomposes at 26°C
Ru	RuF_5	Solid	250	Stable
In	$In(CH_3)_3$	Solid	1.5 at (19°C)	Unknown
Sb	$SbCl_3$	Solid	220	Decomposes under electron bombardment

TABLE 2 (continued)

Commercially Available Gaseous or Easy-to-Evaporate Compounds[1,16,17]

Element	Compound	State (25°C)	Boiling point (°C) at 760 torr	Comments[15] (Check safety regulations!!)
I	I_2	Solid	184	Corrosive
Hf	$HfCl_4$	Solid	319	Stable
Ta	TaF_5	Solid	230	Corrosive
	$TaCl_5$	Solid	242	Corrosive
W	WF_6	Solid	17	Corrosive
Os	OsO_4	Solid	130	Extremely toxic!!
U	UF_6	Solid	56	Corrosive

TABLE 3

Temperatures and Selected Vapor Pressures for Various Compounds and Common Combinations for On-Line Chemical Synthesis[1,3,16,18]

Element	Source feed materials	Typ. operation temp. (°C)	Vapor pressure (torr) at t_e (°C)			Comments[15] (Check safety regulations!!)
			10^{-3}	10^{-2}	760	
Li	LiCl	600	515	582	1380	
Be	$BeCl_2$	220	176	209	481	Toxic
	BeF_2	650	562	632	1159	Toxic
Na	NaCl	675	588	658	1467	
Mg	$MgCl_2$	610	532	596	1412	
K	KCl	650	572	637	1500	
Sc	$ScCl_3$	800	717	787	—	
Fe	$FeCl_2$	500	428	482	1012	
Co	$CoCl_2$	700	612	672	1053	Toxic
Ni	$NiCl_2$	575	491	545	973	Toxic
Se	CdSe	700	588	661	—	Toxic
Rb	RbCl	650	544	614	1380	Can form toxic Rb_2O
Y	YCl_3	750	613	698	—	
Zr	$ZrCl_4$	150	103	128	331	
Ag	$AgCl_2$	725	607	685	1560	
Cd	CdS	725	630	686	1382	Toxic
In	$InCl_3$	275	224	256	470	
I	KI	600	500	565	1323	I is corrosive
Cs	CsCl	600	505	570	1300	Cs ignites in air
Re	Re_2O_7	275	133	157	642	Unstable
Be	Be+CCl_4	400–600	—	—	—	Toxic
Al	Al+CCl_4	600–1000	—	—	—	Toxic
Sc	Sc+CCl_4	600–1200	—	—	—	
Ti	TiO_2+CCl_4	600–1000	—	—	—	
V	V+CCl_4	600–1200	—	—	—	
Cr	Cr+CCl_4	400–600	—	—	—	
Fe	Fe+CCl_4	400–600	—	—	—	
Co	Co+CCl_4	400–600	—	—	—	Toxic
Ni	Ni+CCl_4	400–600	—	—	—	Toxic
Cu	Cu+CCl_4	400–600	—	—	—	
Ga	Ga+CCl_4	600–1200	—	—	—	Toxic
Y	Y_2O_3+Cl_4	600–1200	—	—	—	Toxic
Zr	Zr+CCl_4	400–600	—	—	—	
Nb	Nb+CCl_4	400–600	—	—	—	
Mo	Mo+CCl_4	600–1200	—	—	—	
Rh	Rh+ClF_3	≤1100	—	—	—	Corrosive

TABLE 3 (continued)

Temperatures and Selected Vapor Pressures for Various Compounds and Common Combinations for On-Line Chemical Synthesis[1,3,16,18]

Element	Source feed materials	Typ. operation temp. (°C)	Vapor pressure (torr) at t_e (°C)			Comments[15] (Check safety regulations!!)
			10^{-3}	10^{-2}	760	
Pd	$Pd+ClF_3$	≤1100	—	—	—	Corrosive
In	$In+CCl_4$	600–1200	—	—	—	
La	$La_2O_3+CCl_4$	600–1200	—	—	—	Unknown
Lu	All lanthanides similar		—	—	—	Unknown
Hf	$Hf+CCl_4$	600–1200	—	—	—	
Ta	$Ta+CCl_4$	600–1200	—	—	—	
W	WO_3+CCl_4	600–1200	—	—	—	
Re	$Re+O_2$	600–1200	—	—	—	
Os	$Os+O_2$	600–1200	—	—	—	Very toxic
Ir	$Ir+ClF_3$	≤1100	—	—	—	Corrosive
Pt	$Pt+ClF_3$	≤1100	—	—	—	Corrosive
Th	ThO_2+CCl_4	≤1100	—	—	—	Radioactive
U	UO_2+CCl_3	≤1100	—	—	—	Radioactive
Al	$Al+Cl_2$	800	—	—	—	Corrosive
Si	$Si+Cl_2$	~800	—	—	—	Corrosive
Ti	$Ti+Cl_2$	750	—	—	—	Corrosive
Cr	$Cr+Cl_2$	830	—	—	—	Corrosive
Y	$Y+Cl_2$	~700	—	—	—	Corrosive
Nb	$Nb+Cl_2$	500	—	—	—	Corrosive
Mo	$Mo+Cl_2$	~650	—	—	—	Corrosive

Note: All chlorine and fluorine compounds are corrosive inside the ion source.

1.4 Sputtering

Cathodic sputtering is the most universal method to feed metal atoms into an ion source discharge. Depending on the source construction, however, it is often difficult to expose a large enough area of sputter electrode to the ion source plasma. The sputter yield increases with sputter electrode potential and saturates above ~10 kV (see Figure 2a).[10,19,20] The sputter yield increases with the mass of the ions for the same velocity. Because the sputter voltage in ion sources usually is below, 1 kV the faster Ar^+ gives better results than the slower Xe^+ except for the heaviest target elements. The sputter yield depends on the binding energy of the target element and has a maximum for Cu, Ag, and Au and a minimum for Ti, Nb, and Ta (Figure 2b). The sputter yield has a minimum at normal incidence and increases with sliding incidence. In ion sources the sputter angle is between 30 and 90°. Table 4 gives the sputter yield for various elements at 0.1, 0.5, 1, and 10 keV for argon and xenon ions and normal incidence.

High sputter electrode potential, however, influences the source discharge and may lead to plasma instabilities with negative effect on the extracted ion beam. Sputtering is ideal for materials with high melting point such as W, Ta, Mo, Pt, etc. Low-melting-point materials require proper cooling of the sputter electrode, especially in dc or high-duty cycle operation of the ion source. Semiconductor materials are less suited and insulators can only be used with additional conducting admixtures.[21] Support gas is always necessary for stable operation, argon, krypton, and xenon giving the best results. The amount of metal ions can reach 10 to 50% of the total ion current in stable operation of the ion source. Metal condensation is minimized as metal vapor is produced only when the discharge is switched on. Thus, the

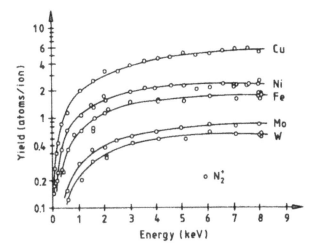

FIGURE 2A
Sputter yield of Cu, Ni, Fe, Mo, and W versus ion energy for N_2.[10]

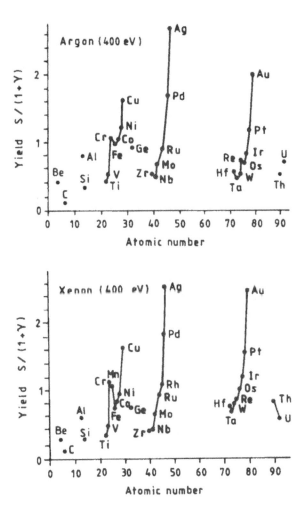

FIGURE 2B
Sputter yield of various elements bombarded by 400-eV Ar or Xe ions.[10]

TABLE 4

Sputter Yield (Atoms per Ion) for Various Elements and Energies of Argon and Xenon Ions (Normal Incidence)[10,19,20]

Element	Argon				Xenon			
	100 V	500 V	1 kV	10 kV	100 V	500 V	1 kV	10 kV
Al	0.2	1	2	—	0.06	0.8	1.2	—
Si	0.06	0.5	0.7	—	—	0.4	0.6	—
Ti	0.08	0.5	0.75	—	—	0.35	0.75	—
Cr	0.3	1	2	—	0.08	1.1	2.5	—
Fe	0.2	1	1.9	3.5	0.06	1	2	7
Cu	0.4	2	3	6.5	0.25	2	3.5	13
Mo	0.09	0.6	1	2	0.03	0.65	1.3	5
Pd	0.4	1.9	2.5	—	0.3	2	4	—
Ta	0.05	0.5	0.7	1.5	0.05	0.7	1.3	5
W	0.07	0.6	1	—	0.05	1	1.8	6
Pt	0.2	1	1.7	4	0.2	2	3	10
Au	0.4	2	3.5	—	0.6	3	5	—
U	0.15	0.8	1.1	2.3	0.25	1.3	2	8

sputter method is especially adapted to pulsed operation with low-duty cycle of the ion source. A comprehensive report on sputter phenomena can be found in Reference 22.

The shape of the sputter electrode depends on the ion source design. Figure 3 shows some typical examples. Materials that are readily available and reasonable in price are machined from the solid material or cut from crystals. To allow intensive cooling, low melting point materials such as Sn, Pb, or Bi are used on a copper backing. The backing technique is also applied to more expensive metals such as Au, Pt, Ag, and rear earth metals. For ferromagnetic materials, special care is necessary if the ion source uses a magnetic field. Small amounts of the material or the use of nonmagnetic alloys keep distortions of the magnetic field small. Copper and molybdenum are preferred backing materials because of their good thermal conductivity. In some cases strips or discs of the material can be soldered to the backing material, but usually melting of the material into an appropriately shaped crucible will give best results for most elements.[21] Whether the crucible can be used directly as sputter electrode or needs some additional machining depends on the ion source design. The melting process usually needs rare gas atmosphere or vacuum environment.

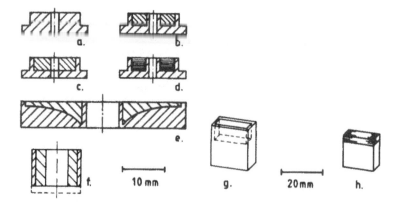

FIGURE 3

Typical shape of sputter electrodes. Disc for duopigatron; a, solid material; b and c, molten with and without middle pin; d, pressed in form; e, disc for CHORDIS; f, cylinder, material molten and machined; g, and h, block electrode for PIG crucible and final shape.

For some elements machining is difficult because of their hardness or brittleness. Hard metals such as W, Ge, and Si can be ground to final shape; for others, ductile alloys may be prepared. Compressed powder is best used in some cases after heat treatment for a degas and sinter process. For rare elements or isotopes or short running time of the ion source, galvanic or vacuum vapor deposition on a copper electrode may be used.[23]

For elements with low melting point, electrodes cut from binary crystals can be used in many cases, such as GaAs for Ga and As, InSb, NaI, KaBr, CsCl, and ZnSe. A comprehensive list of recommended initial substances and electrode preparation techniques is given for various elements in Table 5. Suitable crucible materials and operating temperatures are listed together with remarks on electrode machining or special procedures.

Sputtering is also widely used for negative-ion production where the support gas (Cs, Ba, Xe) is also the electron donor.[24] (For details see Chapter 2, Section 16.) Another ion source where sputtering is very successful is the rf ion source.[25] Here the massive rf antenna acts as a sputter electrode. (For details see Chapter 2, Section 7.)

1.5 Vacuum Arc and Laser Evaporation

Metal evaporation by vacuum arc[6] or laser beams[7,8] are new possibilities for metal ion production, especially the MEVVA ion source developed by I. Brown,[26] which shows very promising results (see Chapter 2, Section 12). A high-voltage vacuum arc or a laser pulse creates a metal plasma, which expands and can be further ionized by an additional discharge. The vacuum arc cathode can replace in some ion sources the hot filament and acts as cathode and metal source at the same time. The change will also change the ion source behavior, but good results have been reported.[50]

This method is as universal as method 4 but does not require support gas. Thus, high amounts of metal ions can be obtained. Usually, the MEVVA ion source design allows only low-duty cycles (up to a few percent), limited by the cooling and size of the cathodes. High-duty cycle or dc operation however, is possible.[27]

For the preparation of vacuum arc cathodes, the same techniques as for sputter electrodes apply.

2 MATERIALS FOR OVEN AND HIGH-TEMPERATURE ION SOURCE DESIGN

Materials for the construction of evaporators for ion sources have to meet some general requirements such as:

- High-temperature stability in vacuum
- Low vapor pressure
- No chemical reaction with other oven or charge material
- Machinable or formable to final shape at reasonable costs

There are many books with data on high-temperature materials available; a small selection includes References 13 and 28 to 34. For every element there are also specialized metallurgical books with detailed data available, which may help in complicated cases.

TABLE 5

Preparation Techniques for Sputter Electrodes

Element	Comp. alloy	Melting point, °C	Technique used	Crucible material	Process temp., °C	Ar pressure (mbar)	Remarks
Li	LiF	850	c(p)	Cu	—	—	Sublimes, 10^{-2}
Be	—	1280	sm	(Cu)	—	—	Toxic
B	LaB$_6$	2210	c	—	—	—	
C	—	3550s	sm	—	—	—	Pyrocarbon
Na	NaI	655	c	—	—	—	
Mg	—	651	sm	—	—	—	
Al	—	660	sm	—	—	—	
Si	—	1420	c	—	—	—	
K	KBr	742	c(p)	Cu	—	—	
Ca	—	850	sm	—	—	—	Machine with kerosene
Sc	—	1397	m	Ta	1500	10	
Ti	—	1610	sm	—	—	—	
V	—	1857	s(sm)	Cu	—	—	40% Ag, 19% Cu, 21% Zn, 20 Cd
Cr	—	1903	s(b)	Cu	1950	250	Machining difficult
Mn	—	1244	m	Mo	1300	200	Ductile alloy, 97% Mn, 2% Cu, 1% Ni
Fe	—	1539	sm(s)	Cu	—	—	Magnetic, 80% Fe, 18% Mn not
Co	—	1492	s(m)	Cu(Mo)	1550	10	Alloy 50% Fe, 50% Co possible
Ni	—	1453	sm(s)	Cu	—	—	Magnetic, carcinogen
Cu	—	1084	sm	—	—	—	Alloy, 80% Ni, 20% Cr, not magnetic
Zn	—	420	m	Cu	630	100	
Ga	GaAs	1238	c	—	—	—	As toxic
Ge	—	937	c	—	—	—	
As	GaAs	1238	c	—	—	—	As toxic
Se	ZnSe	220	m	Fe	300	100	Machining difficult
Se	—	1100	c	Cu	—	—	Toxic
Br	KBr	742	c(p)	Cu	—	—	
Sr	—	770	sm	—	—	—	Machine with kerosene
Y	—	1552	m	Mo,Ta	1550	10	
Zr	—	1852	sm	—	—	—	
Nb	—	2468	sm	—	—	—	
Mo	—	2577	sm	—	—	—	Machine with coolant

Element							Notes
Ru	—	2427	b(sm)	Cu	—	—	Machining difficult
Rh	—	1965	b(sm)	Cu	—	—	Machining difficult
Pd	—	1555	m(s,b)	Ta(Cu)	1650	10	
Ag	—	961	m	Ta	1550	100	Crucible > 400°C
Cd	—	321	m	Cu	380	500	
In	InSb	535	c(P)	(Cu)	—	—	
Sn	—	232	m	Mo	800	0.01	
Sb	InSb	535	c	—	—	—	
I	NaI	655	c	—	—	—	
Cs	CsCl	642	c	—	—	—	
Ba	—	710	sm	—	—	—	Machine with kerosene
La	—	920	m	Mo	930	0.01	
Ce	—	804	m	Mo	1100	200	Pure material necessary
Pr	—	935	m	Ta	950	200	Wets poorly
Nd	—	1024	m	Mo,Ta	1100	20	
Sm	—	1052	m	Mo	1150	700	
Eu	—	825	m	Ta	850	500	Machine with kerosene
Gd	—	1311	m	Mo	1350	10	
Tb	—	1356	m	Mo	1370	1	
Dy	—	1407	m	Mo	1450	700	
Ho	—	1461	m	Mo	1480	500	
Er	—	1500	m	Mo	1500	300	
Yb	—	824	m	Ta	850	500	Soft material
Hf	—	2222	sm	—	—	—	
Ta	—	3000	sm	—	—	—	Machine with coolant
W	—	3380	sm	—	—	—	Grinding necessary
W	WCuNi	3380	sm	—	—	—	Machinable, Cu, Ni 3%
Re	—	3180	sm(b)	(Cu)	—	—	Grinding possible
Ir	—	2454	s(b)	Cu	—	—	
Pt	—	1773	s(m)	Cu(Ta)	1800	0.01	Machining difficult
Au	—	1063	s(m)	Mo	—	—	
Pb	—	328	m(b)	Cu	800	10	Toxic
Bi	—	271	m	Cu	500	0.01	
Th	—	1827	sm	—	—	—	Machined by deliverer
U	—	1132	sm	—	—	—	Machined by deliverer

Note: c, cut from crystal; b, bonding; p, pressed powder; s, soldered; m, molten into electrode; sm, solid material.

2.1   High-Temperature Metals

There is no universal metal for oven design. The most important selection criteria are

- Maximum operation temperature
- No chemical reaction with other oven or charge material
- Mechanical stability and brittleness
- Machinable to final shape at reasonable costs

Materials commonly used are

- Copper is suited for low-temperature operation (600°C) and for cooled ion source parts. It is easy to machine but gets soft at moderate temperatures. The corrosion resistance of copper is poor, which additionally limits its application.

- Titanium has a low thermal conductivity and can be used to reduce the thermal losses, especially in electrical connections. Its melting point is relatively high (1610°C) and it can be used up to 800°C before it reacts with nitrogen. Hydrogen makes Ti brittle and it reacts with fluorine at about 150°C and also with other halogens. Titanium is easy to machine (with a cooling liquid) and has excellent mechanical behavior.

- Stainless steel is an excellent material up to about 1000°C, but its corrosion resistivity decreases with increasing temperature. It forms low-melting alloys with tantalum and molybdenum above 900°C. Its heat conductance is relatively low so one should be aware of local overheating by the plasma or electron beams.

- Molybdenum can be used up to 2000°C. It is relatively resistant to corrosive gases. It is the cheapest of the refractory metals and easy to machine. For parts formed out of plates one has to use a special quality rolled in all directions because ordinary plate material tends to be brittle. Mo gets very brittle after exposure to higher temperatures, but there is a special alloy, TZM, which contains less than 1% Ti, Zr, and C, which can be used in critical cases.

- Tantalum is the easiest to machine of the refractory metals and can be used up to 2600°C for source and oven parts and 2800°C for filaments. Ta gets brittle at high temperatures due to crystallization and also due to reaction with nitrogen and hydrogen. Corrosion resistivity improves with a carbon or carbide layer formed at the surface.

- Tungsten is difficult to machine but can be formed by grinding or spark erosion. It has the highest melting point of about 3400°C and is best suited for filaments. W is also used as emitting surface for surface ionization sources and electron guns and also as converter electrode in negative-ion sources. There are compound materials available containing a few percent copper and nickel that are easy to machine and also a W-sponge containing 25 to 40% Cu. The copper part can be evaporated after machining, leaving the porous tungsten body.

- Rhenium is very expensive but much less brittle than tungsten. Its maximum temperature is 3150°C. It is a good filament material and less affected by corrosive gases and oxygen. W-Re alloy is an easily shaped filament material. Rhenium is also used for surface ionizers and has a higher work function than tungsten.

- High-density graphite is a low-cost material and can be used to high temperatures (3500°C). Its mechanical stability is poor but it is easy to machine. It can be used for high-temperature source parts and directly heated ovens. The corrosion behavior of graphite is excellent, but it tends to absorb many elements and compounds, showing a long memory effect. It outgasses badly after being in air. Graphite may react with ceramic materials (reduction process) and forms carbides with refractory metals at higher temperatures.

- Pyrolytic carbon has a layer structure and, therefore, different behavior in two directions, especially the resistivity. It is not porous, unlike graphite, so does not absorb gas or chemicals. It is harder but still machinable.

- Glassy carbon has the advantage of being gas-tight but is hard and difficult to machine, as with the other ceramics. Its resistivity is as low as other graphites and can be used for directly heated devices. It can be used up to 2500°C in vacuum and has even better chemical resistivity.

Table 6 summarizes the physical and mechanical data of refractory metals.

2.2 High-Temperature Insulators

As with high-temperature metals there is no universal insulator material for oven design. The most important selection criteria are

- Maximum operation temperature
- No chemical reaction with other oven or charge material
- Thermal stability and heat conductance
- Thermal expansion
- Mechanical stability and hardness
- Machinable or formable to final shape at reasonable costs

Materials commonly used are

- Quartz (SiO_2) is a traditional vacuum material and can be used up to 1000°C. It can be formed, similarly to glass, at 1660°C. Quartz is available in many pipe dimensions and is relatively cheap.

- Mica was used for insulation plates and contact insulation. It can be used up to 400 to 600°C, depending on the consistency. Due to its thin layer structure it tends to outgas considerably, which can be reduced by heat treatment (675°C for 16 min).[13] Mica can be cut easily and holes can be drilled.

- Alumina can be used in vacuum up to about 1400°C. It is available in many dimensions (pipes, rods, plates) and can be formed in a green state and burned to the final dimensions quite accurately. It can be cut and ground with diamond cutters. It does not absorb much gas and is chemically stable against many elements and compounds.

- Magnesia is similar to alumina but is more difficult to process and more expensive. It can be used for some materials with which alumina reacts. It can be used for higher temperatures than alumina but has a similar vapor pressure.

- Zirconia is similar to alumina but is more difficult to process and more expensive. It can be used for some materials with which alumina reacts. It can be used for higher temperatures (1600°C) than alumina and has a lower vapor pressure. It changes structure and color when heated, which is reversible under certain conditions. Its resistance drops below 100 Ω above 1000°C.

- Macor or glass ceramic can be easily machined, but is very brittle. It has good vacuum behavior and can be used up to about 1000°C. There are different qualities and mixtures on the market that have to be selected carefully.

- Boron nitride is an excellent material for most applications for temperatures up to 1200°C. It can be used up to 1500°C but starts to decompose and releases large quantities of nitrogen. It outgasses badly and tends to absorb water, which can destroy the parts when heated too fast. There are many qualities on the market and

TABLE 6

Data on Refractory Metals[12,13,35]

	Melting point (°C)	Density (g/cm³)	Therm. expansion (10⁻⁶/K)	Spec. el. resist. (μΩm)	Therm. conduct. (W/m·K)	Emissivity (ε)	Rad. power dissipation (W/cm²)	Vickers hardness (HV 10)ᵃ
Ti	1690	4.5	8.2	0.5	17	50		160–250
Temp. (°C)		20	20	20	20	1200		
S. st.	1400	7.9	18	0.75	20		1.3	190
Temp. (°C)		20	20	20			727	
Mo	2630	10	5.5	0.05	159/109	35	19	155/270
Temp. (°C)		20	20	20	20/1473	1727	1727	
Ta	3000	16.6	6.5	0.125	54.4/74	42	21	90/200–300
Temp. (°C)		20	20	20	20/1527	1727	1727	
W	3395	17–19	4.4	0.055	130/100	42	64	350/450
Temp. (°C)		20	20	20	20/1527	2200	2200	
Re	3176	21	6.8	0.21	—	40	—	250/800
Temp. (°C)		20	20	20		2000		
H-d. graph.	3600	1.8	2–6	10–50	1–80	70–88		30–80
Pyro-C ∥		2.2	0.1	4.0	300	75		
Pyro-C ⊥		2.2	25	10³	3			
Glassy C		1.55	3.1	47	5.5			~10³
Temp. (°C)		20	20	20	20	1727		

ᵃ Annealed/unannealed

TABLE 7

Data on Insulator Materials[13,36,37]

	Max. temp.[a] (°C)	Max. temp.[b] (°C)	Density (g/cm³)	Therm. expansion (10⁻⁶/K)	Specific el. resistance (Ωcm)			Therm. conduct. (W/mK)	Knoop 100 hardness (N/mm²)
					20°C	500°C	1000°C		
SiO_2	1000	1200	2.2	0.6	10^{17}	$<10^9$	—	1.5	820
Mica	500	900	2.8	~15	10^{16}	10^8	—	0.6	135
Al_2O_3	1400	1950	3.8	8.5	10^{14}	10^{10}	10^7	30	23000
MgO	2000	2500	3.4	13.5	10^{14}	10^{10}	10^7	35	7500
ZrO_2	1600	2300	5.5	11	10^{10}	10^3	50	2.5	17000
BN	1200	2400	2	1.8	10^{15}	—	10^9	35	15
Si_3N_4	1400	1500	3.2	3.2	10^{11}	—	10^7	40	15500
Temp. (°C)	—	—	20	20	20	500	1000	20	20

[a] In vacuum $\leq 10^{-4}$.
[b] Inert gas atmosphere.

TABLE 8

Maximum Temperature
for Metal-Ceramic Combinations[13]

Material	Max. temp. (°C) in contact with				
	W	Mo	Ta	Graphite	Ni80Cr20
Al_2O_3	1900	1900	1700	2000	1000
MgO	2000	1800	1800	1800	1000
ZrO_2	1600	1800	1600	1600	1000
BN	2000	1600	1200	2000	1000
BeO	2000	1900	1600	2300	1000
ThO_2	2200	1900	1900	2000	1000
Graphite	1400	1200	1000	—	1000

just the highest density should be used for ion source or oven parts. BN is relatively soft and easy to machine and also to clean from metal condensation. Its surface is smooth, similar to talcum, and it can be used for sliding parts. BN is relatively stable against chlorine and other aggressive chemicals. At temperatures above 1200°C, the nitrogen starts to react with a tantalum heater, which gets extremely brittle or even destroyed. This does not happen with tungsten.

• Silicon nitride is a hard material and difficult to machine. It can be used up to ~1400°C. It is mainly used for insulation with good thermal conductivity and has a high thermal shock resistance. It can be produced cheaply in a wide variety of shapes with high accuracy. Si_3N_4 is stable against most molten metals.

Table 7 summarizes the physical and mechanical data of insulator materials. Table 8 gives the maximum operation temperature for various heating filament and ceramic material combinations.

2.3 Corrosion and Alloying

Two of the major problems confronted in high-temperature application of materials are corrosion and alloying, which have to be checked in any single case of material combination. Some useful information can be found in References 13 and 28 to 34. Table 9 shows the resistance of refractory metals to corrosion by commonly used gases and compounds and Table 10 shows the resistance to liquid metals.

TABLE 9

Resistance of Refractory Metals to Corrosive Gases[35]

Material	Max. temp. (°C) in contact with			
	W	Mo	Ta	Graphite
O_2	800	500	450	500
N_2	2000	2000	900	3000
H_2	3400	2600	500	1500
CO_2	1200	1200	1200	500
CO	900	800	1600	2500
SO_2	700	700	350	500
H_2S	1000	1200	1000	500
F_2	Unstab.	Stab.	Unstab.	Unstab.
Cl_2	250	250	200	Stab.
Br_2	250	400	200	Stab.
I_2	800	400	200	Stab.

Note: All values for 1 atm; at lower pressures higher temperatures are often possible.

TABLE 10

Resistance of Refractory Metals to Liquid Metals[35]

Material	Max. temp. (°C) in contact with			
	W	Mo	Ta	Graphite
Li	1620	1400	1000	900
Na	900	1030	1000	700
K	Stab.	1200	1000	600
Cs	1400	870		500
Be		Unstab.		2000
Mg	600	1000	1150	Stab.
Re.E.		1100	1400	1000
Al	680	700	Unstab.	500
Ga	800	400	450	Stab.
Ti	1600	1600	1000	1000
Fe	Unstab.	Unstab.	Unstab.	1000
Ni	Unstab.	Unstab.	Unstab.	1200
Co	Unstab.	Unstab.	Unstab.	1200
Cu	Stab.	1300	Stab.	Stab.
Au	Stab.	Stab.	1100	Stab.
Zn	750	400	500	900
Cd		—	500	Stab.
Hg	600	700	600	Stab.
Sn	700	500	—	2200
Pb	328	1200	1000	2000
Bi	980	1400	900	600
U		Unstab.	1400	1800
UO_2	3000			1800

Re. E.: Rear earth metals.

2.4 Oven Design Criteria

There are no universal designs of evaporators for ion sources, but there are several considerations to be respected before designing an oven for an ion source:

1. Space available for the oven? Requested operation time before recharging and by that volume of the evaporator?

2. Temperature range needed for evaporation of the desired elements? Are pure elements needed or are compounds or chemical synthesis allowed (is there an analyzing system or is the ion beam directly used from the source)?

3. Maximum power dissipation allowed to other ion source parts? Are cooling shields necessary?

4. Is a high-current feed-through possible for the heating power or does a higher voltage have to be used?

5. Is a precise temperature measurement and control necessary or is the control of the power flow to the oven sufficient?

In the following, the above points are discussed in more detail.

1. The possible shape and position of the oven is mainly defined by the ion source design. Sources in separate housings tend to have less space available than sources in bigger vacuum vessels. For example, for internal cyclotron ion sources space on axis is limited by the first orbit. Duoplasmatron, Nielsen or hollow-cathode sources (see Chapter 2, Sections 2 and 3) have small discharge chambers surrounded by a magnetic coil. Freeman sources and calutrons have plenty of space besides the plasma chamber inside the vacuum chamber.

The best position for an oven would be close to the discharge chamber, in a vertical position to avoid the outflow of liquid material, allowing easy access for maintenance and change of material, and it should be large enough for long-term runs.

2. The temperature necessary for a specific element is given in Table 1, or for compounds or chemical synthesis, Tables 2 and 3. One can define four temperature ranges for different oven designs.

   a. Up to 100°C, one can use external evaporators and heated gas pipes to run the ion source as with a gas. Usually, the power dissipation from the discharge and cathode is enough to maintain a sufficient temperature inside the ion source. Possibly, the cooling of the discharge chamber has to be reduced or it needs a hot liner. Because the vapor pressure is high, outgassing from other parts play no dominant role.

   b. Between 100 and 600°C, alloying rarely occurs and standard materials can be used (Cu, stainless steel) and also insulators are not critical (even glass fiber or Kapton for cable insulation can be used). Even low-power ion sources can reach this temperature for the discharge chamber without additional heating. Heat transfer from the discharge to the oven, however, may cause problems of controlling the oven temperature. To overcome this problem, sufficient thermal insulation of the oven from the discharge chamber has to be provided. The biggest problem in this temperature range is cleaning of the evaporation material by outgassing at low temperatures. The temperature can be controlled with iron constantan thermocouples or resistance thermometers.

   c. Between 500 and 1200°C, radiation cooling gets more and more predominant and careful design of oven and heat shields is necessary. Also, alloying becomes more severe and the right choice of materials is essential for a reliable design. High-temperature metals and insulators presented in Sections 2.1 and 2.2 have to be used for oven and ion source parts. The temperature can still be controlled with thermocouples or resistance thermometers (Pt 100), but photometers and pyrometers can also be used, especially above 800°C. Nearly all elements can be handled in this temperature range as pure element or as compounds (chemical synthesis).

   d. Above 1200°C, the design gets more delicate and specialized for a few pure elements. Up to ~1700°C, conventional resistance heaters and insulators can still be used and directly heated metal or carbon heaters to ~2000°C. Above 1500°C, electron bombardment heating of the sample is a very efficient way to reach the necessary temperatures. Extremely careful design of heat shields and temperature control of the whole ion source and the extraction system are important at this high-temperature level. The temperature has to be controlled by photometers and pyrometers. Losses by heat conduction and unshielded radiation have to be minimum to allow a slightly homogeneous temperature distribution in the oven.

3. In general, all ion source parts exposed to the vapor should be at the same temperature. In the case of high-temperature evaporators, the radiation losses to the environment can be too high and additional cooling of sensitive parts or a cooled shield around the evaporator can become necessary. Also, current feed-throughs may need additional cooling.

4. Directly heated crucibles or carbon heaters can be used for higher temperatures if high-current feed-throughs are available at the ion source. In that way no insulating carrier for the heating wire is necessary. High currents can create disturbing magnetic fields, the influence of which have to be taken into account.

5. In the case of mass analysis of the ion beam, it is possible to control the oven temperature by monitoring the ion current and using the heating power for orientation. In this way one avoids problems with the sensitive electronics at high voltage and problems with metalized windows in the case of optical temperature measurements. Direct temperature control, however, is always advantageous for precise operation of the evaporator of an ion source.

2.5 Examples of Oven Design

The best design is always the simplest that can fulfill the specifications.

- Turned parts are the easiest to produce and are therefore mainly used for oven design. They have also the advantage of smaller surfaces, which means smaller power losses. Cylinders have the advantage over balls of easy application of heating wires, especially if the oven cylinder is made of insulating material, where the metal wire can be directly wound on.

- Heat shields are made of thin foils of reflecting metals and should have several layers for hotter ovens. They can be mounted on a special frame or just wound around the oven cylinder. In the later case, the foil has to be stamped by a point matrix or just structured with a rough wedge grip of a vice. The mechanical and electrical connections of the oven should have low heat conductance to keep losses small.

- The different thermal expansions of the materials used for an oven have to be taken into account. For example, groves in insulators for heating wires have to be larger than the wire diameter; mechanical attachments of oven parts or the oven itself have to have large tolerances or have to be flexible to allow expansion.

- Connections between metal parts tend to bake together at higher temperatures. Bayonet joints are better to handle than screwed parts, and if screws are unavoidable, they should not be too fine and should have large tolerances. Also, the selection of the right material combinations is essential for screwed joints.

Figure 4 shows an oven designed for a Freeman source. It is made with two, easily machined, boron nitride parts with a tungsten or tantalum wire heater around it. The wire is wound in a spiral groove in the BN body and fixed in position with two rings. The rings also serve as the electrical connection. The oven is screwed directly into the source discharge chamber, or for low temperatures, connected through a short pipe. Reactive gas can be fed into the oven through the rear end of the oven where a thermocouple is also installed. This oven can be used up to about 1200°C with pure substances or with compounds and chemical synthesis. For temperatures above 500°C, heat shields are used to reduce power consumption. The material to be evaporated can be placed directly inside the oven or in a separate boat. This oven needs about 100 W to reach 1000°C with proper heat shields.

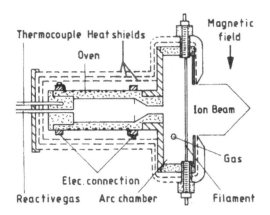

FIGURE 4
Boron nitride oven for up to 1200°C.[16,18]

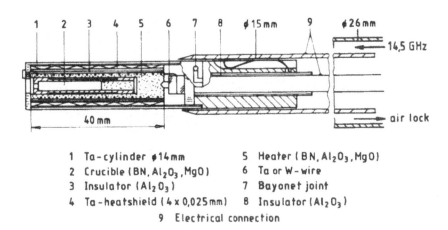

1 Ta-cylinder ⌀14mm   5 Heater (BN, Al₂O₃, MgO)
2 Crucible (BN, Al₂O₃, MgO)   6 Ta or W-wire
3 Insulator (Al₂O₃)   7 Bayonet joint
4 Ta-heatshield (4 x 0,025mm)   8 Insulator (Al₂O₃)
                  9 Electrical connection

FIGURE 5
Minioven for ECR ion source for up to 1500°C.[40]

Figure 5 shows the design of an oven with a small diameter used for an ECR ion source that has limited access to the discharge on axis. Similar ovens are used in Nielsen ion sources (see Chapter 2, Section 2).[29] It is similar in design to the previous one but has an outer diameter of just 14 mm and is mounted on a long coaxial pipe, which allows air- or water-cooling of the support and removal of the oven, without breaking the vacuum, through an air lock for recharging or replacement. The electrical connection of the oven is also through these two pipes. The material of the heater can be boron nitride, alumina, or magnesia, depending on the element to be vaporized and the temperature necessary. The crucible can be made of suitable refractory metal, ceramic, or graphite. The power needed to reach 1500°C is just 80 W. The temperature of the oven is controlled by a pyrometer through the extraction hole of the ECR ion source, but the installation of a thermocouple with connection through the pipe is also possible. This oven can deliver iron or nickel for ion source operation of more than a week without exchange. A similar oven with just 5-mm outer and 1-mm inner diameter was built by Sortais (see Chapter 2, Figure 9.12).[39]

Figure 6 shows an example of a directly heated oven used at the ECR at LBL.[41] It consists of two tantalum rods, one partly machined to a cylindrical crucible and the other to a lid. Both rods are fixed in solid copper bars for mechanical and electrical

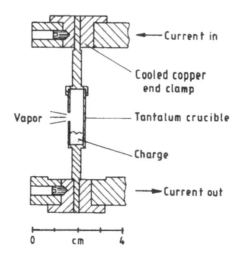

FIGURE 6
Directly heated compact oven for up to 2000°C.[41]

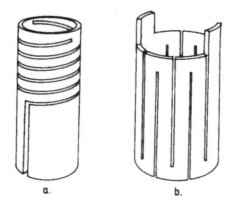

FIGURE 7
Possible shape of graphite heaters: a, bifilar structure; b, meander structure.

connection. At high temperatures, heat shields can be applied to reduce the power flow to the environment. The oven is heated with 340 A and 2.2 V or 700 W at 2000°C. The same design is possible using other refractory metals or graphite. In the case of graphite, the limited mechanical stability could cause problems in this design and the use of a separate graphite heater seems to be more reliable. Graphite heaters are available in various shapes, two of which are shown in Figure 7, and can also be easily prepared from tubes.

Figure 8 shows a high-temperature oven for a high-current source to produce chromium ions.[42] It consists of an oven made of carbon fiber material surrounded by carbon fiber filaments and a multilayer tantalum heat shield. Inside the oven is the crucible cylinder, which is open at the top and which reaches to the discharge chamber through additional heat shields. The crucible is made of tantalum for chromium evaporation. For other elements, a suitable material has to be chosen. The temperature is measured with an optical pyrometer. Insulators for the filaments can be away from the hottest part of the oven, and, therefore, made from usual ceramic material such as alumina or BN.

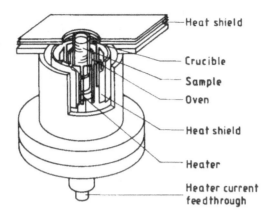

— Heat shield

— Crucible

— Sample

— Oven

— Heat shield

— Heater

Heater current
feedthrough

FIGURE 8
Radiative heated oven for up to 1800°C.[42]

If this design is changed in a way that an additional voltage could be applied between filament and oven, electron beam heating could be used in this arrangement.

Figure 9 shows an electron bombardment heated evaporator that can temperatures up to 2500°C and can be used to evaporate even uranium.[5] This evaporator is incorporated into an electron bombardment ion source. The discharge chamber consists of a 25-μm tungsten cylinder of 15-mm diameter. This cylinder is heated by a direct current flow of 250 A at 10 V to about 2500°C. Inside the cylinder is a tungsten crucible with the material to be evaporated and a perforated half-cylinder connected to it that serves as anode of the source. Electrons from the cathode cylinder heat the crucible and the anode and some of the electrons pass through the holes in the anode and ionize the evaporated atoms, which can be extracted through a slit in the cathode cylinder. For lower temperatures, the ion source can also be used with pure radiative heating of the crucible by reversing the polarity of the discharge and using the outer cylinder as anode and the inner half-cylinder as cathode.

An other ion source using electron bombardment for evaporation is the one by Wilbure and Wei[43] described in Chapter 2, Section 2.3.5. Figure 10 shows a cross-sectional view of the source. Electrons emitted by the cathode can be accelerated toward the crucible, which is then heated to up to about 2500°C. Because of the vertical arrangement of the source, the crucible can have the shape of an open flat bowl, which gives a large evaporation surface.

3 TEMPERATURE CONTROL

To monitor the temperature of an evaporator, one can use direct temperature measurement with thermocouples or resistance thermometers or optical methods with pyrometers or electronic photometers. Thermocouples can be used for lower temperatures up to 2300°C, up to 1300°C with the more common Ni-CrNi ones; optical pyrometers can be used above 600°C and infrared thermometers from –50 to 3000°C, depending on the sensor material.

Thermocouples can be made by preparing a wire pair of the selected material combination or one can use manufactured metal shielded items. The latter have the advantage of being much less affected by noise and high-voltage breakdowns on the small (μV) signal. They are also safe from additional contact potential with other materials, which can falsify the measurement with open wire pairs. There are special

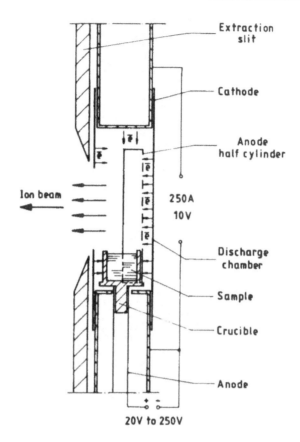

FIGURE 9
Electron bombardment heated oven for up to 2500°C.[5]

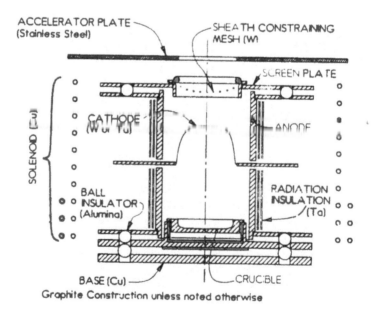

FIGURE 10
Ion source of Wilbur and Wei with electron bombardment heated crucible for up to 2500°C. (From Wilber, P. J. and Wei, R., *Rev. Sci. Instr.*, 63, 2491, 1992. With permission.)

TABLE 11

Data of Thermocouples and Resistance Thermometers[29,44-47]

Material combination	Code	Sensitivity (μV/°C)	Range (°C)
Fe-CuNi	L (J)	50–70	–200–900
Cu-CuNi	U (T)	20–70	–270–400
Ni-CrNi	K	40–35	–270–1375
Chromel-Alumel	N	40–41	–270–1300
PtRh-Pt 10%	S	6–12	–50–1765
PtRh-Pt 13%	R	6–14	–50–1765
PtRh-Pt 18%	B	3.3–11	0–1820
W-WRe 26%	W	10–18.5	0–2300
W3%Re-W26%Re		12–18.5	0–2300
W5%Re-W26%Re		12–18.5	0–2300
Pallaplat		19–52	0–1300
Platinel II			0–1300
Pt 100	Pt	0.44–0.29 Ω/°C	–210–850

extension wires for the different types of thermocouples on the market. To avoid false signals these extension wires have the same thermoelectric forces as the thermocouple; the contacts between thermocouple and extension wire, however, have to be at the same temperature and so does the input to the temperature meter. If no special extension wire is available, copper wires at a stable temperature can be used. Vacuum feed-throughs should be gold-plated to avoid oxide layers at the connection.

Because thermocouples and resistance thermometers are positioned on the high voltage of the ion source, their amplifying electronics have to be at the same potential and have to be designed to resist high-voltage breakdowns. Table 11 gives the thermoelectric potential and the maximum temperature of various thermocouples and resistance thermometers. Iron-constantan (Fe-CuNi), Ni-CrNi, Pt-RhPt, and Pt 100 are the most commonly used.[29]

Pyrometers use the light emitted from the hot object to determine the temperature. This temperature is correct for blackbody radiation. Metal surfaces have an emission coefficient <100% and the measured graybody temperature has to be corrected to get the real blackbody temperature. The correction factor depends not only on the material, but also on the surface condition, which gives in reality an uncertainty for the correction factor. Table 12 gives emission coefficients for refractory metals and ceramics.

The traditional pyrometer compares the emission of a hot tungsten wire with the color of the object, at higher temperatures through a red 0.65-μm filter, to calculate the temperature. Modern electronic pyrometers can be calibrated by the emissivity and window absorption factors and give direct readings of the true temperature of the object. The measurement is done on a small spot that can be just 1 mm in diameter. Other modern pyrometers use two wavelengths (e.g., 0.45 and 0.65 μm) to measure the color temperature of the object. Using the quotient signal of the two measurements, the emissivity and also the absorption in a metallized window are eliminated as long as they are independent of the wavelength. Another option is infrared thermometers using photodiodes or transistors to measure the temperature in the 1- to 20-μm range.[29] In most cases of application to monitor the oven temperature, the absolute value of the measurement is of minor importance compared to the reproducibility of a certain temperature measurement and long-term stability for automatic temperature control of the oven.

TABLE 12

Emission Coefficients of Various Materials at 6650 Å[13]

Material	Temp. (°C)	Emissivity	Material	Temp. (°C)	Emissivity
Ag	940	0.05	Ni	1000	0.37
Al	—	0.18–0.28	NiO	1000	0.89
Al_2O_3	1000	0.10–0.15		1450	0.69
	2000	0.22–0.38	Nb	1730	0.37
Au	1000	0.15	Nb_2O_5	—	0.71
Be	1200	0.5–0.6	Pd	—	0.35
BeO	1470	0.37	Porcelain	—	0.25–0.5
C, smooth	1250	0.75	Pt	1000	0.3
C, rough	1250	0.9	Rh	—	0.24
CaO	—	0.1–0.4	Ta	930	0.47
Co	—	0.36		1730	0.41
Cr	—	0.39	TaC	20–2700	0.55
Cr_2O_3	1200	0.8	Th	800	0.37
Cu	930	0.096	ThO_2	—	0.57
CuO	—	0.7	Ti	880	0.59
Fe	1050	0.38–0.44		1130	0.69
Fe_2O_3	—	0.8	TiO_2	—	0.52
Ir	1750	0.3	W	800	0.45
MgO	1200	0.07–0.15		2200	0.42
	2000	0.13–0.46	Zr	<1230	0.48
Mo	730	0.39		>1230	0.43
	1730	0.35	ZrO_2	—	0.06–0.09

In the case of feedback control of the temperature, the control electronics should be at ground potential, which is naturally the case for optical measurements. In the case of thermocouples, the preamplified signal should be transferred to ground potential by fiber optics to avoid interference with high-voltage breakdowns.

4   EXAMPLES OF METAL ION SOURCES

A few examples of metal ion sources are presented in the following section.

The cold or hot reflex discharge ion source (CHORDIS)[48,49] (see Chapter 2, Section 6) is a modular ion source system, designed for producing high-current beams of all elements in the 10- to 100-mA range. The ion source system comprises four different versions:

1. A cold version for noncondensing gases. The other source versions are built up by adding a few extra components to the basic gas version.

2. A hot-running version for vapors and compound gases with condensable components. In this version, a hot-running liner inside the anode, heated by the discharge plasma, prevents condensation at the cooled anode.

3. A version with internal oven for materials with at least 10 Pa vapor pressure at 1000°C (Figure 11). The oven version has a hot pipe to guide the vapor from the oven through the cathode stem into the discharge chamber. Along with the hot liner, the extraction aperture plate is running hot because limited heat conduction to the flange reduces condensation at the outlet electrode. Ion currents of some 10 mA have been measured for Li, Ca, J, Cs, and Bi with the oven version of CHORDIS.[48,50]

4. A version with a sputter target inside the discharge chamber (Figure 12). The sputter electrode in front of the extraction aperture is insulated from the anode and can be biased to potentials up to –1 kV with respect to the cathode. Pulsed or dc operation

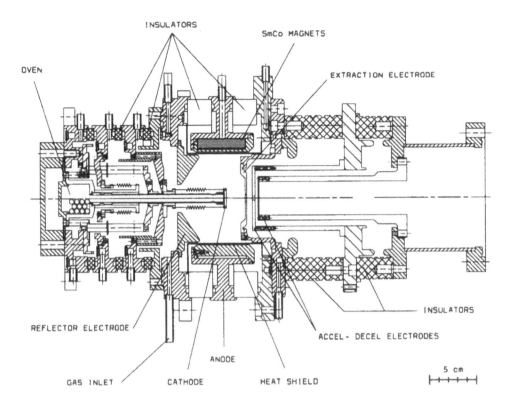

CHORDIS ION SOURCE

FIGURE 11
CHORDIS oven version.

of the sputter electrode is possible, but some power restrictions have to be respected to prevent melting of the sputter electrode. Improved cooling of the electrode extends these limits.[51] Ion currents of a few milliamps have been achieved with the sputter version of CHORDIS for various elements such as Al, Ti, Fe, Cr, Cu, Mo, Ta, W, and rare earth metals.[49,50,52] The sputter electrode is 5 to 8 cm in diameter and 3 to 5 mm thick. It is sheetlike or the material is melted into Cu, Mo, or Ta plates, the shape of which can be seen in Figure 3.

The MEVVA ion source (see Chapter 2, Section 12) has become a powerful metal ion source in ion implantation even though the duty cycle is usually limited to a few percent. The MEVVA ion source is also well suited for synchrotron accelerators, which are naturally low-duty cycle machines.[6,53,54]

The MEVVA can quite often be easily incorporated into existing ion sources by replacing the cathode by a MEVVA cathode with trigger delivering the metal plasma.[54] Depending on the source design, the anode chamber can be used as anode for the MEVVA discharge or as plasma expansion area. Figure 13 shows a combination of a multicathode MEVVA with the CHORDIS ion source as example.

The Freeman ion source (see Chapter 2, Section 4), which has been widely used for ion implantation for almost 30 years, can also deliver high-current beams. Figure 4 shows the Freeman ion source with oven.

This source was optimized for use with corrosive gases and with chlorination. The gas is fed into the source through the oven, which is close to the discharge

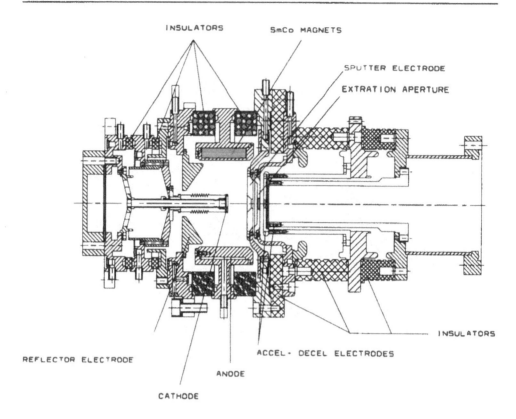

FIGURE 12
CHORDIS sputter version.

chamber. The massive cathode (2-mm-diameter tungsten) enables 20 to 40 h operation with corrosive gases. Source and oven materials are tantalum, boron nitride, and graphite. Ion currents of up to 30 mA have been achieved for B^+ or As^+ out of an extraction slit 90×5 mm.[16,53] Instead of the oven, a sputter electrode can be introduced to the discharge chamber and adjusted by a movable support.[16]

The Penning ion source with radial extraction (PIG; see Chapter 2, Section 5), used at cyclotrons and linear accelerators, produces milliamp beams and also highly charged ions. In particular, the sputter version of the PIG source (Figure 14) can deliver a wide variety of metal ions with high intensity.[50,55] The beam quality of the PIG source, however, is poor compared to the other sources discussed here.

A multicusp, high-current ion source (see Chapter 2, Section 6) for the evaporation of Al, Cr, and Si with temperatures up to 1800°C was developed by Nissin Electric Co. (shown in Figure 15; the oven is shown in Figure 8). A careful arrangement of heat shields and hot extraction electrodes enables prolonged stable operation of the source. Typical ion yields are 50 to 100 mA with extraction voltages below 20 kV.[57] The total heating power is up to 13 kW.

In rf ion sources (see Chapter 2, Section 7), internal antennas are quite often used to couple the rf power into the plasma. Usually, this antenna is coated with some insulating material to protect it from erosion by sputtering. But to get metal ions one can use an uncoated antenna of the specific metal and run the discharge in a mode that leads to high erosion rates. Figure 16 shows the LBL rf ion source, which was developed in this direction by Leung.[25] Ion currents of several milliamps of Cu^+ have been reported and about 20% of the total extracted current was from the antenna material. Other metals can be easily produced by coating the antenna with the respective material.

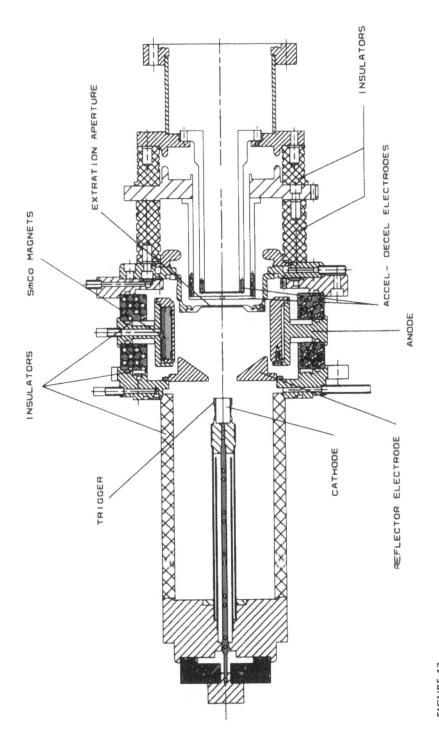

FIGURE 13
MEVVA-CHORDIS combination.

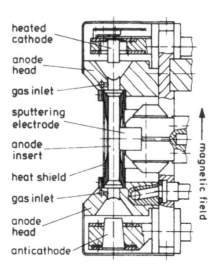

FIGURE 14
Penning sputter ion source.

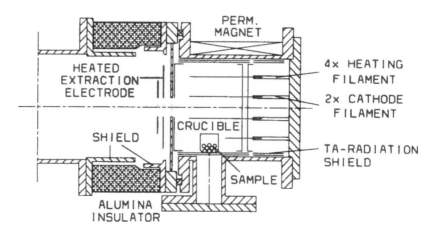

FIGURE 15
Nissin high-temperature ion source.[57]

Microwave ion sources are more frequently used to produce high-current ion beams, most of them using a frequency of 2.4 GHz because of the relatively cheap and reliable generators available (see Chapter 2, Section 8). Using metal ions needs special care to protect the microwave window. The Hitachi slit extraction source developed by Sakudo[58] is used for several semiconductor doping materials like B^+, Al^+, P^+, As^+, and Sb^+, and also Ti^+ and Hf^+ with currents up to 15 mA (Figure 17). Again, reactive gases are used with beneficial effect on the lifetime of the boron nitride plasma chamber.

The hollow-cathode ion source (see Chapter 2, Section 3.3 and Figure 18) uses the cathode filament to heat a crucible that can be moved into the hollow cathode.[59] The temperature is controlled by the position of the crucible and can reach 2000°C.

The production of high charge state metal ions in ECR ion sources is described in Chapter 2, Section 9.2.6 in detail. In addition to evaporation by an oven (see Figures 5 and 6), evaporation from a probe heated by the source plasma is a possibility for

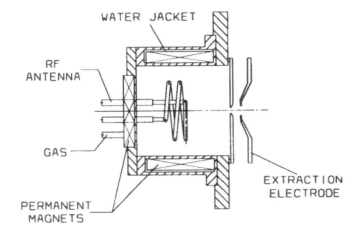

FIGURE 16
LBL rf ion source. (Courtesy of K. N-Leung, Lawrence Berkeley Laboratory, Berkeley, CA.)

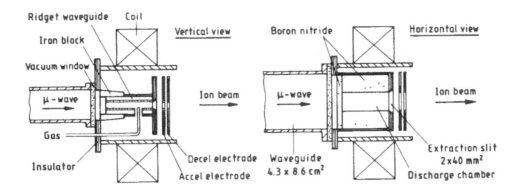

FIGURE 17
Microwave ion source of Sakudo.[58]

ECR ion sources. If this probe (Figure 19, position 2) is biased to a negative potential of 0.1 to 3 kV, metals can be produced using the sputter effect.[60] Figure 19 shows the possible probe positions and the position of an oven for comparison.[61]

REFERENCES

1. J. H. Freeman and G. Sidenius, *Nucl. Instrum. Methods*, 107, 477, 1973; G. Dearnaly, J. H. Freeman, R. S. Nelson, and J. Stephen, *Ion Implantation*, North Holland, Amsterdam, 1973, 292 ff.
2. R. G. Wilson and G. Brewer, *Ion Beams with Applications to Ion Implantation*, R. Krieger, Malabar, FL, 1979.
3. G. Alton, *Nucl. Instrum. Methods*, 189, 15, 1981.
4. B. H. Wolf and R. Keller, in *Ion Implantation into Metals*, V. Ashworth, W. A. Grant, and R. P. M. Procter, Eds., Pergamon Press, Oxford, 1982, 302.
5. M. Bouriant, M. Boge, et al., *Proc. Int. Conf. Ion Sources*, Saclay, 1969, 329; Vienna, 1972, 763; *Proc. Int. Conf. Electromagnetic Isotope Separators and the Technique of their Applications*, Marburg, 1970, 381.

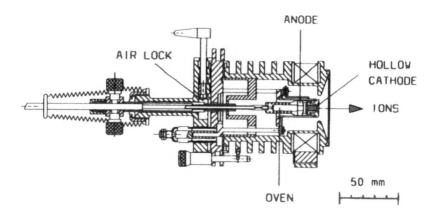

FIGURE 18
Hollow-cathode ion source. (Courtesy of Danfysik A/S, Jyllinge, Denmark.)

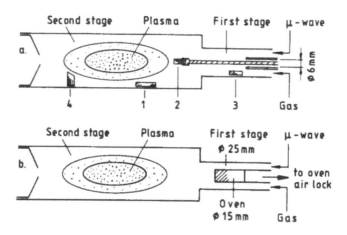

FIGURE 19
Possible sample positions for plasma evaporation in ECR ion sources (a) and position of an oven (b).[61]

6. I. G. Brown et al. *Rev. Sci. Instrum.*, 57, 1069, 1986; *Rev. Sci. Instrum.*, 63, 2351, 1992; Ref. 5, p. 331.
7. R. H. Hughes and R. J. Anderson, in Ref. 5, p. 299.
8. T. R. Sherwood, *Rev. Sci. Instrum.*, 63, 2789, 1992.
9. R. E. H. Honig and D. A. Kramer, *RCA Rev.*, 30, 285, 1969; Ref. 5, p. 427 ff; Ref. 13, p. 880 ff.
10. G. Carter and I. S. Colligon, *Ion Bombardment of Solids*, Heinemann, London, 1968, 38.
11. G. Ranc, *La Technique des Couches Minces et son Application a la Microscopie Electronique*, CEA report number 1117, CEA, Saclay, France, 1959.
12. M. v. Ardenne, *Tabellen zur Angewandten Physik I*, VEB Deutscher Verlag der Wissenschaften, Berlin, 1956 and 1962.
13. W. Espe, *Werkstoffe der Hochvacuumtechnik*, VEB Deutscher Verlag der Wissenschaften, Berlin, 1959.
14. S. Dushman, *Scientific Foundations of Vacuum Technique*, John Wiley & Sons, New York, 1962, 691 ff.
15. N. I. Sax, *Dangerous Properties of Industrial Materials*, Van Northstrand Reinhold, New York, 1975.
16. D. J. Chivers, *Rev. Sci. Instrum.*, 63, 2501, 1992.
17. Company catalogs.
18. H. Ferber, private communication, 1994.

19. G. K. Wehner, *Phys. Rev.*, 102, 690, 1956; *Phys. Rev.*, 108, 35, 1957; *Phys. Rev.*, 112, 1120, 1958; *J. Appl. Phys.*, 30, 1762, 1959; *J. Appl. Phys.*, 32, 887, 1961.
20. N. Matsunami et al., Rep. IPPJ-AM-32, Inst. Plasma Physics, Nagoya University, 1983; *Radiat. Eff. Lett.*, 50, 39, 1980.
21. K. Leible and B. H. Wolf, *Proc. Int. Conf. Low Energy Ion Beams*, Inst. Phys. Conf. Series No. 38, 1978, 96.
22. R. Behrisch, Ed., *Sputtering by Particle Bombardment I*, Springer, Berlin, 1981.
23. M. Müller et al., Rep. GSI-82-11, GSI, Darmstadt, 1982.
24. G. Alton, Y. Mori, et al., *Rev. Sci. Instrum.*, 61, 372, 1990; *Rev. Sci. Instrum.*, 63, 2357, 1992.
25. K. N. Leung et al., *Nucl. Instrum. Methods*, B74, 291, 1993.
26. I. G. Brown, in Ref. 5, p. 331.
27. I. G. Brown et al., *Rev. Sci. Instrum.*, 63, 2417, 1992.
28. C. J. Smithells, Metals Reference Book, Vol. III, Butterworths, London, 1967, 737 ff.
29. I. E. Campbell and E. M. Sherwood, *High Temperature Materials and Technology*, John Wiley, New York, 1967.
30. W. H. Kohl, *Handbook of Materials and Techniques for Vacuum Devices*, Reinhold, New York, 1967.
31. Y. S. Touloukian, *Thermophysical Properties of High Temperature Solid Materials*, MacMillan, New York, 1967.
32. R. Kieffer, G. Jangg, and P. Ettmayer, *Sondermatalle*, Springer, New York, 1971.
33. R. Kieffer and F. Benesovsky, *Hartstoffe*, Springer, Vienna, 1963.
34. E. Meckelburg, *Seltene Metalle*, Technische Rundschau TR 111, Hallwag, Bern-Stuttgart, 1974.
35. Technical Information Plansee Nb, Mo, Ta, W Metallwerke Plansee, A-6600 Reutte, Austria.
36. Technical Information Frialith, Friedrichsfeld, P.O. Box 710261, D-68222 Mannheim, Germany.
37. Technical Information ESK Kempten, Elektroschmelzwerk Kempten, Herzog-Wilhelm-Strasse 16, D-80331 München, Germany.
38. K. O. Nielsen, *Nucl. Instrum. Methods*, 1, 289, 1957.
39. M. P. Bourgarel et al., *Rev. Sci. Instrum.*, 63, 2854, 1992.
40. J. Bossler et al., to be published.
41. D. Clark and C. M. Lyneis, *Proc. Int. Conf. ECR Ion Sources*, Grenoble 1988, *J. Phys. Colloq.*, C1 50, C1–707, 1989.
42. T. Yamashita et al., *Rev. Sci. Instrum.*, 65, 1269, 1994.
43. P. J. Wilbur and R. Wei, *Rev. Sci. Instrum.*, 63, 2491, 1992.
44. Technical Information Degussa, Degussa, Bereich Messtechnik, Leipziger Strasse 10, D 6345 Hanau, Germany.
45. Technical Information Heraeus, W. C. Heraeus, Dep. Temperaturmesstechnik, D 63405 Hanau, Germany.
46. F. Kohlrausch, *Praktische Physik 1*, Teubner, Stuttgart, 1968.
47. Technical information of various companies.
48. R. Keller et al., *Vacuum*, 36, 833, 1986; *Vacuum*, 34, 32, 1984.
49. D. M. Rück et al., *Vacuum*, 39, 1191, 1989.
50. B. H. Wolf et al., *Rev. Sci. Instrum.*, 61, 406, 1990.
51. F. Geiger, private communication.
52. B. Torp et al., *Rev. Sci. Instrum.*, 61, 595, 1990.
53. I. G. Brown, Ed., *The Physics and Technology of Ion Sources*, John Wiley & Sons, New York, 1989.
54. B. H. Wolf et al., *Rev. Sci. Instrum.*, 65, Oct., 1994; Preprint GSI-92–71, GSI, Darmstadt, 1992.
55. V. B. Kutner et al., *Rev. Sci. Instrum.*, 61, 487, 1990; *Rev. Sci. Instrum.*, 65, 1039, 1994.
56. R. P. Vahrenkamp and R. L. Seliger, *IEEE Trans. Nucl. Sci.*, NS-26, 3101, 1979.
57. Y. Inouchi et al., *Rev. Sci. Instrum.*, 61, 538, 1990; *Rev. Sci. Instrum.*, 63, 2478, 1992; Rev. Sci. Instrum., 63, 2481, 1992.
58. N. Sakudo et al., *Rev. Sci. Instrum.*, 63, 2444, 1992; Ref. 5, p. 229.
59. G. Sidenius, *Proc. Int. Conf. Electromagnetic Isotope Separators and the Technique of Their Applications*, Marburg, 1970, 426.
60. R. Harkewitz et al., *Rev. Sci. Instrum.*; Report PHY-7894-Hi-94, Argonne Nat. Lab., Argonne, IL, 1994.
61. R. Geller et al., *Rev. Sci. Instrum.*, 63, 2795, 1992.

Chapter 4

BEAM FORMATION AND TRANSPORT

P. Spädtke

CONTENTS

1 BEAM FORMATION

The extracted current from a charged particle source is limited either by the emission capability or by space charge forces. In the latter case the maximum current density j can be calculated by the Child-Langmuir equation:[1,2]

$$j\left[A/m^2\right] = \frac{4 \cdot \varepsilon_0}{9} \sqrt{\frac{2 \cdot q}{m}} \cdot \frac{\phi^{\frac{3}{2}}}{d^2} \tag{1}$$

with permittivity $\varepsilon_0 = 8.85434 \times 10^{-12}$[F/m], electric charge $q = q^* \times 1.602 \times 10^{-19}$[As], q^* is the charge state, atomic mass $m = u \times 1.6724 \times 10^{-24}$[g], u is the mass in atomic mass units, ϕ[V] is the potential drop across the gap d[m].

In more practical units, this equation can be rewritten:

$$j[mA/cm^2] = 1.72 \cdot \sqrt{\frac{q^*}{u}} \cdot \frac{\phi[kV]^{\frac{3}{2}}}{d[mm]^2} \tag{2}$$

However, for beam transport, not only the extracted current, but also the beam quality is important. The beam quality is measured by the emittance, which describes the particle distribution δ in phase space $\delta(x, \dot{x}, y, \dot{y}, z, \dot{z})$. x, y, z are the spatial coordinates, with z in the direction of the beam, and $\dot{x}$, $\dot{y}$, $\dot{z}$ are the Cartesian velocity components. The velocity component perpendicular to the longitudinal direction is often written as an angle: $x' = \dot{x}/\dot{z}$, $y' = \dot{y}/\dot{z}$. The longitudinal phase space is normally given by the energy W and energy spread ΔW, or by momentum p and momentum spread Δp. The definition of the phase space quantity is given in Reference 3. The longitudinal (ε_L) and transverse emittances (ε_x, ε_y) are given separately. Further examples of applications of the emittance concept are given in Reference 4.

The size of the emittance can be given as

- Real area   physical quantity
- RMS (root mean square)   statistical quantity
- Ellipse enclosing the distribution   practical quantity

In all cases the fraction of the enclosed beam must be specified. The meaning of the RMS quantity is described in more detail in Reference 5.

According to the law of Liouville, the six-dimensional volume stays constant if conservative forces only are applied. For a complete explanation, see Reference 6. Examples for the case where this rule does not apply are

- Collisions of beam particles with residual gas
- Cooling of ion beams by electron cooling or stochastical cooling

A measure of the density of the four-dimensional phase space density is the brightness D of a beam.

$$B[A/(m \cdot rad)^2] = \frac{I[A]}{\pi^2 \cdot \varepsilon_x[m \cdot rad] \cdot \varepsilon_y[m \cdot rad]} \tag{3}$$

In words, the brightness is a measure of the current density per unit solid angle. Techniques for measurement of the beam emittance are described in Chapter 5, Section 4.

When the beam is accelerated, the transverse phase space will shrink with the velocity (see Section 2). The normalized emittance is therefore:

$$\varepsilon_{n,x} = \varepsilon_x \cdot \beta^* \tag{4}$$

$$\varepsilon_{n,y} = \varepsilon_y \cdot \beta^* \tag{5}$$

1.1 Extraction from Fixed Emitters

For fixed emitters, the particle starting coordinates are given by the geometry of the emitter itself. The emitance of the beam is determined by the emitting area, its shape, and the temperature of the emitted particles. The longitudinal emitance is given by the temperature of the particles and the stability of the extraction voltage. Therefore, it is negligible in most cases, especially for extraction voltage above several kilovolts.

1.2 Extraction from Plasma Sources

In plasma sources the ions are generated in a discharge chamber. From that point of generation they drift until a fraction of them reaches the extraction region. The saturation value of ion current density, j_s, which can be extracted from a plasma is given by:[7,8]

$$j_s\left[A/m^2\right] = n_i \cdot q \cdot \sqrt{\frac{kT_e}{m}} \tag{6}$$

with $k = 8.616 \times 10^{-5}[eV/K]$, which is the Boltzmann constant, T_e is the electron temperature, and n_i is the ion density in the plasma. The emitance of the extracted beam is given by the plasma parameter (such as temperature), existing magnetic fields, and aberrations in the extraction. As for extraction from a fixed emitter, the longitudinal emitance is negligible as long as the particle temperature and the stability of the extraction voltage is small compared to the extraction voltage.

In contrast to sources with fixed emitters (see Chapter 2, Sections 15 and 17), the emitting surface (the plasma boundary) of all other ion sources described in Chapter 2 is not fixed. There is only one potential for a given geometry and plasma conditions that will create a beam with desired properties. This is shown in Figure 1. This can also be formulated as: for a given geometry and extraction field strength, the particle density must be matched to generate a beam with desired properties.

Some useful quantities have been defined to describe the beam quality, to compare different ion sources, and to describe scaling laws. The Poissance Π[9,10] has been defined as an electrical quantity:

$$\Pi = \frac{I}{\phi^{\frac{3}{2}}} \tag{7}$$

with I the extracted current, whereas the perveance P_c[9,10] has been defined as a geometric quantity:

$$P_c = \frac{4 \cdot \varepsilon_0}{9} \sqrt{\frac{2 \cdot q}{m}} \cdot \frac{A}{d^2} \tag{8}$$

with A the emitting area.

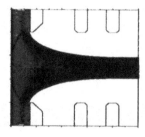 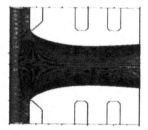 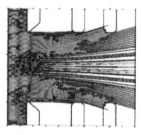

FIGURE 1

Influence of the plasma density on the beam properties. Left: with low plasma density the beam is overfocused; center: the matched condition, where the beam is almost parallel, and right: with too high plasma density.

The minimum theoretical divergence angle ω_0 of the beam for planar extraction systems (slit) has been estimated as:[9,10]

$$\omega_0 = 1.41 \cdot \frac{a}{d} \cdot \left| 1 - \frac{1.47 \cdot \Pi}{P_c} \right| \tag{9}$$

with a[m] the half-width of the slit and d[m] the gap width. For round apertures with radius r a similar expression can be estimated:

$$\omega_0 = 0.5 \cdot \frac{r}{d} \cdot \left| 1 - \frac{1.67 \cdot \Pi}{P_c} \right| \tag{10}$$

According to these equations round apertures have smaller divergence angles; however, zero divergence cannot be achieved because of:

- Thermal energy spread in the plasma
- Aberrations within the extractor

For the cylindrically symmetric case the maximum current can be estimated from References 11 and 12 and the assumption of a certain aspect ratio (aperture radius to electrode separation). A good aspect ratio is on the order of $S \sim 0.5$[9]

$$I[mA] = 0.703 \cdot \sqrt{\frac{q^*}{u}} \cdot \phi[kV]^{\frac{3}{2}} \tag{11}$$

The voltage breakdown limit determines the necessary gap width. The empirically determined limit (valid for clean, flat surfaces) is

$$d[mm] \geq 1.41 \times 10^{-2} \cdot \phi[kV]^{\frac{3}{2}} \tag{12}$$

The emittance of the extracted beam can be roughly estimated as:

$$\varepsilon_{x,y} \approx \omega_0 \cdot r_{opt} \tag{13}$$

if the emittance has a waist at that location. ω_0 is the maximum divergence angle. Therefore, the expected emittance can be rewritten as:

$$\varepsilon_{x,y} \geq 0.7 \times 10^{-2} \cdot \omega_0 \cdot \phi \left[kV \right]^{\frac{3}{2}} \tag{14}$$

The upper limit of the brightness can then be given as:

$$B \leq 1.5 \times 10^3 \cdot \frac{\sqrt{\dfrac{q^*}{u}}}{\omega_0^2 \cdot \phi^{\frac{3}{2}}} \tag{15}$$

This estimate neglects several effects, for example:

- Aberrations created in the extraction system
- The presence of different charge states
- The presence of magnetic fields
- Collisions of the beam particles
- Charge-exchange effects

In general all, $\phi^3/_2/d^2$ scaling laws are applicable only locally. If, for example, the geometry lowers the field close to the emitting surface, the above estimates are not valid anymore.

Whereas the maximum extractable current increases with $\phi^3/_2$, the brightness decreases with $\phi^3/_2$.

Axial Diode System

This simple extraction scheme can be applied if space charge compensation within the extracted ion beam is not important. Two electrodes determine the extraction system: an outlet electrode and a ground electrode. An example is shown in Figure 2.

Axial Triode System

The extraction system consists of three electrodes: an outlet electrode, a screening electrode, and a ground electrode. The following are reasons for using a triode extraction system:

- Preserving the space charge compensation of the extracted ion beam (accel-decel system)
- Having the possibility to change the extraction field strength without changing the beam energy[13,14]

Multigap Extraction Systems

Extraction systems with more electrodes such as tetrodes[15] or pentodes[16] have been used for special purposes. Possible applications are special beam-forming

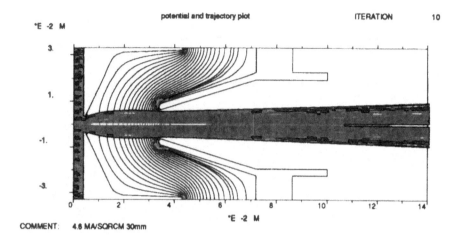

COMMENT: 4.8 MA/SQRCM 30mm

FIGURE 2
Trajectory plot of extraction from an electron cyclotron resonance source. Argon charge states from 2 to 8, 12.5-kV extraction voltage, 1.7 mA total current.

electrodes to improve the beam quality, post-acceleration to higher voltages, or intermediate electrodes in H⁻ extraction systems to separate the electrons from the negative ions.

Multiaperture Extraction System

If the emitting area is too small to deliver enough current, it cannot be increased above a certain limit (due to the optimum aspect ratio) without reduction in beam quality. One possible solution to this problem is to increase the number of extraction holes. In practice, up to several thousand holes have been realized. The emitting area for the emittance estimation must be replaced by the geometric area of the multiaperture system in such a case. The highest achievable transparency can be estimated as ≈60%.

Mesh Extraction System

Advantages of mesh extraction systems are

- Mesh extraction systems provide a stable, well-defined emitting surface
- Emission control capability[17]
- High transparency

The disadvantage is the limited lifetime of the mesh due to sputtering. A mesh in front of the extraction system can also be used to minimize the noise on the beam.[18,19]

Postacceleration

If the extraction voltage is not high enough for the desired beam energy, postacceleration is necessary. This requires that the charged particle source, including all power supplies, be installed on a high-voltage platform, from which the beam can be postaccelerated.

Mass and charge state separation can be done on the high-voltage platform. This increases the power requirements of the terminal (for the bending magnet); on the other hand, it decreases the current requirements for the high-voltage power supply and reduces the space charge problem.

The postacceleration can be done with a multi- or a single-gap acceleration column. In both cases the focusing strength of the gap should be matched to the beam emittance, the rigidity of the beam, and the defocusing strength of the space charge.

Multigap structures normally have low focusing power, best suited for lower ion currents. For high beam current, stronger focusing single-gap devices are used. These devices require good matching because the focusing strength has to compensate the space charge force exactly.

A negative screening electrode can be used behind the acceleration system to preserve the space charge compensation of the accelerated beam.

1.3 Extraction in Axial Magnetic Fields

According to References 20 and 21 the minimum emittance can be estimated as:

$$\varepsilon_{x,y} = 1.60 \times 10^{-5} \cdot \frac{q}{m} \cdot |\vec{B}| \cdot r^2 \cdot \pi [\text{mm} \cdot \text{mrad}] \tag{16}$$

If Equation 12 and the optimum aspect ratio is applied, this can be rewritten as:

$$\varepsilon_{x,y} = 6.44 \times 10^{-9} \, I^2 \cdot |\vec{B}| \cdot \pi [\text{mm} \cdot \text{mrad}] \tag{17}$$

Here $\vec{B}$ is the magnetic flux density in gauss and r the aperture radius in millimeters. Different charge states have to be taken into account separately, since the ions will be influenced by the magnetic field. This is because the beam envelope might be different for different charge states, and the space charge contribution of different charge states will decrease during acceleration differently. An example of such a case is the extraction from an ECR ion source (see Figures 2 and 3).

1.4 Extraction in Transverse Magnetic Fields

For the case when the magnetic field is perpendicular to the extraction system, charge separation takes place already in the extraction. An example of such a situation is the radial extraction from a Penning ion source (see Figure 4).

1.5 Extraction of Negative-Ion Beams

If negative ions are extracted from a plasma source, the electrons are extracted as well. To avoid a huge electron current mixed in with the ion beam, magnetic filters can be applied that bend the electrons on small radii, but not the ions, onto electrodes that are close to the source potential, to keep power losses as low as possible. Such an electrode system is shown in Figure 5.

1.6 Electrode Alignment

If the extraction electrodes are not perfectly aligned, the beam will be steered. For round apertures with a displacement of Δ[mm] of two electrodes with a spacing of d[mm], this steering angle Θ can be estimated to first order by:[9]

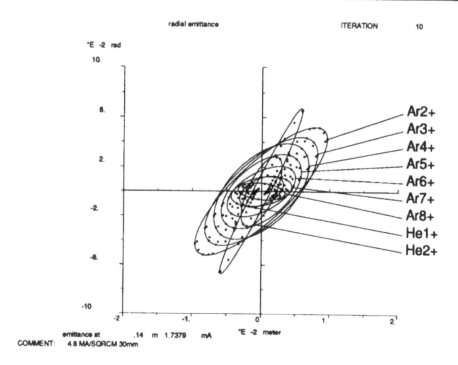

FIGURE 3

Emittance plot of the extracted beam from an ECR ion source. The drawn emittance figures are for the different charge states and different masses (argon and helium).

$$\Theta = \frac{\Delta}{3 \cdot d} \tag{18}$$

For slit apertures the steering angle can be estimated to first order by:

$$\Theta = \frac{2 \cdot \Delta}{3 \cdot d} \tag{19}$$

(see also Section 2.7). For high-current applications computer simulations are recommended for the determination of the correct steering angle (see Chapter 7).

2   BEAM TRANSPORT

The velocity of the particle beam is described as a fraction of the speed of light:

$$\beta^* = \frac{v}{c} \tag{20}$$

If β^* is greater than about 0.5, relativistic effects must be taken into account.

$$m = \frac{m_0}{\sqrt{1 - \beta^{*2}}} \tag{21}$$

KOBRA3-INP VERSION 3.31

3D representation

COMMENT: slit extraction

FIGURE 4

2D projection of a 3D simulation of the extraction from a Penning source; argon charge states from 2 to 6, 10-kV extraction voltage, 10 mA total current. Magnetic flux density in perpendicular direction is 0.3 [T]. Horizontal scale is 10 cm.

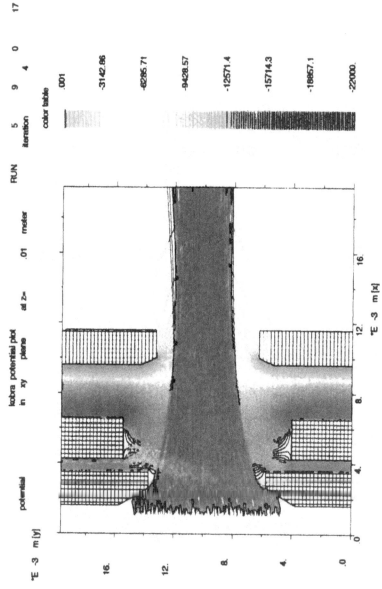

COMMENT: negative ion source extraction

FIGURE 5

2D projection of a 3D simulation of the extraction of a H- beam from a plasma source. The plasma electrode is at –22 kV, the intermediate electrode is at –20 kV, and the third electrode is at ground potential. Perpendicular to the drawing a magnetic field of 0.12 T flux density is applied to bend the electrons out of the beam. The ion beam is slightly deflected as well. To minimize the ion beam deflection, the magnetic field is normally restricted to the region of the electron collector electrode.

The particle motion in electric and magnetic fields is given by:

$$\frac{\partial(mv)}{\partial t} = q \cdot \left(\vec{E} + \vec{v} \times \vec{B} \right) \tag{22}$$

If the field along the path of the charged particles is known, this path can be found by integration. This method is exact but in most cases is only possible using computer codes (see Chapter 7).

If the optical behavior of beam line elements is known, the transport can be described by matrices. This is a much easier method of description than integration of the above equation; however, it neglects higher-order effects. Therefore, integration of the path is recommended if higher accuracy is desired.

The action of a 2×2 matrix for the transverse coordinates can be described as:

$$\begin{pmatrix} \text{size to size} & \text{angle to size} \\ \text{size to angle} & \text{angle to angle} \end{pmatrix} \tag{23}$$

2.1 Drift Space

The matrix that transforms one charged particle with perpendicular angle coordinates x_0', y_0' from point $(x, y, z)_0$ with the velocity $\dot{z}$ to $(x, y, z)_1$ which is $d[m]$ downstream is given by:

$$\begin{pmatrix} 1 & d & 0 & 0 & 0 & 0 \\ 0 & 1 & 0 & 0 & 0 & 0 \\ 0 & 0 & 1 & d & 0 & 0 \\ 0 & 0 & 0 & 1 & 0 & 0 \\ 0 & 0 & 0 & 0 & 1 & d \\ 0 & 0 & 0 & 0 & 0 & 1 \end{pmatrix} \begin{pmatrix} x \\ x' \\ y \\ y' \\ z \\ \dot{z} \end{pmatrix}_0 = \begin{pmatrix} x \\ x' \\ y \\ y' \\ z \\ \dot{z} \end{pmatrix}_1 \tag{24}$$

2.2 Electrostatic Focusing

Acceleration

With the input energy $q^* \cdot U_0$ and the output energy $q^* \cdot U$, the matrix describing the acceleration can be written:

$$\begin{pmatrix} 1 & \dfrac{2 \cdot d}{\sqrt{U/U_0} + 1} & 0 & 0 \\ 0 & \dfrac{1}{\sqrt{U/U_0} + 1} & 0 & 0 \\ 0 & 0 & 1 & \dfrac{2 \cdot d}{\sqrt{U/U_0} + 1} \\ 0 & 0 & 0 & \dfrac{1}{\sqrt{U/U_0} + 1} \end{pmatrix} \begin{pmatrix} x \\ x' \\ y \\ y' \end{pmatrix}_0 = \begin{pmatrix} x \\ x' \\ y \\ y' \end{pmatrix}_1 \tag{25}$$

The longitudinal velocity $\dot{z}$ depends on the total potential drop only:

$$\dot{z} = \sqrt{\frac{2qU}{m}} \tag{26}$$

Einzel Lens

The matrix for a lens that focuses one charged particle with focal length f is given by:

$$\begin{pmatrix} 1 & 0 & 0 & 0 & 0 & 0 \\ -\dfrac{1}{f} & 1 & 0 & 0 & 0 & 0 \\ 0 & 0 & 1 & 0 & 0 & 0 \\ 0 & 0 & -\dfrac{1}{f} & 1 & 0 & 0 \\ 0 & 0 & 0 & 0 & 1 & d \\ 0 & 0 & 0 & 0 & 0 & 1 \end{pmatrix} \begin{pmatrix} x \\ x' \\ y \\ y' \\ z \\ \dot{z} \end{pmatrix}_0 = \begin{pmatrix} x \\ x' \\ y \\ y' \\ z \\ \dot{z} \end{pmatrix}_1 \tag{27}$$

Quadrupoles

$$\begin{pmatrix} \cos\phi & \dfrac{d}{\phi}\sin\phi & 0 & 0 & 0 & 0 \\ -\dfrac{\phi}{d}\sin\phi & \cos\phi & 0 & 0 & 0 & 0 \\ 0 & 0 & \cosh\phi & \dfrac{d}{\phi}\sinh\phi & 0 & 0 \\ 0 & 0 & \dfrac{\phi}{d}\sinh\phi & \cosh\phi & 0 & 0 \\ 0 & 0 & 0 & 0 & 1 & d \\ 0 & 0 & 0 & 0 & 0 & 1 \end{pmatrix} \begin{pmatrix} x \\ x' \\ y \\ y' \\ z \\ \dot{z} \end{pmatrix}_0 = \begin{pmatrix} x \\ x' \\ y \\ y' \\ z \\ \dot{z} \end{pmatrix}_1 \tag{28}$$

with the aperture a of the quadrupole, the applied voltage V_a, and the beam energy q·U:

$$\phi = \frac{d}{a} \cdot \sqrt{\frac{V_a}{U}} \tag{29}$$

2.3 Magnetic Focusing

The bending radius ρ of a charged particle that has been accelerated through a voltage U within a magnetic field is given by:

$$\frac{1}{\rho} = \sqrt{\frac{q}{2m}} \frac{B}{\sqrt{U\left(1 + \dfrac{q}{2m_0c^2}U\right)}} \tag{30}$$

This is a specific quantity for magnetic focusing or deflection in the bending plane.

Solenoid

This transformation was derived in Reference 22. With the abbreviations $c = \cos$ $d/2\rho$ and $s = \sin d/2\rho$ the transfer matrix reads:

$$
\begin{pmatrix}
c^2 & 2\rho sc & sc & 2\rho s^2 c & 0 & 0 \\
-2\rho sc & c^2 & -2\rho s^2 & sc & 0 & 0 \\
-sc & -2\rho s^2 & c^2 & 2\rho sc & 0 & 0 \\
2\rho s^2 & -sc & -2\rho sc & c^2 & 0 & 0 \\
0 & 0 & 0 & 0 & 1 & d \\
0 & 0 & 0 & 0 & 0 & 1
\end{pmatrix}
\begin{pmatrix} x \\ x' \\ y \\ y' \\ z \\ \dot{z} \end{pmatrix}_0
=
\begin{pmatrix} x \\ x' \\ y \\ y' \\ z \\ \dot{z} \end{pmatrix}_1
\tag{31}
$$

Rotating the transverse coordinates about the z-axis by a specific angle, decouples the x and y first-order terms:

$$
\begin{pmatrix}
c & 2\rho s & 0 & 0 & 0 & 0 \\
-2\rho s & c & 0 & 0 & 0 & 0 \\
0 & 0 & c & 2\rho s & 0 & 0 \\
0 & 0 & -2\rho s & c & 0 & 0 \\
0 & 0 & 0 & 0 & 1 & d \\
0 & 0 & 0 & 0 & 0 & 1
\end{pmatrix}
\begin{pmatrix} x \\ x' \\ y \\ y' \\ z \\ \dot{z} \end{pmatrix}_0
=
\begin{pmatrix} x \\ x' \\ y \\ y' \\ z \\ \dot{z} \end{pmatrix}_{1, \text{ roatated}}
\tag{32}
$$

Dipole

The total transfer matrix for an ideal sector magnet is given by:

$$
\begin{pmatrix}
\cos\phi & \rho\sin\phi & 0 & 0 & 0 & \rho(1-\cos\phi) \\
-\dfrac{1}{\rho}\sin\phi & \cos\phi & 0 & 0 & 0 & \sin\phi \\
0 & 0 & 1 & \phi\cdot\rho & 0 & 0 \\
0 & 0 & 0 & 1 & 0 & 0 \\
\sin\phi & \rho(1-\cos\phi) & 0 & 0 & 1 & \rho(\phi-\cos\phi) \\
0 & 0 & 0 & 0 & 0 & 1
\end{pmatrix}
\begin{pmatrix} x \\ x' \\ y \\ y' \\ z \\ \dot{z} \end{pmatrix}_0
=
\begin{pmatrix} x \\ x' \\ y \\ y' \\ z \\ \dot{z} \end{pmatrix}_1
\tag{33}
$$

No special effects such as profiles of the magnet are assumed in this simple model. A more detailed description is given in Reference 23.

Quadrupoles

Focusing in the x-plane and defocusing in the y-plane by a quadrupole lens is given by:

$$\begin{pmatrix} \cos\phi & \dfrac{d}{\phi}\sin\phi & 0 & 0 & 0 & 0 \\ -\dfrac{\phi}{d}\sin\phi & \cos\phi & 0 & 0 & 0 & 0 \\ 0 & 0 & \cosh\phi & \dfrac{d}{\phi}\sinh\phi & 0 & 0 \\ 0 & 0 & \dfrac{\phi}{d}\sinh\phi & \cosh\phi & 0 & 0 \\ 0 & 0 & 0 & 0 & 1 & d \\ 0 & 0 & 0 & 0 & 0 & 1 \end{pmatrix}\begin{pmatrix} x \\ x' \\ y \\ y' \\ z \\ \dot{z} \end{pmatrix}_0 = \begin{pmatrix} x \\ x' \\ y \\ y' \\ z \\ \dot{z} \end{pmatrix}_1 \qquad (34)$$

with

$$\phi = d\cdot\sqrt{\left|\frac{1}{B\cdot\rho}\cdot\frac{dB_z}{dz}+\frac{1}{\rho^2}\right|} \qquad (35)$$

Quadrupoles are commonly used in beam transport as doublet or triplet lenses. A more detailed description is given in Reference 24.

2.4 Comparison of E and B Deflection

For small deflection angles, electric and magnetic deflections are equivalent if:

$$300\cdot\beta^*\cdot B[T] \approx E[MV/m] \qquad (36)$$

However, space charge compensation is destroyed more easily by electric fields.

2.5 Combinations of E and B Focusing

In some cases it might be useful to combine electrostatic and magnetic forces. Such devices are applicable if space charge compensation is not present or not important.

Wien Filter

This device consists of homogeneous electric and magnetic fields, which are perpendicular to each other. The magnetic force is compensated by the electrostatic force for a specific velocity. It is used as a velocity spectrometer.

Quadrupoles

A voltage can be applied to the insulated magnetic poles of a quadrupole. The principal sections are inclined by 45°. Such a device is described in Reference 25.

2.6 Plasmaoptical Devices*

Plasmaoptical devices are most advantageous for the production, formation, and control of intense ion fluxes that can only exist in the charge–compensated state. The operation of plasmaoptical systems is governed by the principle of equipotentialization of the external magnetic field lines of a compensated ion beam penetrating the plasma medium.

High electron mobility is the main factor causing breakdown of space electric field patterns in a plasma. The electron dynamics within a compensated ion beam can be quite accurately described by the following equation:

$$\vec{E} = -\frac{1}{c}\left[\vec{v} \times \vec{H}\right] - \frac{\nabla p_e}{en_e} \tag{37}$$

where v is the electron directional velocity, H is the external magnetic field strength, and $p_e = n_e k T_e$ is the pressure provided by the electron component.

A steady-state electric field $\vec{E}$ within an intense ion beam plasma can exist due to either the Lorentz force or the electron pressure gradient. Often, however, the electron pressure can be neglected ($T_e = 0$); hence

$$\vec{E} = -\frac{1}{c}\left[\vec{v} \times \vec{H}\right] \tag{38}$$

This means that each of the magnetic field lines is at a constant potential. Furthermore, since the field lines cross the walls, one has the fundamental possibility of controlling the potential of the lines using dedicated electrodes to vary the control electric field within the ion beam.

The plasma lens is the simplest plasmaoptical system; it can be used as an example of the principle of plasmaoptics.

Let us consider a ring of radius r. Once the ring is fed with a current I, it is a thin magnetic lens with focal length:

$$\frac{1}{f_H} = \frac{3\pi^2}{16} \frac{q^2}{mc^2} \frac{I^2}{Erc^2} \tag{39}$$

where E is the beam energy. The ring can be operated as electrostatic lens by applying the potential ϕ_L:

$$\frac{1}{f_E} = \frac{3\pi}{128} \frac{e^2 \phi_L^2}{\left(E^2 R\right)} \tag{40}$$

The total focal length in vacuum, when the space charge is negligibly small, can be derived from the expression:

$$\frac{1}{f_0} = \frac{1}{f_H} + \frac{1}{f_E} \tag{41}$$

* This section was provided by A. A. Goncharov, Institute of Physics of the Academy of Sciences of Ukraine, Kiev, Ukraine.

This situation changes drastically when the lens volume is filled with electrons. If this is the case, the principle of equipotentialization of magnetic field lines takes the lead to cause a rearrangement of the electric field. Under these circumstances, the beam ions are deflected by the electric field in one direction only while passing through the lens, and the lens acquires a high focusing capability, given by:

$$\frac{1}{f_{Pl}} = \frac{2\phi_L}{E} \frac{\theta}{R} \tag{42}$$

where θ is a geometric factor.

High-current plasma lenses can be operated in both focusing and defocusing mode. More theory is described in References 26 to 28.

2.7 Beam Uniformity

The current distribution in charged particle beams depends on the properties of the emitting medium and the beam-forming system, the extraction system.

For multiaperture extraction systems the beam uniformity can be influenced by electrode displacements[29-31] and by electrode shaping.[32]

2.8 Space Charge Effects

The space charge force acts as a diverging force because particles of the same charge repel each other. The potential of an infinitely long and parallel beam is given by:

$$\Phi(r) = \frac{I}{4\pi\varepsilon_0 \beta^* c} \cdot \left(1 + 2 \cdot \ln\frac{r_s}{r} - \frac{r^2}{r_b^2}\right) r \le r_b \tag{43}$$

$$\frac{I}{2\pi\varepsilon_0 \beta^* c} \cdot \ln\frac{r_s}{r} r_b \le r \le r_s \tag{44}$$

The radial potential drop across a homogeneous beam with cylindrical cross section can be estimated as:

$$\Delta\Phi[V] = \frac{30 \cdot I[A]}{\beta^*} \tag{45}$$

For sheet beams the transverse potential drop can be estimated as:

$$\Delta\Phi[V] = \frac{50 \cdot I[A]}{\beta^*} \tag{46}$$

The effect of the space charge force is shown in Figure 6. Further investigations of space charge effects are reported in References 33 to 35.

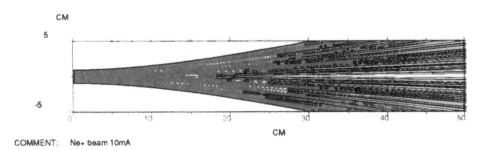

COMMENT: Ne+ beam 10mA

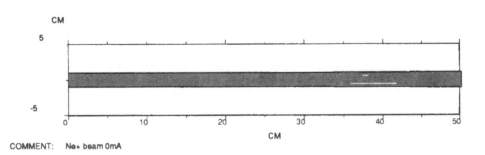

COMMENT: Ne+ beam 0mA

FIGURE 6
Drifting ion beam: Ne+, 15 keV, 10 mA. Top: uncompensated beam; bottom: compensated beam.

An inhomogeneous distribution always tries to change to a distribution that minimizes the internal electric field. This means that optical elements producing nonlinear distributions cause emittance growth. This has been shown in Reference 36.

2.9   Beam Compensation

There are two different kinds of compensation effects: space charge compensation and current compensation.

Space Charge Compensation

Ions of the primary beam i_b collide with residual gas atoms a_r. If the transferred energy is high enough, the residual gas atom will be ionized.

$$i_b + a_r \rightarrow i_b + i_s^+ + e_s^-$$ (47)

The electrons e_s are trapped in the space charge potential of the primary beam, whereas the secondary ion i_s are repelled by this potential. If the potential dip of the primary beam is small compared to the electron temperature, this model will fail. However, in such a case the space charge compensation is close to complete, and the effect of the remaining potential becomes negligible.

Beam transport can be affected if the space charge compensation is not constant in time or if the beam intensity changes with time (noise).[37,38] The effect of insulators or undefined potentials in beam lines, even of insulating layers created by the ion beam itself, has been described in Reference 39. Insulators can influence the space charge compensation. Experimental results are summarized in Reference 40.

A theoretical approach that describes the compensation of an ion beam has been given in Reference 41. In this model the Vlasov equation is solved numerically.

Different approaches have been made to measure the degree of space charge compensation:

- Langmuir probes
- The secondary ion energy distribution
- The deflection of a secondary charged particle beam

Current Compensation

The beam creates its own magnetic field due to the transported current:

$$B_\phi = \frac{\mu_0 \cdot I}{2\pi r} \qquad (48)$$

with the permeability $\mu_0 = 4 \cdot \pi \cdot 10^{-7}$ [Vs/Am], this equation can be rewritten in practical units:

$$B_\phi[T] = 0.2 \times 10^{-4} \frac{I[A]}{r[cm]} \qquad (49)$$

If the focusing force of this magnetic field becomes too large, the ion beam will focus itself, and in the limit can pinch itself off. To avoid this effect a secondary beam of opposite sign would be necessary, or a defocusing lens system (such as the space charge force is) must be present.[42] Applications can be found, for example, in fusion reactors.

REFERENCES

1. C. D. Child, *Phys. Rev. (Ser. 1)*, 32, 492, 1911.
2. I. Langmuir and K. T. Compton, *Rev. Mod. Phys.*, 3, 251, 1931.
3. E. D. Courant and H. S. Snyder, *Ann. Phys.*, 3, 1958.
4. S. Humphries, Jr., *Charged Particle Beams*, John Wiley & Sons, New York, 1990.
5. M. J. Rhee and R. F. Schneider, The root-mean-square emittance of an axisymmetric beam with a maxwellian velocity distribution, *Part. Accel.*, 20, 1986.
6. H. Goldstein, *Classical Mechanics*, Addison-Wesley, Reading, MA, 1950.
7. J. R. Coupland, R. S. Green, D. P. Hammond, A. C. Riviere, *Rev. Sci. Instrum.*, 44, 9, 1973.
8. E. Thompson, *Physica*, 1040, 199, 1981.
9. T. S. Green, *Rep. Prog. Phys.*, 37, 1257, 1974.
10. C. Lejeune, *Advances in Electronics and Electron Physics*, Academic, New York, 1983, Vol. 1, Suppl. 13A.
11. R. Keller, *Nucl. Instrum. Methods*, A298, 247, 1990.
12. R. Keller, Ion extraction, in *The Physics and Technology of Ion Sources*, I. G. Brown, Ed., John Wiley & Sons, New York, 1989.
13. K. Tinschert and W. Zhao, Low energy high-intensity extraction system for Chordis, *Rev. Sci. Instrum.*, 63(4), 2782, 1992.

14. P. Spädtke, I. G. Brown, and P. Fojas, Low energy ion beam extraction and transport, *Rev. Sci. Instrum.*, 1994, to be published.
15. J. H. Whealton, Primary ion tetrode optics for high transparency multibeamlet neutral injectors, *J. Appl. Phys.*, 53 (4), 1982.
16. R. Keller, P. Spädtke, and K. Hofmann, Optimization of a single-aperture extraction system for high-current ion sources, in *Ion Implantation: Equipment and Techniques*, H. Ryssel and H. Glawischnig, Eds., Springer, New York, 1983.
17. R. True, Techniques for the design of guns, focusing systems, and multi-stage depressed collectors, *3rd Workshop EBIS Sources and Their Applications*, 1985.
18. S. Humphries, Jr., C. Burkhart, and L. K. Len, Ion sources for pulsed high-brightness beams, in *The Physics and Technology of Ion Sources*, I. G. Brown, Ed., John Wiley & Sons, New York, 1989.
19. E. Oks, P. Spädtke, H. Emig, and B. H. Wolf, Ion beam noise reduction method for the Mevva ion source, *Rev. Sci. Instrum.*, 1994, to be published.
20. W. Krauss-Vogt, H. Beuscher, H. L. Hagedoorn, J. Reich, and P. Wucherer, *Nucl. Instrum. Methods*, A268, 5, 1988.
21. T. Taylor, High-current dc microwave sources, *Rev. Sci. Instrum.*, 63 (4), 2507, 1992.
22. R. Helm, SLAC-4.
23. H. A. Enge, Deflecting magnets, in *Focusing of Charged Particles*, A. Septier, Ed., Academic Press, 1967.
24. E. Regenstreif, Focusing with quadrupoles, doublets, and triplets, in *Focusing of Charged Particles*, A. Septier, Ed., Academic Press, 1967.
25. F. W. Martin, Electrical rotation of quadrupole lenses, *Nucl. Instrum. Methods*, 189, 1981.
26. D. Gabor, Space Charge Lens for Focusing Ion Beam, Vol. 160, p. 89, 1947.
27. A. I. Morozov, S. V. Lebedev, *Rev. Plasma Phys.*, 247, 1975.
28. A. A. Goncharov and I. M. Protsenko, Formation and transportation of ion beams by plasma-optical systems, *Ukr. Fiz. Zh.*, 36, 11, 1991.
29. J. R. Conrad, Beamlet steering by aperture displacement in ion sources with large acceleration-deceleration ratio, *Rev. Sci Instrum.*, 51(4), 1980.
30. A. J. T. Holmes and E. Thompson, Beam steering in tetrode extraction systems, *Rev. Sci. Instrum.*, 52(2), 1981.
31. S. Tanaka, M. Araki, and Y. Okumura, Application of aperture displacement technique to producing flat beam distributions, *Rev. Sci. Instrum.*, 63(4), 2779, 1992.
32. H. Wituschek, M. Barth, W. Ensinger, G. Frech, D. M. Rück, K. D. Leible, and G. K. Wolf, Alligator — an apparatus for ion beam deposition with a broad-beam ion source, *Rev. Sci. Instrum.*, 63(4), 2411, 1992.
33. L. R. Evans and D. J. Warner, Space Charge Neutralization of Intense Charged Particle Beams, CERN/MPS/LIN 71-2.
34. A. J. T. Holmes, Theoretical and experimental study of space charge in intense ion beams, *Phys. Rev.*, 19, 1, 1979.
35. M. D. Gabovich, Kompensierte Ionenstrahlen, GSI-TR-79/15.
36. O. A. Anderson, Internal dynamics and emittance growth in uniform beams, *Proc. Lin. Acc. Conf.*, Stanford University, 1986.
37. M. V. Nezlin, On the mechanism of space charge fluctuations in quasi-compensated ion beams, *Sov. Phys. Tech. Phys.*, 5, 154, 1960.
38. M. V. Nezlin, Plasma instabilities and the compensation of space charge in an ion beam, *Plasma Phys.*, 10, 1968.
39. J. H. Freeman and W. A. Bell, *Nucl. Instrum. Methods*, 22, 1963.
40. J. E. Osher, The Role of Space Charge in Beam Transport, Low Energy Ion Beams, 1977.
41. R. W. Müller, Ansätze zur numerischen Behandlung des Strahlneutralisationsproblems, GSI internal report, 1985.
42. L. Brillouin, A theorem of Larmor and its importance for electrons in magnetic fields, *Phys. Rev.*, 67, 1945.

Chapter 5

ION BEAM DIAGNOSIS

P. Strehl

CONTENTS

1   THE MOST IMPORTANT ION BEAM PARAMETERS

There are numerous applications of ion beams in the fields of science and technology including industrial production development and applied scientific research. To value and optimize an ion source and to match the beam transport system to the requirements of the application, the following beam parameters are the most relevant ones:

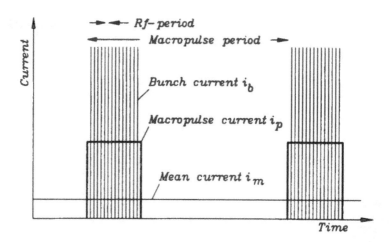

FIGURE 1
Various types of beam currents. (From Strehl, P., *Kerntechnik*, 214-223, 1991. With permission.)

- The current of the desired kind of ions, including also the time dependence of the intensity distribution. In case of a pulsed, rf-accelerated beam one has to distinguish between
    - Macropulse current i_p
    - Bunch current i_b
    - Average (or mean) current i_m, which obviously, corresponds to the dc current in case of dc beams

    Figure 1 illustrates the various types of beam currents one has to deal with.
- Beam profile, which means the intensity distribution in both transverse directions over the beam cross section
- Emittance and brilliance, defined in the transverse phase space
- Beam energy, energy spread, and emittance in the longitudinal phase space

Obviously, there are more parameters for the description of particle beams.[17,89,90,135,162,163] But, in the following we consider the definitions and importance of the beam parameters summarized above, and describe some methods for their measurement.

2 BEAM CURRENT MEASUREMENTS

Precise measurements of beam current may be necessary for the following purposes:

- Accurate measurement of particle flux to an experiment, as, for example, needed for the exact determination of ion implantation doses
- Estimation of transmission through a beam transport system
- Optimization of ion source parameters
- Determination of the ion charge state distribution
- To separate isotopes
- Observation and optimization of time structure in case of pulsed ion sources to minimize plasma oscillations as typical for PIG and other ion sources

The design of a device to measure the beam current depends very much on the beam energy, on the intensity range that has to be covered, on the beam power, and the time structure that has to be monitored. In principle one has to distinguish between destructive and nondestructive measuring techniques. The Faraday cup is the most widely used destructive diagnostic element and the beam current transformer represents a nondestructive device. Both methods allow the absolute determination of beam current. Very sensitive nondestructive measurements can be performed for bunched beams, using resonant tuned capacitive pickups[131] or coaxial resonators,[45] with the shortcoming that the signal output depends on the bunch shape. For low currents, secondary electron monitors (SEMs) or channeltrons may also be used.

2.1 Faraday Cup

A Faraday cup collects the ions of an ion beam in an electrically insulated cup, and the charge is measured using a calibrated resistor in an appropriate electronic circuit. The design of such a cup becomes more complicated if one takes into account the following effects:

- Emission of secondary electrons or other charged particles
- Generation of electrons and ions from the ionization of residual gas
- Leak currents due to the deterioration of insulating material by sputtering or high temperatures
- Formation of galvanic elements due to the use of different materials in case of water-cooled cups
- Leak currents arising from the conductance of the cooling water
- Heating up by high beam power

In the energy range below 2 MeV/u the secondary electron coefficients are of the order of 1 to 10.[1] Early measurements were performed for nitrogen ions having energies up to 300 keV.[2]

Since the flux of secondary electrons varies with $\cos \theta$, where θ is the angle of the secondary electron trajectory to the beam axis, the portion f of electrons passing out of the cup depends on the ratio of cup aperture R to cup length L_{PC}. A simple estimation leads to $f \sim \sin^2 \theta_1/2 = R^2/2(R^2 + L_{PC}^2)$, where $\theta_1 = \arctan R/L_{PC}$.[4] Therefore, it implies $L_{PC} > R$, which will not always be possible. The distortion by secondary electrons can be reduced by a repelling electrical field realized by a dedicated suppressor electrode in front of the cup or by biasing the cup together with the measuring electronics. Figure 2 shows the results of a current measurement in dependence of the suppressor voltage for a 150-keV/u Xe beam. The efficiency of the repelling electrical field can be improved considerably by combining it with a magnetic field that forces the electrons to circular orbits inside the aperture of the cup. By nonrelativistic approximation the bending radius ρ_e is given as:

$$\rho_e = \frac{\sqrt{2m_e T}}{eB} \approx 3.37 \cdot \frac{\sqrt{T[eV]}}{B[mT]} \ [mm] \tag{1}$$

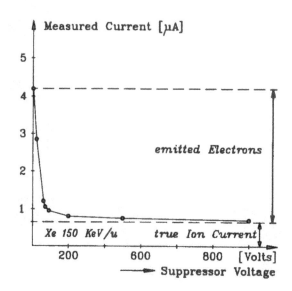

FIGURE 2
Effect of a suppressor on the current measurement.[3]

where T is the kinetic energy of the secondary electrons and B is the magnetic field strength. Using a well-designed yoke in combination with modern permanent magnets (for example cobalt-samarium) a field strength in the order of 50 mT on axis may be realized even for an aperture of 50 mm.[3] Therefore, the radii are in the millimeter range. Obviously, the temperature in the adjacent zone of the magnets has to be limited. Figure 3 shows schematically the layout of an uncooled Faraday cup provided for beam powers on the order of some watts, equipped with an electric suppressor electrode in combination with magnets and having a reasonable R-to-L_{FC} ratio. Figure 4 shows another uncooled tantalum cup, and its mounting onto a standard compressed air actuator[191] for moving the cup into and out of the beam.

In the final design of a cup the effect of sputtering should also be taken into account. To avoid leak currents, ceramic insulators have to be shielded against the deposition of conducting materials. Because the sputtering rate shows maxima at low energies, it may be necessary to estimate also the excavation of material by sputtering. Assuming a sputtering rate of N_s (atoms per incident ion), the amount of material that will be removed by sputtering figures out to be:

$$R_s = 3.6 \times 10^{-1} \cdot \frac{N_s \cdot A}{q \cdot \rho} \cdot \frac{i}{\Delta F} \, [\mu m/h] \qquad (2)$$

with A as the atomic weight and ρ as the density (g/cm^3) of the bombarded material. q is the charge of the ions and $i/\Delta F$ is the electrical current density (mA/cm^2). The sputtering rate N may differ considerably for various materials as shown in Figure 5 for a 45-keV krypton beam. The data given in Figure 5 have been confirmed by measurements[18] for targets of Al, Cu, Mo, and Ta with argon ions between 20 to 35 keV. Table 1 gives the R_s values for some typical stopper materials assuming q = 1 and $i/\Delta F$ = 1 mA/cm^2, taking the sputtering rates N_s from Figure 5.

In the case of higher beam power, the effectiveness of water cooling has to be considered before starting the design of a cooled device. Sometimes cooling will not help since the material will be melted or even vaporized instantaneously, or by only

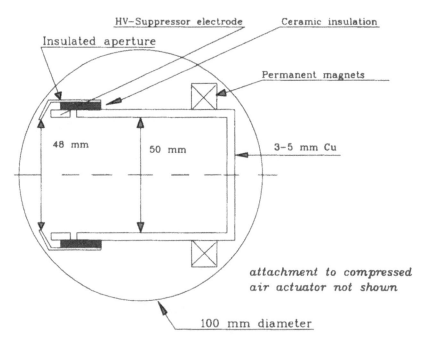

FIGURE 3
Scheme of a uncooled Faraday cup.

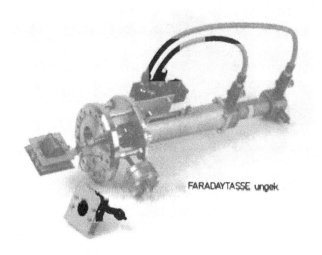

FIGURE 4
Uncooled Faraday cup and compressed air actuator.

a few beam pulses in the case of pulsed operation. This may happen also in the low-energy region because the ranges $\rho\Delta x$ and Δx, respectively, of the particles are very small and, therefore, the beam energy will be deposited in a very small volume, defined by $V = \pi R_0^2 \Delta x$ with the beam radius R_0. The energy needed to melt the material within the penetration range volume is determined by the energy (ΔW_1) to heat up this volume from T_0 to the melting point plus the heat for transition (ΔW_2) from the solid to the liquid phase:

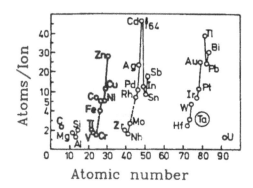

FIGURE 5

Measured sputtering rate for a 45-keV Kr beam. (From Carter, G. and Colligon, J., *Ion Bombardment of Solids*, Heinemann, London, 1969, 310. With permission.)

TABLE 1

R_s Values (μm/h)

Al	Ti	Fe	Cu	Mo	Ta	W
5.0	7.7	10.2	30.9	8.5	11.8	15.4

$$\Delta W_1 = mC_p(T)(T_m - T_0) = \rho V C_p(T)(T_m - T_0) \tag{3}$$

$$\Delta W_2 = W_{tr} \cdot m \tag{4}$$

where m is the mass (g), $C_p(T)$ is the heat capacity (W s/g·K) ρ is the specific weight (g/cm³), V is the volume (mm³), and W_{tr} is given in watt seconds per gram. Table 2 summarizes the relevant properties of mostly used construction materials and the calculated energies $\Delta W = \Delta W_1 + \Delta W_2$ with $T_0 = 20°C$ using interpolated C_p values. Taking energy loss data from the literature[9,11-35,195] the calculation of the range volume as defined above is straightforward. The results are shown in Figure 6 for the ion-target combinations Ne → Cu, Ne → Ta, U → Cu, and U → Ta. With the given energy ΔW for melting from Table 2, the required number of particles to melt the range volume has been calculated, and Figure 7 gives the results for the discussed ion-target combinations.

TABLE 2

Relevant Properties of Some Elements

Element	ρ (g/cm³)	λ (W/mm·K)	T_m (°C)	C_p(25°C) (W s/g·K)	C_p(2000°C) (W s/g·K)	W_{tr} (W s/g)	ΔW (W s/mm³)
Be	1.87	0.450	1278	1.83	3.276	1090	7.42
Al	2.70	0.213	660	0.903	1.09	390	2.67
Ti	4.51	0.020	1660	0.525	0.790	436	6.66
Fe	7.87	0.0803	1535	0.454	0.827	272	9.254
Cu	8.96	0.396	1083	0.386	0.496	210	5.84
Mo	10.22	0.140	2617	0.252	0.374	280	11.68
Ta	16.60	0.054	2996	0.140	0.168	177	10.90
W	19.30	0.180	3410	0.134	0.168	192	14.38

From References 6, 7, and 8.

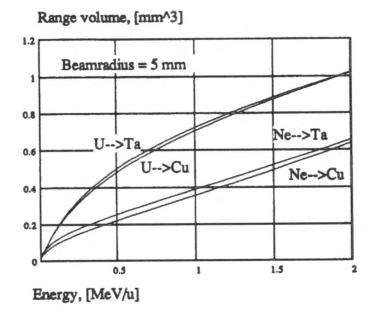

FIGURE 6
Calculated range volume for ion energies up to 2 MeV/u.

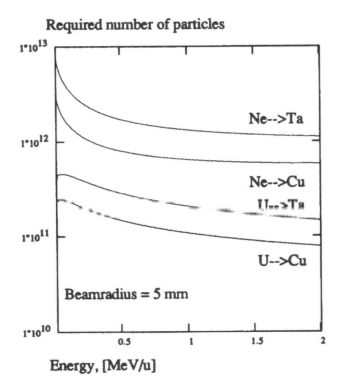

FIGURE 7
Number of particles required to melt the range volume.

Obviously, these numbers hold only if all particles hit the target within a time during which nearly no heat transfer to the material around the range volume takes place. Therefore, some information about the time constant of heat transfer is needed. For that purpose let us consider the partial differential equation of heat:

$$\frac{\partial T}{\partial t} = a^2 \Delta T + f(x, y, z, t) \tag{5}$$

where a is defined by

$$a = \sqrt{\frac{\lambda}{c \cdot \rho}} \tag{6}$$

Because the energy is deposited within a very thin layer at the surface, this equation can be solved numerically taking into account the energy flow $\dot{Q}/F$ from the beam by the relation:

$$\frac{\dot{Q}}{F} = -\nabla T \tag{7}$$

To demonstrate the time dependence of heat transfer, calculations were performed[16] with the following parameters:

Beam radius	5 mm
Beam pulse power	10 kW
Beam pulse length	1 ms
Radius of the target	25 mm
Thickness of the target	0.5 mm
Target materials	Cu and Ta

Figures 8 and 9 show the calculated temperature distributions in the center of the target ($r = 0$) along the z-axis after $t = 1$, $t = 2$, and $t = 5$ ms, fixing the temperature to $T = 0°C$ on the backplane of the target, and assuming complete heat isolation of the target disks mantle surface. As expected from the thermodynamic properties, there is a clear difference of the time constants for the two considered target materials. Similar calculations for thin targets and high-power pulsed beams have been performed.[92,93] The partial differential equation (Equation 5) shows that there must be also a dependence of the time constant on the geometry. Therefore, this has to be taken into account, estimating the time of heat transfer from the range volume. The time of heat transfer can be estimated by considering a solution of the partial differential equation (see, for example, Reference 14)

$$T(z, t) = \frac{1}{2a\sqrt{\pi \cdot t}} \cdot \exp\left(-\frac{z^2}{4a^2 \cdot t}\right) \tag{8}$$

which describes the time dependence of the temperature along one coordinate axis (z). Here the thermodynamic properties of the material are described by a as given above, while the term z^2/t mainly describes the time and geometry dependence. Figure 10 gives the shape of the function for copper and tantalum in dependence on

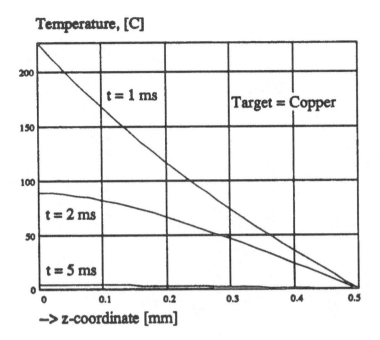

FIGURE 8
Calculated distribution of temperature for the copper target.

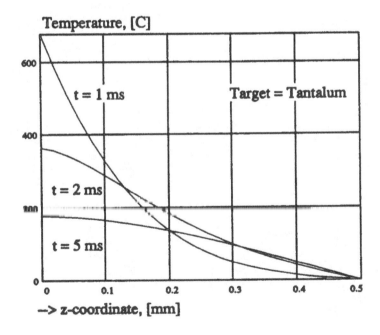

FIGURE 9
Calculated distribution of temperature for the tantalum target.

the parameter t/z^2. Applying the result to the numerical examples for copper and tantalum shows that there is nearly no heat transfer to the cooled end plate ($z = z_{max}$ = 0.5 mm) if $t \ll 1.25$ ms for Cu, and $t \ll 5$ ms for Ta. Assuming the same target thickness in both cases, the difference, of course, is determined by the values of a. Table 3 gives a^2 (mm²/s) at T = 20°C for all materials considered in Table 2. Figure 11

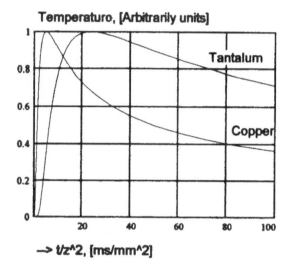

FIGURE 10
Estimation of the time and geometry dependence of heat transfer.

TABLE 3

Values of a^2 (mm²/s)

Be	Al	Ti	Fe	Cu	Mo	Ta	W
94.26	79.17	6.75	15.93	100.22	43.77	21.12	61.76

demonstrates the geometry dependence of heat transfer for two tantalum targets of different thicknesses (L = 1 mm, L = 0.2 mm). In the numerical calculations a beam pulse length of 1 ms was assumed. Both targets are heated up near to the melting point. Keeping in mind the discussed time dependence of heat transfer, the calculated particle numbers to melt the range volume of Figure 7 have been converted in Figure 12 to more relevant units of watt seconds and kilowatt milliseconds. Now, if the estimations lead to the conclusion that cooling of a stopping device makes sense, some important points have to be considered when designing a cooled Faraday cup:

- Selection of the stopping material has to be optimized with respect to the heat conductivity and the maximum allowed temperature, which is related to the melting point of the material.

- Shape and dimensions of the cup have to take into account contradictory demands concerning the material thickness in beam direction: while a low maximum temperature requires a thin layer of material between the stopping range volume and the cooled surface, the effect of film boiling[15] requires the distribution of heat across a large area, which only can be realized if there is enough material around. This has been discussed, e.g., in Reference 17. Figure 13 gives the results from a model calculation with the following parameters:

Beam radius	5 mm
Beam power	3.5 kW, c.w.
Radius of the target	25 mm
Thickness of the target	variable, 5 to 30 mm
Target material	Cu
Cooling	cylinder mantle and back plate T = 0

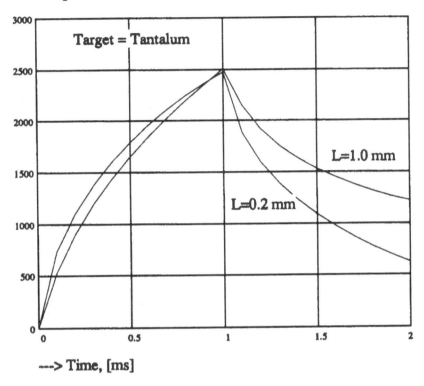

FIGURE 11
Time and geometry dependence of heat transfer for a tantalum target.

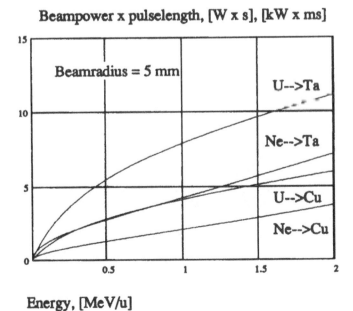

FIGURE 12
Estimation of the required product of beam power × pulse length to melt the range volume.

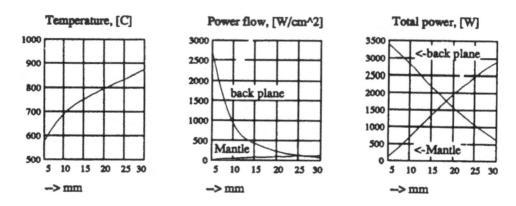

FIGURE 13
Results of the model calculation; see text for parameters.

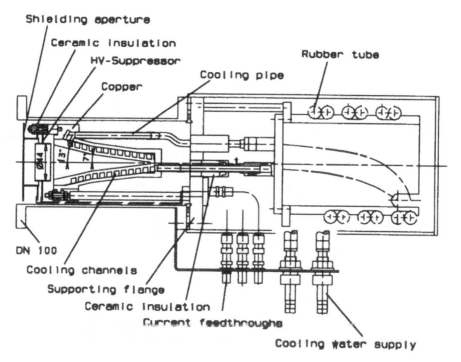

FIGURE 14
High-power Faraday cup.

Figure 14 shows the scheme of a Faraday cup provided for a proton beam of 25 MeV at 25 kW beam power with a maximum power flow of 10 kW/cm². To reduce the power flow on the surface, and at the same time to increase the range volume, the cup has a conical shape. The cup has been designed for laminar flow of the cooling water, and, therefore, a maximum power flow of ≈120 W/cm² was another condition for the design. The maximum power flow can be increased considerably by applying turbulent coolant flow in the tubes, having Reynolds numbers ≥3000.[15,19] Figure 15 shows the head of a cooled tantalum cup, provided for heavy ions up to 20 MeV/u, and a maximum beam power of 6 kW. Figure 16 is a picture of a complete diagnostic unit for beam current measurements, combining a cooled Faraday cup with a compressed

FARADAYTASSE gek.

FIGURE 15
Head of a 6-kW Ta-Faraday cup.

FIGURE 16
Cooled Faraday cup provided for a maximum beam power of 4.5 kW.

air actuator.[192] The cup is designed for a maximum beam power of 4.5 kW (45 MeV protons) with a minimum beam radius of 4 mm.

In some cases there is a need to monitor fast plasma oscillations of ion sources or even an rf bunch structure with a Faraday cup. Because the required band width for signal detection and signal transmission will be on the order of some 100 MHz, the impedance of the stopping device has to be matched to the broadband signal processing system to avoid reflections. The exact dimensions of a 50-Ω broad-band Faraday cup are given in Figure 17 and Figure 18 shows the parts of the 50-Ω head with the dimensions given in Figure 17. The effect of signal broadening due to the longitudinal electrical field of the bunches depends very much on the velocity of the particles and has been considered in References 3 and 135. The signal of a coaxial cup can also be falsified by escaping secondary electrons as demonstrated in the two oscillograms of Figure 19. The picture with the 20-mV vertical scale shows the true bunch signal of positive ions, supplying about -500 V to the suppressor grid in front of the cup's collector plate. Without suppression the measured signal strength (note the 100-m V vertical scale) is enhanced by the escaping electrons, and considerable signal broadening results from the energy spread of the emitted electrons. Figure 20 gives shape and dimensions of a head in 50-Ω geometry, provided for larger beam diameters and beam power up to 10 kW. For the design of stoppers with a given impedance the well-known books on rf techniques[20,94] are useful. In the case of direct water-cooling of the stopper material, there are limitations for the detection of low currents arising from the formation of galvanic elements and caused by the conductance of the cooling water. Assuming a reasonable conductance of the cooling water of 1 μS/cm < κ < 10 μS/cm a bypass of 1 MΩ·cm·l/F > R_w > 0.1 MΩ·cm·l/F will disturb the correct current reading. F and l are cross section and length of the insulation gap bridged by the water. Neglecting bias current and offset voltage of a common operational amplifier, the simplified electronic circuit for a current measurement looks like Figure 21 and the influence of the bypass can be estimated from:

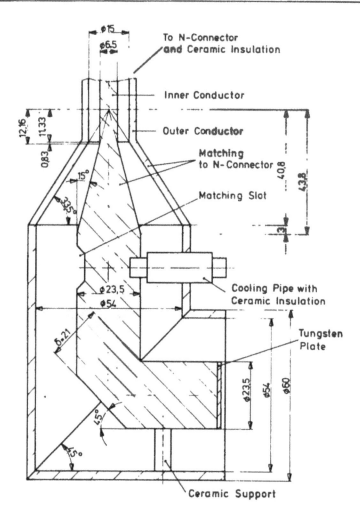

FIGURE 17
Head of a 50-Ω broad-band Faraday cup (dimensions in mm).[191]

$$U_a = \frac{-is \cdot R}{1 - \frac{1}{A}\left(1 + \frac{R}{R_w}\right)} \qquad (9)$$

with is as the beam current that has to be measured. In general, the open loop gain A is complex and $|A|$ will be greater than 100.000. The ratio l/F is on the order of 1/cm. Therefore, the error will be less than 1% even for R = 100 MΩ as usually needed for the measurement of currents in the nanoamp range. On the other hand, experience has shown that currents induced by the formation of galvanic elements are on the order of some nanoamps and, therefore, measurements of such small currents with a cooled Faraday cup should be avoided.

A compromise avoiding the discussed effects is introduced by indirect cooling using appropriate insulation material such as BeO or AlN and taking advantage of their high thermal, but low electrical conductivity. The dependence of λ on the temperature is given in Figure 22 for BeO and AlN and Figure 23 gives the specific resistance of BeO, which is about the same for AlN. For comparison, the corresponding values for copper are λ = 396 W/m⁻¹·K for the thermal conductivity (see Table 2)

COAXIAL FARADAY CUP

FIGURE 18
50-Ω parts of the head of a broadband Faraday cup.[191]

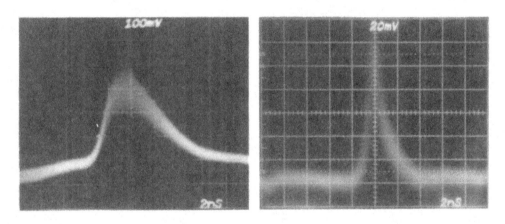

FIGURE 19
Bunch signal of positive ions measured with an 50-Ω broad-band Faraday cup (100-mV scale: without secondary electron suppression, 20-mV scale: −500 V on the grid).[90]

and $\rho = 1.7 \times 10^{-6}$ Ω·cm, neglecting the dependence on the temperature. Figure 24 shows the scheme of an end cup provided for heavy ions of small penetrating range, and a maximum beam power of 1 kW using an AlN insulation.[134] The effect of the shunt impedance arising from the AlN insulation can be estimated by using the simplified diagram of Figure 21 and the relation given by Equation 9 with R_w determined by the AlN plate dimensions. Taking $\rho = 3 \times 10^7$ Ω·cm, about 1 MΩ results for the cup of Figure 24.

2.2 Calorimetric Measurements

The beam current may be determined indirectly by a calorimetric measurement of the beam power and the deposited energy. The use of calorimetry has been discussed by a number of authors.[21-26] Since the method determines the balance of energy, which does not depend on the electrical charge of the particles, a variety of

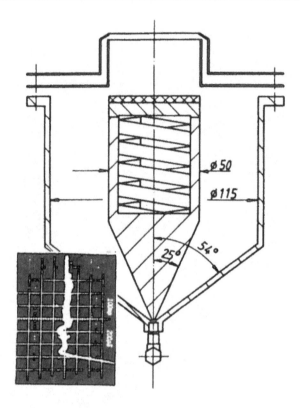

FIGURE 20
50-Ω head of a Faraday cup. The small insert shows the result of a test with a time domaine reflectometer (TDR).

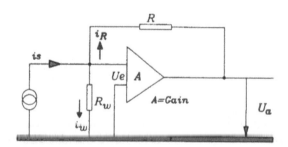

FIGURE 21
Simplified circuit diagram of the current measurement.

other applications like determination of neutral particle flux, radiometric calorimetry, and estimation of dose in case of ion implantation are also possible. The design of a calorimetric system for beam diagnosis depends very much on the beam parameters and the requirements on the response time of the device. In principle there are two viewpoints influencing the design and use of a calorimetric diagnostic device. The first one is the measurement of deposited energy, which is determined by Equation 3 with ΔW_1 as the deposited energy and $T_m - T_0 = \Delta T$ as the resulting rise of temperature. Obviously, in this application the time of temperature change is not very important and may be on the order of some seconds. The aim will be an exact determination of ΔQ by measuring ΔT. Therefore, the calorimetric device has to be thermally isolated by adapting the geometric size as well as the selection of the

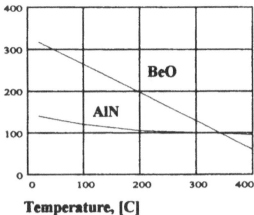

FIGURE 22
Heat conductivity of BeO and AlN.

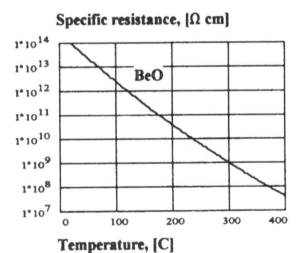

FIGURE 23
Specific resistance of BeO.

material to the energy deposition which has to be detected. Figure 25 illustrates the method. The characteristics of the calorimeter will be determined by the product of $C_p(T) \cdot \rho$ if one compares devices of the same volume. Table 4 gives the $C_p(T) \cdot \rho$ values for the materials listed in Table 2 taking $C_p(25)$.

Because the variation is rather small, the relation between energy deposition and resulting change of temperature is approximately given by:

$$\Delta T[K] \approx \frac{1}{3} \frac{\Delta Q[Ws]}{V[cm^3]} \tag{10}$$

On the other hand, there may be the demand to use a calorimetric device for monitoring and optimizing the continuous particle flow from an ion source or even

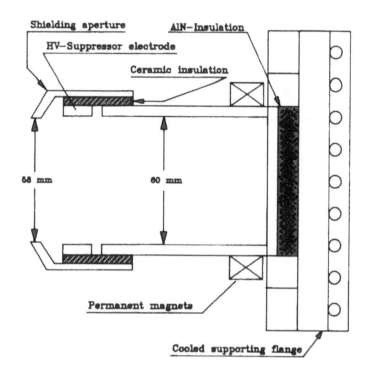

FIGURE 24
Scheme of an indirectly cooled Faraday cup.

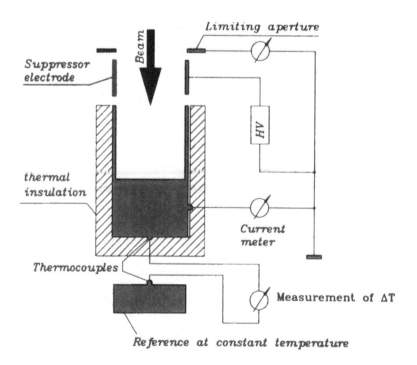

FIGURE 25
Scheme for a calorimetric measurement of a given amount of beam energy.

TABLE 4

Values of $C_p(25) \cdot \rho$ (Ws/K cm^3)

Be	Al	Ti	Fe	Cu	Mo	Ta	W
3.42	2.44	2.37	3.57	3.46	2.58	2.32	2.59

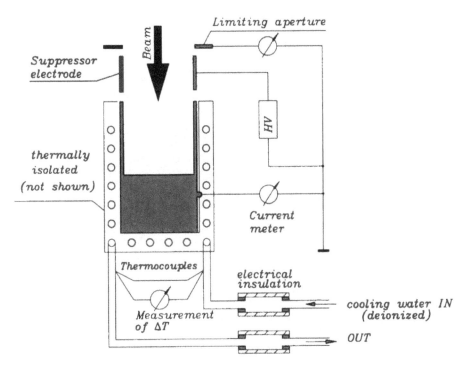

FIGURE 26
Principle of calorimetric beam power measurement.

from an accelerator. In this case the calorimeter has to be cooled and the fluctuations of input power can be detected by changes of the cooling water output temperature. From the operational aspect of this application a response time of about 100 ms seems to be an adequate requirement. Therefore, the design of a calorimeter has to be considered especially from this point of view. Figure 26 shows schematically the principle of continuous beam power measurement using a calorimeter. The following relations have to be taken into account to meet the requirements with respect to response time and thermal characteristics of a calorimetric measuring system:

- From the time derivation of Equation 3, applied to water

$$\dot{W}_1 = P_{beam} = \rho_w \dot{V}_w C_w \Delta T_w \tag{11}$$

with $\rho_w = 1$ g/cm^3 and $C_w = 4.18$ W s/g K (at 293 K), the response time with respect to changes in beam power can be estimated as:

$$\frac{d\Delta T}{dt} = \frac{\Delta T}{P_{beam}} \cdot \frac{dP_{beam}}{dt} = \frac{\Delta T}{i_{beam}} \cdot \frac{di_{beam}}{dt} \tag{12}$$

assuming constant cooling water flow $\dot{V}_w$. Although a large ΔT between cooling water input and output results in a better response time, it should be kept in mind that also the dependence of ρ_w and C_w on the temperature has to be taken into account.

- For a precise measurement of the temperature the coolant flow should be as turbulent as possible, which means one has to optimize the ratio of the velocity v_w and ΔT in the product $A_w \cdot v_w \cdot \Delta T$, with $V_w = A_w \cdot v_w$ and A_w as the cross section of the cooling channels. Therefore, Reynolds number has to be checked for the selected geometry and velocity (Re > 2300 for circular tubes).[5,19]

- The transfer of heat from the stopping material of the calorimeter to the water is given by

$$P_{beam} = \alpha \cdot A \cdot \delta T_{w-c} \tag{13}$$

where A is the cooled surface, and δT_{w-c} is the temperature difference between water and calorimeter. The coefficient of heat transfer α for water is given by[18]

$$\alpha = 350 + 210\sqrt{v_m[m/s]} \ [W/m^2\ K] \tag{14}$$

- The response time of the calorimeter itself is determined by the value of a^2 as given in Table 3, and the geometric dimensions of the calorimeter. A rough estimation of the time t_{trans} that is required to transfer a change of P_{beam} into a change of δT_{w-c} is given by

$$t_{trans} \approx \frac{z^2}{a^2} \tag{15}$$

where z is the material thickness in direction of the heat transfer to the cooled surface.

The relations and conditions given above show very clearly that the design of a calorimetric device for continuous monitoring of ion beams depends on many parameters, which have to be optimized with respect to the beam power and the desired overall response. Otherwise, a well-designed and carefully calibrated calorimeter may be used in combination with an independent current measurement to determine even the energy of ions.[18]

2.3 Beam Current Transformers

The use of transformers for beam current measurement has the following advantages:

- Nondestructive signal extraction and, therefore, most of the problems discussed in Chapter 2, Section 1 will not arise. Because the ion beam will not be distorted by the measurement, the beam current transformer is well qualified for on-line closed-loop feedbacks to control the ion beam intensity (for example, in the case of ion implantation).

- Assuming careful design of the transformer system, the measurement will be nearly independent of beam position and beam size.

- Obviously, the most important advantage is the direct proportionality of the output signal to the beam current, which, in turn, gives a means of absolute calibration by using an external current source.

One basic shortcoming of beam current transformers is the strong dependence of sensitivity and time constant on the time structure of the beam. Because there is a great variety in the characteristics of beams, a broad spectrum of solutions concerning the following is discussed in the literature:[27-30,32-44,46,48,49,53,127,161]

- Mechanical design
- Type of core material, for example, Vitrovac,[51] Ultraperm,[51] Ni-Zn-Ferrites[193]
- Winding schemes, for example, bifilar windings to reduce common mode noise signals
- Electronics following the current transformer
- Applications including presentation of experimental results

Although all information needed for the design of beam current transformers, including sophisticated electronic signal processing, can be found in the references, a summary of the most important relations may be helpful:

- Figure 27 shows a physical model of a current transformer and Figure 28 is a simplified equivalent electrical circuit of the transformer (see Reference 37 for more details). The transfer function of the electrical circuit, according to Figure 28, with s as the Laplace variable is given by

$$U_a = -i_{beam} \cdot \frac{sRL}{N} \cdot \frac{1}{R + sL + R_L + sRC_L(sL + R_L)} \qquad (16)$$

Here L is the transformer inductance, N is the number of secondary windings, R is the load resistance of the system, R_L is the resistance of the secondary circuit, and C_s is the stray capacitance.

- The required bandwidth df results from rise time, time structure, and length of the beam pulse. Considering beam monitoring for ion source development and applications in the discussed low-energy region, the high-frequency limit will be determined in most cases by plasma oscillations of the ion source, which are normally below 1 MHz. The required low-frequency limit is determined by the length of the beam pulse, which includes even dc current.

- The droop and, as a consequence, the drop of the output signal are related to the low-frequency response of the transformer. Neglecting the stray capacitance C_s, and assuming $i_{beam} = i_p$, which corresponds to a step function, the solution of Equation 16 is

$$U_a = -i_p \cdot \frac{R}{N} \cdot \exp\left(-\frac{(R + R_L)}{L} \cdot t\right) \qquad (17)$$

and the time constant of the exponential drop is given by

$$\tau = L/(R + R_L) \approx L/R \qquad (18)$$

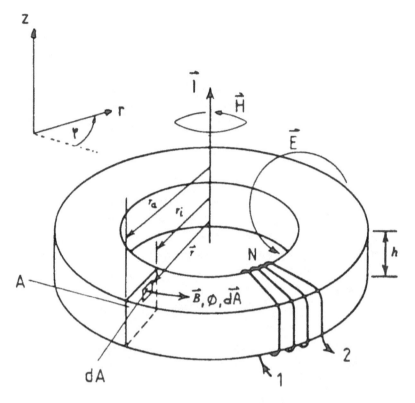

FIGURE 27
Physical model of a current transformer.

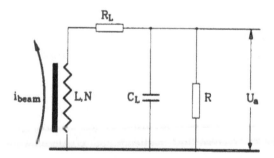

FIGURE 28
Simplified equivalent circuit of the beam current transformer. (From Strehl, P., *Proc. 9th Int. Conf. Cyclotrons and Their Applications*, Caen (France), 1981, 545–554. With permission.)

- The transformer inductance L is given by

$$L = \frac{N^2 \mu_0 \mu_r h}{2\pi} \ln \frac{r_a}{r_i} \tag{19}$$

Here $\mu_0 = 1.2566 \times 10^{-6}$ Vs/Am is the permeability of the vacuum, μ_r is the relative permeability, and the core dimensions are (see Figure 27) core height h, inner core radius r_i, and outer core radius r_a. The effective inductance and, therefore, the time constant τ can be reduced by eddy currents if the relation $\tau_w \ll \tau$ is not fulfilled, as

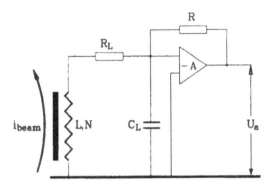

FIGURE 29
Modified electrical equivalent circuit of the beam current transformer, including an operational amplifier with feedback loop.

discussed, for example, in References 38 and 51. The time constant τ_w is determined by the eddy currents and is determined by:[51]

$$\tau_w = \frac{\mu_0 \mu_i d^2}{\pi^2 \rho} \tag{20}$$

where μ_i is the relative initial permeability, ρ is the specific electric resistance of the core material, and d is the thickness of the lamination.

- According to Equation 17, the achievable sensitivity is proportional to R/N, while the resolution is influenced by the bandwidth df and the load resistance R due to the thermal noise voltage U_{eff} of R, given by:

$$U_{eff} = \sqrt{4k_b \cdot T \cdot R \cdot df} \tag{21}$$

where k_b is Boltzmann constant and T is the temperature (K).

Therefore, a conflict situation arises with respect to the selection of N since the output voltage is proportional to 1/N while the transformer inductance, which should be high for a good low-frequency response, is proportional to N^2. Fortunately, the development of beam current transformers has followed the evolution of particle accelerators. One important step in this development was the extension of the low-frequency range by placing the current transformer in the feedback loop of an operational amplifier[28,49] as shown in the modified electrical equivalent circuit of Figure 29.

The resulting transfer function of this active-passive transformer is

$$U_a = -i_{beam} \cdot \frac{sRL}{N} \cdot \frac{1}{\dfrac{R}{A} + sL + R_L + sRC_L(sL + R_L)} \tag{22}$$

Here A is the complex gain of the operational amplifier. The most important effect is a considerable increase of the time constant τ, which has changed to $\tau = L/(R/A + R_L) \approx L/R_L$. Therefore, the droop rate can be reduced considerably in comparison to a nonfeedback system, which, in consequence, allows lower N values with the result

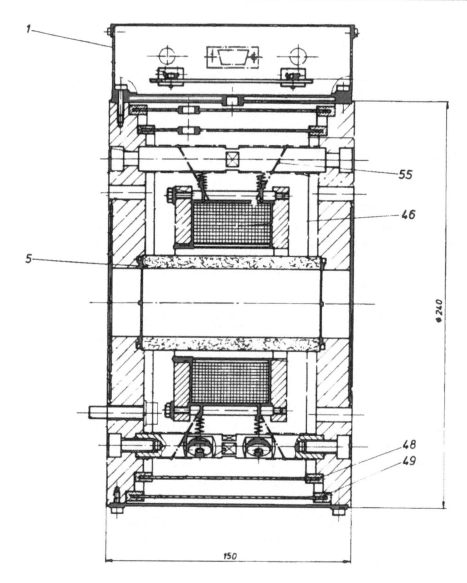

FIGURE 30

Simplified design drawing of a beam current transformer (measurements in mm, sealing flange: 40"
– CF). The most important parts are 1 — shielded housing for electronics, 5 — insulating ceramic
tube, 46 — secondary windings, 48, 49 — Mu-metal shields, 55 — elastic suspension system to damp
mechanical vibrations.

of higher sensitivity. To give an example, Figure 30 shows a simplified design
drawing of a beam current transformer, which has been designed first as a pure
passive system. Using an Ultraperm core with 400 windings and a load resistance of
470 Ω, a sensitivity of about 3 μA (s/n = 1) with a drop of ≤3% for a 5-ms beam pulse
could be achieved. By adding a feedback loop as discussed above, the number of
windings could be reduced to 20, and, therefore, the sensitivity could be improved
to about 150 nA. Figure 31 is a picture of the transformer.

For the design of a passive beam current transformer, calculations assuming the
parameters given in Table 5 may be used to estimate the achievable resolution and
to scale the system parameter. In the first step the required number of windings for

FIGURE 31
Picture of the beam current transformer shown as design drawing in Figure 30. The two Mu-metal shields and the last cover plate have been removed.

a pure passive system is determined. Figure 32 shows the results in dependence of the beam pulse length T_p. Knowing the number of windings N, the required beam current to get a signal-to-noise ratio s/n = 1. can be estimated from Equation 21, assuming $i \cdot R/N = U_{eff}$ and neglecting additional noise from the signal processing electronics. This leads to:

$$i_{min} = 10^6 \cdot \frac{N(T_p)}{R} \cdot \sqrt{4k_b \cdot T \cdot R \cdot df} \quad [\mu A] \tag{23}$$

The results for df = 10 kHz, 100 kHz and 1 MHz are given in Figure 33 in dependence on T_p, keeping in mind the determination of N by T_p and the allowed droop. Converting now the electronic circuit into an **active-passive system** by introducing a feedback loop, the calculated number of windings may be reduced up to about a factor of 20 (not going below 10) improving the resolution. The resolution can be even more improved by the use of a noise-optimized amplifier input stage.

In addition to the aspects summarized above, the following viewpoints are also important for the design of a beam current transformer system:

TABLE 5

Parameters of a Beam Current Transformer

Core material	Vitrovac 6025 F, Ref. [51]
Tape thickness	25 μm
Inner core radius	30 mm
Outer core radius	45 mm
Core height	28 mm
Relative permeability	>80.000, taken for the calculation
Load resistance	75 Ω
Allowed droop	1%

Number of windings, N

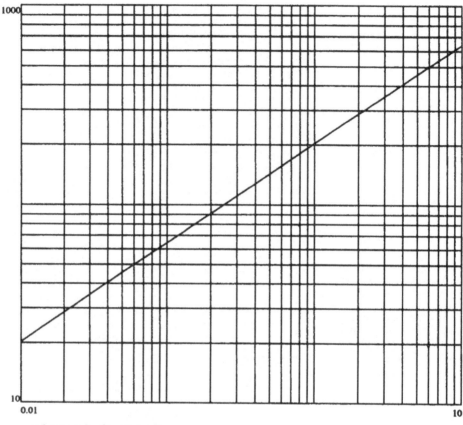

---> beam pulse length,[ms]

FIGURE 32
Required number of windings; see text.

- Shortcomings of a beam transformer are its high sensitivity to mechanical vibrations, as well as to electric and magnetic stray fields. Therefore, the location of installation along the beam line should be selected very carefully, and the mechanical layout also has to take these aspects into account. One has to keep in mind that the magnetic flux density produced by 1-μA beam current is in the order of 10^{-11} T, which is about five orders of magnitude below the magnetic earths field.

- The measured current may be falsified if beam plasma electrons are moving with or against the ions.

- Differential signal transmission is essential if line interference lies inside the signal passband.

- Use of ferromagnetic cores with low magnetostriction and reduced remanent induction minimize microphonic effects.

- Because the beam line has to be cut by an insulating gap the mechanical design has to provide an external bypass for the image currents, taking into account the required bandwidth determined by the beam pulse structure.

- With respect to the mechanical layout of the insulating gap, it is also important to decide about the needed vacuum sealing at the connection between gap and beam pipe. Figure 34 shows the simplified mechanical layout of a beam transformer

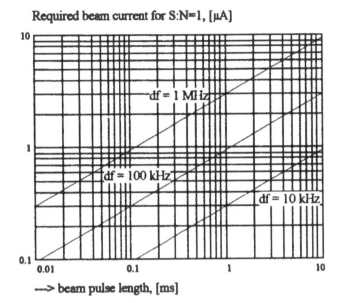

Required beam current for S:N=1, [μA]

---> beam pulse length, [ms]

FIGURE 33
Required beam current to get s/n = 1; see text.

sealed with O-rings (FPM, NBR), while the transformer shown in Figure 35 has metallic sealing using copper gaskets. There are even solutions where the whole transformer device has been placed inside the vacuum chamber.[44]

In many cases of ion source development and accelerator operation, dc currents have to be monitored. The development of beam transformers having a frequency range down to dc was influenced very much by the operation of storage rings, especially the intersecting storage ring (ISR) at CERN [195]. The basic principles of the so-called magnetic modulator have been described, for example, in References 28, 41, 52, and 127. The principle of operation may be explained referring to Figure 36, which shows a two-core magnetic modulator. Two toroidal cores are modulated in opposite sense (typical: 1 kHz). Assuming perfectly identical magnetic characteristics of both cores the detector signal, as shown in the scheme, should be exactly zero when there is no beam current through the cores. On the other hand, an asymmetric shifting of the hysteresis curve results if dc beam is fed through the toroids. The dc current is measured by means of the current generated in the automatic compensation circuit, which forces the output signal to zero again. The detector can even use a harmonic of the modulation frequency, which results in higher sensitivity and an improvement of the signal-to-noise ratio. Due to the extremely specific requirements concerning the identity of the magnetic characteristics for a core pair, the design of a magnetic modulator with high resolution and dc stability is rather complex and the success depends very much on the selection and treatment of the core material as discussed more in detail in Reference 41. The specifications of a typical dc system that has been developed[47] for the heavy ion synchrotron (SIS)[194] are given in Table 6. Figure 37 is a picture showing the mechanical combination of three types of beam current transformers in one housing, which even can be baked out. Here the two-core magnetic modulator (two cores at the left hand side) has been combined with an L/R integrator type of transformer (core in the center) and a fast transformer systems with the specifications given in Table 7.

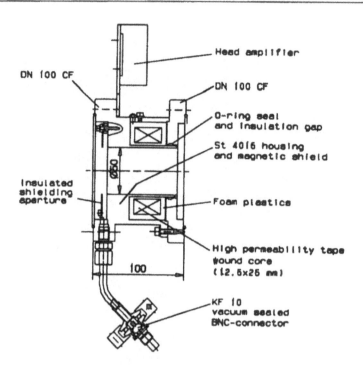

FIGURE 34
Example of a O-ring sealed beam transformer (measurements in mm).[50]

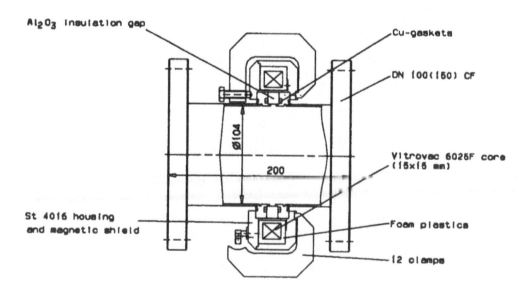

FIGURE 35
Example of a metallically sealed beam transformer (measurements in mm).[50]

To extend dc current measurements to the nanoamp region the application of cryogenic current comparators is discussed in References 55, 57, and 196. The principle of this current meter is based on the use of superconducting interference devices (SQUIDs) for the detection of the magnetic field of the ion beam, which switches on a surface current on the superconducting material (Meissner Ochsenfeld effect). Figure 38 shows schematically the main components of such a device, developed at GSI.[54]

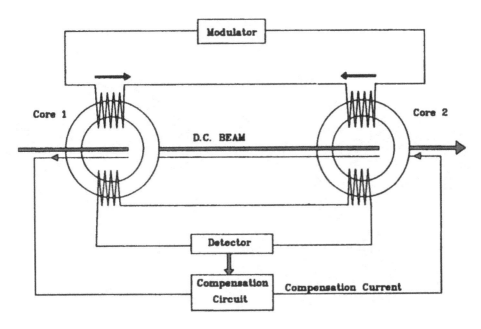

FIGURE 36
Principle of the two-core magnetic modulator.

TABLE 6

The dc Beam Transformer

Core material	Vitrovac 6025 F, Ref. 51
Aperture	>200 mm
Insulation Gap	Al_2O_3
Ranges (FSR)	300 µA ... 1 A
	1 ... 3 ... 10, bipolar
Resolution	2 $µA_{pp}$,
	s/n = 1, full bandwidth
1/f noise threshold	ca. 2 Hz
Bandwidth	dc-20 (200) kHz,
	Rise time: ca. 20 µs,
	Overshoot < 1%
Gain error	<.3%
Offset compensation	±1.5 µA,
	in auto zero mode
Linearity error	<0.1%

3 BEAM PROFILE MEASUREMENTS

With the implementation of computer-controlled measuring procedures and the introduction of computer-aided operation tools the importance of beam profile measurements has been increased. As a consequence there is a great variety of profile measuring devices,[44,89,90,136–143,146–152,154–159,165,168,177,181] which are mainly based on:

- Viewing screens, including digitizing electronics to perform computerized evaluation of the optical images
- Profile grids or harps
- Beam profile monitors based on residual gas ionization

FIGURE 37
Combination of three beam current transformers; see text. The cover plate has been removed. The insulating ceramic gap is the ring in white color at the right-hand side.

TABLE 7

Specifications of the Fast Beam Transformer (see text)

Core material	Vitrovac 6025 F ([51])
Aperture	>200 mm
Insulation Gap	Al_2O_3
Ranges (FSR)	100 µA ... 300 mA,
	1 ... 3 ... 10
Resolution	5 μA_{pp}, s/n = 1,
	100 µA-range, full bandwidth,
Other ranges	<3% of full scale
Bandwidth	700 kHz
Rise time	10 ... 90% = 500 ns
Droop	1% within 5 µs, 100 µA range
	<3% in all other ranges.
Gain error	<3%

- Scanning devices such as wire scanners or ionization beam scanners, which are, of course, special residual gas monitors
- Slit-Faraday cup combinations

Considering ion implantation, material research, and other industrial applications of ion beams, the measurement and monitoring of beam profiles and the control of their uniformity is essential.

3.1 Viewing Screens

The combination of a scintillating screen and the observation by a video camera is a very simple, reliable profile monitor. Since this combination does not require a computer control or computerized signal processing like more sophisticated monitors, it may be very useful during commissioning periods and for troubleshooting. Therefore, viewing screens should be installed in addition to other monitors at all

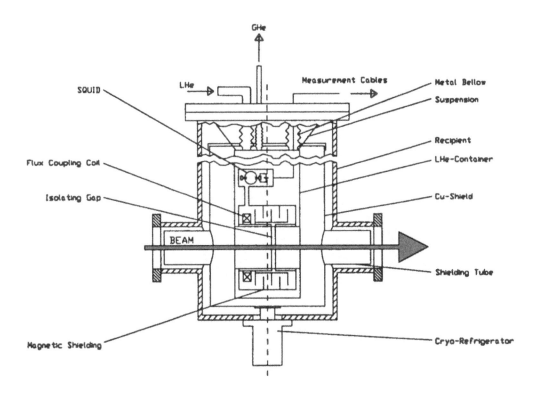

FIGURE 38
Scheme of a cryogenic current measurement device for the detection of very low dc currents.[54]

critical points along a beam transport system. Of course, the scintillation material has
to be matched to the kind of ions and their energy, as well as to the expected beam
currents. Furthermore, the screen material has to be adapted to the characteristics of
the video camera and the optical detector.

Various scintillating materials are used. In case of low intensities the old ZnS
screen may be a solution. Its lifetime is limited, especially in case of low energies, and
has been estimated to about 10^{14} protons/mm^2.[163] More radiation resistant screens are
made of ZnCdS/Ag (P20) and ZnS/Cu (P31), which have been tested in Reference
149. The decay time (down to 10%) of those materials is in the order of some hundred
microseconds. Most popular materials are cerium-activated lithium glasses with a
very short decay time of about 100 ns and Al_2O_3 doped with chromium (red, decay
time $\approx$ 1s) or cobalt (blue).[163,164] From the viewpoint of vacuum qualification, screens
made of Al_2O_3 or Li-glass are more suitable. Most of the tests concerning lifetime and
sensitivity were performed with minimally ionizing protons and, therefore, scaling
to low energies and to various ion species is difficult. Assuming a minimal number
of 2×10^6 protons/mm^2 required within 1 s, as typical for Chrolox 6® material with a
decay time of $\approx$ 1 s,[164,167] a rough estimation for other particles is possible by scaling
the data $\sim z^2$.[166] Figure 39 shows the results for ions up to uranium for beam radii of
3, 5, and 10 mm. The sensitivity can be improved by use of special videocameras or
amplifiers. In case of high beam powers, metallic foils may be used to observe the
thermal heating by the beam. Figure 40 is a picture of a complete diagnostic unit
consisting of a large Chrolox 6® viewing screen (diameter = 75 mm) and a compressed
air actuator.

Although the handling of viewing screens is very simple they have some disadvantages:

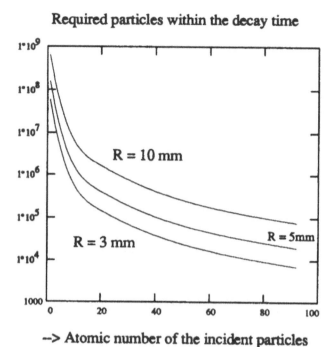

Required particles within the decay time

--> **Atomic number of the incident particles**

FIGURE 39
Estimated number of required ions for a Chrolox 6[R] viewing screen[167], R is the beam radius.

FIGURE 40
Viewing screen mounted onto a compressed air actuator. The small flange is a window for observation of the beam spot using a video camera and a mirror.

- In the energy region previously discussed the beam will be completely stopped, and, therefore, profile measurements at the same time at different positions downstream are not possible.

- The intensity range that can be covered is very limited.

- In the combination screen/video camera there is no data signal available for computer-aided signal analysis, e.g., automatic beam alignment or emittance measurements. This restriction can be solved using video digitizers and frame grabbers, such as those available for laser beam analysis,[189] which allows the extraction of more quantitative information. Figure 41 shows a simplified block diagram of a digitizing electronics that has the following specifications:[137]

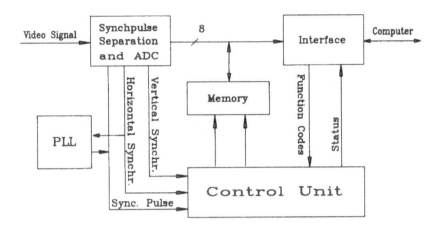

FIGURE 41
Simplified block diagram of a digitizing electronics for viewing screen observation by video cameras.[137]

Resolution	64 × 64 pixels
Size of the digitized image	64 × 61 mm
Pixel size	1.0 × 0.92 mm
Video input voltage range	0.3–1.0 V_{pp}
ADC	8 bits, linear
Sampling frequency	1.76 MHz
Antialiasing low-pass filter	Butterworth, 11th order, $f_g = 600$ kHz
Memory capacity	32 KB, static
Used memory	4 KB
Power supplies	+12, +5, –12 V

The digitizing system also includes a software procedure for correct alignment and focusing of the camera system. The resolution of 64 × 64 pixels corresponds to 4096 measured profile points, and, therefore, gives more quantitative information than, e.g., a 2 × 64-wire profile grid, which integrates along the wire. In comparison to a profile grid electronic this electronic is cheaper by about a factor of two, including even the video camera. On the other hand, the very limited intensity range of a viewing screen is a serious disadvantage.

3.2 Profile Grids, Scanning Devices

From the viewpoint of signal acquisition profile grids, scanning wires and slow-moving slit-Faraday cup combinations are very similar. Both deliver an electrical signal that is proportional to the beam intensity at the wire and the slit position. In the energy range below 2 MeV/u, the particles will be stopped completely in the wires and, therefore, the electrical signal is proportional to the collected charge of the stopped ions. Since in most cases there is no secondary electron suppression, the signal may be amplified or attenuated by the emitted secondary electrons, depending on the sign of the ion charge. Figure 42 shows a standard profile grid with 2 × 16 wires of 0.1-mm diameter and a spacing of 1.5 mm. The grid can be mounted onto a standard compressed air actuator.[191,192] Figure 43 is a picture of a complete diagnostic unit for beam profile measurements. The unit is a very special one.[192] The aperture of the grid itself is 170 × 170 mm, and to save costs for the electronics by reducing the

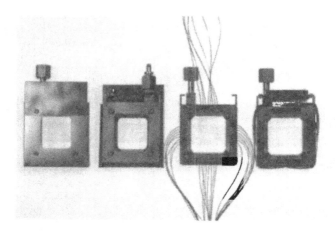

FIGURE 42
GSI-standard profile grid. Left: front face, showing also the metallic shields, next: rear; next two: some parts removed, to show the Kapton-insulated signal wires and the Duratherm springs, to tighten each wire.

number of channels, the spacing of the most outside wires is larger. Table 8 summarizes the most important parameters of a profile grid measuring system. Figure 44 is the picture of a rotating wire scanner, which has to be mounted in a 45° position with respect to the horizontal and vertical coordinate, to measure both transverse profiles during one half revolution as described in Reference 147. The scanner is provided to rotate with ≤750 rpm, whereby 750 rpm corresponds to a scanning speed of 1.344 mm/ms. The maximum beam spot size can be 35 mm. The diameter of the scanning wire, made of tantalum, is 1 mm, and, therefore, the resolution that can be achieved is on the order of 1 mm. Both devices, profile grids or scanning wires, have their advantages and disadvantages, and the following points should be taken into account when deciding about application:

- With a profile grid, the beam intensity is sampled at the same time, while a moving wire will sample the profile at different locations at different times. Therefore, longitudinal variations of the beam intensity will be mixed with transverse intensity variations.

- In the case of pulsed beams further complications may arise from the demand for exact synchronization, which can be solved without any problems in case of profile grid application.

- For a profile grid system, the signal integration improves the signal-to-noise ratio of the acquired beam signal.

- Considering a rotating wire scanner as, for example, described in Reference 147, the geometric duty factor, generated by the rotation, allows a higher beam power than a static grid.

- Falsifications of profiles due to changes of the secondary emission coefficient are minimized by using only one moving wire.

- The electronics for data acquisition is cheaper for a scanning system. A harp requires one channel per wire, as shown schematically in Figure 45. But, in both cases, if there is no need to digitize the signals for computer control the profiles may be monitored by using only an oscilloscope.

When using a grid system, the maximum beam current is limited by the applicable beam power-time product to avoid wire melting, and the minimum required beam

FIGURE 43
Large profile grid mounted onto a special compressed air actuator.

current has to be estimated from the geometrical data of the grid, taking into account the beam spot size as well as the noise voltage of the signal processing electronics.

3.2.1 Maximum Applicable Beam Current

As already discussed in Chapter 2, Section 1, the maximum applicable beam current can be estimated considering the energy that is needed to melt the range volume. Figure 46 shows the results for pulsed beams assuming a parabolic shape of the intensity distribution, taking the center wire as the critical one. The graph is valid for tungsten wires of 0.1-mm diameter. On the other hand, the maximum allowed continuous power loss for such a wire has been proven experimentally[17] to be about 0.5 W/mm wire. Taking this value into account, the curves of Figure 47 show the

TABLE 8

Important Parameters of a Profile Grid Measuring System

Diameter of the wires	0.05–0.5 mm
Spacing	0.5–5 mm
Transparency	≥80% (typically)
Length	40–100 mm
Material	W-Re alloy (typically)
Insulation (frame)	Glass ceramics
Number of wires	15–127
Maximum power rating	0.5–1 W/mm
Sensitivity (lowest range)	
Integration time = ≤5 ms	2–5 nA/V
Integration time = ≤5 s	2–5 pA/V
Dynamic range	$1:10^3–1:10^6$
Number of ranges	4–16
Maximum output voltage	10 V

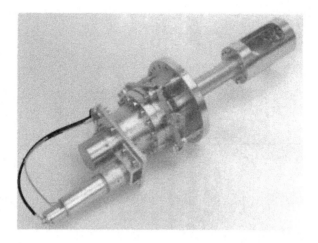

FIGURE 44
Rotating wire scanner.[191,192]

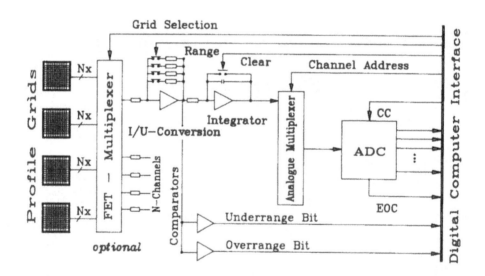

FIGURE 45
Block diagram of a profile grid measuring electronics.[17]

results for dc beams under the same assumptions as for Figure 46. In the case of low energies and high beam currents, the destruction of the wires by sputtering, as estimated in Chapter 2, Section 1, should also be considered.

3.2.2 Minimum Required Beam Current

For the calculation of the minimum required beam current to get a profile measurement with a profile grid or a moving wire, the specifications of the mentioned amplifier-integrator system ACF2101 (see Chapter 4, Section 1) can be used. Assuming a minimum output signal of ≥ 50 mV for the wire at the center of the beam, the required current-time product has been calculated for a wire thickness of 0.1 mm. Figure 48 shows the results in dependence on the beam radius. The diagram holds for

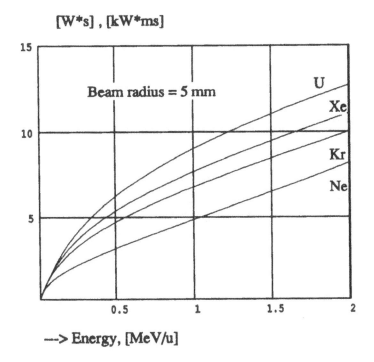

[W*s] , [kW*ms]

Beam radius = 5 mm

U
Xe
Kr
Ne

---> Energy, [MeV/u]

FIGURE 46
Estimation of the maximum allowed product of beam power·pulse length using a profile grid for various ion species.

positive ions and the ratio $S = q/(q + N_e)$ has been introduced to take into account the signal amplification by the emitted secondary electrons per incident ion (N_e).

Due to the integration of current along the wire, strange beam profiles, as, for example, the profile of a hollow beam, cannot be resolved. One possibility to improve resolution is the rotation of a harp around the longitudinal axis, taking data at various angles and applying reconstruction techniques. This technique has been analyzed and tested in References 169 to 175 and 190. Due to requirements for the vacuum system, the design of a grid rotatable around the beam axis gets rather complicated. Furthermore, the mathematical procedures for data evaluation and profile display are much more sophisticated than for a static grid system. A compromise may be the use of two complete profile grid units, installed at different angles with respect to the both transverse coordinates.

3.3   Residual Gas Ionization Monitors

The first devices for nondestructive profile measurements using residual gas ionization were developed for proton beams.[150,158] Since then the sensitivity has been improved by use of electron multipliers, microchannelplates (MCP), and Chevrons, whereof the first use of a MCP has been reported in Reference 178. The most sophisticated device seems to be the ionization beam scanner,[142] which also has been developed for profile measurements of proton beams.

The use of residual gas ionization monitors for ion beams in the considered energy region has the advantage of high ionization cross sections. Therefore, it may be sufficient to collect the liberated electrons or ions on metal strips using a simple profile grid electronics for signal amplification as shown in Figure 45. Using the

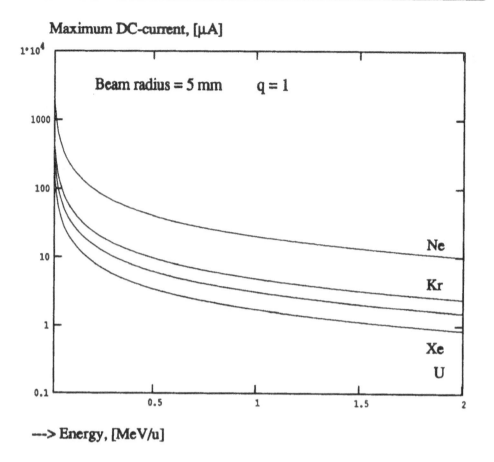

Maximum DC-current, [μA]

--> Energy, [MeV/u]

FIGURE 47
Estimation of the maximum allowed dc current using a profile grid for various ion species.

stopping power data given in Reference 180, the ratio between primary ion beam current and current of charged particles generated by residual gas ionization has been estimated with the parameters given in Table 9.

Figure 49 shows the results for hydrogen and Figure 50 for nitrogen assuming all liberated charged particles are collected by applying an appropriate electrical field. Note the scaling of the ordinate axis by a factor of 10^{-1} and keep in mind the higher charge state of the heavier ions.

Figure 51 is the scheme of a monitor, developed for future high beam current operation of the heavy ion accelerator Unilac at GSI.[188] The monitor fits into a beam pipe of 100-mm diameter as shown in Figure 52. The most important specifications are given in Table 10.

Use of residual gas ionization monitors in the considered energy region also has some disadvantages:

- Due to the relatively low longitudinal momentum the deflection of the beam by the collecting electrical field E_c can be rather high, and has to be compensated for. The deflection is given by:

$$x'[\text{mrad}] = 1.074 \cdot \frac{q}{A} \cdot \frac{E_c[\text{kV/cm}] \cdot l[\text{cm}]}{\beta^2[\%]}$$

(24)

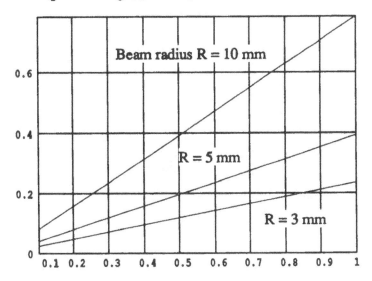

Required charge, [nA*ms]

Beam radius R = 10 mm

R = 5 mm

R = 3 mm

--> Ratio S, see text

FIGURE 48
Estimation of the minimum required beam current-time product to allow a profile measurement using wires of 0.1 mm, $S = q/(q+N_e)$; see text.

TABLE 9

Parameters to Calculate the Current Ratio (see text)

Length (l) of the monitor	100 mm
Residual gas pressure	10^{-6} torr
Ionization energy/electron-ion pair	60 eV
Residual gas	Hydrogen, nitrogen
Ion species	Ne^{1+}, Kr^{5+}, Xe^{7+}, U^{9+}

This deflection may be compensated for with the addition of a steerer pair behind the monitor or the application of a magnetic field along the monitor, which is at right angles to the beam direction as well as to the collecting electrical field. The necessary magnetic field strength B_m depends only on the velocity v_i of the primary ions and the electrical field strength and is given by $B_m = E_c/v_i$. Considering electrical fields in the order of ≤ 1 kV/cm, the required magnetic field strengths will be some Teslas for ion energies of about 1 keV/u and some milliteslas for energies of about 1 MeV/u.[181]

- The fringe fields of the collecting plates act like a lens on the primary beam, which has to be taken into account especially in the case of low energies.[181]

- Due to the proportionality of the beam potential to i_{beam}/β, the momentum transfer to the generated residual gas ions and electrons perpendicular to the collecting electrical field may result in a considerable distortion of the measured profiles.[153,156,181-183]

- A further effect is the recoil caused by the ionization process, which also leads to deterioration of the profiles.[153,156,183]

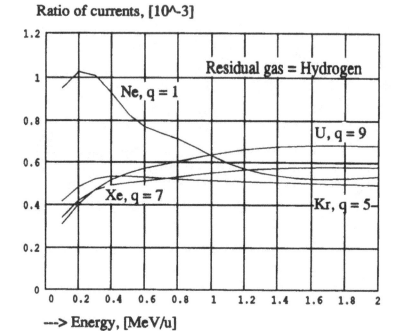

FIGURE 49
Estimated current ratio for collecting charged particles from hydrogen residual gas ionization; see Table 9 for parameters.

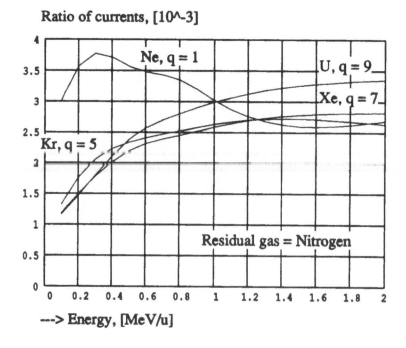

FIGURE 50
Estimated current ratio for collecting charged particles from nitrogen residual gas ionization; see Table 9 for parameters.

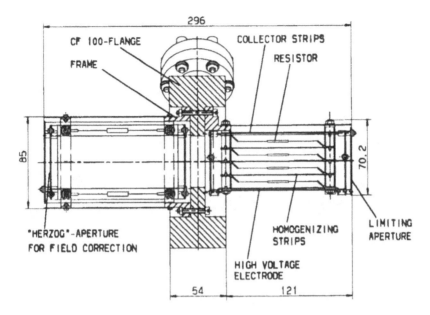

FIGURE 51
Scheme of a residual gas monitor (measurements in mm); see Table 10 for parameters.

FIGURE 53
Picture of the residual gas monitor shown schematically in Figure 51.

TABLE 10

Most Important Parameters of the Residual Gas Monitor Shown Schematically in Figure 51

Number of measuring planes	2, horizontal and vertical
Number of collecting strips	16 per plane
Spacing between the strips	1.6 mm
Maximum profile width	25 mm
Length of the strips	100 mm
Overall length of the monitor	300 mm
Supporting flange	NW 100 CF
Provided collecting field strength	1 kV/cm

Obviously, the two latter effects are much smaller for the residual gas ions than for the liberated electrons, and, therefore, for better resolution, the collection of ions should be preferred. The measured profiles may also be distorted by secondary electrons generated by the impinging residual gas ions. The distortion will be even more serious when collecting electrons, because the generated ions also will hit the common high-voltage plate. As a consequence, secondary electrons will be emitted and accelerated from there to the detector strips, which results in a distortion of the profile signals. But, still having in mind the mentioned disadvantages by detecting the electrons, an interesting improvement of a residual gas profile monitor is discussed in Reference 153, proposing the collection of electrons and ions at the same time, giving both transverse beam profile data with only a one-plate system. As mentioned above, considerable refinements concerning sensitivity as well as transversal resolution of residual gas ionization can be achieved with the use of MCPs, Chevrons, etc. in combination with wedge-and-strip anodes, resistive anodes, and time-of-flight techniques as discussed in References 153 and 184 to 187.

4   MEASUREMENTS IN THE TRANSVERSE PHASE SPACE

Formation, acceleration, transport, and focusing of charged particles depend very much on the optical properties of the beam. Real ion sources are extended and have geometric imperfections. Furthermore, the extracted beams have a thermal velocity spread and aberrations due to space charge. Therefore, the analytical approach of description by the well-known paraxial formalism[66,104,110-112] had to be improved by introducing the terms of emittance and, especially, brilliance. The basic definitions, invariance conditions, relations between the parameters, and methods of measurement are summarized and discussed in detail in References 58 and 59. Therefore, we will consider here only a short summary of the most important definitions and relations.

In general, a two-dimensional projected phase space area is defined as

$$A_i = \int dx_i dp_i \tag{25}$$

where x_i, p_i are the conjugate coordinate pairs in a Hamiltonian system. According to Liouville's theorem,[197,198] the areas A_i are maintained, provided only nondissipative forces act on the particles, and the motion in each of the three subspaces are decoupled from other motions. In most cases of beam transport these conditions are fulfilled, and the three projected phase space areas are defined by the two transverse phase planes and the longitudinal one. Now, describing the motions in the three phase planes by the coordinates z (longitudinal motion, i = 3 in Equation 25), x (horizontal motion, i = 1), and y (vertical motion, i = 2), and neglecting relativistic effects in the both transverse directions ($\dot{x} \ll \dot{z}$, $\dot{y} \ll \dot{z}$) the momentums p_x and p_y are given by

$$p_x = m\dot{x} = m_0\gamma\frac{x'}{\dot{z}} = m_0\gamma\beta cx'$$

$$p_y = m\dot{y} = m_0\gamma\frac{y'}{\dot{z}} = m_0\gamma\beta cy' \tag{26}$$

Here m is the relativistic mass, given by $m = \gamma m_0$ with $\gamma = 1/\sqrt{1 - \beta^2}$, $x' = dx/dz$, $\dot{z} = dz/dt = v_z = \beta c$, ($c = 2.997925 \times 10^8$ m/s). Introducing the transverse particle momentums p_x and p_y of Equation 26 into Equation 25 it is evident that for constant β the areas

$$E_x = \int dy' dx \quad \text{and} \quad E_y = \int dx' dy \qquad (27)$$

are conserved. Because in most cases the distribution of the particles in the two transverse phase planes has an elliptical shape, and the area of an ellipse contains a factor of π, it is convenient to define the horizontal and vertical emittances as

$$\varepsilon_x = \frac{1}{\pi} \int dy' dx = \frac{1}{\pi} E_x \quad \text{and} \quad \varepsilon_y = \frac{1}{\pi} \int dx' dy = \frac{1}{\pi} E_y \qquad (28)$$

Because Liouville's theorem holds only for the area defined by Equation 25, the emittances defined in Equation 28 will change with β. Therefore, the so-called normalized emittances ε_x^n, ε_y^n have been defined as

$$\varepsilon_x^n = \beta \gamma \varepsilon_x \quad \text{and} \quad \varepsilon_y^n = \beta \gamma \varepsilon_y \qquad (29)$$

Obviously, to determine the emittance in one of the two transverse phase planes one has to measure the angular divergence of all particles in dependence on their coordinate, and, therefore, the area has the dimension of radian meters. Figure 53 illustrates the principle of an emittance measurement, and Figure 54 shows how it can be realized to perform such a measurement. Figure 55 gives an example of the data one obtains by such a measurement, while Figure 56 is a surface plot of the data demonstrating the usefulness of describing emittance areas by ellipses.

For practical reasons the brilliance B (sometimes called brightness), which is the average value of the density in trace space, has been defined as:[60,61]

$$B = i_{beam} / \pi^2 \varepsilon_x \varepsilon_y \qquad (30)$$

which obviously has the dimensions of amperes per squared radian meter, and can be determined by measuring the beam intensity together with the emittances in the two transverse phase planes.

4.1 Destructive Measurement Methods

There is a great variety of solutions and according to References 58 and 59 the methods applied may be classified as follows:

- The pepper-pot method, which consists of an array of small holes in front of a viewing screen, film, or another recording device. Here, the x,y-coordinates are defined by the holes, while the divergence can be determined from the spacing of the spots on the recording device. Figure 57 shows a complete cooled unit provided for beam powers up to 6 kW.

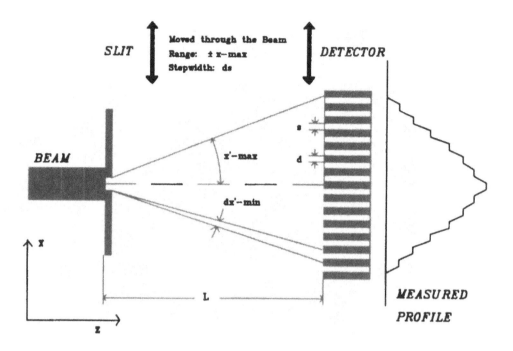

FIGURE 53
Principle of an emittance measurement. (From Strehl, P., *Kerntechnik*, 214-223, 1991. With permission.)

EMITTANZ –
–MESSEINRICHTUNG

FIGURE 54
Emittance measuring device. A slit and a sandwich-detector are attached to a stepping motor-driven linear feed-through, to move them together through the beam.

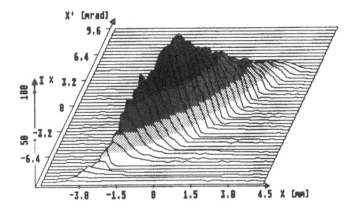

FIGURE 55

Typical example of the data obtained by an emittance measurement.[199] The gray scale represents various percentages of the maximum wire current (50, 25, and 10%).

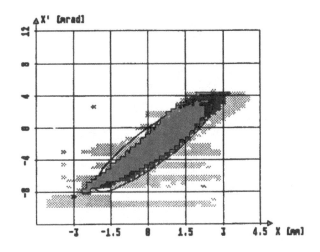

FIGURE 56

Surface plot of the data shown in Figure 55. The gray scale marks various percentages of the total beam current (95, 90, 85, and 80%).

- Two slits combined with a current measuring device such as a Faraday cup or beam current transformer
- One slit in front of a profile scanner
- One slit in combination with a multichannel profile measurement, for example, a detector-sandwich or a movable profile grid
- One hole in front of a slit and a current meter
- One crossed slit plus a photographic film
- Two pairs of crossed slits combined with a current measuring device

The specific characteristics of the various methods as well as the evaluation and interpretation of measured data are discussed in detail in References 58 and 59. Numerous kinds of realizations are described in the literature where the essential differences are characterized by:[69–76,85,86,89,90,115,116,144]

FIGURE 57

Pepper pot for emittance measurements,[191] the unit is equipped with a mirror for observation of the beam spots from outside using a video camera.

- The requirements with respect to the angular range and the beam spot size which have to be covered, as well as with respect to the desired angular resolution and resolution of position
- The penetrating of the particles in material and the beam power one has to deal with
- The measuring algorithm and the electronic signal processing
- The time spent for a measurement
- The mathematical extraction of relevant beam parameters for the valuation of an ion source and the calculation of proper settings for beam transport systems

Taking the slit-multichannel detector as a typical example, the most important relations for the design of a complete emittance measuring system can be derived and some general rules concerning the mechanical design, the achievable sensitivity, resolution, and ranges can be discussed. The calculated data may easily be scaled to other systems and beam parameters. Since a harp mounted on a movable vacuum feed-through has been proven to be very flexible we will consider this device as the multichannel detector. The maximum divergence x'_{max} of the beam influences the choice of the distance L between slit and detector, the number of detector channels N, their spacing, and the stroke of the feed-through that supports the movable detector. Mostly, there are also geometrical restrictions. If the influence of the space charge can be neglected, the determination of L is straightforward. Otherwise, the distance between slit and detector should be taken as small as possible to minimize the effects of space charge.[80-82,88] Whether the emittance or space charge spread is dominating can be determined by checking the quantity S given by[62-64,67]

$$S = \frac{q}{A} \cdot \frac{e \cdot I_{beam}}{2\pi \cdot \varepsilon_0 \cdot m_u c^2 \cdot (\beta\gamma)^3 \cdot c} \tag{31}$$

which is derived from the well-known paraxial envelope equation[64,66,67] incorporating the effect of space charge. q and A are charge state and mass number of the ions, e = 1.6×10^{-19} As, $\varepsilon_0 = 8.85 \times 10^{-12}$ V s/Am is the dielectric constant of the vacuum, and $m_u c^2$ = 931.502 MeV. Space charge will dominate if

$$S \cdot r_m^2 = 64.4 \cdot \frac{q}{A} \cdot \frac{I_{beam}[mA] \cdot r_m^2[mm^2]}{(\beta[\%] \cdot \gamma)^3} \geq \varepsilon^2 [mm^2 \cdot mrad^2] \tag{32}$$

where a circular beam with a minimum radius of r_m at the waist has been assumed and ε is the emittance of the beam that has to be measured. The divergence growth due to space charge during a drift can be determined from the derivation of the well-known beam-spreading curve[62,65] or also from the envelope equation. For the assumed circular beam, starting with a waist radius r_m this divergence is determined by

$$\frac{dr}{dz} \approx \frac{2}{2.09^2} \cdot \frac{q}{A} \cdot \frac{e \cdot I_{beam}}{\pi \cdot \varepsilon_0 \cdot m_u c^2 \cdot (\beta\gamma)^3 \cdot c} \cdot \frac{z}{r_m} \tag{33}$$

$$\frac{dr}{dz}[mrad] \approx \frac{0.059}{(\beta[\%] \cdot \gamma)^3} \cdot \frac{q}{A} \cdot \frac{z}{r_m} \cdot I_{beam}[mA] \tag{34}$$

Because the divergence rises proportionally to the drift in the z direction, the distance L between slit and detector has to be optimized with respect to the space charge effect. Figure 58 shows the calculated divergence according to Equation 34, in dependence on L for some typical β values.

Taking into account that

$$\frac{dr}{dz} \sim \frac{q}{A} \cdot \frac{I_{beam}}{r_m} \tag{35}$$

the results given in Figure 58 can be easily scaled to other ions and currents. The following points have to be kept in mind when calculating S by Equation 32 and r′ by Equation 34:

- The divergence induced by the space charge results from the radial electric field inside the beam and may be different for the two transverse coordinates, for example, in case of elliptical beams. The correction of r′ is straightforward.[63,64]
- No compensation of space charge by fractional neutralization of electrons trapped in the beam has been considered.[77-79,83,84,91,114,117]
- The calculations hold for a dc beam; therefore, a considerable increase of r′ may result with bunched beam operation.

The last two effects can be taken into account replacing I_{beam} by I^*_{beam}[63] where

$$I^*_{beam} = \frac{I_{beam} \cdot (1-f)}{B} \tag{36}$$

with 1 − f as the fractional neutralization and the bunching factor B = 4c/3d, where c is the semiaxis of the phase space ellipsoid in the longitudinal direction and d is the distance between the bunches. Obviously, B = 1 for a dc beam and may be calculated for a bunched beam in geometrical terms as well as in terms of bunchlength and bunch repetition period.

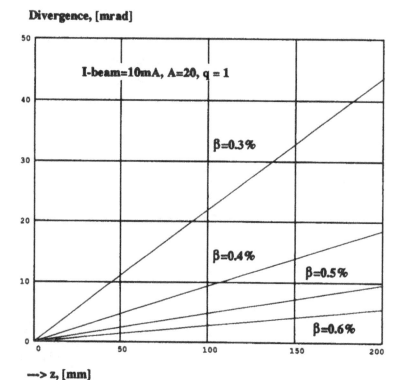

FIGURE 58
Calculated divergence growth due to space charge.

Once the drift space L has been determined, the relevant design parameters of the emittance measuring system can be fixed as follows:

$$x_{max} = \frac{1}{2} \cdot S_{slit} \tag{37}$$

$$\delta x = W_{slit} \tag{38}$$

$$\Delta x = N_{slit} \cdot \delta S_{slit} \tag{39}$$

$$x'_{max} = \frac{N_d}{2} \cdot \Delta d + \frac{1}{2} \cdot S_{det} \tag{40}$$

$$\delta x' = D_{det} \tag{41}$$

$$\Delta x' = \frac{\Delta d}{ZS + 1} \tag{42}$$

with the definitions:

X_{max} half of the maximum beam width to be measured (30 mm)
S_{slit} stroke of the slit feed-through (100) mm
δx resolution in x-coordinate (0.1 mm)
W_{slit} width of the slit (0.1 mm)
Δx separation of measured points in x (0.25 mm)
N_{slit} number of steps for each slit displacement (10)
δS_{slit} slit-displacement per step (0.025 mm)

x'_{max} maximum measurable divergence (50 to 150 mrad)
N_d number of detector channels (16 to 60)
Δd separation of detectors (1 mm, harp; 0.1 mm, sandwich)
S_{det} stroke of the detector feed-through (100 mm)
$\delta x'$ resolution in divergence (1 mrad)
D_{det} diameter of the harp wires (0.1 mm)
$\Delta x'$ separation of points in x' (1 mrad)
ZS number of intermediate steps for the detector movement (5)

with some typical values given in parentheses. Figure 59 is a block diagram of a typical emittance measuring electronic device provided to process the signals of a profile grid with 61 wires.

All the methods of destructive emittance measurements summarized above result in a considerable reduction of current intensity distributed across the profile detectors. Since the designer of the electronics for an emittance measuring system needs information about the expected intensities, an estimation of the fraction of current passing the slit and the current distribution over the profile measuring detectors becomes essential. Let us assume a circular beam with transverse emittances, which can be described by ellipses, characterized by the so-called Twiss parameters α, β, and γ and the emittance parameter ε (area = $\pi \cdot \varepsilon$)[107] as illustrated in Figure 60. Furthermore, let us assume typical emittance parameters ε = 120 mm $\cdot$ mrad, β = 0.533 mm/mrad, γ = 1.88 mrad/mm and α = 0, which corresponds to a beam waist having a radius of R_0 = 8 mm and a maximum divergence of x'_{max} = 15 mrad. For intensity distribution with parabolic shape the radial current density is

$$N_{data} \approx \frac{2 \cdot x_{max}}{\Delta x}(1+ZS)(1+N_2) \tag{43}$$

$$I(r) = \frac{2\,I_{beam}}{\pi R_0^2} \cdot \left[1 - \left(\frac{r}{R_0}\right)^2\right] \tag{44}$$

Moving the slit along one of the transverse axis gives a current through the slit

$$I_s(R_s) \approx \frac{8}{3} \cdot \frac{I_{beam} \cdot W_{slit} \cdot \sin(\alpha(R_s))}{\pi R_0} \cdot \left[1 - \left(\frac{R_s}{R_0}\right)^2\right] \tag{45}$$

where R_s is the position of the slit and $\alpha(R_s)$ = arccos(R_s/R_0). Figure 61 shows the percentage of beam passing the slit for the selected beam parameters. A further reduction of intensity results from the current distribution across the detectors due to the divergence of the beam at the slit.

The current density I_d as a function of slit position R_s and detector position x is given by

$$I_d(x,R_s) \approx \frac{4}{3} \cdot \frac{I_s(R_s)}{x_{max}(R_s)} \cdot \left[1 - \left(\frac{x}{x_{max}(R_s)}\right)^2\right] \tag{46}$$

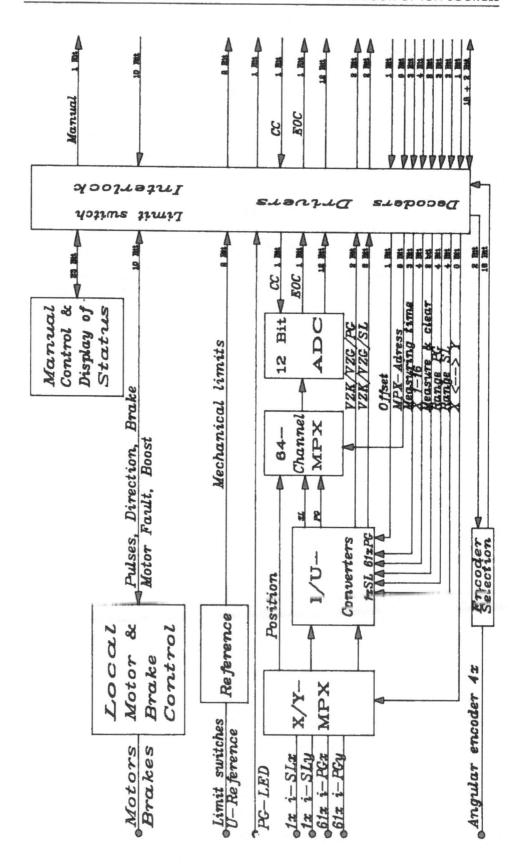

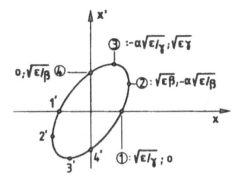

FIGURE 60
Characteristic points of emittance ellipses described by the Twiss parameters.

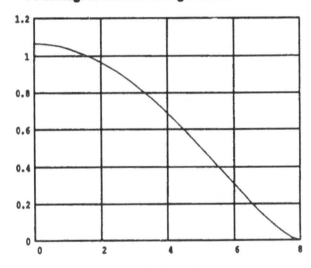

FIGURE 61
Percentage of beam passing a slit; see text for parameters.

where $x_{max}(R_s) = L \cdot x'_{max} \cdot \sin(\alpha(R_s))$ is the maximum extension of the beam behind the slit, which, obviously, depends on the drift length L and the maximum divergence at position R_s. To obtain the current on the detectors, $I_d(x, R_s)$ has to be multiplied by D_{det}, the diameter of the detector wires. Taking a typical distance of L = 200 mm and d = 0.1 mm, the fraction of current was calculated. Figure 62 shows the result for three detector positions in terms of x_{max}.

The conversion of the relations for other beam parameters and different specifications of the emittance measuring device is straightforward. But, mostly, the reduction of current will be on the same order of magnitude as shown in Figures 61 and

FIGURE 59
Block diagram of a typical emittance measuring electronic device.[192]

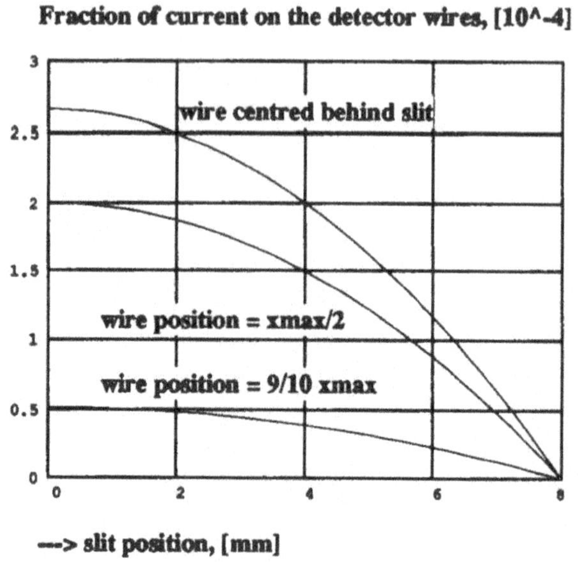

Fraction of current on the detector wires, [10^-4]

---> slit position, [mm]

FIGURE 62
Fraction of beam current on the detectors behind a slit; see text for parameters.

62. In the case of low beam currents the situation can be improved by integrating the detector currents over an appropriate time. Especially in case of dc currents, an integration over one or more periods of the line frequency will reduce the hum contribution considerably. A suitable amplifier-integrator system is the ACF2101 switched integrator,[68] which gives an output voltage of 10 V for input current-integration-time products between 0.01 $\mu A \cdot 100$ ms and 10 $\mu A \cdot 100$ μs.

Some emittance measurement systems use a crosslike slit and a profile grid instead of a harp as detector. By moving the slit and grid in a 45° direction through the beam, both transverse emittances can be measured at the same time. Obviously, there will be a contribution of current from one cross arm to the other, which will falsify the results. The contribution can be estimated using the equations given above. The additional pick-up current on the wire in position x is given by

$$I(x, R_i) = D_{det} \cdot W_{det} \cdot \frac{2 I_{beam}}{\pi R_0^2} \cdot \left[1 - \left(\frac{R_i \perp x}{R_0} \right)^2 \right] \qquad (47)$$

Taking L = 200 mm and D_{det} = 0.1 mm the fraction of false current was calculated. Figure 63 shows the result for the same three detector positions as shown in Figure 62. A comparison of the two figures leads to the conclusion that the contribution from the second cross-arm can be considerably high. Therefore, this method, which will save money and measuring time, will not be recommended for precise emittance measurements. Other effects such as slit scattering and secondary particle emission may distort the measurement and have to be considered in dependence on the system parameters.

4.2   Nondestructive Methods

The methods discussed above are based on direct measurements of the intensity distribution in the transverse phase space. Therefore, the boundary contours of the

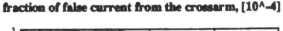

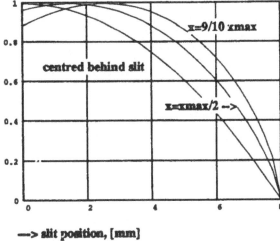

---> slit position, [mm]

FIGURE 63
Fraction of false current from a second perpendicular slit; see text for parameters.

emittance pattern may be of any shape, and extraction of the emittance area and brilliance from the measured data will be possible in most cases. But, depending on the system parameters, there may be the following drawbacks:

- In most cases the measurement completely destroys the beam.
- High-energy beams are not stopped by a thin knife-edged slit.
- In the case of very high beam power the slit and detector materials may be melted.
- The measuring time can be very long.

If particle transfer through a system of linear optical elements can be treated by the well-known matrix formalism,[104,105] which is the case for the transport of a normal beam[58] having elliptical phase space contours, the Twiss parameters as defined in Figure 60 can be derived from profile measurements. Keeping in mind that the parameter β is correlated to the width of the beam according to Figure 60, two variants are discussed in the literature.[39,44,47-49,100,109,105,125]

- The gradient variation method, illustrated in Figure 64, and which, in principle, requires only three different settings of a quadrupole in front of a profile measuring device
- The three position method, which is very similar to the gradient variation method, except that the three profiles have to be measured at various positions without the necessity of quadrupole variation.

Obviously, both methods may be improved by measuring more than three profiles.

Taking the 2×2 matrix representation of the Twiss parameters[108] the transformation of α, β, and γ through a linear optical system between any two points 0 and s is given by

$$\begin{pmatrix} \beta(s) & -\alpha(s) \\ -\alpha(s) & \gamma(s) \end{pmatrix} = M(s) \cdot \begin{pmatrix} \beta_0 & -\alpha_0 \\ -\alpha_0 & \gamma_0 \end{pmatrix} \cdot M(s)^T \qquad (48)$$

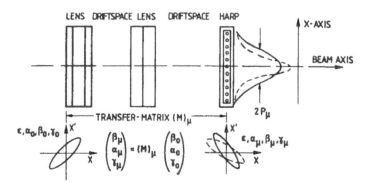

FIGURE 64
Scheme to illustrate the determination of transverse emittances by the gradient variation method. (From Strehl, P., *Proc. 9th Int. Conf. Cyclotrons and Their Applications*, Caen (France), 1981, 545–554. With permission.)

where M(s) is the 2×2 matrix of the single-particle transformation between 0 and s, defined by

$$\begin{pmatrix} y(s) \\ y'(s) \end{pmatrix} = M(s) \cdot \begin{pmatrix} y(0) \\ y'(0) \end{pmatrix} \qquad (49)$$

and $M(s)^T$ is the transpose of M(s). In a quadrupole system, taking the notation of Reference 104, M(s) is determined by the so-called cosinelike trajectory C(s), the sinelike trajectory S(s) and their derivations C'(s) = dC(s)/ds, S'(s) = dS(s)/ds:

$$M(s) = \begin{pmatrix} C(s) & S(s) \\ C'(s) & S'(s) \end{pmatrix} \qquad (50)$$

Therefore, the interesting transformation of β is given by

$$\beta(s) = \left(C(s)^2 \quad -2C(s)\,S(s) \quad S(s)^2 \right) \cdot \begin{pmatrix} \beta_0 \\ \alpha_0 \\ \gamma_0 \end{pmatrix} \qquad (51)$$

With respect to the two variants of measuring algorithms under discussion, only the transfer matrices for a drift space M(D), for a focusing quadrupole M(k), and their combination M(D) · M(k) have to be considered. Extending both variants to more than the needed three measurements, a least squares fit can be applied to the relation[89]

$$F = \sum_\mu \left[P_\mu^2 - \left(C_\mu^2 \beta_0 \varepsilon - 2C_\mu S_\mu \alpha_0 \varepsilon + S_\mu^2 \gamma_0 \varepsilon \right) \right]^2 \qquad (52)$$

where the $P_\mu = \sqrt{\varepsilon \beta_\mu}$ are determined by the measured profile width (see Figure 60) and μ is the number of settings with different transfer matrices. Solution of the equations resulting from

$$\frac{\partial F}{\partial \beta_0 \varepsilon} = 0 \quad \frac{\partial F}{\partial \alpha_0 \varepsilon} = 0 \quad \frac{\partial F}{\partial \gamma_0 \varepsilon} = 0 \tag{53}$$

is straightforward keeping in mind $\beta_0 \gamma_0 - \alpha^2 = 1$. Because the method is very sensitive to the accuracy of the profile widths and the matrix elements, some work has been devoted to analyze the conditions for most accurate determination of the ellipse parameters. Obviously, the number of profile measuring devices that can be involved in the improved three-position method is limited and their geometric arrangement may not be optimal with respect to the emittance measurement. In each case optimum results can be obtained if one of the profile width measurements is performed in a waist position and the other profile measuring devices are arranged symmetrical around the device in waist position. There are also detailed investigations about the best ratio between profile width in the waist position and the other widths.[95,100,109] In this sense the gradient variation method is more flexible due to the possibility of varying the profile width in a wide range, which improves the available accuracy. But, before varying the transfer matrix by changing the quadrupole gradients, the beam should be aligned carefully because steering effects can falsify the results. Practical experience has also shown that the implementation of a semiautomatic algorithm[113] offers more control for the physicist, with respect to the range of profile variation and the location of the measured points in the $P_\mu = f(k_\mu)$ diagram (k_μ are the quadrupole gradients), than a fully automatic procedure. Figure 65 shows the data represented to the operator using the gradient variation method in a semiautomatic mode.

As discussed in Reference 97, detailed beam-profile measurements contain far more information about the distribution in the transverse phase space than necessary for the determination of one elliptical emittance contour using the methods discussed so far. Taking the entire profile shape, even the percentage of beam contained within a calculated phase-space area may be extracted from the collected profile data. This has been also discussed by various authors. In Reference 101 a new method is shown that obtains a so-called percent emittance even without the requirement for a Gaussian profile shape.

The accuracy of the gradient variation method can be improved by using more than one profile if there are other detectors in the line. Furthermore, extracting the exact location of the minimum profile width in the $P_\mu = f(k_\mu)$ diagram gives one more independent equation for the determination of the Twiss parameters. Remembering that for a focusing quadrupole of effective length l_q in front of a drift space D the matrix elements C and S are given by

$$C(k, D) = \cos(kl_q) - Dk \sin(kl_q) \tag{54}$$

$$S(k, D) = \frac{1}{k} \sin(kl_q) + D \cos(kl_q) \tag{55}$$

the location of the minimum profile width is determined by $\partial P / \partial k = 0$, and from Equation 52 it follows immediately

$$\left(C\frac{\partial C}{\partial k}\right) \cdot \beta_0 - \left[\left(S\frac{\partial C}{\partial k}\right) + \left(C\frac{\partial S}{\partial k}\right)\right] \cdot \alpha_0 + \left(S\frac{\partial S}{\partial k}\right) \cdot \gamma_0 = 0 \tag{56}$$

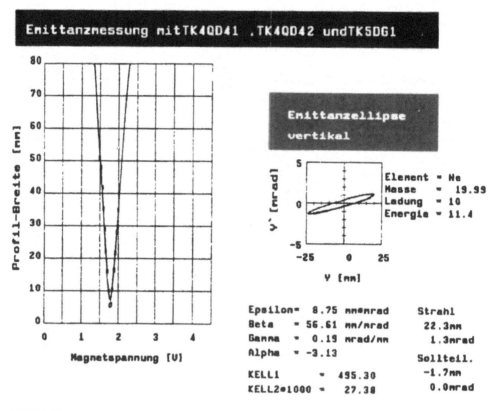

Emittanzellipse
vertikal

Element = Ne
Masse = 19.99
Ladung = 10
Energie = 11.4

Epsilon= 8.75 mm•mrad   Strahl
Beta = 56.61 mm/mrad   22.3mm
Gamma = 0.19 mrad/mm   1.3mrad
Alpha = -3.13   Sollteil.
KELL1 = 495.30   -1.7mm
KELL2•1000 = 27.38   0.0mrad

FIGURE 65
Computer display of an emittance measurement applied to a neon beam with 11.4 MeV/u, using the quadrupole gradient variation method. The diagram at the left-hand side shows the measured profile width (axis of ordinates in mm) in dependence of the quadrupole gradient (abscissa normalized to 10 Volts). The calculated emittance ellipse (representing ca. 90% of the total beam) is shown in the small insert (axis of ordinates in mrad, abscissa in mm).

The evaluation of the partial derivations leads to

$$\left(\frac{\partial C}{\partial k}\right) = -l_q \sin\left(kl_q\right) - D\sin\left(kl_q\right) - Dkl_q \cos\left(kl_q\right) \tag{57}$$

$$\left(\frac{\partial S}{\partial k}\right) = \frac{-1}{k^2}\sin\left(kl_q\right) + \frac{l_q}{k}\cos\left(kl_q\right) - Dl_q \sin\left(kl_q\right) \tag{58}$$

One of the unknown Twiss parameters β_0, α_0, or γ_0 can now be determined from Equation 56 to reduce the number of variables in the least squares fit of Equation 52 by one.

5   MEASUREMENTS IN THE LONGITUDINAL PHASE SPACE

For bunched beams the analogous description of emittances is defined by the momentum spread that replaces the divergence in the transverse phase space and the phase deviation that replaces the transverse coordinate. Both coordinates have to be measured with respect to a fictitious reference particle. Due to the proportionality

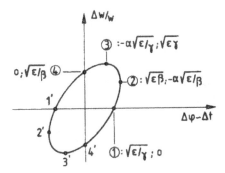

FIGURE 66
Emittance ellipse definition in the longitudinal phase space. (From Strehl, P., *Kerntechnik*, 214–223, 1991. With permission.)

between momentum and energy, a representation as shown in Figure 66, using the relative energy spread instead of absolute momentum spread, has proven to be very practical. Obviously, for dc beams the definition of a phase deviation does not make sense. But in many cases there will be a prebuncher, buncher, or even a small rf-accelerating structure available, which allows the use of capacitive pickups, coaxial cups, and the application of measuring procedures suitable for bunched beams.[119–122,126,127,130] Although energy can easily be determined with spectrometers, and energy spread may be measured with a Thomson-parabola spectrometer, Wien-filter,[118] or by using a chopper cavity,[128] the application of methods for bunched beams can be useful. A good example is the precise energy determination applying the time-of-flight (TOF) technique to bunches generated by a prebuncher in the injection area of an accelerator. The important parameters are explained in Figure 67. Figure 68 is the picture of a capacitive pickup. Due to the low energies under consideration the geometrical distance $\beta \cdot \lambda$ between bunches is on the order of only some centimeters for typical λs, where $\lambda = c/f = c \cdot T_f$ and f is the frequency of the bunching system. The particle energy is determined by the measured time difference δt between the signals at the first and the second pickups (see Figure 67) and the number N_b of bunches by:

$$v = \beta \cdot c = \frac{L_s}{N \cdot T_f + \delta t} \tag{59}$$

$$\gamma = \frac{1}{\sqrt{1-\beta^2}} \tag{60}$$

$$W = m_u c^2 (\gamma - 1) \quad [\text{MeV u}] \tag{61}$$

The number N_b has to be determined from

$$N_b = \frac{L_s}{\beta_1 \cdot \lambda} \tag{62}$$

where β_1 results from a first, rough determination of energy, taken, for example, from a coarse readout of the high voltage or from the measured magnetic field of a

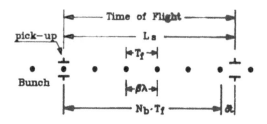

FIGURE 67
Precise energy measurement using the TOF technique.

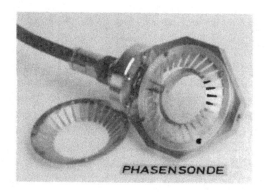

FIGURE 68
Capacitive pickup.

deflecting magnet. Obviously, the relation $\Delta N_s/N_s < 1$ has to be fulfilled. The accuracy that can be achieved can be estimated nonrelativistically and figures out to be

$$\left|\frac{\Delta W}{W}\right| = \frac{2 \cdot \Delta \delta t}{N_b \cdot T_f + \delta t} \tag{63}$$

neglecting the error in the determination of L_s, with $\Delta \delta t$ as the measuring error in δt. Taking some typical values of $L_s = 1$ m, $\beta = 0.5\%$, $f = 27$ M H z, $\Delta \delta t = 100$ µs, an accuracy of $\Delta W/W \approx 3 \times 10^{-4}$ can be realized. Clearly, the improvement in accuracy results from the large number of bunches between both pickups, taking into account the very high precision of T_f. From Figure 67 it becomes evident that β_1 can also be obtained by installing a third pickup in a short distance to one of the other two pickups.[89,119] The technique may be used for the very precise calibration of the terminal voltage in an injector system. Taking advantage of a buncher in the beam transport system, the longitudinal emittance can be determined in analogy to the methods described for nondestructive transverse emittance measurements. Again, there are two possibilities:[132]

1. Measuring the bunch width Δt at three or more positions along a drift space
2. Varying the bunch width with the help of an rf resonator (buncher, debuncher) operated at $\varphi_s = \pm 90°$, which corresponds to no energy change for the center of the bunch

Figure 69 is a simplified schematic diagram illustrating the algorithm according to the second method.

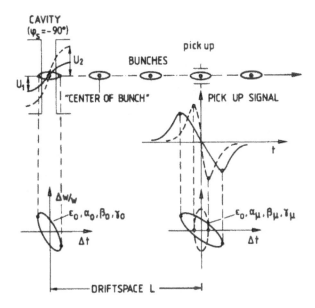

FIGURE 69
Simplified schematic diagram illustrating the algorithm according to method 2. (From Strehl, P., *Kerntechnic*, 214-223, 1991. With permission.

Referring to Equation 50, the change of the bunch width ($\sim\sqrt{\varepsilon} \cdot \beta$; see Figure 66) is determined by

$$\Delta t_\mu^2 = C_\mu^2 \cdot \beta_0 - 2C_\mu S_\mu \cdot \alpha_0 + S_\mu^2 \cdot \gamma_0 \tag{64}$$

$$C_\mu^2 = \left(1 + D_C \cdot U_\mu \cdot K\right)^2 \tag{65}$$

$$2C_\mu S_\mu = 2 \cdot \left(1 + D_C \cdot U_\mu \cdot K\right) \cdot K^2 \tag{66}$$

Here K is defined by the driftspace between the resonator and the pickup and D_C contains the relevant beam parameters as well as the characteristics of the rf resonator according to:

$$\frac{\Delta W}{W} = \frac{1}{W} \cdot \frac{q}{A} \cdot T_{transit} \cdot U \cos\varphi_s = D_C \cdot U \cdot \cos\varphi_s \tag{67}$$

where U is the rf gap-voltage and $T_{transit}$ is the transit time factor. For small $\Delta\varphi$ around $\varphi_s = \pm 90°$ (see Figure 69 for the definition of φ_s) and positive ions, Equation 67 simplifies to

$$\frac{\Delta W}{W} = D_C \cdot U \cdot \Delta\varphi \tag{68}$$

According to Equation 67 the constant D_C can be determined experimentally by switching the rf phase to accelerate or decelerate the bunches and measuring the energy change for the selected rf voltage U by the TOF technique. From practical experience[113,132] the following dimensions are recommended for calculating the relevant parameters:

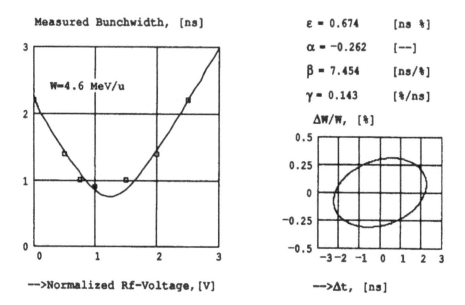

FIGURE 70
Results from a longitudinal emittance measurement at the Unilac, using method 2.

$$K[ns/\%] = -\frac{0.005 \cdot L[m]}{\beta \cdot c[m/ns]} \qquad (69)$$

$$D_C[\%/V \cdot ns] = \frac{200}{U[V] \cdot \cos\varphi} \cdot \frac{L[m] \cdot \delta t[ns]}{\beta \cdot c[m/ns]} \cdot \frac{2\pi f[MHz]}{1000} \qquad (70)$$

where δt is the measured change in the time of flight t_f for $\varphi_s \neq \pm 90°$, determined in nonrelativistic approximation by $\Delta W/W = -2\delta t/t_f$, with $t_f = L/\beta c$. Figure 70 shows the results of a measurement using method 2.

Obviously, the relations given above hold also for method 1, replacing K by K_μ with $\mu \geq 3$ and $D_C \equiv 0$.

The energy spread of charged particles may be also measured by the use of semiconductor detectors.[129] According to the scheme in Figure 71, the pulse height distribution from the detector, analyzed with the aid of a multichannel analyzer, represents the energy distribution of the particles. Semiconductors of barrier type[133] with a typical electrode thickness of 40 µg/cm² and a sensing depth of 100 µm are suitable for the energy region of interest. This method does not require a bunched beam. But in applying this technique one has to keep in mind that the particle flux has to be attenuated considerably to avoid detector destruction. The achievable resolution is in the order of 1% for $\Delta W/W$. In the case of bunched beams, the TOF technique, taking advantage of the very stable rf reference frequency according to Figure 72, allows precise measurements of bunch structure and also determination of longitudinal emittances.

ACKNOWLEDGMENT

The author is very grateful to the members of his group, the beam diagnostics group of GSI, for many years of good collaboration. The author wishes especially to thank Peter Moritz, Hansjörg Reeg, and Helgi Vilhjalmsson for the corrections and valuable suggestions. Most of the photographs are from A. Zschau, GSI.

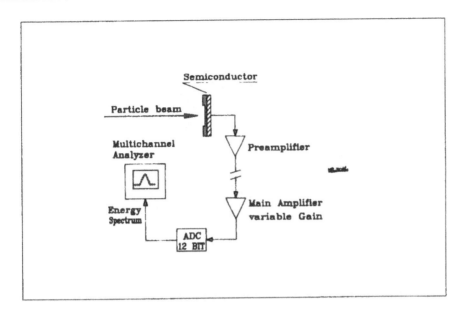

FIGURE 71
Simplified scheme for the measurement of energy spread with semiconductors.

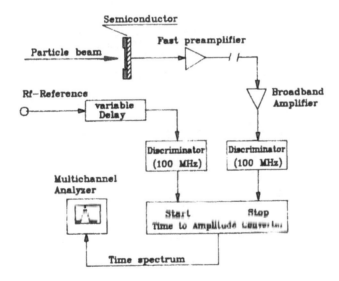

FIGURE 72
Simplified scheme for measuring the time structure of bunches with semiconductors.

REFERENCES

1. Sanborn, C. and Brown, *Basic Data of Plasma Physics*, MIT Press, Massachusetts, 1966.
2. Jamba, D., Secondary particle collection in ion implantation dose measurement, *Rev. Sci. Instrum.*, 49, 5, 634, 1978.
3. Kraus, H., Störmer, J., and Strehl, P., Konstruktive Auslegung von Faraday-Tassen, GSI-Bericht PB-1–75, pp. 27–33, 1975.
4. McKenna, C. M., High Current Dosimetry Techniques, *Radiation Effects*, Vol. 44, Gordon and Breach, Holland, 1979, 93–110.

5. Carter, G. and Colligon, J., *Ion Bombardment of Solids*, Heinemann, London, 1969, 310 ff.
6. *Physikhütte, Band II*, Verlag von Wilhelm Ernst & Sohn, Berlin, 1971, 308 ff.
7. *Handbook of Chemistry and Physics*, 54th ed., CRC Press, Cleveland, OH, 1973–1974, D56 ff.
8. Goodfellow Cambridge limited, Cambridge Science Park, Cambridge, C84 4DJ, Great Britain (Katalog 1993/94).
9. Northcliffe and Schilling, Nuclear Data Tables, Vol. 7, No. 3–4, Jan 1970, p. 233–463.
10. CERN Report 72–19, Nov. 1972.
11. Hubert, F., Fleury, A., Bimbot, R., and Gardes, D., *Ann. Phys.*, Suppl., 5, 1, 1980.
12. Hubert, F., Bimbot, R., and Gauvin, H., *At. Data Nucl. Data*, 46, 1, 1990.
13. Ziegler, J. F., *Handbook of Stopping Cross-Sections for Energetic Ions in All Elements, Vol. 5, The Stopping and Ranges of Ions in Matter*, Pergamon Press, 1980.
14. Smirnow, W. L., *Lehrgang der Höheren Mathematik, Teil II*, VEB-Verlag der Wissenschaften, Berlin, 1960, 547–559.
15. Gröber, E. G., *Die Grundgesetze der Wärmeübertragung*, Springer-Verlag, Berlin, 1981, 330–336.
16. Strehl, P., Thermische Berechnungen zum Hochstrombetrieb, GSI-Report 94–03, 1994.
17. Strehl, P., Beam diagnostics, *Rev. Sci. Instrum.*, 63 (2), 2652–2659, 1992.
18. Kandler, Th., Konstruktion, Bau und Inbetriebnahme einer Apparatur zur Bestimmung der Strahlleistung von Ionenstrahlen, Diplomarbeit im Fachbereich Physikalische Technik der Fachhochschule Wiesbaden, durchgeführt bei der GSI-Darmstadt, 1984.
19. Hell, F., *Grundlagen der Wärmeübertragung*, VDI-Verlag GmbH, Düsseldorf, ISBN 3–18–400529–1, 1982.
20. Meinke, G., *Taschenbuch der Hochfrequenztechnik*, Springer-Verlag, Berlin, 1986.
21. Christodoulides, C. E. and Freeman, J., H., Ion beam studies. Part II, Nucl. *Instrum. Methods*, 135, 13–19, 1976.
22. Sanders, J. H., *J. Sci. Instrum.*, 26, 36, 1949.
23. Harrison, E. R., *J. Sci. Instrum.*, 34, 242, 1957.
24. Andersen, H. H., *Rad. Effects*, 3, 51, 1970.
25. Van de Runstraat, C. A., Wijnaendts, van Resandt, R., and Los, J., *J. Phys. E*, 3, 575, 1970.
26. Gunn, S. R., Radiometric calorimetry: a review, *Nucl. Instrum. Methods*, 29, 1, 1964.
27. Williams, F. C. and Nobel, S. W., The fundamental limitations of the second harmonic type of magnetic modulator as applied to the amplification of small DC signals, *J. IEE London*, 97, 445–459, 1950.
28. Unser, K. B., Beam current transformer with DC to 200 MHz range, *IEEE Trans. Nucl. Sci.*, NS-16, 934–938, 1969.
29. Gardiner, S. N., Matthews, J. L., and Owens, R. O., An accurate non-intercepting beam current integrator for pulsed accelerator beams, *Nucl. Instrum. Methods*, 87, 285–290, 1970.
30. Steiner, R., Merle, K., and Andresen, H. G., A high precision Ferrite-Induction beam current monitoring system, *Nucl. Instrum. Methods*, 127, 11–15, 1975.
31. Reimann, R., Methode zur rückwirkungsfreien Messung eines Ionenstrahls, *Nucl. Instrum. Methods*, 137, 385–386, 1976.
32. Appelo, H. C., Groenenboom, M., and Lisser, J., The zero flux DC current transformer, a high precision wide-band device, *IEEE Trans Nucl. Sci.*, NS-28, 3, 1810–1811, 1977.
33. Groenenboom, M. and Lisser, J., Accurate measurements of DC and AC by transformer, *Electron. Power*, 52–55, 1977.
34. Dunn, P. C., Absolute beam charge measurements with toroid monitors: experience at the Bates Linac, *Nucl. Instrum. Methods*, 165, 163–167, 1979.
35. Tokuda, N. and Watanabe, S., Electrostatic and ferrite core monitors, INS, University of Tokyo, report, INS-NUMA-21.
36. Unser, K. B., A toroidal DC current transformer with high resolution, *IEEE Trans. Nucl. Sci.*, NS-28, 3, 2344–2346, 1981.
37. Unser, K. B., Toroidal AC and DC transformers for beam intensity measurements, *Atomenergie-Kerntechnik*, 47, 1, 48–52, 1985.
38. Reeg, H.-J., Dimensionierung, Konstruktion, Bau und Erprobung eines elektromagentisch abgeschirmten Strahltransformators mit einer Zeitkonstanten von etwa 0, 5 μs, Diplomarbeit, FH-Wiesbaden, GSI-Darmstadt, 1986.
39. Loyer, F., Andre, T., Ducoudret, B., and Rataud, J. P., New beam diagnostics at Ganil: very sensitive current transformers in beam lines and counting system of beam turns in cyclotrons, *IEEE Trans. Nucl. Sci.*, NS-32, 5, 1938–1940, 1985.
40. Sato, Y., Yamada, T., Ogawa, H., and Fujii, R., A non-intercepting current monitor, *Nucl. Instrum. Methods*, 228, 576–578, 1985.
41. Unser, K. B., Design and preliminary tests of a beam intensity monitor for LEP, *Proc. IEEE Particle Accelerator Conference*, March 1989, Chicago, Vol. 1, pp. 71–73.

42. Unser, K. B., Measuring bunch intensity, beam loss and bunch lifetime in LEP, *Proc. Second European Particle Accelerator Conference*, Nice, June 12–16, 1990, Vol. 1, pp. 786–788.

43. Burtin, G., Colchester, R., Fischer, C., Hemery, J. Y., Jung, R., Vanden Eynden, M., and Vouillot, J. M., Mechanical design, signal processing and operator interface of the LEP beam current transformers, *Proc. Second European Particle Accelerator Conference*, Nice, June 12–16, 1990, Vol. 1, pp. 794–796.

44. Degueurce, L. Beam diagnostics for the Ebis Ion Source at Saturn, *Nucl. Instrum. Methods*, A260, 538–542, 1987.

45. Reimann, R. and Rüede, M., Strommonitor für die Messung eines gepulsten Ionenstrahls, *Nucl. Instrum. Methods*, 129, 53–58, 1975.

46. Schneider, N. and Walter, H., Beam current measurements, GSI Scientific Report 1990, GSI 91-1, March 1991, ISSN 0174–0814, p. 386.

47. Reeg, H. and Schneider, N., The beam current transformers at SIS and ESR, GSI Scientific Report 1990, GSI 91-1, March 1991, ISSN 0174–0814, p. 392.

48. Unser, K. B., The parametric current transformer, Particles and fields, Series 46, Conference proceedings No. 252, *Third Annual Workshop on Accelerator Instrumentation*, CEBAF, Newport News, 1991, pp. 266–275, ISBN 0–88318–934–8.

49. Internal Cern-Report, The induction type beam Monitor for the PS, ("Hereward Transformer"), MPS/Int. CO 62–15, 1962.

50. Reeg, H.-J. and Störmer, J., Gessellschaft für Schwerionenforschung mbH, Planckstraße 1, Postfach 11 05 52, D-64220 Darmstadt, Germany, 1993, private communication.

51. Boll, R., *Weichmagnetische Werkstoffe, Vacuumschmelze GMBH*, Verlag: Siemens AG, Berlin und München, 1965, ISBN 3–8009–1546–4.

52. Unser, K. B., Recent advances in beam current transformer technology and avenues for further developments, *Proc. 1st European Workshop on Beam Diagnostics and Instrumentation for Particle Accelerators*, Montreux, May 3–5, 1993, CERN PS/93–35 (BD), CERN SL/93–35 (BI), pp. 105–109.

53. Pruitt, J. S., Electron beam current monitoring system, *Nucl. Instrum. Methods*, 92, 285–297, 1971.

54. Peters, A., Vodel, W., Dürr, V., Koch, H., Reeg, H., and Schroeder, C. H., A cryogenic current comparator for low intensity ion beams, *Proc. 1st European Workshop on Beam Diagnostics and Instrumentation for Particle Accelerators*, Montreux, May 3–5, (1993), CERN PS/93–35 (BD), CERN SL/93–35 (BI), pp. 100–104.

55. Kuchnir, M., McCarthy, J. D., and Rapidis, P. A., Squid based current meter, *IEEE Trans. Magn.*, 21, 2, 997–999, 1985.

56. Grohmann, K. and Hechtfischer, D., Krystromkomparatoren als Präzisionsstandards für rationale Gleich und Wechselstromverhältnisse, PTB-Mitteilungen 92, 5/82, pp. 328–344.

57. Harvey, K. I., A precise low temperature dc ratio transformer, *Rev. Sci. Instrum.*, 43, 1626, 1972.

58. Lejeune, C. and Aubert, J., Emittance and brightness: definitions and measurements, *Adv. Electron. Electron Phys.*, Suppl. 13A, 159–259, 1980.

59. Van Steenbergen, A., Evaluation of particle beam phase space measurement techniques, *Nucl. Instrum. Methods*, 51, 245–253, 1967.

60. Van Steenbergen, A., *IEEE Trans. Nucl. Sci.*, NS-14, 746.

61. Walcher, W., *Proc. Int. Conf. Ion Sources*, 2nd, Vienna, 1972, p. 111.

62. Hutter, R., Beams with space-charge, in *Focusing of Charged Particles*, Vol. II, A. Septier, Ed., Academic Press, 1967, 3–22.

63. Müller, R. W., Handwerkszeug für Raumladungs-Rechnungen, GSI, 1979, internal note.

64. Lawson, J. D., Beams and sources, *Nucl. Instrum. Methods*, 139, 17–24, 1976.

65. Pierce, J. R., *Theory and Design of Electron Beams*, Van Nostrand, Princeton, NJ, 1954, chap. 6.

66. Lawson, J. D., *The Physics of Charged Particle Beams*, Clarendon Press, Oxford.

67. Kapchinskij, I. M. and Vladimirskij, V. V., *Proc. Conf. High Energy Acc.*, CERN, Geneva, 1959, 274.

68. Fradj, M., GSI-Darmstadt, Gesellschaft für Schwerionenforschung mbH, Planckstraße 1, Postfach 11 05 52, D-64220 Darmstadt, Germany, 1993, private communication.

69. Witkover, R. L., Automatic electronic emittance device for the BNL 200 MeV Linac *Proc. 1970 Proton Linear Acclerator Conference*, Batavia, Illinois, Vol. 1, 1970, pp. 125–142.

70. Wulf, F., Egelhaaf, C., Homeyer, H., Das Strahldiagnosesystem am Schwerionenbeschleuniger VICKSI, HMI-Berlin, Report Nr. HMI-B 300 (D 86), June 1979.

71. Kuibida, R. P. and Stolbunov, V. S., Gerät zur Messung der Strahlemittanz am Eingang und Ausgang des Linearbeschleunigers I-2, translated from Russian, internal report, DESY L-Trans-305, June 1985.

72. Riehl, G., Untersuchung der mehrdimensionalen transversalen Phasenraumverteilungen von intensiven Ionenstrahlen, Dissertation, Institut für Angewandte Physik, J. W. Goethe Universität Frankfurt, 1992, to be published as GSI-report.

73. Tanaka, J., Baba, H., Sato, I., Inagaki, S., Anami, S., Kakuyama, T., Takenaka, T., Terayama, Y., and Matsumoto, H., Development of the KEK injector Linac, IEEE Trans. Nucl. Sci., NS-24, 3, 1124–1126, 1977.

74. Wenzler, F., Konstruktion und Aufbau eines Strahldiagnosesystems zur gleichzeitigen Messung der horizontalen und vertikalen Emittanz, Diplomarbeit, Fachhochschule Wiesbaden, Physikalische Technik, 1988.

75. Kuhlmann, W., Bojowald, J., Mayer-Böricke, C., Reich, J., and Retz, A., An automatic beam-emittance-measuring device for the Jülich isochronous cyclotron (JULIC), Nucl. Instrum. Methods, 80, 89–94, 1970.

76. Wroe, H., Some emittance measurements with a Duoplasmatron ion source, Nucl. Instrum. Methods, 52, 67–76, 1967.

77. Gabovich, M. D., Kompensierte Ionenstrahlen, Ukr. Fiz. Zh., 24, 1979; translated as GSI-report, GSI-tr-79/15.

78. Nezzlin, M. V., Plasma instabilities and the compensation of space charge in an ion beam, Plasma Phys., 10, 337–358, 1968.

79. Humphries, S., Jr., Anderson, R. J. M., Freeman, J. R., Lockner, T. R., Poukey, J. W., and Ramirez, J. J., Pulselac program: space charge neutralized ion beams for internal fusion applications, Nucl. Instrum. Methods, 187, 289–294, 1981.

80. Bozsik, I. and Hofmann, I., Space charge effects in the focusing of intense ion beams.

81. Hofmann, I., Emittance growth of ion beams with space charge, Nucl. Instrum. Methods, 187, 281–287, 1981.

82. Holmes, A. J. T., Theoretical and experimental study of space charge in intense ion beams, Phys. Rev. A, 19, 1, 389–407, 1979.

83. Evans, L. R. and Warner, D. J., Space-charge neutralisation of intense charged particle beams: some theoretical considerations, CERN/MPS/DIN. 71-2 (August 1971).

84. Humphries, S., Jr. and Poukey, J. W., Proposed method for the transport of ions in linear accelerators utilizing electron neutralization, Part. Accel., 10, 107–123, 1980.

85. Wroe, H., Emittance measurements with a single gap accelerator and the production of very low emittance proton beams, Nucl. Instrum. Methods, 58, 213–222, 1968.

86. Ames, L. L., Rapid measurement of emittance and brightness, Nucl. Instrum. Methods, 151, 363–369, 1978.

87. Tajima, S., Yoshida, T., and Kikuchi, S., Measurements of the negative ion beam emittance for the JAERI tandem accelerator, Nucl. Instrum. Methods Phys. Res., A326, 407–415, 1993.

88. Struckmeier, J., Klabunde, J., and Reiser, M., On the stability and emittance growth of different particle phase-space distributions in a long magnetic quadrupole channel, Part. Accel., 15, 47, 1984.

89. Strehl, P., Beam diagnostic devices for a wide range of currents, Proc. 9th Int. Conf. Cyclotrons and Their Applications, Caen (France), 1981, pp. 545–554.

90. Strehl, P., Beam diagnostics, CAS Cern accelerator school, Aarhus (Denmark), September 1986, CERN 87–10, 1987, pp. 99 104.

91. Holms, A. J. T., Neutralization of ion beams, CAS-Cern accelerator school, Aarhus (Denmark), September 1986, CERN 87–10, 1987, pp. 79–92.

92. Nickel, F., Marx, D., and Ewald, H., Temperature of thin targets in a pulsed electron beam, Nucl. Instrum. Methods, 134, 11–14, 1976.

93. Nickel, F., Festkörpertargets bei hohen Strahlbelastungen, Inaugural-Dissertation, Justus-Liebig-Universität Giessen, Fachbereich 13, 1976.

94. ITT Reference Data for Radio Engineers, 5th ed., Howard W. Sams & CO., Inc., Indianapolis/Kansas City/New York.

95. Metzger, C., Mesures des emittances et du centrage des faisceau dans la ligne de mesure "800 MeV" du PSB, CERN/SI/Int. DL69–10.

96. Baribaud, G. and Metzger, C., The 800 MeV measurement line of the CERN P.S. Booster, IEEE Trans. Nucl. Sci., NS-20, 3, 659, 1973.

97. Goodwin, R. W., Gray, E. R., Lee, G. M., and Shea, M. F., Beam diagnostics for the NAL 200 MeV line, Proc. 1970 Proton Linear Accelerator Conference, 1970, pp. 107–113.

98. Franczak, B., Emittanzmessung mit Quadrupolen — Auswertung, internal note, GSI-Darmstadt, AN-F-060673-3, 1973.

99. Schaa, V., Emittance Measurements with Profile Grids, Thesis, TH-Darmstadt, GSI-Darmstadt, 1978.

100. Qian, Y. L., Riedel, C., Schaa, V., Strehl, P., Determination of beam emittance by profile measurements, internal report, GSI-Darmstadt, 1990.

101. Ebihara, K., Tejima, M., Kawakubo, T., and Takano, S., Non-destructive emittance measurement of a beam transport line, *Nucl. Instrum. Methods*, 202, 403–409, 1982.

102. Birukov, I. N., Mirzojan, A. N., Ostroumov, P. N., and Petronevich, S. A., Transverse beam parameter measurements at the INR Proton Linac, *Proc. 3rd European Particle Accelerator Conference*, EPAC, Berlin, March 1992, ISBN 2-86332-114-5, pp. 1109–1111.

103. Kester, O., Rao, R., and Rinolfi, L., Beam emittance measurement from CERN thermionic guns, *Proc. 3rd European Particle Accelerator Conference*, EPAC, Berlin, March 1992, ISBN 2-86332-114-5, pp. 1017–1019.

104. Steffen, K. G., *High Energy Beam Optics*, Wiley-Interscience, New York, 1965.

105. Banford, A. P., *Transport of Charged Particles*, Spon, London, 1966.

106. Bedfer, Y., Farvacque, L., Milleret, G., and Soudan, J. M., Emittance measurements at the experimental areas of the synchrotron Saturne II, *Kerntechnik*, 56, 4, 224–228, 1991.

107. Courant, E. D. and Snyder, H. S., *Ann. Phys.*, 3, 1–48, 1958.

108. Joho, W., Representation of beam ellipses for transport calculations, SIN-report, TM-11-14, 1980, pp. 1–31.

109. Bovet, C. and Guignard, C., Mesure de l'emittance d'un faisceau primaire, Rep. CERN/DL/68-7.

110. Grivet, P. A., *Electron Optics*, Pergamon, Oxford, 1965.

111. Grivet, P. A., Electron Optics, 2nd ed., Pergamon, Oxford, 1972.

112. Septier, A., Ed., Focusing of Charged Particles, Vol. II, Academic Press, 1967, 123.

113. Scheeler, U., GSI-Darmstadt, Gesellschaft für Schwerionenforschung mbH, Planckstraße 1, Postfach 11 05 52, D-64220 Darmstadt, Germany, 1993, private communication.

114. Holmes, A. J. T., Space charge neutralisation of ion beams, *Inst. Phys. Conf. Ser.*, 38, 222–227, 1978.

115. Smith, H. V. and Allison, P., H⁻ beam emittance measurements for the Penning and the asymmetric, grooved magneton surface-plasma sources, *Rev. Sci. Instrum.*, 53(4), 405–408, 1982.

116. Haddock, Ch., Anderson, J., De Jong, M., Konopasek, F., and Smith, Ch. A., Neutral beam profiles and a computer-controlled emittance measuring device for a cyclotron beam, *Rev. Sci. Instrum.*, 53(8), 1185–1188, 1982.

117. Sun, B. H. and Cheng, Q., Space-charge neutralization lens and its application, *Rev. Sci. Instrum.*, 64(6), 1437–1441, 1993.

118. Matsuzawa, T., Takahashi, A., Masugata, K., Ito, M., Matsui, M., and Yatsui, K., Time-resolved measurement of energy and species of an intense pulsed ion beam, *Rev. Sci. Instrum.*, 56(12), 2279–2289, 1985.

119. Strehl, P., Klabunde, J., Schaa, V., Vilhjalmsson, H., and Wilms, D., Das Phasensondensystem am Unilac: Sondendimensionierung und Signalauswertung, GSI-Report 79–13, 1979, translated for the Los Alamos Scientific Laboratory by Leo Kanner Associates, Redwood City, 1980, LA-TR-80-43.

120. Feinberg, B., Meaney, D., Thatcher, R., and Timossi, C., On-line velocity measurements using phase probes at the Superhilac, *Nucl. Instrum. Methods*, A270, 1–5, 1988.

121. Bongardt, K. and Kennepohl, K., Determination of 100 ps short Linac pulses by broad-band pickups and reconstruction technique, *Proc. European Particle Accelerator Conference*, EPAC, ROME, June 1988, ISBN 9971-642-4, pp. 976–978.

122. Van der Hart, A., Schmidt, F., Kennepohl, K., Bongardt, K., and Uhlemann, R., Nondestructive beam monitors for the SNQ-Linac, *Proc. 1986 Linear Accelerator Conference*, Stanford, June 1986, SLAC-report-303, pp. 56–58.

123. Gilpatrick, J. D. and Grant, D. L., Multiple-measurement beam probe, *Proc. 1986 Linear Accelerator Conference*, Stanford, June 1986, SLAC-report-303, pp. 154–156.

124. Allison, P. W., Sherman, J. D., and Holtkamp, J. D., An emittance scanner for intense low-energy ion beams, Proc. 12th Int. Conf. High-Energy Accelerators, IEEE Trans. Nucl. Sci., Vol. NS-30, (1983), pp. 2204–2206.

125. Miller, R. H., Clendenin, J. E., James, M. B., Sheppard, J. C., Nonintercepting emittance monitor, Proc. 12th Int. Conf. High-energy Accelerators, *IEEE Trans. Nucl. Sci.*, NS-30, 602–605, 1983.

126. Cuperus, J. H., Monitoring of particle beams at high frequencies, *Nucl. Instrum. Methods*, 145, 219–231, 1977.

127. Reimann, R., Kapazitive Sonde für Phasenmessungen an einem gepulsten Ionenstrahl, *Nucl. Instrum. Methods*, 136, 397–398, 1976.

128. Bongardt, K., Mitra, A. K., Sauer, M., and Stockhorst, H., The deflector-pickup system: a novel approach to measure online the energy spread of DC ion beams, *Nucl. Instrum. Methods*, A287, 273–278, 1990.

129. Ziem, A. and Borisch, P., Unilac beam diagnostics with particle detectors, GSI-Scientific report, 84–1, 1984, p. 305.

130. Vilhjalmsson, H., Capacitive pick-up versus inductive loop monitor, GSI-Scientific report, 84–1, 1984, p. 306.

131. Schneider, N., The beam structure demodulator, a nondestructive current monitoring system, GSI-Scientific report, 84–1, 1984, p. 307.

132. Strehl, P., A new method for longitudinal emittance measurement, Proc. 1983 Particle Accelerator Conference, *IEEE Trans. Nucl. Sci.*, NS-30, 2198–2200, 1983.

133. EG&G ORTEC, 100 Midland Road, Oak Ridge, Tennessee, 37831–0895, U.S.A.

134. Sintek Keramik GmbH, Romantische Strasse 18, D-87642, private communication, 1993.

135. Strehl, P., Measurement of particle beam parameters, *Kerntechnik*, 214–223, 1991.

136. Harris, M., Maurer, A., and Turner, S., Mechanical design of beam monitors for the ISR and beam transfer systems, CERN-ISR-OP/71–30, Geneva, August 1971.

137. Knoll, F., Digitalisierungselektronik für die Auswertung von Intensitätsverteilungen auf Leuchttargets, Diplomarbeit D2607 ÜT, GSI-Darmstadt, April 1991.

138. Steinbach, Ch. and van Rooij, M., A scanning wire beam profile monitor, *IEEE Trans. Nucl. Sci.*, NS-32, 5, 1920–1922, 1985.

139. Hoffman, E. W., Macek, R. J., van Dyck, O., Lee, D., Harvey, A., Bridge, J., and Caine, J., High intensity beam profile monitors for the LAMPF primary beam lines, *IEEE Trans. Nucl. Sci.*, NS-26, 3, 3420–3422, 1979.

140. Jung, R. and Colchester, R. J., Development of beam profile and fast position monitors for the LEP injector Linacs, CERN/LEP-BI/85–16, Prevessin, May 1985.

141. Wegner, H. E. and Feigenbaum, I. L., High current beam scanner, Brookhaven National Laboratory, Upton, New York, for: Physicon Corporation, Boston, Mass. 02138, U.S.A.

142. Johnson, C. D. and Thorndahl, L., Non-destructive proton beam profile scanners, based on the ionization of residual gas in the beam vacuum chamber, *IEEE Trans. Nucl. Sci.*, NS-16, 3, 227, 1969.

143. De Luca, W. H., Beam detection using residual gas ionization, *IEEE Trans. Nucl. Sci.*, NS-16, 813, 1969.

144. Kobayashi, H., Yamazaki, Y., Kurihara, T., Sato, I., Otani, S., and Ishizawa, Y., Emittance measurement for high-brightness electron guns, *Proc. 1992 Linear Accelerator Conference*, Ottawa, Ontario, Canada, August 1992, pp. 341–343.

145. Gilpatrick, J. D., Marquez, H., Power, J., and Yuan, V., Design and operation of a bunched-beam, phase-spread measurement, *Proc. 1992 Linear Accelerator Conference*, Ottawa, Ontario, Canada, August 1992, pp. 359–361.

146. Hornstra, F., Jr., and Simanton, J. R., A simple nondestructive profile monitor for external proton beams, *Nucl. Instrum. Methods*, 68, 138–140, 1969.

147. Hortig, G., A beam scanner for two dimensional scanning with one rotating wire, *Nucl. Instrum. Methods*, 30, 355–356, 1964.

148. Michaelsen, R. and Seidel, E., Modification of the VICKSI beam profile monitor system for application at low beam intensities, *Nucl. Instrum. Methods*, A268, 503–505, 1988.

149. Samulat, G., Strahlprofilmessungen in den HERA-Protonen-Beschleunigern unter Ausnutzung der Restgasionisation, Diplomarbeit, Universität Hamburg, September 1989.

150. Budker, G. I., Dimov, G. I., and Dudnikov, V. G., Experiments on producing intense proton beams by means of the method of charge-exchange injection, *Sov. At. Energy*, 22, 5, 441–448, 1967.

151. Dimov, G. I. and Dudnikov, V. G., Determination of the current and current distribution in a proton beam, *Instrum. Exp. Techn.*, 3, 553, 1969.

152. Krider, J., Residual gas beam profile monitor, *Nucl. Instrum. Methods*, A278, 660–663, 1989.

153. Hochadel, B., Ein Strahlprofilmonitor nach der Methode der Restgasionisation für den Heidelberger Testspeicherring TSR, Report MPI Heidelberg, MPI H-1990-V-19, 1990.

154. Witkover, R. L., Proposal for a beam profile monitor for the booster ring, Booster Technical Note, No. 122, Brookhaven National Laboratory, (BNL), Upton, NY, 11973, May 1988.

155. Kawakubo, T., Ishida, T., Kadokuraa, E., and Ajima, Y., Fast data acquisition system of a non-destructive profile monitor for a synchrotron beam by using a microchannel plate with multi-anodes, *Nucl. Instrum. Methods*, A302, 397–405, 1991.

156. Wittenburg, K., Experience with the residual gas ionisation beam profile monitors at the DESY Proton accelerators, *Proc. 3rd European Particle Accelerator Conference*, EPAC, Berlin, March 1992, ISBN 2–86332–114–5, pp. 1133–1135.

157. Schippers, J. M., Kiewiet, H. H., and Zijlstra, J., A beam profile monitor using the ionization of residual gas in the beam pipe, *Nucl. Instrum. Methods*, A310, 540–543, 1991.

158. Hornstra, F. and Deluca, W. H., Nondestructive beam profile system for the zero gradient synchrotron, *Proc. 6th Int. Conf. High Energy Accelerators*, M. I. I., Cambridge, MA, 1967, p. 374.

159. Weisberg, H., Gill, E., Ingrassia, P., and Rodger, E., An ionization profile monitor for the Brookhaven AGS, *IEEE Trans. Nucl. Sci.*, NS-30, 4, 2179–2181, 1983.

160. Ishimaru, H., Igarashi, Z., Muto, K., and Shibita, S., Beam profile measurements for KEK 12 GEV proton synchrotron, *IEEE Trans. Nucl. Sci.*, NS-24, 3, 1821–1823, 1977.

161. Degueurce, L., Armani, J. M., and Launay, F., A new beam current monitor for the MIMAS storage ring, *Proc. 3rd European Particle Accelerator Conference*, EPAC, Berlin, March 1992, ISBN 2–86332–114–5, pp. 1103–1105.

162. Borer, J. and Jung, R., Diagnostics, CAS-Cern accelerator school, Aarhus (Denmark), October 1983, CERN 84–15, 1984, pp. 385–467.

163. Koziol, H., Beam diagnostics for accelerators, CAS-Cern accelerator school, Jyväskylä, Finland, September 1992, to be published.

164. Jung, R., Beam intercepting monitors, Joint US-CERN school on particle accelerators; frontiers of particle beams, lecture note in physics, No. 343, Anacapri, Italy, October 1988, pp. 403–422.

165. Neet, D. A. G., Beam observation in the transfer channels of the intersecting storage ring, CERN/ISR-CO/68–47, October 1968.

166. Jung, R., CERN, Geneva, private communication, 1989.

167. Morgan Matroc, Ltd., D-53721 Siegburg, Germany, private communication.

168. Calabretta, L., Cuttone, G., Giove, D., and Raia, G., Beam diagnostic elements at the laboratorio nazionale del sud, *Nucl. Instrum. Methods*, A268, 496–502, 1988.

169. Chamberlin, D. D., Jameson, R. A., Minerbo, G. N., and Sander, O. R., Beam tomography in two and four dimensions, *Proc. 1979 Linac Accelerator Conference*, Los Alamos, 1979, pp. 314–318.

170. Colsher, J. G., Iterative three-dimensional reconstruction from tomographic projections, *Comput. Graph. Image Process.*, 6, 513–537, 1976.

171. Fraser, J. S., Beam tomography or ART in accelerator physics, Los Alamos Scientific Laboratory Report, LA-7498-MS, Nov. 1978.

172. Gilbert, P., Iterative methods for the three-dimensional reconstruction of an object from projections, *J. Theor. Biol.*, 36, 105–117, 1972.

173. Gordon, R. and Herman, G. T., Three-dimensional reconstruction from projection: a review of algorithms, *Int. Rev. Cytol.*, 38, 111–151, 1974.

174. Minerbo, G. N. and Sanderson, J. G., Reconstruction of a source from a few (2 or 3) projections, Los Alamos Scientific Report, July 1978.

175. Oppenheim, B. E., More accurate algorithms for iterative 3-dimensional reconstruction, *IEEE Trans. Nucl. Sci.*, NS-21, 72–77, 1974.

176. Galvin, J. E. and Brown, I. G., Ion beam profile monitor, *Rev. Sci. Instrum.*, 55(11), 1866–1867, 1984.

177. Toshiyuki, Iida, Yoshimi, Maekawa, Ryohich, Taniguchi, Masami Byakuno, and Kenji Sumita, Simple profile monitor for low intensity, low duty cycle pulse ion beams, *Rev. Sci. Instrum.*, 52(9), 1328–1331, 1981.

178. Bosser, J. and Burnod, L., Proposal for a profile monitor, CERN SPS/ABM/JB/Report 78-3, August 1978.

179. Northcliffe, L. C., Passage of heavy ions through matter, *Annu. Rev. Nucl. Sci.*, 13, 67, 1963.

180. Ziegler, J. F., *Stopping Cross-Sections for Energetic Ions in All Elements: The Stopping Power and Ranges of Ions in Matter 5*, Pergammon Press, 1984.

181. Gabel, A., Entwicklung eines Restgas-Monitors für Strahlprofilmessungen im Bereich höherer Teilchenströme, Durchführung von Messungen, Diplomarbeit im Fachbereich Physikalische Technik der Fachhochschule Wiesbaden, durchgeführt bei der GSI-Darmstadt, 1990.

182. Venica, E., Utilisation d'un detecteur base sur le principe de l'ionisation d'un gaz pour la mesure du profil d'un faisceau d'ions dans la domaine des hautes energies, GSI-Darmstadt, Germany, Report, June 1993.

183. Schotmann, Th., Das Auflösungsvermögen der Restgas Ionisations Strahlprofil Monitore für Protonen in PETRA und HERA, Diplomarbeit, DESY-Hamburg, Germany, DESY HERA 93–09, 1993.

184. Lapington, J. S., Breeveld, A. A., Edgar, M. L., and Trow, M.-W., Span — a novel high speed high resolution position readout, Optical and Optoelectronic Applied Science and Engineering, July 1990.

185. Lapington, J. S., Breeveld, A. A., Edgar, M. L., and Trow, M.-W., A novel imaging readout with improved speed and resolution, 2nd London PSD Conference, September 1990.

186. Unverzagt, M., Entwicklung eines Strahlprofilmonitors für den ESR, Diplomarbeit, Institut für Kernphysik der Johann-Wolfgang-Goethe-Universität Frankfurt, September 1992.

187. Berg, H. and Schmidt-Böcking, H., Residual gas beam profile monitor for SIS beams, GSI-Scientific Report 1989, GSI 90–1, March 1990, p. 280.

188. Reeg, H.-J., Gesellschaft für Schwerionenforschung mbH, Planckstraße 1, Postfach 11 05 52, D-64220 Darmstadt, 1993, private communication.

189. Spiricon, Advanced laser beam analyzer, Model LBA-100A, technical data, May 1991, Spiricon, Laser beam diagnostics, U.S.A.

190. Eckert, G. and Winter, J., Strahlrekonstruktion aus Profilmessungen, Diplomarbeit, Fachhochschule Darmstadt, Fachbereich Informatik, durchgeführt bei GSI-Darmstadt, Germany, 1986.

191. Neue Technologien Gesellschaft mbH & Co KG Am Spielacker 12–14, D-63571 Gelnhausen, Germany.

192. PET, Physik Elektronik Technik, Smetanaweg 6, D-64291 Darmstadt, Germany.

193. Siemens AG, Germany.

194. Gesellschaft für Schwerionenforschung mbH, Planckstraße 1, Postfach 11 05 52, D-64220 Darmstadt, Germany.

195. European Organization for Nuclear Research, CH-1211 Geneva 23, Switzerland.

196. Grohmann, K. and Hechtfischer, D., Kryostromkomparatoren als Präzisionsstandards für rationale Gleich- und Wechselstromverhältnisse, PTB-Mitteilungen 92 5/82, 1992, p. 328.

197. Tolman, R. C., The principles of statistical mechanics, University Press, Oxford, 1938, chaps. II, III.

198. Wilson, E., Transverse beam dynamics, CAS-Cern accelerator school, Gif-sur-Yvette, Paris, France, 1984, CERN 85–19, 1985, pp. 64–95.

199. Riehl, G., Gesellschaft für Schwerionenforschung mbH, Planckstraße 1, Postfach 11 05 52, D-64220 Darmstadt, Germany, private communication.

Chapter 6

Ion Source Electronics and Microwave Generators for Ion Sources

H. Horneff
F. Bourg

CONTENTS

0-8493-2502-1/95/$0.00+$.50
© 1995 by CRC Press Inc.

1   INTRODUCTION

To run an ion source several power supplies and other electronic equipment are necessary. For many dc-operated ion sources commercial power supplies can be used. For a duoplasmatron, for example, or for other low-voltage arc ion sources, commercially available current-regulated power supplies are well suited for maintaining a stable arc plasma even without an additional series resistor. But inexpensive power supplies without electronic regulation can also be used to run ion sources in a dc mode. Such power supplies are discussed in Section 2.

If, on the other hand, pulsed mode operation of the ion source is required, then the necessary pulse switches have to be laboratory-made. Several examples of pulse switch designs are presented in the following sections.

For ion extraction at voltages of up to several tens of kilovolts, commercially available power supplies are well suited. For a few special applications, pulsed extraction is needed. Some examples are presented in Section 8.

The environment of the ion source electronics is dominated by high voltages of up to several hundred kilovolts. In many cases the entire ion source equipment package is on a platform at several hundred kilovolts and the ion source power supplies themselves are at an additional potential of several tens of kilovolts (see Figure 1). High-voltage sparks and breakdowns, which are sources of voltage spikes that can interfere with the electronics in the area, are unavoidable. Additionally, ion

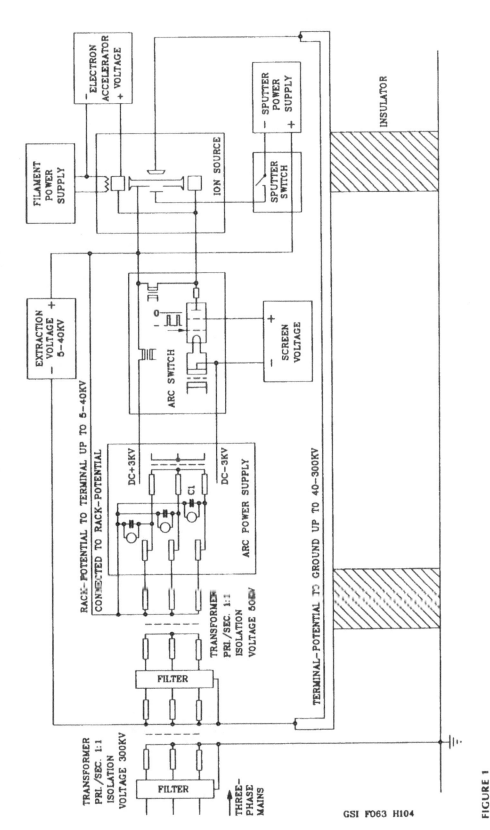

GSI F063 H104

FIGURE 1
Arrangement for Penning ion source.

sources tend to generate quite energetic plasma oscillations over a wide range of frequencies. For all of these reasons it is obvious that certain rules and precautions have to be respected when constructing an ion source supply arrangement. Special care has to be taken about the base potential, i.e., the ground potential of the HV platform. Ground loops have to be carefully avoided or, better still, all ground connections should be done at one single point. The main ground connections should be fabricated from solid copper bars. It is preferable to design all command inputs and all control outputs to be floating so as to avoid ground loops. The same rules as for the supply arrangement apply to every supply component itself.

High-voltage sparking and plasma oscillations within the ion source can cause severe problems in modern electronics. Malfunctioning of the electronic devices and destruction of electronic components may result. To avoid problems and to guarantee trouble-free operation of the equipment conservative techniques with passive electronic components should preferably be used in critical areas. Thus, we present here a number of quite simple but sturdy and reliable designs that do not have the precision and finesse of modern electronics, but are virtually indestructable. Experience has shown that all transformers should be designed for a permanent isolation voltage of 10 kV and should have their windings sealed in epoxy.

For switching high voltages and high pulse power in many cases the good old vacuum tube is still the best choice, especially if constant current pulse operation is necessary, for example, for PIG ion sources. We also present some modern electronic designs, which are neither conservative nor passive, but have specific and novel circuit configurations that protect them from high-voltage and plasma interference.

In Figure 1 the complete infrastructure of supplies for a Penning ion source is shown, together with the high-voltage platform. Clearly shown is which neutral point of the transformers is connected to ground, which to platform potential, and which to rack potential. Note also that the rack potential is connected to the anode and to the positive output of the sputter power supply, and is defined by the extraction potential with respect to the platform potential. It is important to install filter elements at the input to the isolation transformers. After the regulation transformer each phase of the arc power supply is connected to rack potential via a capacitor C_1 of 200 pF and an overvoltage protector.

2   POWER SUPPLIES

Various power supplies are necessary to operate an ion source terminal as, for example, for the Penning source shown in Figure 2. Along with the main arc power supply for the vacuum tube pulse switch, there is a filament supply and the acceleration supply for the cathode heating, a screen grid supply, and one supply for the sputter electrode. All these power supplies are put at extraction potential (≤ 50 kV).

Figure 2 shows the basic circuitry of these power supplies. The different parameters of the various power supplies are summarized in the table in Figure 2. The power supplies are designed using sturdy and conservative techniques. They are constructed floating with an isolation voltage of 10 kV, and, thus, all transformers shown in Figure 2 should have an isolation voltage of 10 kV. The installed test converters for monitoring the current and voltage are of a standardized design presented in Section 10 and Figure 17. They may be designed with the values given in the list in Figure 17 and should be insulated to 10 kV as the transformers used inside.

GSI F063 H98

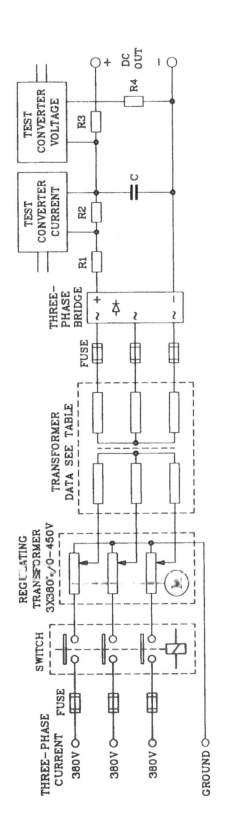

TABLE POWER SUPPLIES FOR ION SOURCE OPERATION	FILAMENT HEATING	SPUTTER VOLTAGE	SCREEN VOLTAGE	ACCEL VOLTAGE	ARC VOLTAGE	ARC VOLTAGE
PRIMARY VOLTAGE	450V AC	450V AC	450V AC	450V AC	450V AC	450V AC
SECONDARY VOLTAGE	9.4V AC	1480V AC	1100V AC	1480V AC	2300V AC	222V AC
DC VOLTAGE	10V DC	2000V DC	1500V DC	2000V DC	3100V DC	300V DC
CURRENT	100A	1A	0.33A	2A	3A	20A
CAPACITANCE	4.7mF	20µF	40µF	40µF	600µF	50mF
R1	0	100 OHM	500 OHM	0	INDUCTOR 1HENRY	0
R2 SHUNT	0.006 OHM	1.5 OHM	4.5 OHM	0.75 OHM	0.5 OHM	0.16 OHM
R3	0	100 OHM	125 OHM	250 OHM	0	0

FIGURE 2
Power supplies.

The output voltage of the power supplies is controlled by a regulating transformer adjusted by a small motor. There are fuses before and after the transformer. Three resistors R_1, R_2 and R_3 are positioned in the output line of the power supply; R_1 and R_3 are used for stabilizing of the output voltage and for limiting the output current. Their values are calculated individually and may be left out under certain conditions. For the case of the arc power supply, R_1 is replaced by an inductor (see list in Figure 2). R_2 is the shunt for the current transformer and R_4 is the instrument multiplier for the voltage transformer. Its value is given in Section 10, Figure 17.

The charging capacitance C (Figure 2) can be calculated as an approximation as follows:

C   = charging capacitance in farads
R   = load-resistance in ohms
U_{dc} = output voltage in volts
I_{dc} = output current in amperes
p   = ripple in percent

The load resistance is given by

$$R = \frac{U_{dc}}{I_{dc}} \left(\frac{V}{A} \right)$$

and the time constant

$$T = \frac{U_{dc}}{I_{dc}} \cdot C(s)$$

It follows for the capacitance

$$C = \frac{I_{dc}}{U_{dc}} \cdot T(F)$$

It is approximately true for the discharge of C with p ≈ 1% if a three-phase bridge circuit is used

$$C \approx \frac{I_{dc}}{U_{dc}} \cdot T \cdot \frac{60}{p}$$

and for 50 Hz, T is given by

$$T = \frac{1}{6 \cdot f_{mains}} = 0{,}0033(s)$$

For the power supplies shown in Figure 2, with $R_1 \approx 100$, the simplified formula for C changes to

$$C \approx \frac{I_{dc}}{U_{dc}} \cdot T \cdot \frac{12}{p}$$

All the above formulas are valid only for dc load. For pulse load and an average current below I_{dc} one can calculate

$$C \approx \frac{I_{pulse}}{U_{dc}} \cdot T \cdot \frac{60}{P}$$

To calculate the output voltage for a three-phase bridge-circuit one can simplify as follows, if the voltage drop at the rectifier and along the windings and cables is neglected:

$$U_{dc} \approx \frac{U\sim}{0.74}$$

where $U\sim$ = voltage at the input of the rectifier.

Figure 3 shows the circuitry of a power supply of very special design. Eight transformers with an input voltage of 220 V and output voltages of 4, 8, 16, 32 V, etc. are connected in series. Each transformer can be activated or shorted. A damping resistance in front of each transformer is shorted soon after activation of the particular transformer. For switching we use Triacs, which are controlled by separate electronics. With a test converter and some electronics the power supply can be operated as voltage- or current-controlled. With such a design it is possible to construct a power supply that can deliver a higher current for lower voltages without increasing the size of the power supply for higher voltages. The low-voltage transformers are just designed for the higher current and the high-voltage ones for lower current and a very compact supply is the result. The time constant of the control, on the other hand, is relatively slow (t ≈ 20 ms for 50 Hz), a fact that is even an advantage in our application.

C_1 is a small capacitance that equalizes the output voltage for the test converter. C_2 is a large storage capacitor for pulse operation. A diode is connected between C_1 and C_2, which prevents C_2 from influencing the power supply regulation.

The power supply is short-circuit-proof since the transformers are shorted in that case. The power supply always starts from zero when switched on without any reset procedure. This type of power supply is especially suited for operation in connection with a transistor switch because one does not need any short circuit protection except for a small current-limiting resistor.

3 PULSE SWITCH WITH VACUUM TUBES

Figure 4 shows the basic circuitry of a vacuum-tube-operated pulse switch. The operation voltage is delivered by an arc power supply (3.1 kV, 3 A) of the type discussed in the previous section (Figure 2). There is a resistor R_1 between the anode of the tube and the ion source, the value of which is between 10 and 100 Ω. The connection from the screen power supply to the screen grid is done via an inductor, which together with the resistors R_1 and R_2 prevents the circuit from oscillating. The resistor R_2 can be chosen between 100 and 500 Ω. A test transformer is in series between the ion source and the positive output of the arc power supply, providing a monitor of the pulse current off ground potential. A second test transformer is connected in parallel with the ion source to monitor the arc voltage off ground

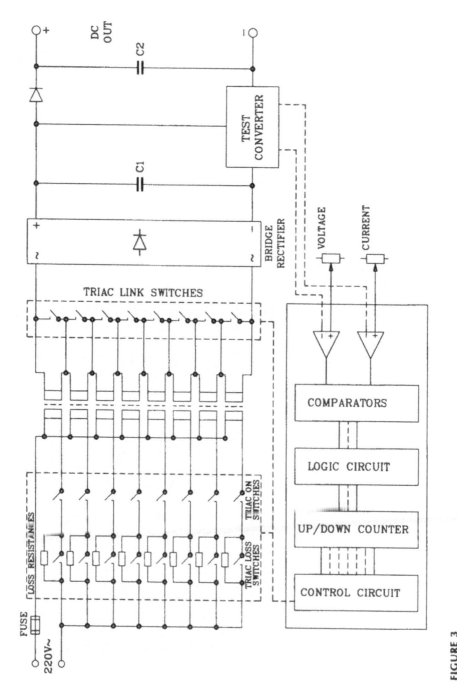

FIGURE 3
Basic diagram — power supply with transformer cascade.

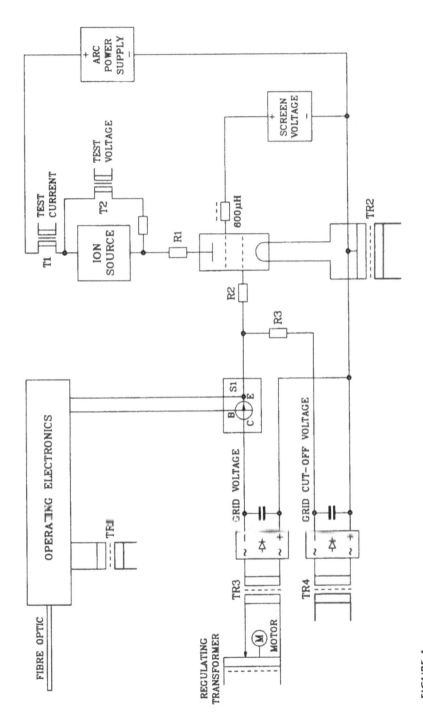

FIGURE 4

Basic diagram — tube pulse switch with regulating transformer.

potential. The design of the test transformers is discussed in Section 10, Figure 16 in more detail.

The cathode of the vacuum tube is heated by a center-tapped transformer TR_2. All components connected to tube cathode potential have to be connected to this center tap to avoid interference of the power line ripple.

The tube is switched off by a negative potential at the grid. This potential is maintained by a small transformer TR_4 with rectifier and charging capacitance via R_3 and R_2; R_3 should be about 100 kΩ. It could be chosen larger, but a slower pulse edge would be the result. An additional small transformer with rectifier and charging capacitance delivers the grid control voltage. A motor-driven transformer controls the voltage of TR_3 and, thus, the grid potential. The switch S_1 connects this voltage periodically to the grid of the vacuum tube via R_2. Because R_3 is large compared to the source resistance of the grid power supply, the grid cutoff voltage is dropped across R_3 and is not active while the pulse is on. The switch S_1 is controlled by the operating electronics, which are designed floating and supplied by the transformer TR_1. Fiber optics link the pulse control signal to the operating electronics. A high-voltage transistor, as used in the horizontal deflection of a TV set, can be used for the switch S_1. Also, a MOSFET or an insulated-gate bipolar transistor (IGBT) could be used. The transformers TR_1 to TR_4 should have a high isolation voltage to guarantee reliable operation of the device.

Figure 5 shows a different design for a vacuum tube pulse switch with a motor-driven potentiometer for the grid control. All other parts follow the design of Figure 4. With two (or more) potentiometers P_1, P_2, ... and corresponding switches S_1, S_2, ..., double (or multiple) pulses can be easily created. The value of the potentiometer should be about 20 kΩ and, of course, the operating electronics also have to be double (or multiple). With a pulse switch such as this, one can run the ion source at a lower level in a standby mode and then switch it on for short periods of time. This may improve ion source lifetime or, in the case of a PIG ion source, help to maintain a set cathode temperature. Finally, an example of a vacuum tube pulse switch for 3 kV and 10 A is given. We choose the tube 4CW25,000A (English Valve Company Ltd.), which needs a screen voltage of 0 to 1500 V. The data sheet gives a cut-off voltage of about 500 V. The grid voltage will be 0 to 500 V and for a maximum current of 10 A, R_1 will be 10 Ω. The tube is water-cooled and allows a power of 25 kW at maximum. In our case we are far below this value even with 50% duty cycle and 50 Hz (a part of the voltage drop is at the ion source). Even so, one has to use water-cooling because the heating of the cathode alone creates enough heat to destroy the metal-ceramic soldering, which usually has an upper temperature limit of 200°C.

4 PULSE SWITCH WITH TRANSISTORS

A high-voltage transistor pulse switch is shown in Figure 6. A high-voltage transistor, such as used in the horizontal deflection of a TV set, serves as the switch. Each transistor carries at maximum 2.5 A, a current at which the current gain of the transistor is just 2 or 3. Therefore, a high driver current is necessary. To overcome this inconvenience the switch current is also used as driver current. To prepare a minimum operating voltage for the driver transistor, a network of 12 resistors of 0.22 Ω is used. This trick reduces the necessary driver current for each transistor to ~100 mA, delivered by the floating driver electronics. Each of the n stages switches at maximum 600 V and gets its power from a transformer carrying n independent secondary windings. For 3000 V one then needs five stages. A zinc oxide varistor of 650 V

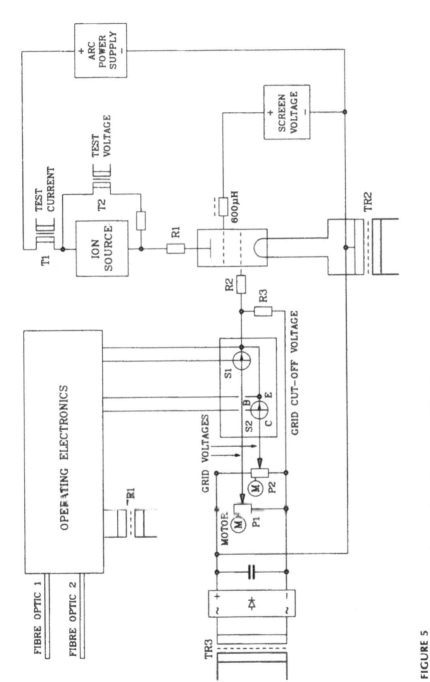

FIGURE 5

Basic diagram — tube pulse switch with motor potentiometer.

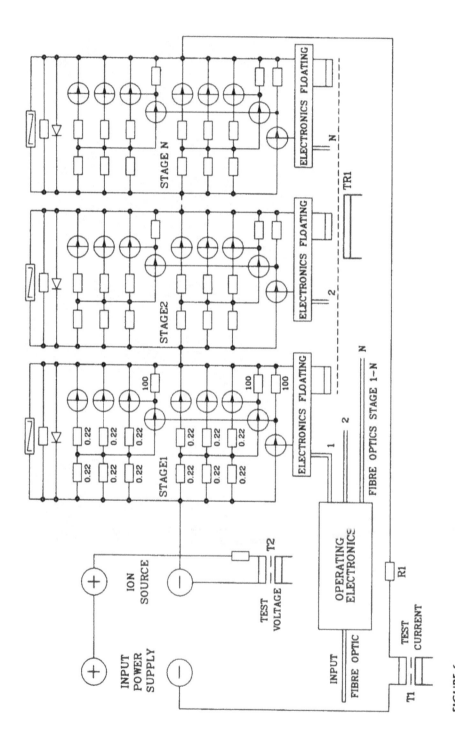

FIGURE 6

Basic diagram — HV transistor pulse switch.

protects the transistor from high voltages and a diode protects it from reverse currents. A resistor serves to equalize the potential of the different stages. This design leads to a low saturation voltage for the transistors, and for this reason air-cooling with a small fan is sufficient to operate the device. With five stages one can construct a pulse switch for 3 kV and 15 A. The operating electronics are controlled by fiber optics and each stage is also driven through a fiber-optic system.

Figure 7 shows a high-current pulse switch with transistors. A power module of 1 kV and 200 A is used as the switching device. We have limited the maximum current to 50 A, which gives us the advantage of low saturation voltage and low heat production, which can be easily cooled by a fan. The driver current is lower and the current gain larger if the operating current is kept low. Also, the driving electronics can be of simple design. In case of a short circuit at the ion source this pulse switch can hold a much higher current for a period of time until the power supply is discharged. Again, the operating electronics are controlled via fiber optics and connected floating through a 15-kV optocoupler to the driver electronics. The test transformers are described in Section 10.

With six modules such as QM200HA-2H (Mitsubishi) one can design a compact pulse switch for 600 V and 300 A, which needs three units of a 19-in. rack only. For short pulses one could even increase the current to 500 A without any problem. A zinc oxide varistor of 650 V protects the transistor from overvoltage. An R/C-combination of 4.4 Ω and 5 μF prevents oscillations, which may occur with long cable connections to the ion source.

5   PULSE SWITCH WITH IGBTS

An insulated-gate bipolar transistor (IGBT) is a power-switching device that combines both high-speed capabilities and high-impedance characteristics of MOSFETs, and low saturation voltage characteristics of bipolar transistors. A new kind of driver electronics, which can be used for all transistors with a gate — such as IGBTs or field-effect transistors — is shown in Figure 8. The control for the driver is connected to a ring core, which acts as a current-voltage transformer, by a current loop through the center. Positive and negative current pulses are sent through the loop. A negative signal is related to the start of the control pulse and a positive one to its end. There are two turns with the same orientation as the control loop on the ring core. In these two turns a positive voltage pulse is generated by the trailing edge of the negative signal. The beginning of the coil is connected to the emitter of the IGBT and its end to the gate via a diode (1N4151). The gate is charged up to 8 V by the positive voltage pulse. The ring core carries another winding of just one turn with the same orientation as the control loop. The trailing edge of the positive signal generates a negative voltage pulse in this turn. The end of the turn is connected to the emitter of the IGBT and the start is connected via resistor R_1 to the switching transistor (2N3904). By the reverse connection of this turn the negative voltage pulse becomes positive and switches the transistor (2N3904), which discharges the gate of the IGBT.

At the gate there will be a control voltage, which corresponds to the initial signal but is delayed by about 30 μs because of the use of the trailing edge of the command signals. The advantage, on the other hand, is that the method is independent of the number of ring cores involved. The initial current loop in Figure 8 is connected to ground on the left side and to the operating electronics on the right side. The operating electronics are shown in more detail in Figure 14. There are several advantages of this control design. It is easy to build up cascades of transistors and one needs

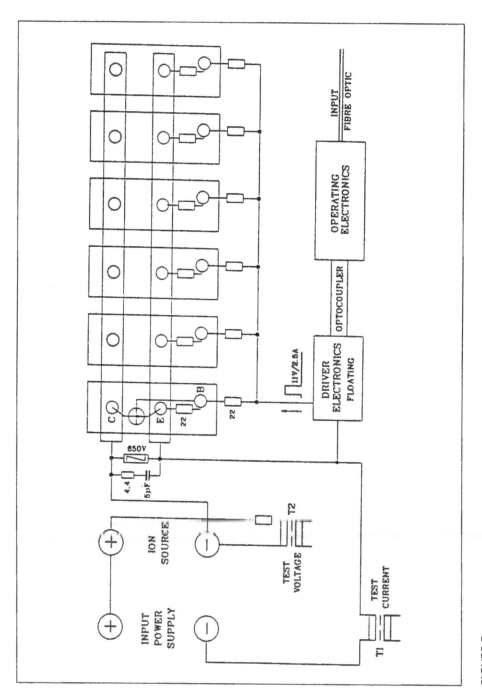

FIGURE 7
Basic diagram — HC transistor pulse switch.

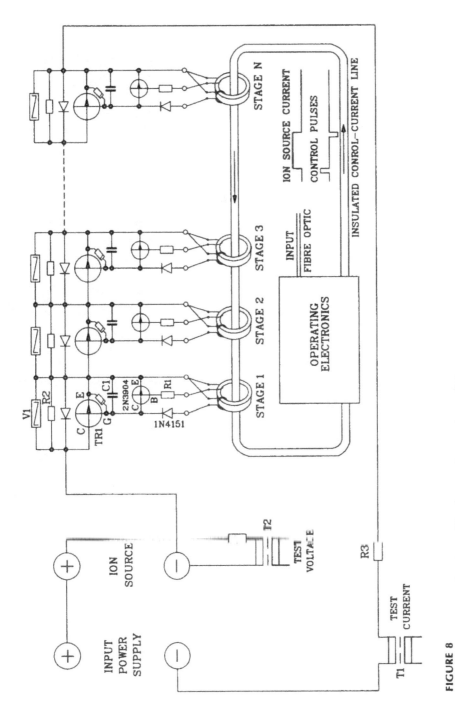

FIGURE 8

Basic diagram — insulated gate bipolar transistor pulse switch.

no separate power supplies for the operating electronics of each stage. Additionally, one can insulate the control loop easily for high voltages and can put the switch at any potential without danger to the operating electronics, which remains at ground potential. The inductive coupling of the loop is negligible and the capacitance below 0.5×10^{-12} F per stage.

As an example we define a sputter pulse switch (2.4 kV, 2 A) for a PIG ion source following the above-described design. As transistor we select an IGBT of 1000 to 1200 V, and a current maximum of 35 A. A BUP307 (Siemens) with a body TO-218AA could be chosen, for example. The resistor R_1 is then 2.2 Ω and C_1 10 nF. R_2 is 300 kΩ and V_1 is a zinc oxide varistor, which takes a current of 1 mA at 450 V. The protection diode in parallel with the transistor is about 1200 V and 2 A. R_3 is a stabilizing and short-circuit-protection resistor of 50 Ω. For reliability reasons, a IGBT of 35 A dc or 50 A pulse is chosen. Between gate and emitter a transient voltage suppressor 1N 6273A (General Semiconductor Industries, Inc.) is installed. The necessary test transformers are described in Section 10, Figure 16.

6 CONSTANT-CURRENT PULSE GENERATOR WITH SEMICONDUCTORS

Figure 9 shows the basic circuitry for a constant-current pulse switch with semiconductors, which allows, for example, Penning ion source operation in a similar way as with the vacuum tube switch described in Figures 4 and 5. A very compact arrangement can be designed for high pulse power but low average power. One does not need the peripherical power supplies for the tube heating and the screen grid; the efficiency is much higher and one has to handle less heat removal.

The whole current source consists of a series of semiconductor current sources, which need a special stabilization to make the design operational. The resistor R_1 generates a negative feedback and a zinc oxide varistor V_1 is designed to take over the current if an asymmetry occurs. So the feedback signal of R_1 is amplified. Small current deviations of the whole cascade, initiated by voltage changes of the ion source, are controlled by a follow-up control unit.

The circuit is realized by switching cascades controlled by ring core transformers as used in Figure 8. As opposed to Figure 8, we now use one ring core to control two transistors via separate windings; the feedback resistor R_1 is also new. The cascade is controlled through a current loop similar to Figure 14 in Section 10, but instead of a constant operating voltage for the switch of the positive current pulse, we now use an adjustable one. The positive current pulse is now defined by the leading edge of the control signal and the negative one by the trailing edge. That is contrary to the operation in Figure 14. To distribute the heat production it is preferable to use more stages than necessary for the design voltage.

As an example we take a constant pulse device for 5 kV and 4 A with a maximum pulse width of 1 ms with a repetition rate of 5 Hz (these are the values needed for a cold cathode PIG ion source for synchrotron accelerators). We choose 32 stages for the cascade. Each stage consists of two parallel MOSFET transistors, which should have a drain-source voltage of at least 1000 V and a drain current of 10 A. The gate threshold voltage is supposed to be between 2 and 4 V. The input capacitance may be between 1500 and 4000 pF and C_1 at around 1000 pF. If the shape of the constant-current pulse is not satisfactory one has to adjust C_1 to the input capacitance of the transistor. We take 1 Ω for the feedback resistor R_1. The value of the zinc oxide

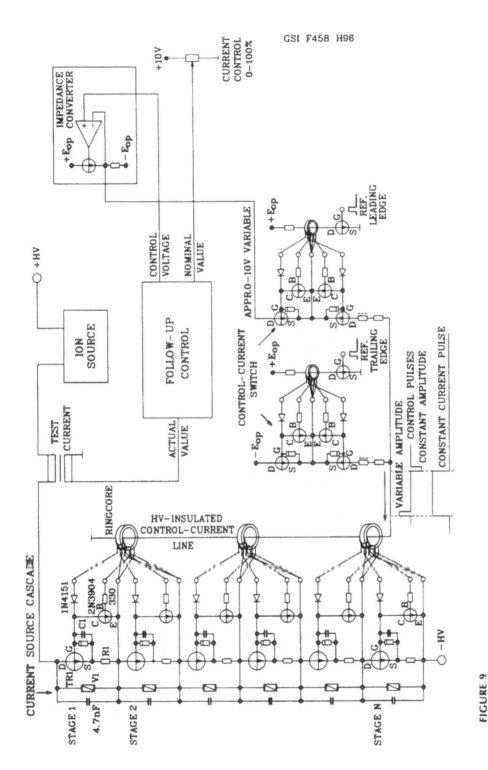

FIGURE 9
Basic diagram — constant current transistor pulse switch.

varistor V_1 has to be chosen carefully. One needs a type that allows a current of 1 mA at 140 V and can tolerate a permanent load of 0.8 W. To protect the transistors, a transient voltage suppressor is connected to gate and source (e.g., 1N 6273A [General Semiconductor Ind.]). The ferrite ring core should have an inductance of 1.5 mH per turn.

7   HIGH-VOLTAGE TRIGGER GENERATOR

For metal-vapor vacuum arc ion sources (MEVVA) a high-voltage trigger supply is needed to start the vacuum arc. Figure 10 shows the circuitry of a device for generating high-voltage, high-current pulses. A thyristor Th_1 charges a capacitor of 20 μF via an inductor of 600 μH. A second thyristor Th_2 discharges the capacitor to the primary winding of a transformer, which delivers from the secondary winding the high-voltage pulse with the necessary power to strike the MEVVA source. The transformer is mounted in a (plastic) box and is connected as close as possible to the ion source so as to avoid long, high-voltage cables and minimize interference with sensitive electronics in the environment. The power supply that delivers voltage and power for Th_1 has an overcurrent protection. If by some accident Th_1 and Th_2 switch at the same time, the power supply is automatically switched off for a short time and then switched on again.

For the generation of the trigger pulses a C-core transformer with an iron cross section of about 20 cm^2 and a power of about 200 W is used. In our case an air gap of 0.25 mm per iron connection is needed. The total air gap is then 0.5 mm. The primary winding has five turns of plastic-insulated wire of 4 mm^2. The secondary winding consists of 200 turns of Teflon wire of about 0.14 mm^2. Between primary and secondary winding there is a layer of 2-mm insulation. The control circuit for the thyristors Th_1 and Th_2 is shown in detail in Figure 15. If a higher current than 11 A is needed one has to reduce the number of secondary turns and increase the operational voltage of Th_1. The supply is controlled via fiber optics and needs just three units of a 19-in. rack.

8   PULSE LINE FOR MEVVA ION SOURCES

To generate short pulses with a fixed pulse length one can use a transmission line. Figure 11 shows a line-simulating network. This pulse-forming network is charged by a power supply and a series resistor. One can use a dc power supply or a voltage pulse transformer system. The pulse-forming network is discharged through a terminating impedance and some kind of switch. In the case of a MEVVA ion source the ion source trigger starts the discharging of the network line via the ion source arc and a current-limiting resistor. The pulse-forming network has to be calculated with the terminating impedance, the pulse length, and the rise time. The ideal case would be to build the transmission line with an unlimited number of parallel L-C elements. In practice one uses just a few elements, depending on the pulse shape necessary. It is possible to compensate for the ripple on the flat top by adding a small resistor (R_{co}) in series with the capacitance of the last L-C element in front of the ion source. Also, it could be necessary to change the last inductance in front of the ion source, for example, L into 0.5L or 1.25L.

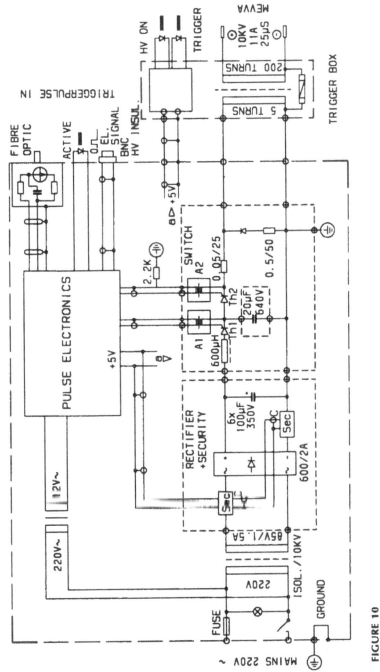

FIGURE 10
Basic diagram — HV trigger generator.

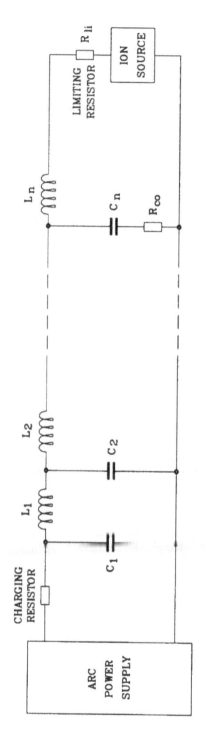

FIGURE 11
Basic diagram — pulse transmission line.

To calculate the pulse-forming network one can use the following simplified formulas:

C = capacitance per element in farads
L = inductance per element in henrys
Z = terminating impedance in ohms
t_w = pulse width in seconds
n = number of elements
t_r = rise time in seconds

The capacitance is given by

$$C \approx \frac{tw}{2 \cdot n \cdot Z}$$

the inductance by

$$L \approx \frac{tw \cdot Z}{2 \cdot n}$$

the terminating impedance by

$$Z \approx \sqrt{\frac{L}{C}}$$

and the pulse width by

$$t_w \approx 2n\sqrt{L \cdot C}$$

The rise time is given by the last L-C element in front of the ion source by

$$t_r \approx \sqrt{L \cdot C}$$

For further information the following books should be consulted: *Theory and Problems of Transmission Lines*, by Robert A. Chipman, McGraw-Hill Book Comp.; *Pulse Generators*, by G. N. Glasoe and I. V. Lebacqz, McGraw-Hill Book Comp.; *Electronic Design Data Book*, by Rudolf F. Graf, van Nostrand Reinhold Company.)

Finally, we calculate an example for a pulse-forming network with a pulse width of approximately 220 µs. We choose an impedance of approximately 2 Ω and the number of elements n = 5. The rise time should be less than 20 µs.

$$C \approx \frac{tw}{2 \cdot n \cdot Z} \approx \frac{220 \times 10^{-6}}{10 \times 2} \approx 11 \times 10^{-6}F$$

we choose a capacitance of 10 µF

$$L \approx \frac{tw \cdot Z}{2 \cdot n} \approx \frac{220 \times 10^{-6} \times 2}{10} \approx 44 \times 10^{-6}$$

we choose an inductance of 50 µH

$$Z \approx \sqrt{\frac{L}{C}} \approx \sqrt{\frac{50}{10}} \approx 2.2 \, \Omega$$

$$t_w \approx 2 \cdot n \cdot \sqrt{L \cdot C} \approx 10 \sqrt{500} \times 10^{-6} \approx 224 \times 10^{-6} \, s$$

$$t_r \approx \sqrt{L \cdot C}$$

for the last element in front of the ion source we choose 0.5 L

$$t_r \approx \sqrt{250} \times 10^{-6} \approx 16 \times 10^{-6} \approx 16 \, \mu s$$

for the resistor R_{co} (see Figure 11) we choose approximately 1 Ω, and for the limiting resistor R_{Li} we choose approximately 0.5 up to 1 Ω. With a charging voltage from 0 to 600 V we get up to a 200-A pulse, depending on the value of R_{Li}

9 EXTRACTION AND HIGH-VOLTAGE POWER SUPPLIES

The use of standard industrial high-voltage power supplies has been proven to be useful. One should select especially reliable power supplies, which withstand the "unfriendly" environment of an ion source. The power supply should be short-circuit proof and should have a reserve in output current to make operation with pulsed loads easier. If there are problems with pulsed loads one may connect a resistor and capacitance between power supply and ion source with the capacitance on the side of the ion source. High-voltage capacitors have to be handled with great care since they can store a lot of energy.

In case one needs to pulse the extraction voltage also Figure 12 shows the design of a vacuum tube high-voltage pulse generator. Vacuum tube, grid supply, screen supply, cathode heating supply, and the automatic control with demodulator are all at high potential. At ground potential are the operating electronics with the modulator, a follow-up control, the test electronics, the ventilating system, and an oil cooling system. The pulse generator is built in a 19-in. rack. Inside the rack there is a PVC housing for all components at high potential, which is designed with shelves and an acrylic door. The different power supplies are positioned on the PVC shelves behind the glass window at high potential. Each power supply has its own connection to the 220-V mains via a 50-kV isolation transformer. This arrangement has the advantage of making the maintenance of a single power supply easy when the high voltage is switched off. The PVC "cupboard" has a hole in the bottom and on top connected to a ventilation system. This is necessary because ozone is produced inside the high-voltage region. The vacuum tube is mounted in a glass container filled with transformer oil. The oil is circulated by an oil pump through an air-cooled radiator (from a car wrecker) at ground potential. The operating electronics are at ground potential as well. The control signal is modulated and sent through fiber optics to the automatic gain control with a demodulator. The test electronics get their signals from two voltage dividers at the input and the output of the pulse switch. The external power supply is controlled by a follow-up control in such a way that the vacuum tube always sees 1000 V — no more, no less. This is not only advantageous for the electric

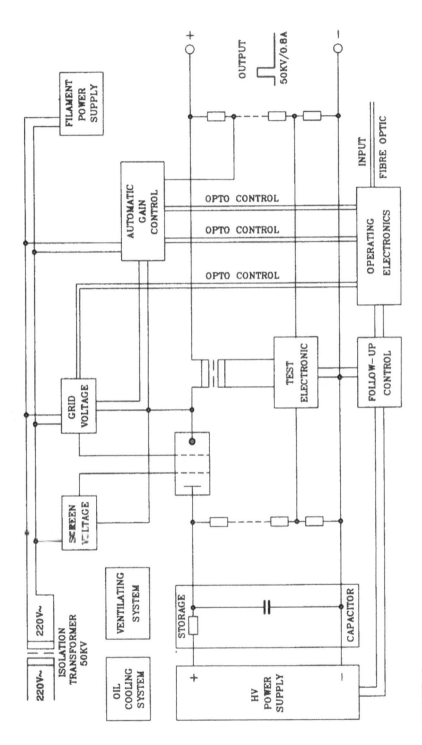

FIGURE 12

Basic diagram — extraction voltage tube switch.

function but also for the cooling system. For X-ray protection the rack has to be a closed metallic box. A storage capacitor is necessary in case of high pulse currents above the dc current maximum of the power supply. This capacitor is connected to the power supply through a charging resistor. The capacitor, typically 5 μF, has to be handled with great care because a lot of power is stored when charged to 50 kV (like a bomb).

A design example: Tetrode TH 5184 (Thomson-CSF); cutoff voltage ~ –400 V; screen voltage ~450 V; cathode heating 7.2 V, dc, 17 A; grid voltage 0 to –370 V pulse; HV power supply 50 kV, 80 mA; storage capacitor 5 μF; oil pump 1 bar with 10ℓ/min, oil radiator compact car (VW). With such a device one can produce pulse currents of up to 0.8 A at 49 kV.

The above-described vacuum tube pulse generator needs a complicated expansive construction. If constant-current operation is not needed one can use a much simpler design with a semiconductor switch. Figure 13 shows such a high-voltage switch with a transistor cascade similar to Figure 8. Differing from Figure 8, one ring core is used to control two transistors. The control circuit is shown in Figure 14 in Section 10. An IGBT is used as switch TR_1. The collector-emitter voltage should be 1200 V and the collector current about 25 A. The gate threshold voltage can be between 4.5 and 6.5 V and the input capacitance between 1000 and 4000 pF. For example, BUP 307 (Siemens) is suitable. In parallel with the transistor is a diode of 1200 to 1500 V and about 2 A and a zinc oxide varistor, which takes a current of 1 mA at 450 V. The total cascade has 96 transistors, so the control current loop has to feed 48 ring cores. In this case the total inductance of the control loop and also the reverse voltages get very high. To overcome this problem one can use two control loops with 24 ring cores each or one has to increase the drain-source voltage of the switching transistors in the control electronics (see Figure 14). Additionally, one has to reduce the inductance of 25 μH or eventually eliminate any inductance at all. A transient voltage suppressor 1N 6273A is connected to gate and emitter as overvoltage protection. The base resistor R_1 of the transistor 2N 3904 is 2.2 Ω. Such a high-voltage switch has to be constructed very carefully. One should divide the cascade into six cards of 16 transistors. They are mounted in a plastic box with six divisions and sealed in silicon rubber, which has to be carefully evacuated to avoid air bubbles inside. If this is not carefully done ozone will be created and cause damage to the device. Such a switch can be used at 40 kV and could theoretically switch some amps. To avoid problems with heat dissipation and coupling the current is limited to 1 A in our arrangement. The heat production, of course, is also influenced by the pulse length and the repetition rate. It is possible to measure the maximum tolerable current at a voltage below 500 V. One should do this measurement before the cascade is sealed in silicon. A current-limiting series resistor is necessary for this switch. One should choose the resistor as large as tolerable for the pulse shape and the voltage drop. Because line and circuit capacitances are discharged when switching high voltages, large (destructive) currents can occur without the use of a current-limiting resistor.

10   CONTROL AND TEST ELECTRONICS

To be independent of the operating voltage of the pulse device and to avoid ground loops, fiber-optic control is advantageous. To simplify maintenance and improve reliability one should try to use a standard system. Commands like power supply on and off, voltage up and down, reset, etc. are also applied via fibre optics, but it is sufficient to use one control system for the whole power supply arrangement.

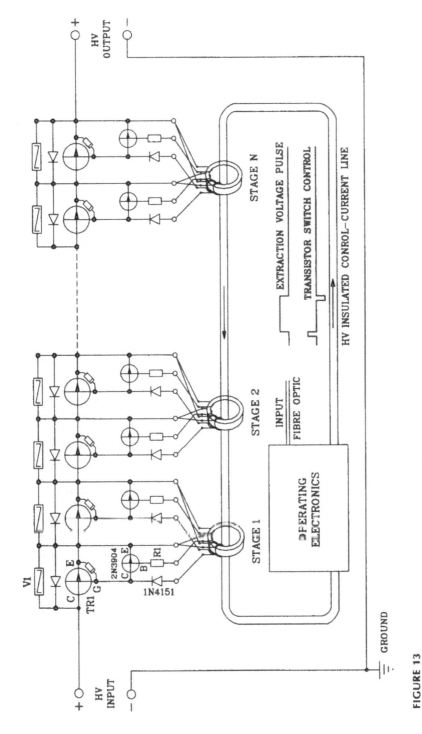

FIGURE 13

Basic diagram — extraction voltage transistor switch.

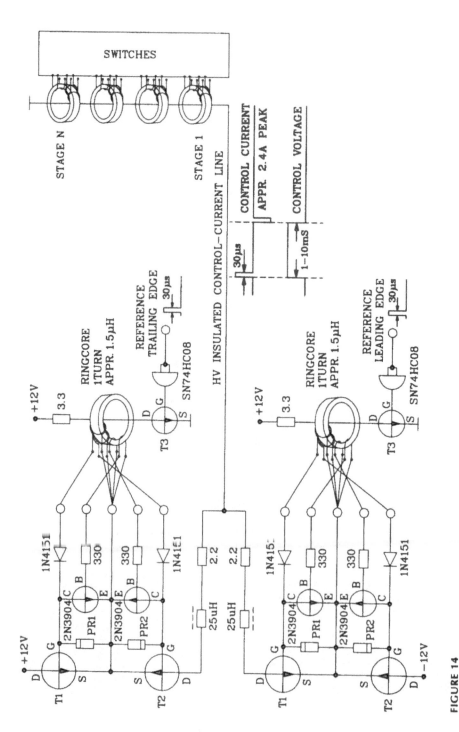

FIGURE 14
Control circuit for ringcore-controlled switches.

Figure 14 shows the control circuit with ring core as used in the designs of Figures 8, 9, and 13. A switch consisting of two transistors, T_1 and T_2, is designed to operate with positive or negative voltages as well. Two of these switches, one connected to +12 V the other to –12V, generate through an inductance of 25 µH and a resistor of 4.7 Ω positive and negative pulses as shown in Figure 14. Because T_1 and T_2 are switched in opposite ways, it is advantageous to control them as described in Section 5, Figure 8 with a ring core, which can be small but should have at least 1.5 µH per turn. On the core there are two coils of two turns each and two additional coils with one turn with the opposite sense. One has to construct and connect these coils with care to avoid malfunctioning. The ring core is driven by a current loop, consisting of the transistor T_3 and a resistor of 3.3 Ω. T_3 is driven directly by an integrated circuit (SN 74 HC 08). Again, one has to consider the direction of the current flow.

The leading edge of the loop current pulse of about 30 µs charges via the coils with two turns and diodes (1N 4151) the gates of T_1 and T_2 to ~4 to 8 V and the switch is on. The trailing edge of this current pulse discharges via the coils with one turn and the resistors of 330 Ω and the transistors (2N 3904), the gate, and the switch are switched off.

The transistors T_1 and T_2 must have an operational voltage of ~200 to 400 V because in the case of a IGBT switch with 16 ring cores reverse voltages of ~150 to 250 V may occur. Because these voltages work in both directions, the switches have to be independent of the operation voltage. T_1 and T_2 should be designed for a pulse current of 3 A (at 30-µs pulse width). The input capacitance should be about 500 to 2000 pF and the threshold voltage between 2 and 6 V. For T_3 the same type could be used but also one with just 50 V, e.g., BSS295 (Siemens), would be sufficient.

For overvoltage protection between gate and source transient voltage suppressors PR_1 and PR_2 are used (e.g., 1N6273 with 15 V [General Semiconductor Industries, Inc.]). The two 30-µs wide pulses, which control the integrated circuit (SN 74 HC 08) in Figure 14 are generated by the leading and the trailing edge of the control signal, respectively. The current pulses generated by the circuit shown in Figure 14 switch the IGBT just with their trailing edge. This has the advantage of being independent of the amount of ring cores controlled by the current loop. A time delay of 30 µs is the price one has to pay. This delay is usually unimportant. The leading edge of the control current pulses is slowed down by the inductance of 25 µH.

Figure 15 shows a circuitry that is suitable for controlling the thyristors of the trigger design in Figure 10. The arrangement is very simple and easy to copy. Such a quasi-bifilar transformer is very fast (some nanoseconds). Figure 15 shows the correct construction procedure.

Wide-band current-voltage transformers are ideal for monitoring pulse currents. They are easily constructed for high isolation voltages, sturdy, and reliable. They can be put in any current-carrying line without interference with the operation voltage of the line. They have a low resistance and are nearly feedback-free. Such a pulse test transformer consists of one primary turn for the input signal and several secondary turns with a terminating impedance at the monitoring output. For the case of high currents an air gap is necessary. The number of turns, iron core, air gap, and the behavior at low frequencies can be calculated in a simplified way. The behavior at higher frequencies needs a more complicated analysis because there are coil capacitances and leakage inductances involved. From theoretical calculations and practical experiments we can say that for secondary windings with less than 2500 to 3000 turns higher frequencies do not need to be taken into account.

In the following, the necessary formulas for a wide-band current-voltage transformer will be developed. Because we are interested only in the real region of the

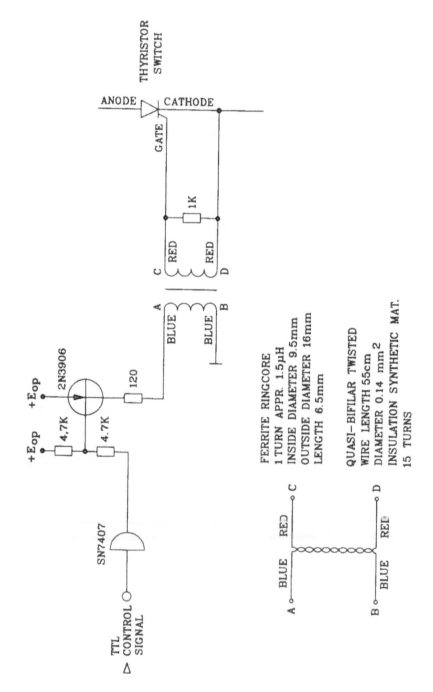

FIGURE 15

Control circuit for HV trigger generator.

transformer, the formulas can be simplified greatly. At the end of the calculation we will find three formulas for easy and quick calculation of a current-voltage transformer with good accuracy.

Simplified transformer in the real range without leakage inductance and capacitance:

Rw_2 = resistance of secondary coil
R_1 = terminating impedance
w_1 = number of primary windings
w_2 = number of secondary windings
Ua = output voltage

It is valid for the output voltage

$$Ua = \frac{i_1}{w_2}\left(Rw_2 + R_1\right)$$

With R for $R_{w2} + R_1$ one gets

$$Ua \approx \frac{i_1}{w_2}\cdot R$$

and

$$R \approx \frac{U_a \cdot w_2}{i_1} \tag{1}$$

Consideration of the flat top (lower frequency limit):

Ua_1 = max. output voltage
Ua_2 = output voltage with pulse droop after time t
L = secondary inductance
R = terminating impedance
t = pulse length
T = time constant $= L/R$

We have

$$Ua_2 = Ua_1 \cdot e^{-\frac{t}{T}}$$

Upon series expansion one gets:

$$\frac{Ua_2}{Ua_1} = e^{-\frac{t}{T}} = 1 - \frac{\left(\frac{t^1}{T}\right)}{1!} + \frac{\left(\frac{t^2}{T}\right)}{2!} - \frac{\left(\frac{t^3}{T}\right)}{3!} + \ldots \pm \frac{\left(\frac{t^n}{T}\right)}{n!}$$

Simplifying for small pulse droops

$$\frac{U_{a2}}{U_{a1}} \approx 1 - \frac{t}{T}$$

or

$$\frac{U_{a2}}{U_{a1}} \approx 1 - \frac{R}{L} \cdot t$$

For a pulse droop P (%) one gets

$$\frac{U_{a2}}{U_{a1}} \approx \frac{100 - P}{100} \approx 1 - \frac{R}{L} \cdot t$$

or

$$R \approx \frac{P}{100} \cdot \frac{L}{t} \qquad (2)$$

For the output voltage Ua one gets from Equations 1 and 2

$$U_a \approx \frac{P}{100} \cdot \frac{L}{t} \cdot \frac{i_1}{w_2} \qquad (3)$$

Iron core with air gap (leakage neglected):

H   = field strength ($^{Aw}/_{cm}$)
l_e = average length of iron core (cm)
l_0 = air gap (cm)
Θ = flux (A · w)
A   = ampere
w   = turns
Aw = ampere turns

The flux is then

$$\Theta = H_{iron} + H_{air} = i_i \cdot w_1 (Aw)$$

with

$$\Theta_{iron} = H_{iron} \cdot l_e$$

$$\Theta_{air} = H_{air} \cdot l_0$$

For large current and large air gap one may simplify:

$$\Theta \approx \Theta_{air}$$

with

$$H_{air} = \frac{B}{\mu_0}$$

B = induction (G)
μ_0 = induction constant
    = 1.257 ($^{G \cdot cm}/_A$)
G = gauss

It follows that

$$\Theta = i_1 \cdot w_1 \approx H_{air} \cdot l_0 \approx \frac{B}{\mu_0} \cdot l_0$$

$$l_0 \approx \frac{i_1 \cdot w_1 \cdot \mu_0}{B}$$

and for $w_1 = 1$

$$l_0 \approx \frac{i_1 \cdot \mu_0}{B} \tag{4}$$

The inductance without air gap:

Fe = iron cross section
He = henry

$$L = 1.257 \times 10^{-8} \cdot \mu_{rel} \cdot \frac{Fe}{le} \cdot w_2^2 \; (He)$$

With the air gap we introduce an effective permeability μ_{eff} and get

$$L = 1.257 \times 10^{-8} \cdot \mu_{eff} \cdot \frac{Fe}{l_e} \cdot w_2^2 \; (He)$$

with

$$\mu_{eff} = \frac{\mu_{rel}}{1 + \frac{\mu_{rel}}{\mu_0} \cdot \frac{l_0}{l_e}}$$

μ_{eff} will be independent of μ_{rel} for large values of μ_{rel}:

$$\mu_{eff} \approx \mu_0 \cdot \frac{l_e}{l_0} \left(\frac{G \cdot cm}{A} \right)$$

and one gets

$$L \approx 1.257 \times 10^{-8} \cdot \mu_0 \cdot \frac{F_e}{l_0} \cdot w_2^2 \tag{5}$$

From Equations 4 and 5 together it follows that

$$L \approx 1.257 \times 10^{-8} \cdot F_e \cdot \frac{B}{i_1} \cdot w_2^2 \tag{6}$$

and from Equations 3 and 6

$$U_a \approx \frac{P}{100 \cdot t} \cdot F_e \cdot B \cdot w_2 \times 1.257 \times 10^{-8} \tag{7}$$

or

$$w_2 \approx \frac{U_a \times 100 \cdot t}{P \cdot F_e \cdot B \times 1.257} \times 10^{+8} \tag{8}$$

To summarize the necessary formulas:

$$w^2 \approx \frac{U_a \times 100 \cdot t}{P \cdot F_e \cdot B \times 1.257} \times 10^{+8}$$

$$R \approx \frac{U_a \cdot w_2}{i}$$

$$l_0 \approx \frac{i_1 \cdot \mu_0}{B}$$

For example, we choose an iron core for a main transformer or a C-core with an iron cross section of at least 20 cm². We assume:

i_1 = 500 A
U_a chosen = 5 V (corresponding to 500 A)
F_e = 20 cm²
p chosen = 0.44%
t = 10 ms (pulse width ~50% at 50 Hz)
B = we assume 18,000 G (maximum value)

and get for

$$w_2 \approx \frac{U_a \times 100 \cdot t}{p \cdot F_e \cdot B \times 1.257} \times 10^{+8} \approx \frac{5 \times 100 \times 1 \times 10^{-2}}{0.44 \times 20 \times 18000 \times 1.257} \times 10^8 \approx 2500 \text{ turns}$$

for

$$R \approx \frac{U_a \cdot w_2}{i_1} \approx \frac{5 \times 2500}{500} \approx 25\,\Omega$$

and for

$$l_0 \approx \frac{i_1 \cdot \mu_0}{B} \approx \frac{500 \times 1.257}{18,000} \approx 0.0349\,\text{cm}$$

Primary and secondary windings are insulated for at least 10 kV.

The transformers described are very efficient for currents of 500 A and more. For a pulse test transformer for 5 A the output voltage is just 50 mV. If one has to measure smaller currents or needs higher signals at the output one can use more than one primary turn. Additionally, one can increase the terminating resistor and compensate the resulting pulse droop with a complex impedance consisting of an ohmic resistor and a capacitor in series. An example of a pulse test transformer is shown in Figure 16. We use a core from a standard 30-W power transformer. The primary winding has 10 turns and the secondary 2680 turns. Both windings are insulated for 10 kV with respect to each other and to the core. The resistor and capacitor network at the output corrects the flat top of the current pulse with the potentiometer P_1. The potentiometer P_2 is used to adjust the output level. Both potentiometers have to be adjusted together because they influence each other. An input current of 10 A generates a test voltage of 5 V in this device.

For measurement of the voltage of a pulsed load a voltage transformer can be used (see Figure 16). A 30-W power transformer core has 5000 turns for the primary and the secondary windings. The primary winding is connected by a resistor cascade to the high voltage to be monitored. The secondary coil together with resistor-capacitor network delivers the output signal. P_3 adjusts the flat top of the pulse and P_4 its level. Again, both potentiometers have to be adjusted simultaneously. An input voltage of 3000 V generates a test signal of 3 V at the output of the voltage transformer.

The test transformers described here are all wide-band pulse transformers. Figure 17 shows a conservative design for a test converter for potential-free measurement of dc voltages and dc currents. The circuitry is sturdy and the possible operation voltage depends mainly on the insulation of the transformers. The test transformer is driven by two transistors in push-pull operation. Power-line frequency is used to drive the transistors inversely phased. At the output of the test converter there is a bridge rectifier that delivers a dc signal of 40 V peak, which needs very little smoothing. The voltage was chosen this high to keep the linear deviation of the bridge rectifier small. It is also advantageous to select all semiconductors with higher voltage limits than actually used. A voltage divider at the output is calculated to give 10 V dc at a load of 10 kΩ as full-scale signal. One has to make sure that the input resistance of the instrument or data acquisition system is 10 kΩ. Figure 17 also shows a list with design data of the test converters. The top line shows the necessary input voltage for the test converter, which has to be delivered by a shunt or a voltage divider for current or voltage measurement, respectively. One can also use this converter for direct measurement of ac voltages. In this case, the primary winding is connected directly to the shunt (no center tap) without any input electronics.

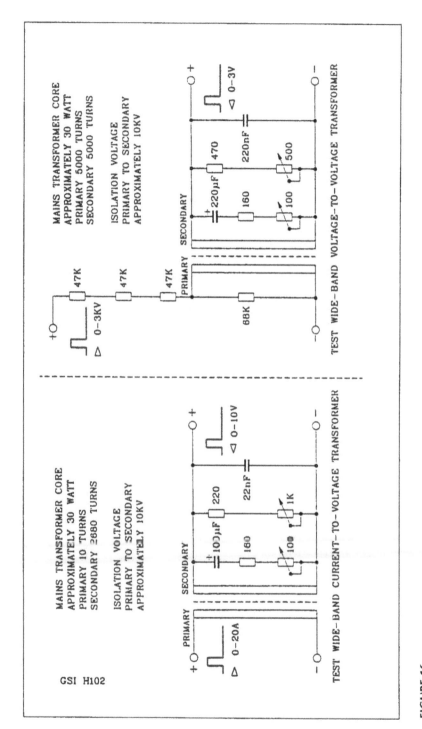

FIGURE 16
Pulse test transformer.

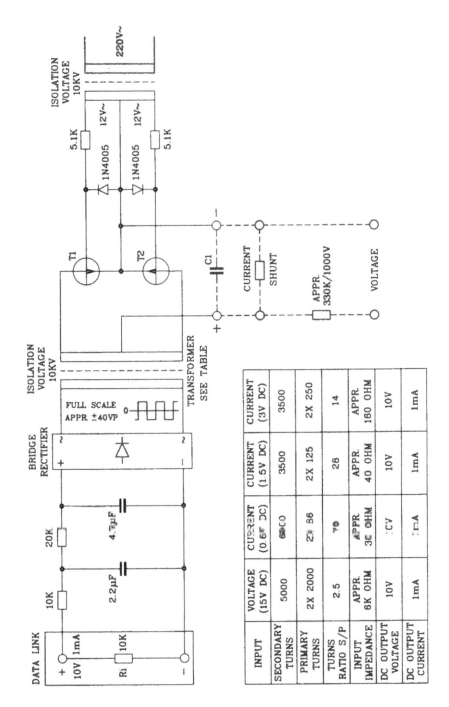

The following data appears in the figure table:

INPUT	VOLTAGE (15V DC)	CURRENT (0.6V DC)	CURRENT (1.5V DC)	CURRENT (3V DC)
SECONDARY TURNS	5000	3500	3500	3500
PRIMARY TURNS	2X 2000	2X 36	2X 125	2X 250
TURNS RATIO S/P	2.5	70	28	14
INPUT IMPEDANCE	APPR. 6K OHM	APPR. 30 OHM	APPR. 40 OHM	APPR. 160 OHM
DC OUTPUT VOLTAGE	10V	10V	10V	10V
DC OUTPUT CURRENT	1mA	1mA	1mA	1mA

FIGURE 17
Circuit dc test converter.

11 MICROWAVE ELECTRONICS: THE HIGH POWER GENERATORS

For a few decades, electron cyclotron resonance ion source (ECRIS) have used microwave device generators in the range between 2.45 and 18 GHz at a power level of several kilowatts to create and heat the plasma. The microwave tubes are able to supply high peak and average power over a larger microwave spectrum. They are used in numerous systems requiring high output power, including radar and telecommunication systems, as well as scientific, industrial, and medical equipment.

11.1 General Description

In the case of ECRIS many products are proposed by microwave equipment manufacturers. The standard designs include rf power output levels from 250 W to 15 kW, cw or pulsed. The frequency range covered is 1.0 to 40 GHz with different band-passes.

The characteristics of the generators are

- Reliability
- Pulsed or cw microwave
- Power stability <1%; for metallic ion
- Frequency stability
- Local control or remote control
- Start in manual or automatic (microprocessor)
- Internal and external security

To make a complete generator, there are a number of other devices required. The amplifier tube is preceded by the low-level circuit, which includes rf devices such as a low-level oscillator, a preamplifier, and a PIN diode attenuator (Figure 18).

This circuit allows the control of the rf level into the amplifier. The PIN diode attenuator can be controlled by a signal, in pulsed or cw operation. While a PIN diode is certainly one way of generating a pulse of rf energy, there are other methods. As an example, the traveling wave tube itself can be pulsed. To prevent energy from being reflected from the tube amplifier back into the driving device, an input isolator can be added. Isolators or circulators have the unique capability of being able to couple energy moving in a particular direction. This is done magnetically. Reflected energy is present in rf systems because of impedance mismatches. If the impedance of the load does not match the impedance of the source, energy will be reflected from the load back to the source. This can potentially damage the source and always degrades the gain linearity. An isolator will divert any energy traveling back toward the source to a resistive load.

The tube amplifier is followed by a directional coupler. Directional couplers are devices that have the ability to couple a small fraction of the energy traveling in a particular direction. A typical coupler can be a 50-dB coupler. This means the energy coupled will be at a power level 100.000 times below the power level in the main path. This lower power level can easily be applied to a crystal detector or converted to heat in a bolometer to determine the power level of the main path. In many cases two directional couplers are used in opposite directions. This allows the measurement of the power level leaving the amplifier and the level of power being reflected back into the generator. The reflected power level can be used to monitor a potential load mismatch. Reflected power level can be converted to an electrical signal that can

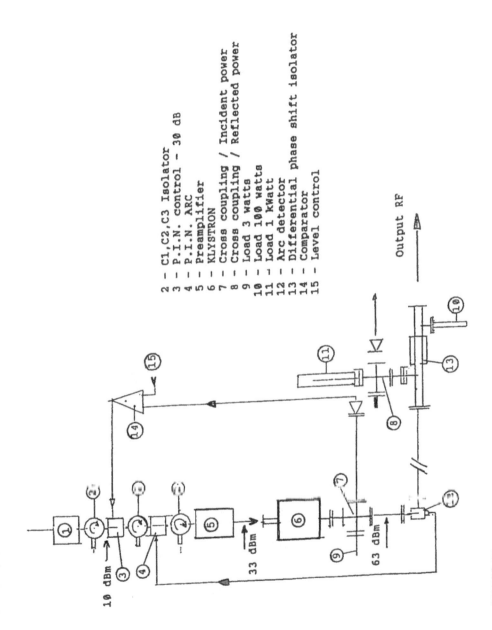

2 — C1,C2,C3 Isolator
3 — P.I.N. control – 30 dB
4 — P.I.N. ARC
5 — Preamplifier
6 — KLYSTRON
7 — Cross coupling / Incident power
8 — Cross coupling / Reflected power
9 — Load 3 watts
10 — Load 100 watts
11 — Load 1 kWatt
12 — Arc detector
13 — Differential phase shift isolator
14 — Comparator
15 — Level control

Output RF

10 dBm
33 dBm
63 dBm

FIGURE 18
Generator, 10 or 14 CHz.

trigger adjustable protection circuits. This protection is called voltage standing-wave ratio (VSWR) protection.

11.2 Microwave Oscillators

Oscillators can be characterized in frequency as fixed and/or tunable. Under the fixed tuned category are the crystal-controlled, surface acoustic wave (SAW) oscillators and also dielectric resonator oscillators (DRO). The tunable oscillators can be subdivided further into voltage-tuned, current-tuned, and mechanically tuned oscillators. A crystal-controlled oscillator can be tuned by incorporating a mechanically tuned capacitor or a varactor as a part of the resonant circuit. The tuning range is normally 0.1% of the center frequency. A dielectric resonator oscillator can also be tuned mechanically by approximately 1.0% of its center frequency and electronically by 0.1% of its center frequency. A mechanically tuned oscillator, on the other hand, can be locked as a fixed tuned oscillator unless one mechanically changes the cavity dimensions. Here again the cavity oscillators can also be finely tuned electronically.

Fixed tuned oscillators normally exhibit higher frequency stability and lower phase noise power spectral density. A new development that emerged about 10 years ago combines all of the desirable features mostly sought for in oscillators, namely, frequency accuracy, frequency stability, frequency tunability, and acceptable phase noise performance. Frequency synthesizers offer all the above features. Even though introduced about 10 years ago, they have recently become commercially available, thanks to improvements both in bipolar transistors and digital circuits.

11.2.1 Mechanically Tuned Oscillators

Mechanically tuned oscillators can be used as stand-alone frequency sources for applications where stability is not critical. They can be controlled with an optional voltage-sensitive frequency-tuning port for integration into phase-locked source assemblies. The quality of the mechanically tuned oscillators depends on the Q of the resonant circuit, the phase noise performance of the active circuit elements, and the sensitivity of the circuit elements to environmental temperature variations. Along with the basic oscillator circuit, the voltage-controlled oscillator assemblies can include voltage regulators, buffer amplifiers, frequency multipliers with bypass filters, and output amplifiers for load isolation or for output power enhancement (Figure 19).

11.2.2 Phase-Locked Frequency Sources

A properly designed phase-locked oscillator (Figure 20) is an accurate and stable frequency source. Its stability and accuracy are directly related to the stability and accuracy of a lower frequency reference. The lower frequency reference can take many forms. It can be a fixed-frequency, high-stability crystal oscillator supplied as an integral part of the phase-locked oscillator assembly, or an externally supplied crystal reference, or a bank of multiple switchable crystal oscillators, or an externally supplied low-frequency synthesizer. The reference source is sometimes a phase-locked oscillator itself, locked to a low-frequency 5- or 10-MHz standard. Fixed-frequency dedicated sources consist of cavity or dielectrically controlled oscillators. Multifrequency sources use cavity oscillators for applications that can tolerate the time required for cavity adjustments, or automatic frequency-agile oscillators where short acquisition time or

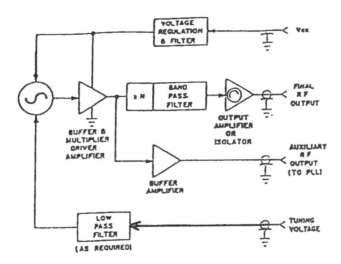

FIGURE 19
Functional block diagram of voltage-tuned oscillator.

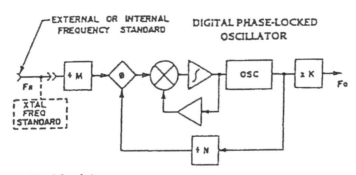

M,N:divider
∅:phase detector
⊗:harmonic mixer
K:multiplier
Fr:input frequency
Fo:output frequency

FIGURE 20
Phase-locked oscillator.

remote frequency control is required. For high-shock and -vibration environments that cannot tolerate mechanical changes, voltage-controlled oscillators are often used.

A phase-locked frequency source offers two distinct advantages to the system designer:

- It stabilizes the microwave oscillator to the stability of a high-performance, high-stability reference.
- It shapes the phase noise power spectral density of the microwave oscillator to approximately that of the reference within the loop bandwidth.

In a practical phase-locked loop there are additional sources of noise that degrade the phase noise performance of the output. In an analog phase-locked loop, the additional sources of noise result from the phase detector and from the operational amplifiers used in the loop and in the active loop filters. In the digital phase-locked loops, digital dividers and the digital phase detector add to the phase noise.

11.2.3 Frequency Synthesizers

The ever increasing complexity and sophistication of today's systems are imposing more stringent performance requirements on the frequency sources that are used as system building blocks. Frequency stability, frequency agility, coherency, and good spectral purity are only some of these new demands. Frequency agility itself requires either the output frequency of the transmitter or the local oscillator frequency of the receiver to be changed in a preprogrammed or predetermined manner. There are two design alternatives:

- A frequency-agile oscillator phase-locked to a low-frequency synthesizer
- A self-contained microwave synthesizer with step size small enough to meet the frequency accuracy and frequency resolution requirements of the application

Frequency synthesizers have a wide range of applications. For instance, they are used in superheterodyne receivers and transmitters that require reception or transmission of multiple carriers over a specified frequency range, frequency-hopping radars, and frequency generators that require extreme frequency stability, settability, and accuracy.

The phase noise power spectral density of the frequency synthesizer depends on the synthesizer frequency coverage. As the tuning range of the synthesizer increases, so does the phase noise power spectral density. This degradation of the phase noise performance is caused by the lower Q of the tuned circuits associated with the more widely tunable oscillators.

Although single-loop synthesizers are very cost-effective frequency sources, they have performance limitations in some more stringent system applications. The basic design limitation of a single loop is in its close in phase noise and relatively high spurious outputs close to the carrier. Even though the final output of the synthesizer is locked to the reference and has the same frequency stability, the narrow loop bandwidth required for small frequency steps makes the output sensitive to outside perturbances, such as line-voltage pickup, vibrations, shocks, etc.

11.3 Microwave Integrated Circuit

In selecting the best and cost-effective preamplifier for a particular application, one has to consider the following amplifier performance parameters: frequency coverage, amplifier bandwidth, gain flatness, noise figure, input and output match, and dynamic range of the output power.

There are applications where all of the above-listed characteristics are important and critical. Moderate bandwidth amplifiers are used for radar, radio astronomy, and communication applications to name a few; 5 to 10% bandwidth is sufficient, while test instrumentation and countermeasure applications require octave and multioctave frequency coverage. Therefore, in most cases, the amplifiers require simultaneous noise and power match at the amplifier input port. Amplifier gains are normally on the order of 30 to 50 dB. The other parameters are

Gain ripple	±0.25 to ±0.5 dB
Input VSWR	1.25:1 to 1.5:1
Output VSWR	1.5:1 to 2.0:1
Output power capability	+10 to 44 dBm

In practice, the amplifiers are noise figure optimized at a given frequency within the bandwidths of interest. To achieve low noise figures, the amplifier input has to be noise matched. However, most of the specifications ask for a simultaneous power match with little or no degradation of the overall amplifier noise figure. The required input return loss varies from 10 to 20 dB, depending on amplifier bandwidth. For the lowest noise figure (noise match), the input VSWR can be as high as 6. Amplifier manufacturers normally design and specify amplifier performance from a 50-Ω source impedance at the microwave frequencies. All performance specifications are referred to the 50-Ω source and load impedances.

In addition, if the source and the amplifiers are physically separated, they often are connected by low-loss, 50-Ω, unmatched impedances. These conditions decrease system sensitivity and introduce gain ripple and group delay distortions as a function of operating frequency. Thus, good system design requires a simultaneous noise and power match. Above 1 GHz, for moderate bandwidth designs, a well-matched, low-loss isolator is used between the source and the amplifier input. The isolator acts as a buffer between the source and the amplifier.

The input signal incident at the isolator input port is transmitted to the amplifier input port through the forward junction of the isolator. Power reflections at the amplifier input pass through the second junction and are absorbed by the isolator termination. Thus, the isolator presents a well-matched 50-Ω impedance to the source and provides a constant 50-Ω source impedance to the amplifier for noise matching purposes, even when the actual source impedance differs from 50 Ω. For the same reasons, an isolator is used to protect the amplifier output.

11.4   Automatic Level Control Loop

In the case of highly charged metallic ion production in ECR ion sources, it is important to keep the power level in a microwave setup constant with good accuracy. It is necessary to have a system consisting of three parts (Figure 21):

- A sensing element
- A control element
- A comparator and amplifier element that detects changes in the output from the sensing element

The sensing element is a standard crystal detector, the control element is a PIN diode, and the comparator amplifier is a unit made especially for this application. The automatic leveling control operates as follows. With the reference signal in the on position, a portion of the output rf signal is sampled by the cross-coupled P1, detected and fed to the noninverting terminal of the operational amplifier. At the same time, a reference dc voltage is applied to the box control input (Vr) terminal. This reference voltage sets the level of the rf output power. The magnitude of this reference is set by adjusting the reference signal. An operational amplifier compares the two input voltages and amplifies the difference voltage, also known as the error voltage. This amplifier error voltage drives the current-controlled attenuator in such a manner as

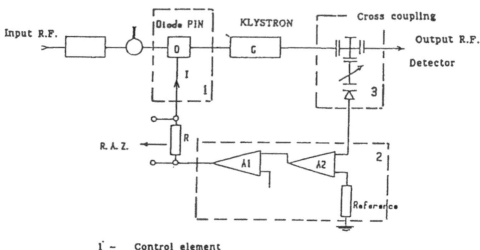

1 –   Control element
2 –   Comparator
3 –   Sensing detector

FIGURE 21
Automatic level control loop.

to continuously adjust its insertion loss, thus controlling the rf input signal level to the klystron. As the error voltage approaches zero, the output rf power level approaches the desired level as set by the control rf. One element of the control loop is the PIN diode attenuator. When the current is maximum, the attenuation of the rf input is maximum (40 dB).

12   MICROWAVE TUBES

The different tubes may be grouped into two main classes:

- Linear-beam tubes, divided into slow- (klystrons, TWTs) and fast-wave (gyrotrons, masers at free electrons) tubes
- Crossed-field tubes: magnetrons, cross-field amplifier, and carcinotrons[1]

Each type has an optimum frequency band where the power output varies little with frequency, and beyond which the power drops more quickly (Figure 22).

12.1   The Magnetron[2]

The magnetron oscillator is the most widely used of a general class of tubes know as crossed-field devices. It consists of a cylindrical cathode surrounded by an anode structure that possesses cavities opening into the cathode-anode or interaction space. A constant magnetic field is applied with flux lines parallel to the axis of the cathode. In these devices the electron beam is kept from reaching the anode under the combined effect of a dc electric field between cathode and anode and a magnetic field at right angles to the electric field. Beyond a certain critical value of the magnetic induction, Bo, depending upon the anode voltage, Vo, and the cathode and anode radius, the electrons can no longer reach the anode. An rf circuit provides an electric

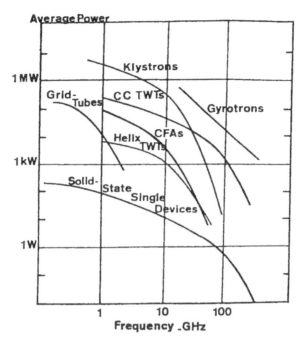

FIGURE 22
Microwave tubes.

field component parallel to the direction of the electron motion that interacts with the beam. When the electrons give up energy to the rf field, they may reach the anode, thus causing a dc beam current to flow. Under normal magnetron operating conditions, this electron cloud extends to about midway between the anode and the cathode.

Several configurations (Figure 23) have been used to make magnetron oscillate.[1] If the circuit contains an even number (N) of cells, there are only (N/2) + 1 distinct resonant frequencies, corresponding to phase shifts of 0, 2·Π/N, 4·Π/N, and Π rad. Oscillations occur when the synchronism condition is obtained. This means that the electrons must have an angular velocity Ωe such that the transit angle is equal to one of the phase shifts cited above, to within 2·n·Π rad. There are therefore a wide variety of oscillation modes possible, among which the one generally chosen corresponds to a Π-rad phase shift. This value is selected because it is the only one not degenerated. In this condition the magnetron is seen to be a higher-efficiency device. This means that only in this case is the field pattern determined by the geometry of the anode.[3]

Today, the circuitry used for the anode are strapped vane structures as shown in Figure 24. Alternatively, the same result can be obtained with alternating long and short slots, as in the rising sun structure (Figure 25). Finally, in the Π mode, a vaned anode can be coupled to an external cavity resonating in the TE 011 mode, whose field lines are circles, by slots in the bottom of every other cavity. This is called a coaxial magnetron.

The different types of magnetron structures can be mechanically tuned in many different ways. For example, a plunger can be inserted at different depths into the cavity. In the case of the coaxial magnetron, the external cavity is tuned by means of a piston, which, for the TE 011 mode, does not need to touch the cavity walls. An attenuating substance, placed on the back of this piston, attenuates the undesirable modes. The band covered by mechanical tuning can attain 10 to 15% of the central

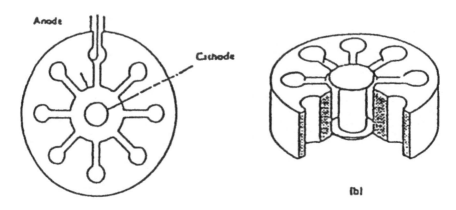

FIGURE 23
Top (a) an perspective (b) views of a simple magnetron.

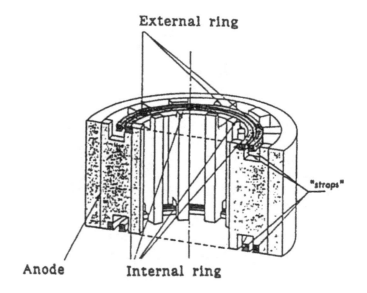

FIGURE 24
Detail of a double-strapped magnetron.

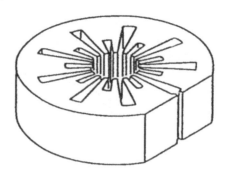

FIGURE 25
Rising-sun magnetron structure.

operating frequency. Tuning devices are also made for very quick changes in frequency over a narrow bandwidth. This is known as frequency agility.

For low-power tubes or low frequencies (f < 4GHz), the output energy is extracted by means of a coaxial line coupled to one of the cavities by a loop (Figure 26) or directly connected to the straps. In other cases, the energy is extracted by means of a waveguide coupled to the magnetron's circuit by an appropriate impedance transformer.

The magnetic field is normally created by a permanent magnet. The pole pieces can be integrated with the tube. Sometimes magnetic shunts are provided, which enable changing the magnetic field without changing the magnet.

The magnetron's cathode is bombarded by electrons that have been accelerated by the rf field. This back-bombardment, whose power can be 5% of the output power, heats the cathode and can necessitate decreasing the heater voltage or even cutting it off completely during magnetron oscillation. This is one source of limitations of magnetron power. Moreover, this back-bombardment frees secondary electrons, which are the essential source of the electrons present in the rotating cloud during oscillations.

12.1.1   Characteristics

Usually, the magnetron characteristics are shown on a current-voltage diagram upon which are traced lines of equal value of the magnetic field and isopowers (lines of equal output power), such as shown in Figure 27. The frequency of the oscillation varies with the current, which is indicated by a numerical frequency-pushing figure, for example, 100 kHz/A. Finally, the load also impacts the frequency and the output power. These considerations are shown on a Rieke diagram (Figure 28). It is possible to characterize the frequency variation as a function of the load by taking the extreme frequency shift encountered on the 1.5:1 VSWR circle. This is called the frequency-pulling figure.

12.2   Klystron

This tube uses a type of modulation called velocity modulation. These are longitudinal interaction tubes, having three to five resonant cavities, separated from each other by a narrow drift space. A constant magnetic field is applied with flux lines parallel to the axis of the cathode and the cavities. Below we describe a three-cavity klystron as illustrated in Figure 29. The construction of this tube is similar to the other amplifier, differing solely in the number of cavities.

The microwave signal impresses velocity modulation on the beam at the input cavity gap. The second cavity is placed a quarter of a plasma wavelength (first drift tube) away at the position of the maximum rf convection current modulation. The induced current in this cavity produces a voltage across its gap. This second cavity voltage, which is considerably larger than the first cavity voltage, impresses velocity modulation on the beam at this point.

This velocity modulation produces current modulation at the output cavity, a quarter plasma wavelength (second drift tube) away. In the second drift tube, the beam, already partially bunched and subjected to velocity modulation by the first and the second cavities, undergoes further bunching of its electrons and thus carries a higher alternating current, which excites the third cavity. The rf convection current passing through the output cavity produces an induced current in the output cavity,

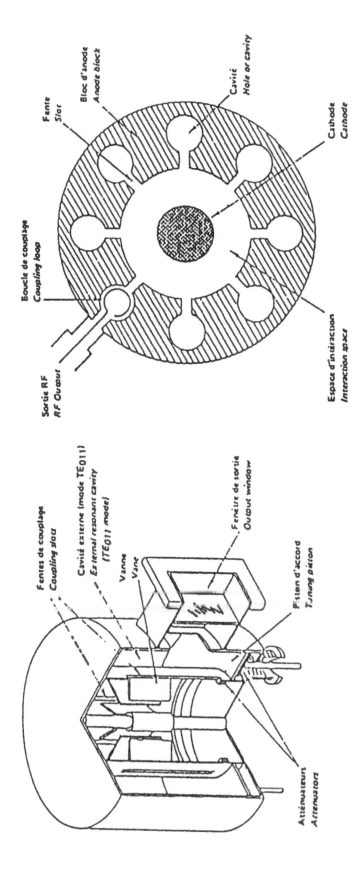

FIGURE 26
Coaxial magnetron with rf output via a coupling loop.

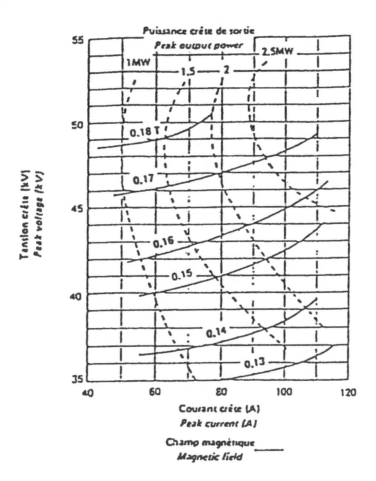

FIGURE 27
Typical diagram of operating characteristics for a high-power magnetron.

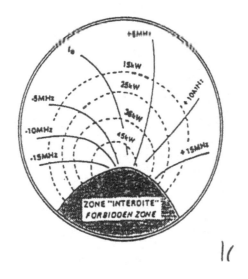

FIGURE 28
Rieke diagram.

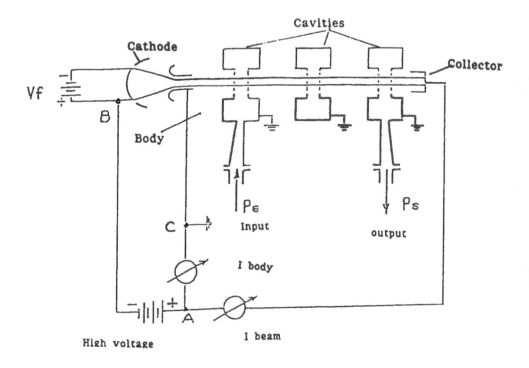

FIGURE 29
Three-cavity klystron.

which causes rf power to be delivered to the load. Most klystrons are equipped with mechanical tuning devices on each cavity. The frequency range of a given klystron will be specified by the tuning range and the bandwidth at each point. Normally, the bandwidth will be fairly constant over the tuning range.

In summary, the klystron amplifiers are characterized by high gain, good efficiency, and freedom from oscillations. On the other hand, their bandwidths are relatively small.

12.2.1 The Power Supply. High Voltage

The power supply has to provide all of the voltages required by the klystron. In addition, the power supply performs all necessary voltage sequencing, tube protection, and monitoring. There is generally a considerable amount of imbedded intelligence in a power supply to protect the tube power supply and amplifier as a whole.

Power supply technology has been advancing at a very fast rate. New devices and topologies are being announced continually. Many of the advances in the power conversion are very applicable to tube amplifier power supplies. Many supplies use linear power supply technology. In this sort of supply the power semiconductors are operated in their linear or dissipate mode. These supplies are characterized by low efficiency, good noise performance, large size, high cost, and good reliability due to the lower number of parts.

An alternative to the linear supply is the switching supply. The switching supply operates its power semiconductors as switches in either the full-on or full-off mode. This reduces power dissipation in the switch to a fraction of the dissipation in the pass element of a linear supply. The net result is a dramatic improvement of power supply efficiency. A typical linear supply will have an efficiency in the 30 to 50%

region. A switching supply will normally operate in the 70 to 90% range. In addition to power saving, the supply itself becomes smaller and less costly. While any improvement in efficiency, size, weight, and cost is certainly welcome, there are other problems. Switching supplies are rich sources of noise. The noise, if not properly controlled, can add directly to the output rf signal. The noise can create an electromagnetic compatibility problem. This is a problem that occurs in complex electrical systems.

What can be done to keep the amplifier power supply from interfering with other elements in the system? Basically, good design practice is required. Filtering on all input power leads is better. This prevents the pulses of power, characteristic of switching power supplies, from appearing on the power lines. The filter serves mainly to attenuate the rf content of the pulses. It should also prevent the chassis itself from becoming a conduction path of high-frequency conducted emissions.

12.3 Gyrotron Tubes[4]

The gyrotron is the microwave generator that has the most significant industrial development. The reason is that the application of these generators for the heating of thermonuclear plasmas is of immediate interest, at least at the experimental level.[5] The physical principle used in gyrotrons has been known since 1961. The gyrotron is a microwave vacuum tube based on the interaction between an electron beam confined by a constant magnetic field and a microwave field, where coupling is achieved by the cyclotron resonance condition. This device requires an electron beam where most of the electron energy is transverse to the axis of the tube. The dc magnetic field is related to the operating frequency by the cyclotron resonance condition. This relation is given by:

$$
\begin{aligned}
\text{Wo} &= \text{N·Wc} \\
\text{Wo} &= \text{operating pulsation} \\
\text{N} &= \text{an integer} \\
\text{Wc} &= \text{angular velocity of the electron} \\
\text{Wc} &= \text{e·B/m·g} \\
\text{B} &= \text{magnetic field} \\
\text{e} &= \text{electron charge} \\
\text{g} &= \text{relativistic mass factor}
\end{aligned}
$$

The interaction occurs only for magnetic fields where N is near integer values.[6] For most microwave field shapes, such as encountered in conventional waveguides and resonators, the fundamental resonance condition with N = 1 has the strongest interaction. With certain special microwave field shapes, useful interactions have the advantage that the magnitude of the dc field for a given frequency can be reduced by 1/N. For the fundamental resonance, a frequency of 2.8 MHz/G is necessary.[7] For example, the magnetic field is 10 kG for a frequency of 28 GHz. For higher frequencies, proportionally higher magnetic fields are needed. This has led to the use of superconducting magnets. Operation at the second harmonic of the cyclotron resonance has been obtained, which allows a reduction of the magnetic field (Figure 30).

Historically, the general name of these devices is electron cyclotron masers (ECMs). In a practical electron tube, electrons are injected by an electron gun along the dc magnetic field into the rf circuit, cavity, or waveguide. They could be collected on the walls of the cavity, but it is, of course, better to let the electrons leave the cavity

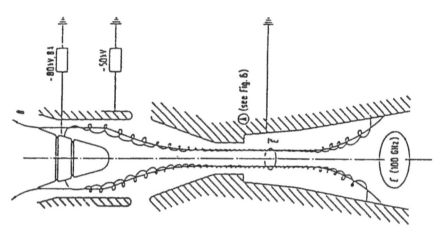

FIGURE 30
Schematic diagram of a gyrotron.

in the homogeneous magnetic field. Then, as the field diverges, the electrons are spread over a large collector. The axial velocity is necessary for a large current, but the corresponding energy is lost. It is difficult to produce an intense electron beam with large transverse and small longitudinal velocity.

The Soviet workers built their efficient and powerful device with a structure that they named the gyrotron, which solves the essential problems.[8] It consists of an electron gun of the magnetron type, which produces an annular electron beam and accelerates it through a growing magnetic field, and an rf circuit through which the beam flows. Often, a circular waveguide is excited near the cutoff of a TE0n mode, or something closely related.[8] All the varieties have been discussed in the literature. Although the concepts used in the more powerful tubes[9] differ in some ways from this model, it can be taken as a basis to explain the theory of electron cyclotron masers.

Gyrotrons very important features for high power in long pulse are cw operation, narrow linewidth, and good spatial beam quality with mode converters. These capabilities have been extended into millimeter and submillimeter wavelength range where other long pulse or cw high-frequency microwave devices and lasers have power levels orders of magnitude weaker at much lower efficiencies.[9]

12.4 Traveling Wave Tube (TWTs)

The broad bandwidth of frequencies attainable with traveling wave amplifiers (Figure 31) make them ideally suited for broad-band communications and as laboratory amplifiers. A bandwidth of 20 to 40% can be achieved.

The traveling wave tube consists of a electron beam associated with a periodic slow-wave structure, not resonant, where the phase velocity of the signal on the interaction structure is approximately equal to the electron velocity in the beam. The velocity modulation created all along the beam by the structure's field induces an alternating current that excites the slow-wave structure in both directions. The energy may be transferred from the electron beam to the rf traveling wave. The synchronism condition is such that all of the partial waves created by the excitation add together in the direction of the beam motion, and are out of phase in the opposite direction.

At low input signal levels the power output of a TWT is directly proportional to the power input. This operating range is called the linear region of the tube. The gain

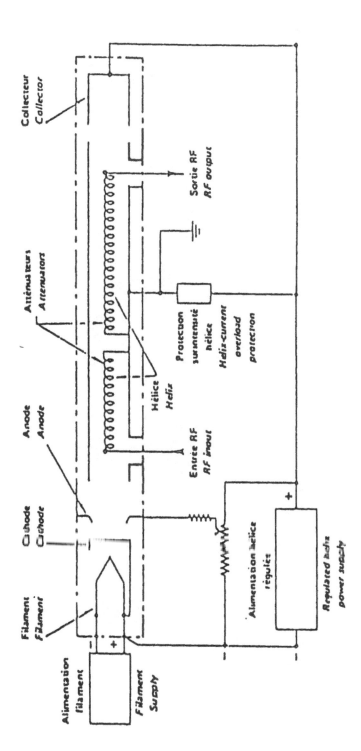

FIGURE 31
Schematic diagram of a helix TWT.

in this region is called the small-signal gain and it is constant. As the input power level is increased, the tube starts to saturate. The gain starts to decrease, so the output is no longer a linear function of the input. The gain at saturation is typically from 5 to 7 dB below the small-signal gain level. Rated power for a traveling wave tube is defined as a suggested operating point that is approximately from 1 to 2 dB below the point of saturation. The linear region of a traveling wave tube and its dynamic range are often defined as the level at which the signal is just equal to the noise power at the point at which gain has compressed by 1 dB. This point is called the 1-dB compression point. Noise in a traveling wave tube amplifier is important because it limits the useful output of the device.

The traveling wave tube is a nonlinear amplifier; its microwave output is not a linear function of its input. It therefore generates harmonics as well as amplifies the fundamental input signal. Harmonics can be reduced by using external filters.

12.4.1 Power Supply

What sets a TWT power supply apart from most other power supply is the voltage. A TWT requires a cathode voltage that can vary from 1 to above 15 kV. The sensitivity of a medium-power traveling wave tube to a change in helix voltage is typically 0.01 dB/V under small-signal gain conditions in the center of the traveling wave tube's microwave passband. The power supply ripple causes both amplitude modulation and pushing modulation in a traveling wave tube amplifier, with the voltage on the helix being the most critical part. Amplitude modulation can be minimized by adjusting the helix voltage for operation at maximum gain. But in a wide-band traveling wave tube, this condition cannot be maintained over the full band. In addition, there may be limits on the amount of helix voltage adjustment because of other tube operating requirements. It is to be emphasized that traveling wave tubes will often have considerable bandwidth outside the published frequency range. The supply for this tube is more complicated than for a klystron, because it is necessary to dispose of several supplies, filament voltage, whenelt, anode, and helix voltage. High regulation and low ripple are required in the helix voltage supply to prevent undesirable phase shift variations.

13 MICROWAVE WINDOW FOR RF POWER INJECTION INTO THE ION SOURCES[10]

The existing materials are quartz, aluminium oxide, magnesia, zirconia, sapphire, beryllium oxide, and boron nitride.[11] In order to inject the microwave power into the ion source, two main positions, axial and radial, may be used. In any case, there are several conditions to fulfill to protect the window from being destroyed.

- The plasma particles should not impinge on the window; thus, a bend in the input waveguide may be used so that the window is not in the line of sight of the plasma.

- No cyclotron resonance can be tolerated (i.e., a local magnetic field B such that $B = m \cdot w / e$) right in front of the window on the vacuum side; such a resonance could locally create a plasma, thus abnormally heating the window.

- The pressure in the source has to be below a minimum value for a normal microwave breakdown (a few thousandths of a millibar for the current source dimensions); otherwise, arcing may occur in the waveguide under vacuum and will usually go back to the generator and end up on the window.[12]

Puncturing of the window is caused by initial internal arcing, which extends to the surface and produces a vacuum leakage path across the window. Failure by cracking is associated with heating of the window. Because the E field can be temporarily strong due to faults in the waveguide and its load, it is essential to operate the system in a very good vacuum, taking all possible precautions to prevent load arcs.[13]

In the case of high-power microwave signals, other phenomena can occur, such as resonances with harmonic frequencies. The existence of ghost modes in dielectric obstacles of arbitrary shape may be made plausible. The insertion of dielectric material into the guide effects an increase in the electrical cross section of the guide, so that a wave may have propagating character within the length L of the obstacle, while it decays outside. Waveguide discontinuities (iris, couplers, bends, etc.) located in the vicinity of the window may also serve this function.[14]

14   TRANSMISSION LINES

The choice of dimensions for a waveguide to be operated at a given wavelength involves the following considerations:

- Mode of operation
- Cutoff wavelength for the operating mode
- Attenuation in the wall surfaces
- Voltage breakdown

14.1   Rectangular Waveguides

Commercially available rectangular waveguides can seldom be considered as perfectly rectangular and the majority of practical waveguide systems are operated in air or with an internal gas whose permittivity is greater than that of free space.

In free space the wavelength λ_o is directly proportional to the speed of electromagnetic waves C and inversely proportional to the frequency involved $\lambda_o = C/F_o$. In media with permittivity greater than that of free space, the speed of electromagnetic waves becomes $(C/\sqrt{\varepsilon_r})$ where ε_r is the relative permittivity. If F_o is stated in hertz and C in centimeters per second then λ_o is in centimeters. In a rectangular waveguide operated above its cutoff frequency in the dominant TE10 mode, the waveguide length λ_g is given by:

$$\lambda_g = \lambda \left[\varepsilon_r - \left(\lambda/\lambda_c \right)^2 \right]^{-1/2}$$

where the waveguide cutoff wavelength, λ_c, measured in the same units as λ_o, is simply equal to twice the broad dimension a, i.e., $\lambda_c = 2 \cdot a$.[15]

Waveguide nomenclature can differ from country to country. For example, the outside-dimensioned rectangular waveguide is included in all present-day nomenclature systems and thus has an IEC number, three RG numbers, one WR number, four British Joint Service numbers, three NATO numbers, a Japanese number, a French number, and two commonly used band letter designations. The classified rectangular waveguides are listed in Table 1 along with cross-references to other classifications systems.

TABLE 1

Waveguide Standard Data Chart

HP band designation	Frequency range, TE_{14} mode (GHz)	IEC R-	EIA WR-	British WG-	JAH RG-	Other Common Usage	Material (B:Brass A:Alum. S:Silver)	Choke UG-	Cover UG-	Inside Width cm (in.)	Inside Height cm (in.)	Inside Tol μm (mils)	Outside Width cm (in.)	Outside Height cm (in.)	Outside Tol μm (mils)	Hom. wall thickness mm (in.)	Cutoff Frequency (GHz)	Theoretical attenuation, low to high frequency, brass/alum/silver, dB/100 ft	Theoretical attenuation, low to high frequency, brass/alum/silver, dB/100 m	Theoretical cw power, low to high frequency, MW (kW)
	1.12–1.70	14	650	6	69	L	B		4178	16.510 (6.50)	8.255 (3.25)	±12.7 (±5)	16.916 (6.66)	8.661 (3.41)	±12.7 (±5)	2.03 (0.080)	0.908	0.412–0.272	1.353–0.894	11.8–17.1
					103		A		4188									0.269–0.178	0.883–0.584	
	1.45–2.20	18	510	7			B			12.954 (5.10)	6.477 (2.55)	±12.7 (±5)	13.360 (5.26)	6.883 (2.71)	±12.7 (±5)	2.03 (0.080)	1.16	0.574–0.390	1.883–1.280	7.5–10.6
							A											0.374–0.255	1.229–0.836	
	1.70–2.60	22	430	8	104	LS,F	B		4358	10.922 (4.3)	5.461 (2.15)	±12.7 (±5)	11.328 (4.46)	5.867 (2.31)	±12.7 (±5)	2.03 (0.080)	1.375	0.759–0.504	2.492–1.655	5.2–7.5
					105		A		4378									0.496–0.329	1.626–1.080	
	2.20–3.30	26	340	9A	112		B		553A	8.636 (3.40)	4.318 (1.70)	±12.7 (±5)	9.042 (3.56)	4.724 (1.86)	±12.7 (±5)	2.03 (0.080)	1.735	1.030–0.716	3.382–2.352	3.4–4.71
					113		A		554A									0.673–0.468	2.207–1.535	
S	2.60–3.95	32	284	10	48		B	54	53	7.214 (2.84)	3.404 (1.34)	±12.7 (±5)	7.620 (3.00)	3.810 (1.50)	±12.7 (±5)	2.03 (0.080)	2.080	1.435–0.982	4.711–3.225	2.18–3.1
					75		A	585A	584									0.937–0.642	3.074–2.105	
	3.30–4.90	40	229	11A	—		B		CMR229	5.817 (2.29)	2.908 (1.145)	±12.7 (±5)	6.142 (2.418)	3.233 (1.273)	±12.7 (±5)	1.63 (0.064)	2.59	1.828–1.296	6.002–4.255	1.56–2.14
							A											1.194–0.346	3.917–2.777	
G	3.95–5.85	48	187	12	49	C,H	B	148C	149A	4.755 (1.872)	2.215 (0.872)	±12.7 (±5)	5.080 (2.00)	2.540 (1.00)	±12.7 (±5)	1.63 (0.064)	3.16	2.695–1.869	8.849–6.134	(941–1317)
					95		A	4068	407									1.760–1.220	5.774–4.003	
C	4.90–7.05	58	159	13	—	C	B		CMR159	4.039 (1.59)	2.019 (0.759)	±10.2 (±4)	4.364 (1.718)	2.344 (0.923)	±10.2 (±4)	1.63 (0.064)	3.71	3.091–2.324	10.15–7.630	(754–983)
							A											2.019–1.518	6.622–4.980	
J	5.85–8.20	70	137	14	50	XH,C,G	B	343B	344	3.484 (1.372)	1.580 (0.622)	±10.2 (±4)	3.810 (1.50)	1.905 (0.750)	±10.2 (±4)	1.63 (0.064)	4.29	3.821–3.018	12.54–9.907	(554–696)
					106		A	440B	441									2.496–1.971	8.187–6.465	
H	7.05–10.00	84	112	15	51	XB,W	B	52B	51	2.850 (1.122)	1.262 (0.497)	±10.2 (±4)	3.175 (1.25)	1.588 (0.625)	±10.2 (±4)	1.63 (0.054)	5.26	5.355–4.161	17.58–13.66	(355–454)
					68		A	137B	138									3.497–2.717	11.47–8.913	
	7.00–11.00	—	102	—	—		B			2.591 (1.02)	1.295 (0.510)	±7.6 (±3)	2.845 (1.12)	1.549 (0.610)	±7.6 (±3)	1.27 (0.050)	6.50	6.939–4.360	22.78–14.31	(280–424)
							A											4.532–2.848	14.87–9.34	
X	8.20–12.40	100	90	16	52		B	40B	39	2.286 (0.90)	1.016 (0.40)	±7.6 (±3)	2.540 (1.00)	1.270 (0.50)	±7.6 (±3)	1.27 (0.050)	6.56	8.362–5.784	27.45–18.99	(206–293)
					67		A	136B	135									5.461–3.778	17.91–12.39	
M	10.00–15.00	120	75	17	—		B			1.905 (0.75)	0.953 (0.375)	±7.6 (±3)	2.159 (0.850)	1.207 (0.475)	±7.6 (±3)	1.27 (0.050)	7.88	9.893–6.909	32.48–22.68	(166–229)
							A											6.461–4.512	21.19–14.80	
P	12.40–18.00	140	62	18	91	XU,Y,U	B	541A	419	1.580 (0.622)	0.790 (0.311)	±6.4 (±2.5)	1.783 (0.702)	0.993 (0.391)	±7.6 (±3)	1.02 (0.040)	9.49	12.46–0.162	40.92–30.08	(119–157)
							A	—	—									8.141–5.984	26.71–19.63	
					107		S											6.165–4.531	20.22–14.87	

14.1.1 Attenuation and Losses in the Waveguide

In the waveguide, a theoretical cw power rating imposes a limitation by a temperature rise resulting from power dissipation within the rectangular waveguide walls. This power rating can be determined by predicting the rise in temperature. The major power rating consideration in present waveguide designs, using the TE10 mode, has been the limitation imposed by voltage breakdown (peak power rating). However, with the advent of high-average-power microwave systems, the average power rating of waveguides should be considered. The normal attenuation of a transmission line is the ratio of the output power to the input power when the load is matched to the characteristic impedance of the transmission line. The common expression for normal attenuation in decibels is

$$\alpha = 10 * \log(P2/P1)$$

P2 power (in watts) at the output transmission line
P1 power (in watts) at the input transmission line

The theoretical attenuation, for a copper waveguide, is found by employing the following formula:

$$\alpha_{dB/m} = \frac{1.47 \times 10^{-4}}{b_m^{3/2}} \frac{\left(\frac{f_c}{f}\right)^{3/2} + \frac{b}{2a}\left(\frac{f}{f_c}\right)^{1/2}}{\left[1 - \left(\frac{f_c}{f}\right)^2\right]^{1/2}}$$

b   width of waveguide, in meters
a   height of waveguide, in meters
f   operating frequency, in hertz
f_c cutoff frequency of waveguide

The attenuation curves (Figure 32) and power handling (Figure 33) are indicated for standard waveguides. The curves shown are traced on 15000 V/cm as the maximum voltage gradient of dry air. For a rectangular waveguide operating in the TE10 mode, the maximum power-handling capability can be expressed by:

$$P_{watt} = 6.66 \times 10^{-4} \frac{\lambda}{\lambda_g} \cdot a_{cm} \cdot b_{cm} E_{0\,volt/cm}^2$$

b   width of waveguide, in centimeters
a   height of waveguide, in centimeters
λ free space wavelength, in centimeters
λ_g waveguide wavelength, in centimeters
E_0 breakdown voltage gradient of the dielectric filling the waveguide, in volts per centimeter

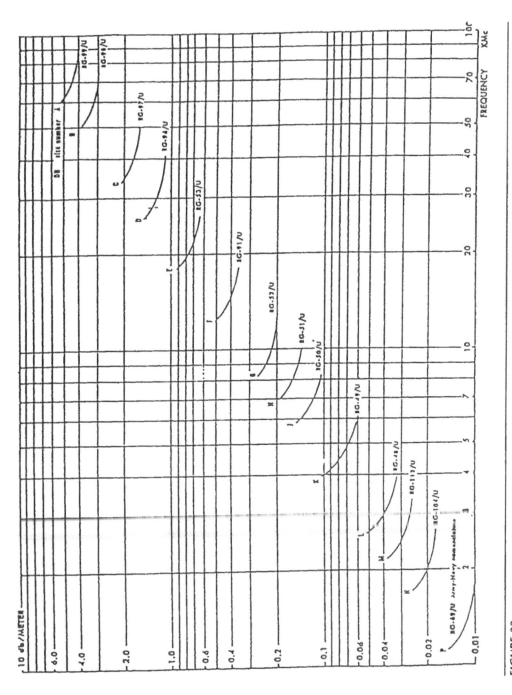

FIGURE 32
Attenuation of standard: waveguides.

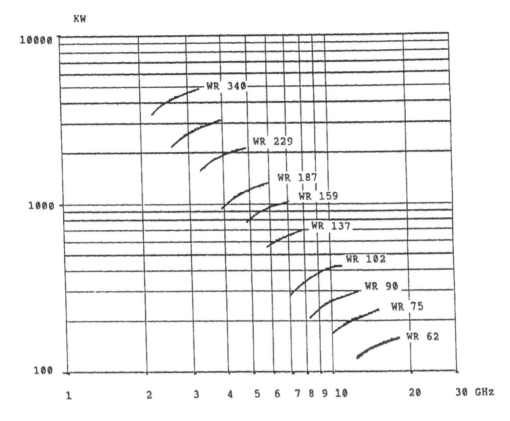

FIGURE 33
Power handling vs. frequency.

14.1.2 Flanges

The contact coupling is usually made of flat flanges that are soldered to the ends of the tubing and bolted together. The contact coupling is not frequency-sensitive. Choke-flange coupling may be used. It consists essentially of a series-branching transmission line whose length is one half wavelength, thus presenting zero series impedance to the main line (Table 2).

14.2 Circular Waveguide

For a circular waveguide operating in the TE11 mode, the maximum power-handling capability can be expressed by:

$$P_{max} = 3.97 \times 10^{-3} \sqrt{1 - \left(\frac{\lambda}{3.41a}\right)^2} \, a^2 E_{max}^2$$

a   radius of waveguide, in centimeters
λ free space wavelength, in centimeters
λ_g waveguide wavelength, in centimeters
E_{max} breakdown voltage gradient of the dielectric filling the waveguide, in volts per centimeter

TABLE 2

Flange Standards

Frequency (GHz)	Designations			Flange UG ( )/U		Choke flange	
	CEI.	E.I.A.	J.A.N.	Brass (flat)	Aluminium alloy	Brass	Aluminium alloy
2.17/3.3	R 26	WR 340	RG 112	553 A	554 A		
2.6/3.95	R 32	WR 284	RG 48	53	584	54 A	585 A
3.22/4.90	R 40	WR 229	RG 341				
3.94/5.99	R 48	WR 187	RG 49	149 A	407	148 B	406 A
4.64/7.05	R 58	WR 159	RG 343				4.64
5.38/8.17	R 70	WR 137	RG 50	344	441	343 A	440 A
6.57/9.99	R 84	WR 112	RG 51	51	138	52 A	137 A
8.2/12.5	R 100	WR 90	RG 52	39	135	40 A	136 A
9.84/15.0	R 120	WR 75	RG 346				
11.9/18.0	R 140	WR 62	RG 91	419	1665	541 A	1666

Other factors to be considered with the above formula are the voltage standing-wave ratio and internal pressure. The VSWR of the waveguide lowers the maximum power-handling capability of the waveguide by the reciprocal of the magnitude of the VSWR. At higher internal pressures, the power is approximately proportional to the square of the density of air.

14.3 Coaxial Transmission

For many applications, flexibility is required and there are many manufacturers[19] of coaxial cable (Table 3).

For a coaxial line operating in the TEM mode, the maximum power-handling capability can be expressed by:

$$P = \frac{E_0^2 a^2 \sqrt{k_{e_1}}}{120} \ln \frac{b}{a}$$

k_{e_1} dielectic constant
b   radius of outer conductor, in centimeters
a   radius of the inner conductor, in centimeters
λ   free space wavelength
E_0 breakdown voltage gradient of the dielectric filling the waveguide, in volts per centimeter

14.4 Ridge Waveguides

The ridge waveguides have a longer cutoff wavelength and a lower characteristic impedance than conventional rectangular waveguides having the same internal dimensions. They also have a wider bandwidth, free of higher-mode interference. Because of these advantages, ridge waveguides have been used as transmission links in systems requiring a wide, free range in the fundamental mode, as matching or transition elements in waveguide-coaxial junctions, as filter elements, and as components for other special purposes. Ridge waveguides are used as slow-wave structures in traveling wave tubes (the phase velocity of the wave is close to an electron beam velocity).

TABLE 3
Coaxial Power Handling

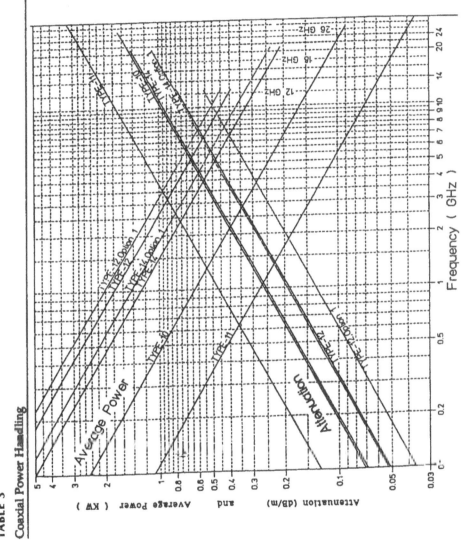

In the development of tunable magnetrons, double-ridge waveguides have been used as external tuning cavities because their reduced cutoff frequency permits a compact cavity section. The electric field is concentrated between the ridges; satisfactory tuning is obtained by means of a plunger that short-circuits the narrow gap. Ridge waveguides are also used as H-type output transformers in magnetrons for the purpose of transforming the high impedance of the waveguide used for power transmission to the low impedance level of the magnetron. In these applications of the waveguides, the cutoff wavelength and the characteristic impedance are important parameters to know for design purposes. Formulas for finding these characteristics are expressed explicitly in terms of the guide dimensions, and charts are constructed to facilitate computations for waveguides having various aspect ratios and having ridges of different widths and depths.[20]

14.5   Isolating Flange

To extract ions from the ion source it is necessary to polarize the body of the source to negative high voltage. A sheet of Teflon is used between two flanges of the feeder to isolate the generator. It appears that the best material is the Teflon FEP:[18]

Material	Teflon
Manufacturer	Du Pont de Nemours
$\tan\delta$ at 3 cm	0.001
ke at 3 cm	2
Dielectric strength	Thickness
240 kV/mm	0.025 mm
70 kV/mm	5 mm

14.6 Transition Waveguide to Coax

The design of a continuous transition for a microwave transmission line requires the use of a symmetrical field configuration. This can be the dominant coaxial mode (TEM mode), the symmetrical TM01 mode, or a circularly polarized TE11 mode in a round waveguide. When rectangular tubing is used for the waveguide transmission line, as is generally the case, transitions from the TE10 mode in the rectangular waveguide to one of the symmetrical modes in the coaxial line or a round waveguide are necessary. Many possibilities exist for the arrangement in construction of a coaxial line-to-waveguide transition with differences depending on the conditions under which the unit has to be operated. Different configurations are shown in Figure 34.

By adjustment of proper dimensions, it is possible to obtain an impedance match. The bandwidth obtained in the units should not be expected at other wavelengths when various standard sizes of waveguides and coaxial lines are used. There are, of course, relative considerations based upon the electrical properties of the transition as a function of its geometrical configuration. This does not alter the fact that a microwave component is more sensitive to changes in dimensions at shorter wavelengths than at longer ones.

15   STUB TUNER

A stub tuner (Figure 35) is used to reduce the VSWRs of equipment. It matches any load impedance very close to the characteristic impedance of the waveguide. It

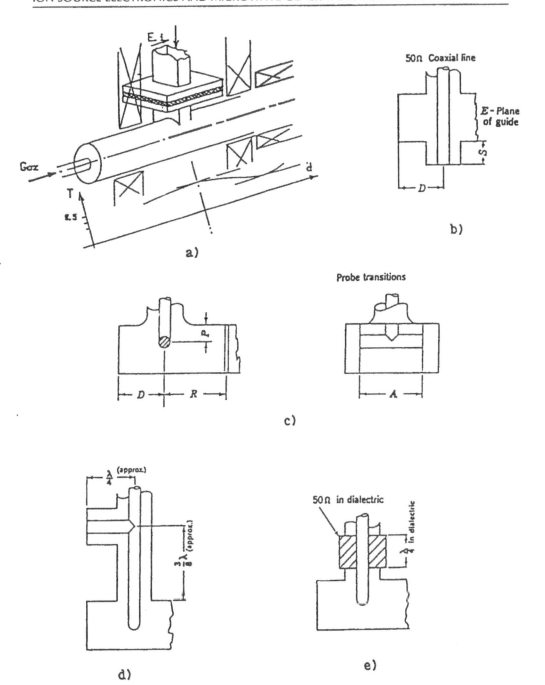

FIGURE 34
a) Rf injection E//Bo., b) crossed transition; c) crossbar transition; d) broad-band probe transition;
e) bead-supported probe.

is used with monomode or multimode cavities whose reflection coefficients are low. These stubs may permit the adjustment of the impedance of the line even with a VSWR as high as 10. But the device is limited in power and reduces the band-pass of the line.

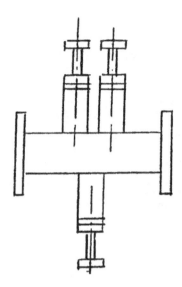

FIGURE 35
Triple-stub tuner used with rectangular waveguide.

16   HAZARDS CAUSED BY MICROWAVE RADIATION[16]

Precautions must be taken to prevent exposure of people to microwaves. In addition, under some tuning conditions, regeneration or oscillations may occur if rf energy from the output line reaches the input cavity because of faulty rf connectors or inadequate shielding. External leakage may be prevented or reduced to a safe level by tightening rf input and output connections.

Although overall heating is the best-documented biological effect of nonionizing radiation, some heating effects that occur only at specific frequencies have been reported. Body resonance occurs when the wavelength of the radiation is about the same as the size of an exposed body part. Also, energy can be concentrated in some areas and cause local hot spots. It is even possible to experience localized burns and electric shocks. These specific heating effects are important because some body organs, particularly the testes and eyes, are especially sensitive to heating.[17] Research has been performed worldwide to support development of standards or guidance that would protect humans from exposure to this energy and to determine what energy levels are safe. The most commonly utilized standard in the U.S. has been the American National Standards Institute (ANSI) C95.1–1990 standard entitled "Safety Levels with respect to Human Exposure to Radio Frequency Electromagnetic Fields, 300 kHz to 100 GHz." (See Figure 36.)

17   POWER MEASUREMENTS

The measurement of microwave power is of fundamental importance in directly establishing the proper functioning of many types of microwave apparatus. Many quantities can be measured indirectly in terms of a quantitative knowledge of power. Thus, electric or magnetic field magnitudes can be measured if one knows power and impedance, and this information may be put to use in a variety of applications including noise measurements and voltage breakdown measurements. Power measurements at microwave frequencies may be divided into low, medium, and high

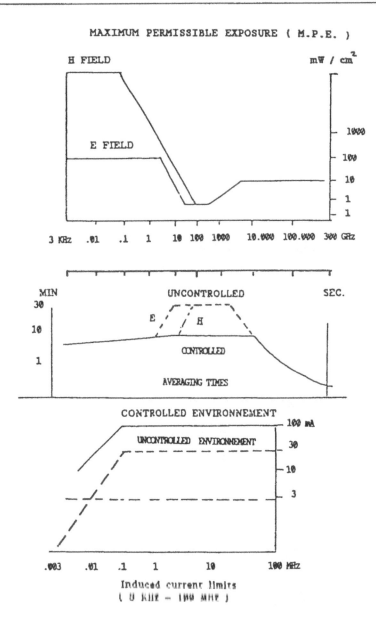

FIGURE 36
ANSI standard C95.1–1990.

power levels. Low power levels (less than 10 mW) are most often measu. .:d using crystal rectifiers, bolometers, thermistors, and, recently, thermocouples as sensing elements. Bolometers and thermistors are devices whose resistance varies in proportion to the incident rf power as a result of the conversion of rf energy into heat energy. Thermocouples are devices that generate a dc voltage proportional to the rf power.

Medium power levels (10 mW to 1 W) are frequently measured by using calibrated attenuators together with low-level power meters. However, thermocouples are presently being used for direct power measurements above 100 mW.

High power is measured by using directional couplers and attenuators together with low-level power meters or by using microwave calorimeters. In a microwave calorimeter, power is determined by measuring the temperature rise as the rf energy is dissipated in a lossy termination. A circulating fluid in a load directly absorbs the rf energy or carries off heat produced by rf power. Inlet and outlet temperature difference plus the fluid flow rate are used to compute the power. A unit (bolometer, detector) mounted on a directional coupler can be calibrated to low power.

To measure power accurately at microwave frequencies, sensing elements must be mounted in a transmission line section in such a way that they reflect as little of the incident power as possible. Power absorbed in the walls of the mount or in the matching structure and connections all degrade the efficiency of the mount. The match of the mount is equally important in determining that accuracy of absolute power measurements. Often, the VSWR of a mount can be improved by external or integral adjustable tuning. When making absolute power measurements, the generator end of the line must be well matched. This will reduce the errors resulting from phase addition or subtraction of secondary reflection from the source end of the transmission line. To reduce the errors resulting from drift of the power bridge, the bolometer or thermistor should be permitted to stabilize thermally before any measurements are attempted.

For applications requiring peak-level determination, a diode is required due to its fast response. Using a diode to perform these measurements has typically been difficult because of higher VSWRs, temperature drifts, and linearity errors.

REFERENCES

1. Technical review, *Thomson Tube Electron.*, 23, 1991.
2. SAIREM, Manufacturer/microwave, Technical review, Vaux en Velin, Lyon, France, 1990.
3. G. Firmain, Technical review, *Thomson Tube Electron.*, 23, 1991.
4. G. Mourier, Technical review, *Thomson Tube Electron.*, 23, 1991.
5. G. Mourier, A theoretical study, Arch. Electron. Ubertragungtech., 34, 1980.
6. V. A. Flyagin et al., The gyrotron, *IEEE Trans. Microwave Theory Tech.*, p. 77, 1977.
7. V. A. Flyagin et al., Gyrotron oscillator, *Proc. IEEE*, p. 345, 1988.
8. G. Mourier, Current gyrotron development in Thomson tubes electronics, *Int. Workshop Strong Microwaves in Plasmas*, Suzdal, Sept. 1990.
9. V. A. Flyagin et al., Gyrotrons. State of the art and projects, *Int. Workshop Strong Microwaves in Plasmas*, Suzdal, Sept. 1990.
10. Microwave tubes, *Proc. 5th Int. Congr.*, Paris, 1964.
11. *Dielectric Materials and Applications*, John Wiley & Sons, New York, 1960.
12. D. H. Priest and R. C. Talcott, On the heating of output windows of microwave tubes by electron bombardment, *IRE Trans. Electron Devices*, ED8, 243–251, 1961.
13. R. C. Talcott, The effects of titanium films on secondary electron emission phenomena in resonant cavities and at dielectric surfaces, *IRE Trans. Electron Devices*, ED-9, 406–410, 1962.
14. M. P. Forrer and E. T. Jayne, Resonant modes in waveguide window, *IRE Trans. Microwave Theory Techn.*, MTT8, 147–150.
15. M. M. Brady, Note on the presentation of the radio refractive index, *Radio Sci. (New Ser.)*, 1523–1524, 1967.
16. Electromagnetic waves and biology — report congress of URSI and CNFRS (French National Scientific Committee of Radioelectronics, Jouy-en-josas, France, 1980.
17. S. F. Cleary, Considerations in the evaluation of the biological effects of exposure to microwave radiation, *Am. Ind. Hyg. Assoc. J.*, 31, 1970.
18. Du Pont de Nemours International S.A., Plastics department, Geneva 24, Switzerland.
19. Spectrum Elektrotechnik GmbH, Munich, Germany.
20. S. B. Cohn, Properties of ridge waveguides, *Proc. IRE*, 35, 783–788, 1947.
21. M. Sucher and J. Fox, *Microwave Measurements*, 3rd ed., John Wiley & Sons, New York, 1963.
22. R. E. Collin, *Field Theory of Guided Waves*, McGraw-Hill, New York, 1966.

23. R. E. Collin, *Foundation for Microwave Engineering*, McGraw-Hill, New York, 1966.
24. Montgomery, *Technique of Microwave Measurements*, McGraw-Hill, New York, 1948.
25. G. L. Ragan, *Microwave Transmission Circuits*, McGraw-Hill, New York, 1948.

Chapter 7

COMPUTER CODES

P. Spädtke

CONTENTS

A compendium of computer codes used in particle accelerator design has been compiled in Reference 1. Here the general approach is described.

The codes described here run on large mainframe computers; however, the hardware improvement of small computers today has reached a level that makes it possible to implement most of these codes on a PC. The memory of these small computers can be expanded to several MB, which was not an easy task just several years ago even on a large computer. CPU performance has been improved as well,

0-8493-2502-1/95/$0.00+$.50

so that execution time on a PC is comparable to the performance of a mainframe divided by the number of users.

For these reasons, today most of the described programs are available commercially.[2-4]

1   GENERAL SOLUTION OF MAXWELL'S EQUATION

The general formulation of Maxwell's equations for the electric field $\vec{E}$, flux density $\vec{D}$, and the magnetic field $\vec{H}$ and flux density $\vec{B}$ is

$$\nabla \times \vec{E} = -\frac{\partial \vec{B}}{\partial t} \tag{1}$$

$$\nabla \times \vec{H} = \frac{\partial \vec{D}}{\partial t} + \vec{J} \tag{2}$$

$$\nabla \cdot \vec{D} = \rho \tag{3}$$

$$\nabla \cdot \vec{B} = 0 \tag{4}$$

where ρ is the charge density and $\vec{J}$ the current. The constitutive equations between the fields and the flux densities are

$$\vec{D} = \varepsilon_0 \varepsilon_r \vec{E} \tag{5}$$

$$\vec{B} = \mu_0 \mu_r \vec{H} \tag{6}$$

$$\vec{J} = \kappa \vec{E} + \rho \vec{v} \tag{7}$$

with the permittivity ε_0, ε_r, permeability μ_0, μ_r, and the conductivity κ, which can depend on the material and on the field itself. A numerical approach for the solution of Maxwell's equations is presented in Reference 5.

2   LAPLACE SOLVER

If there is no magnetic field, only steady-state electric fields, and no space charge, the potential equation is derived from Equation 3:

$$\partial^2 \phi / \partial x^2 + \partial^2 \phi / \partial y^2 + \partial^2 \phi / \partial z^2 = 0 \tag{8}$$

3   POISSON AND VLASOV SOLVER

The same equation with a space charge term added is the Poisson equation:

$$\partial^2 \phi / \partial x^2 + \partial^2 \phi / \partial y^2 + \partial^2 \phi / \partial z^2 = \rho / \varepsilon_0 \tag{9}$$

If the space charge distribution depends on the potential itself, the problem is known as the Vlasov equation.

3.1 Fixed Emitter

The solution of the Vlasov equation depends on a known distribution of the charged particles at the origin, for example, the cathode for the case of electrons. The current distribution of the beam is unknown; however, it can be calculated by applying the Child-Langmuir law (see Chapter 4, Section 1).

3.2 Plasma Sources

In plasma sources the Child-Langmuir law still applies, but the gap distance d is not defined anymore. Normally, the current density is now set by discharge parameter and the plasma boundary sets itself to adjust the space charge limit.

The space charge of the ion distribution is compensated by the electron distribution, described analytically by:[6]

$$n_e = n_{e0} \cdot e^{\frac{\phi_{pl} - \phi}{kT_e}} \tag{10}$$

$$n_{e0} = n_{i0} \tag{11}$$

where n_e and n_i are the electron density and the ion density, which are equal in the plasma. The electron density decreases depending on the electron temperature T_e and the potential with respect to the plasma potential ϕ_{pl} if a Maxwellian energy distribution of the plasma electrons is assumed.

If the plasma contains different charge states, this has also to be taken into account, especially if magnetic fields are present. Not only will the beam envelopes of different charge states differ, but also the velocity will increase differently for different charge states. This may have a strong influence on the total space charge distribution. This has been demonstrated numerically and experimentally for the extraction of ion beams from ECR ion sources.[7] However, in this case one can assume that the electron density will depend on the potential as described in Equation 10. The space charge distribution of the ions can be found by tracing all the different charges through the extraction system.

3.3 Plasma Effects

Plasma effects on the beam transport, such as two-stream instabilities, beam plasma instabilities, and Landau damping, have been described in Reference 8.

4 MAGNETIC FIELDS

In some cases (for example, in ECR sources) the magnetic field plays an important role, not only for the discharge process but also for the extraction optics.[7,9] The magnetic field might be obtained by integration or by a finite difference method.

The contribution of currents can be used to calculate the vector potential $\vec{A}$:

$$\nabla^2 \vec{A} = -\vec{J} \tag{12}$$

This equation can be solved in Cartesian components similar to the Poisson equation. With the relation:

$$\vec{H} = \nabla \times \vec{A} \tag{13}$$

the magnetic field of arbitrary coils can be derived. The contribution from magnetized materials to the magnetic field can be derived from a scalar potential Ψ superimposed on the coil fields,

$$\vec{H} = \nabla \Psi \tag{14}$$

Another possibility for getting the magnetic flux density in the region of interest is by measurement. Because in most cases it is not possible to measure the magnetic flux density with the desired resolution, two different techniques can be applied.

Laplace Method

If the magnetic flux density is known on a closed surface, the Laplace equation can be applied for each component separately.

$$\nabla^2 B_x = \nabla^2 B_y = \nabla^2 B_z = 0 \tag{15}$$

With known boundary conditions (they have been measured), the flux distribution can be calculated with Equation 2.

Extrapolation Method

For the case of cylindrical symmetry, the measurement of the magnetic flux density on axis is sufficient. With a series expansion the B_z and B_r components can be evaluated:

$$B_z(z,r) = B(z) - \frac{1}{4} \cdot r^2 \cdot B''(z) \tag{16}$$

$$B_r(z,r) = \frac{1}{2} \cdot r \cdot B'(z) \tag{17}$$

Higher-order terms are neglected here, because the measured data are not precise enough in most cases. On the other hand, this restricts the application of series expansion to close to the symmetry axis.

5 TRAJECTORY CODES

Trajectory codes are applicable if the beam is steady state. The first trajectory codes were developed for electron guns.[10] After the 1D plasma sheath model by Self[6] had been added, the extraction of ions from a plasma[11-13] was also possible to describe.

6 PARTICLE-IN-CELL CODES

Particle-in-cell codes are applied if the longitudinal interactions of the particles are not negligible or if time-dependent fields are present. A summary of this class of programs is given in Reference 14.

7 NUMERICAL METHODS

7.1 FDM — Finite Difference Method

The region of interest is covered by a mesh. On each of these mesh points the differential equation, for example, the Laplace equation, is replaced by a difference equation.

The first derivative of the potential between the i-th and the (i+1)-th mesh point can be written as:

$$\left(\frac{\partial \phi}{\partial x}\right)_{i+1/2} = \frac{\phi_{i+1} - \phi_i}{\Delta x} \tag{18}$$

Similarly, for the (i–1)-th and i-th mesh point:

$$\left(\frac{\partial \phi}{\partial x}\right)_{i-1/2} = \frac{\phi_i - \phi_{i-1}}{\Delta x} \tag{19}$$

The field is now known between the two mesh points (i – ½ and i + ½); the second derivative can therefore be calculated exactly at the i-th mesh point:

$$\left(\frac{\partial^2 \phi}{\partial x^2}\right)_i = \frac{\left(\frac{\partial \phi}{\partial x}\right)_{i+1/2} - \left(\frac{\partial \phi}{\partial x}\right)_{i-1/2}}{\Delta x} \tag{20}$$

$$\left(\frac{\partial^2 \phi}{\partial x^2}\right)_i = \frac{\phi_{i+1} - 2 \cdot \phi_i + \phi_{i-1}}{\Delta x^2} \tag{21}$$

Similarly to this 1D discretized differential equation, the appropriate 2D or 3D equation can be obtained.

Points that are within an electrode can be set to the known potential. Boundary conditions have to be known. The set of equations can then be solved. Further description is given in References 15 and 16.

7.2 FEM — Finite Element Method

The region of interest is divided into elements. An integral equation is used instead of the differential one used in the FDM method. This method is described in more detail in Reference 17.

The advantage of FEM lies in its better discretization. The disadvantage is the more difficult formulation of shape functions.

7.3 Solver

Both variants (FDM and FEM) lead to a large number of equations that are to be solved numerically.

There are iterative and direct methods for solving the large set of equations.[18] Iterative methods can be accelerated by relaxation methods, for example, as described in Reference 19.

If Φ^n denotes the solution of iteration n

$$\Phi^{n+1} = \omega\Phi + (\omega - 1) \cdot \Phi^n \tag{22}$$

is the result of iteration n + 1 with

$$0 < \omega < 2 \tag{23}$$

If the relaxation parameter ω is less than 1 it is an underrelaxation, while if greater than 1 it is an overrelaxation.

Different solving algorithms, and mesh-relaxation and matrix methods are described and compared in Reference 20.

Special features such as plasma boundary simulation or emission restricted current require iterative methods.

Because the particle distribution depends on the fields, this equation cannot be solved directly; however, an iterative method can be applied to solve the Vlasov equation (see Figures 1 and 2).

7.4 Integration Methods for Particle Motion

The equation of motion is a second-order differential equation. In Cartesian coordinates, three of these systems have to be solved. Such a system can be replaced by six first-order equations, which can then be solved by integration. For the integration the fields have to be calculated.

$$x = \int \dot{x} dt \tag{24}$$

$$y = \int \dot{y} dt \tag{25}$$

$$z = \int \dot{z} dt \tag{26}$$

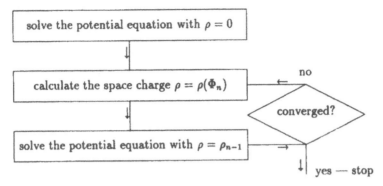

FIGURE 1
Iteration scheme to calculate a self-consistent solution for a particle distribution. Items 2 and 3 are repeated until the difference between two successive iterations becomes small enough. A self-consistent solution has then been found. The convergence behavior of the iteration scheme can be seen in Figure 2.

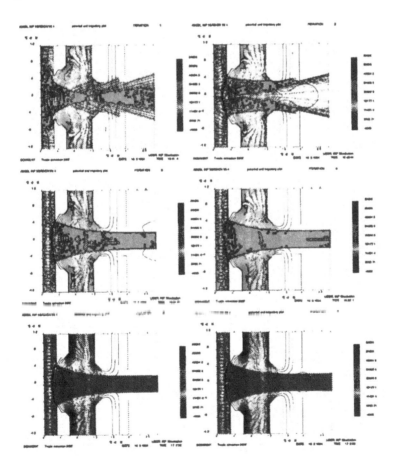

FIGURE 2
Six iterations of an extraction problem, showing the convergence behavior. The first row shows iteration 1 (left) and 2 (right), the second row shows iteration 3 (left) and 4 (right), and the third row shows iteration 5 (left) and 6 (right). Between iterations 5 and 6, only small differences can be observed in the trajectory plot within the extraction system. Some further iterations are necessary to achieve final convergence.

With the abbreviation:

$$k = \frac{q}{m} \cdot \sqrt{1 - \beta'^2}$$

$$\dot{x} = \int k \left(E_x - \dot{x} \cdot \frac{E_x \cdot \dot{x} + E_y \cdot \dot{y} + E_z \cdot \dot{z}}{c^2} + \dot{y} \cdot B_z - \dot{z} \cdot B_y \right) dt \tag{27}$$

$$\dot{y} = \int k \left(E_y - \dot{y} \cdot \frac{E_x \cdot \dot{x} + E_y \cdot \dot{y} + E_z \cdot \dot{z}}{c^2} + \dot{z} \cdot B_x - \dot{x} \cdot B_z \right) dt \tag{28}$$

$$\dot{z} = \int k \left(E_z - \dot{z} \cdot \frac{E_x \cdot \dot{x} + E_y \cdot \dot{y} + E_z \cdot \dot{z}}{c^2} + \dot{x} \cdot B_y - \dot{y} \cdot B_x \right) dt \tag{29}$$

Several integration algorithms are available to solve the above equations. A fast extrapolation method for integration has been described in Reference 21.

8 DIAGNOSIS

Along with the calculation, the presentation of the computed quantities is important for the interpretation of the results. Therefore, it should be possible to display the following quantities:

- Potential
- Space charge
- Magnetic flux
- Electric field
- Magnetic field
- Currents

If particles are incorporated, it should be possible to plot the following information:

- Trajectory plot
- Phase space diagrams
- Profiles (density distributions)
- Energy distribution

REFERENCES

1. Computer Codes for Particle Accelerator Design and Analysis: A Compendium, Los Alamos Code Group, 2nd ed., 1990.
2. Profi Engineering, Otto-Röhm-Str. 26, Darmstadt, Germany.
3. Vector Fields, 24 Bankside, Kidlington, Oxford OX5 1JE, England.
4. INP, Junkernstr. 99, 65205 Wiesbaden, Germany.

5. M. Bartsch, M. Dehler, M. Dohlus, F. Ebeling, P. Hahne, R. Klatt, F. Krawczyk, M. Marx, Z. Min, T. Pröpper, D. Schmitt, P. Schütt, B. Steffen, B. Wagner, T. Weiland, S. Wipf, and H. Wolter, Solution of Maxwell's equation, *Comput. Phys. Commun.*, 72, 1992.

6. S. A. Self, Exact solution of the collisionless plasma-sheath equation, *Phys. Fluids*, 6, 1762, 1963.

7. P. Spädtke, K. Tinschert, and D. Ivens, Simulation of ion beam extraction from an ECR source, *11th Int. Workshop ECRIS*, KVI report 996, 1993.

8. C. K. Birdsall and A. B. Langdon, Plasma Physics via Computer Simulation, 1991.

9. P. Spädtke and H. Wituschek, The influence of axial magnetic fields on the extraction of an ion beam from a plasma source, *Rev. Sci. Instrum.*, 61, 1, 1990.

10. W. B. Herrmannsfeldt, SLAC-266, Stanford Linear Accelerator Laboratory, 1979.

11. E. F. Jäger and J. C. Whitson, Numerical Simulation for Axially Symmetric Beamlets in a Duopigatron Ion Source, ORNL/TM-4990, 1975.

12. J. H. Whealton and J. C. Whitson, Space-charge ion optics including extraction from a plasma, *Part. Accel.*, 10, 1980.

13. J. H. Whealton, Ion extraction and optics arithmetic, *Nucl. Instrum. Methods*, 189, 1981.

14. G. P. Boicourt, Parmila an Overview, AIP conference proceedings, 177, *Linear Accelerator and Beam Optic Codes*, Charles R. Eminhizer, Ed., La Jolla Institute, 1988.

15. C. Weber, Numerical solution of Laplace's and Poisson's equation and the calculation of electron trajectories and electron beams, in *Focusing of Charged Particles*, A. Septier, Ed., Academic Press, 1967.

16. P. Spädtke, Computer modeling, in *The Physics and Technology of Ion Sources*, I. G. Brown, Ed., John Wiley & Sons, 1989.

17. A. Konrad and P. Silvester, A finite element program package for axisymmetric vector field problems, *Comput. Phys. Commun.*, 9, 1975.

18. V. Rokhlin, Rapid solution of integral equations of classical potential theory, *J. Comput. Phys.*, 60, 1983.

19. J. H. Wegstein, *Accelerating Convergence of Iterative Processes*, National Bureau of Standards, Washington, D.C.

20. R. W. Hockney, *Computer Simulation Using Particles*, McGraw-Hill, 1981.

21. R. Bulirsch and J. Stör, Numerical treatment of ordinary differential equations by extrapolation methods, *Numer. Math.*, 8, 1965.

INDEX